D0880977

User Interfaces
for All

Concepts, Methods, and Tools

Human Factors and Ergonomics
Gavriel Salvendy, Series Editor

User Interfaces for All

Concepts, Methods, and Tools

Edited by

Constantine Stephanidis

LAWRENCE ERLBAUM ASSOCIATES, PUBLISHERS

2001   Mahwah, New Jersey   London

Lawrence Erlbaum Associates, Inc., Publishers
10 Industrial Avenue
Mahwah, NJ 07430

Cover design by Kathryn Houghtaling Lacey

Library of Congress Cataloging-in-Publication Data

User interfaces for all : concepts, methods, and tools /
edited by Constantine Stephanidis.
      p. cm.
Includes bibliographical references and index.
    ISBN 0-8058-2967-9 (cloth : alk. paper)
    1. User interfaces (Computer systems)   I. Stephanidis,
Constantine.
    QA76.9.U83 U838   2001
    005.4'28—dc21   99-086232
                                      CIP

Printed in the United States of America
10 9 8 7 6 5 4 3 2 1

*To those who have offered
their commitment towards a vision*

Contents

Part V: Evaluation

Foreword

With the rapid introduction of highly sophisticated computers, (tele)communication, service, and manufacturing systems, a major shift has occurred in the way people use technology and work with it. The objective of this book series on Human Factors and Ergonomics is to provide researchers and practitioners a platform where important issues related to these changes can be discussed, and methods and recommendations can be presented for ensuring that emerging technologies provide increased productivity, quality, satisfaction, safety, and health in the new workplace and the Information Society.

The present volume is published at a very opportune time, when the Information Society Technologies are emerging as a dominant force, both in the workplace, and in everyday life activities. In order for these new technologies to be truly effective, they must provide communication modes and interaction modalities across different languages and cultures, and should accommodate the diversity of requirements of the user population at large, including disabled and elderly people, thus making the Information Society universally accessible, to the benefit of mankind. The present volume provides a comprehensive account of the state of the art in *User Interfaces for All*, as it has been evolving through recent research and development efforts worldwide. Furthermore, it assesses alternative solutions, examines their technological feasibility, practical usability and economic viability, and identifies areas where additional work is needed. The book adopts a multidisciplinary perspective spanning across theory, technology, and policy.

The book consists of thirty chapters written by the world's foremost authorities from industry, academia, research organizations and government from Europe, USA, and Asia. It provides a systematic structured approach to the design, implementation and evaluation of user interfaces in the context of contemporary public research policies, so that the emerging Information Society would be accessible to, and usable by, all individuals irrespective of their abilities, skills, preferences, and educational, cultural, and professional background.

For providing an appropriate platform to achieve these objectives, the editor of the book and the authors of the chapters are to be congratulated.

—*Gavriel Salvendy*
Series Editor

Preface

The proliferation of computer-based systems and applications in every walk of life, and the anticipated widespread use of the emerging telematic services, have introduced new dimensions into the field of Human–Computer Interaction (HCI). Human activities are increasingly becoming mediated by computers; the now common use of computing systems in the workplace and their widespread employment in a variety of business tasks has replaced the initial view of computers as scientific devices with the perception of computers as tools for productivity enhancement. Moreover, with the advent of the information society, the creation of new virtual spaces is bringing about yet another paradigm shift, whereby computers are no longer conceived as mere business tools, but as integrated environments of use, available to *anyone, anytime, anywhere*. Therefore, in the information era, it is important to design high-quality user interfaces, accessible and usable by a diverse user population with different abilities, skills, requirements, and preferences. Citizens/users in the information society are people with different cultural, educational, training, and employment background, novice and experienced computer users, the very young and the elderly, as well as people with different types of disabilities. Human–computer interfaces should not only support more effective and efficient user interaction, but also address the individual end-user requirements and expectations in the variety of contexts of use to be encountered.

In this context, *user interfaces for all* is defined as a systematic approach to the design, implementation, and evaluation of user interfaces that cater to the requirements of the broadest possible user population. The scope of user interfaces for all as a perspective on HCI, is necessarily broad and complex, involving challenges that pertain to issues such as context-oriented design, diverse user requirements, and adaptable and adaptive interactive behaviors. This complexity arises from the numerous dimensions that are involved, and the multiplicity of aspects in each dimension. Accordingly, the fields of science that might provide useful insights toward user interfaces for all are many and diverse. This is not essentially different from the history of HCI itself, in the course of which the field has grown and continues to develop strong links with a variety of scientific disciplines, such as psychology (e.g., information processing and developmental psychology), linguistics and the social sciences (e.g., anthropology, sociology, etc.), and so forth. With regards to technology, the concept of user interfaces for all links with recent advances in user interface software, including software engineering, advanced inter-action platforms, tools for development, and so on.

The present book aims to bring together complementary perspectives spanning across theory, technology, and policy to address a two-fold objective. The first is to provide a comprehensive account of the state of the art in user interfaces for all, as it has been evolving through recent research and development (R&D) efforts. The second objective is to project into the future and identify areas where additional work is needed, as well as to assess alternatives and examine their economic feasibility in terms of adoption and diffusion.

The book is structured in eight parts, each containing a collection of chapters addressing specific research issues. Part I, "Introduction," consists of chapter 1, "User Interfaces for All: New Perspectives Into Human–Computer Interaction" by C. Stephanidis, which introduces the concept of user interfaces for all, and elaborates on its implications on HCI research and practice, at the dawn of a new era, often referred to as the Information Society.

Part II, "Dimensions," unfolds some of the dimensions in HCI that are relevant to the aims and objectives of user interfaces for all, elaborating the sources of diversity that are considered important in HCI design, and postulating the critical role of the study of context. Chapter 2, "Individual Differences and Inclusive Design" by D. Benyon, A. Crerar, and S. Wilkinson, focuses on human diversity and the role of individual differences in designing advanced interactive software. It provides a comprehensive account of the type and range of human parameters that shape inclusive design, and reveals the complexity involved in dealing with the widest possible end-user population.

Chapter 3, "International and Intercultural User Interfaces" by A. Marcus, unfolds another dimension of diversity resulting from culture and language, and highlights the critical turning points in designing interactive software for international use. The chapter points out that studying different users and contexts of use of interactive software is no longer exclusively the premise of dedicated research communities (e.g., assistive technology) but a critical factor of success in a global market situation.

Chapter 4, "Intelligent User Interfaces for All" by M. T. Maybury, overviews the roots and premises of intelligent interface technologies. It highlights recent advances and achievements towards a vision of joint cognitive systems, whereby interactive software exhibits adaptive control, and can cooperate with the human partner in the accomplishment of tasks.

Chapter 5, "Interaction Technologies beyond the Desktop" by L. Bass, summarizes recent results in user interface software technologies and points out new capabilities enabled by novel interactive technologies. The chapter highlights the changing context in the use of interactive devices and forecasts a wide range of new virtual spaces, which extend our current conceptions of the desktop computer and the scope of computer-mediated human activities.

Finally, Part II is complemented by two accounts of special needs and disability access, chapter 6, "Special Needs and Enabling Technologies" by P. L. Emiliani, and chapter 7, "Everyone Interfaces" by G. C. Vanderheiden and S. L. Henry. Chapter 6 reviews a decade of experiences gained through European collaborative R&D efforts aimed at advancing the understanding of users with special needs, their specific accessibility problems and how the latter can be addressed by employing advanced information technology and telecommunications. In a similar fashion, chapter 7 highlights the vision of "everyone interfaces" and provides illustrative examples of how two seemingly different objectives—designing systems that work in a wide range of environments and designing systems for a wide range of user characteristics—have similar solutions. The chapter argues that when we manage to design a system that is truly universal and mobile, we will have created a system that is accessible to almost anyone with a physical or sensory disability.

Part III, "Design," focuses on well-established as well as emerging design approaches, their underlying assumptions and corresponding implications in the design of user interfaces for all. Chapter 8, "Theory and Practice From Cognitive Science" by M. Wilson, provides a comprehensive account of cognitive science as a perspective into HCI design and evaluation. The author reviews the premises of cognitive inquiry, identifies alternative cognitive research strands and their respective deliverables, and contrasts perspectives to justify the role of cognitive science as an influential line of research work in the broad area of HCI, while, at the same time, it briefly introduces many of the assumptions that underpin the cognitive approach.

Chapter 9, "From Programming Environments to Environments for Designing" by T. Winograd, is an adapted version of a previously published article in the *Communications of the ACM* journal [Winograd, T. (1995). Environments for software design. *Communications of the ACM, 38* (6), 65–74.]. The author questions the prevailing programming-intensive ground for HCI, and advances a proposal for establishing a more design-oriented basis. The chapter offers several useful insights into how HCI can be improved with regard to its conduct (i.e., tools for the practitioner) and outcomes (i.e., interactive software for end-users).

Chapter 10, "From Human–Computer Interaction to Computer-Mediated Activity" by L. J. Bannon and V. Kaptelinin, views HCI design from an alternative perspective, namely that of *activity theory*, which is rooted in developmental psychology and in particular in the Vigotskian perspectives on cultural/historical psychology.[1] Activity theory constitutes both a challenge and an opportunity for HCI researchers. The challenge amounts to becoming accustomed to the new footing on which the HCI inquiry is placed and the need for skilled and competent inquiry. The opportunity, according to the advocates of the approach, is that activity theory

---

[1]It should be mentioned that activity theory is not the only research effort that fosters more developmental perspectives into HCI design. Equally competent accounts include distributed cognition [Hutchins, E. (1995). *Cognition in the wild*. Cambridge, MA: MIT Press] and situated action models [Suchman, L. A. (1987). *Plans and situated actions: The problem of human machine communication*. Cambridge, England: Cambridge University Press], which could not be addressed in this book. Activity theory has been selected due to its popularity among HCI researchers, as well as to provide an account of a representative effort rooted in developmental frames of reference.

can provide a powerful prescriptive frame of reference for obtaining richer insights into, and a deeper understanding of a wider range of HCI phenomena.

Chapter 11, "Sociological Issues in HCI Design" by M. Pieper, offers another informative perspective into HCI design, emphasizing the role and potential contributions of the social sciences, and in particular sociology.

Finally, chapter 12, "Generating Design Spaces: An NLP approach to HCI design" by M. Antona, D. Akoumianakis, and C. Stephanidis, presents an on going effort to exploit natural-language-processing techniques, and in particular unification-based formalisms and grammars, to facilitate more analytical insights into HCI design.

Part IV, "Software Technologies and Architectural Models," provides a comprehensive collection of recent technological contributions by leading researchers, academics, and software vendors. These chapters depict not only the state of the art in software technologies and architectural models, but also some of the on going efforts by industry to provide technology for building innovative applications and services. Chapter 13, "The FRIEND21 Framework for Human Interface Architectures" by H. Ueda, offers an overview of the FRIEND21 project. FRIEND21 was one of the early collaborative basic research efforts to address and realize, as early as the mid-1980s, several of the notions that have since prevailed (e.g., media fusion, pluggable look and feel, adaptive change, etc.). A distinctive property of FRIEND 21 was its focus on two of the themes that have been at the core of HCI since its origin, namely HCI design and user interface architecture. The results of the project offer useful insights to both these issues, and are still highly relevant to the HCI research community.

Chapter 14, "User Modeling for Adaptation" by J. Kay, provides a comprehensive account of user modeling, a line of research work that, for more than a decade now, has provided critical tools for implementing interactive software capable of exercising adaptable and adaptive control. The author unfolds the criticality of adaptation as a technique to accommodate diverse user requirements and contexts of use, and highlights some of the outcomes (e.g., techniques and tools) of user-modeling research that can be used to implement systems exhibiting intelligent and co-operative interactive behavior.

Chapter 15, "Interface Agents: A New Interaction Metaphor and Its Application to Universal Accessibility" by A. Waern and K. Höök, provides a review of an emerging interaction paradigm, namely agent-based interaction. Interface agents constitute an alternative, more cooperative and distributed insight into how computer-mediated human activities can be conceived, organized, and implemented. The authors review recent practice and experience in an attempt to extrapolate and assess the relevance of this emerging perspective on universal accessibility. Interestingly, their conclusions impinge on two prevalent themes throughout the book, namely HCI design and architectural models for interactive software.

Part IV concludes with a review of the efforts of a major industrial software vendor, namely Sun Microsystems, to provide accessibility hooks into mainstream technologies. Chapter 16, "Accessibility in the Java™ Platform" by P. Korn and W. Walker, describes the accessibility hooks available in the Java™ platform. This chapter also depicts the evolution in the prevailing industrial practices, and the greater awareness of the challenges and opportunities resulting from the capability to address diverse user requirements through standardized, well-documented, and reliable software tools.

Part V, "Evaluation," unfolds issues related to evaluation in the context of user interfaces for all, and is composed of two chapters that highlight indicative challenges related to evaluating interactive systems for diverse user groups and contexts of use. Chapter 17, "User Interface Adaptation: Evaluation Perspectives" by D. Akoumianakis, D. Grammenos, and C. Stephanidis, provides an account of the evaluation challenges posed by interactive computer-based systems exhibiting adaptable and adaptive behavior. In particular, the authors review past and recent experiences in the evaluation of adaptable and adaptive systems, and point out some of the reasons why the traditional performance-oriented comparative evaluation paradigm may not provide the insights needed. The authors' conclusions link with those of other researchers who identify the compelling need for context-oriented assessments, which, in turn, reinforces the quest for more analytical and developmental insights into HCI design to reveal, unfold, model, and prescribe context in meaningful and measurable terms.

Chapter 18, "Quality in Use for All" by N. Bevan, reviews recent work on quality in use, examines the interplay between quality concepts and usability engineering, and highlights their potential relevance to the study of user interfaces for all. The discussion reveals that, whereas from a process-oriented perspective much of what is "preached" and practiced by usability professionals can directly contribute to advancing user interfaces for all, the currently available instruments do not suffice to accommodate the peculiarities of accessibility. In other words, the research challenge is to develop, validate, and deploy cost-effective techniques enabling designers to assess their tentative and final outcomes with regard to adherence to well-defined accessibility principles.

Part VI, "Unified User Interfaces," presents unified user interface development as one possible pathway towards user interfaces for all. The chapters in this part introduce and elaborate upon different aspects of unified user interfaces. In particular, chapter 19, "The Concept of Unified User Interfaces" by C. Stephanidis, discusses some distinctive properties that qualify unified user interface development as a promising HCI strand. Chapter 20, "The Unified User Interface Software Architecture" by A. Savidis and C. Stephanidis, proposes an architectural framework for structuring unified user interface software. Chapter 21, "The Unified User Interface Design Method" by A. Savidis, D. Akoumianakis, and C. Stephanidis, presents a user interface design method tailored to the development requirements introduced by unified user interfaces. Chapter 22, "Development Requirements for Implementing Unified User Interfaces" by A. Savidis, and C. Stephanidis, presents the required and recommended properties that user interface development tools should support, so as to enable the effective and efficient implementation of unified user interfaces. Chapter 23, "USE-IT: A Tool for Lexical Design Assistance" by D. Akoumianakis and C. Stephanidis, and chapter 24, "The I-GET UIMS for Unified User Interface Implementation" by A. Savidis and C. Stephanidis, present two tools that facilitate, respectively, the design and implementation phases of unified user interface development. Finally, chapter 25, "A Case Study in Unified User Interface Development: The AVANTI Web Browser" by C. Stephanidis, A. Paramythis, M. Sfyrakis, and A. Savidis, presents a comprehensive case study of unified user interface development in the user interface of the AVANTI system.

Part VII, "Support Measures," complements the technological flavor of the previous chapters by providing a policy perspective that identifies some of the nontechnical issues pertaining to user interfaces for all. Specifically, Part VII provides an account of possible support measures that are deemed necessary to accompany technological developments in order to facilitate the wider diffusion and adoption of the principles and practices of user interfaces for all.

The first chapter in this part, chapter 26, "Making the Web Accessible" by D. Dardailler, J. Brewer, and I. Jacobs, focuses on the World Wide Web as an infrastructure, and provides an account of recent efforts aiming to increase its accessibility. Particular reference is given to the efforts initiated by the World Wide Web Consortium's Web Accessibility Initiative (W3C), describing some of the key results to date.

Chapter 27, "Industrial Policy Issues" by C. Stephanidis et al., elaborates some of the industrial policy issues that underpin further deployment and the successful adoption of the principles of user interfaces for all. To this effect, a brief description of the opportunities offered by the Assistive Technology sector and the mainstream industry are discussed, pointing to some of the impediments that need to be addressed by each sector in order to appropriate the benefits of user interfaces for all. Additionally, alternative strategies are presented, through which accessibility can be promoted.

Finally, chapter 28, "Economics and Management of Innovation" by N. Vernardakis, D. Akoumianakis, and C. Stephanidis, provides an alternative insight into the concept of user interfaces for all, from the perspective of innovation diffusion theory. A parallel is drawn with recent experiences in the diffusion of Java so as to provide a more informative account of the issues being discussed. One issue that emerges very strongly is that, independent from other intrinsic properties of innovation, the role of nonmarket institutions is critical to promoting a culture that is more sensitive and respectful of diversity—irrespective of its source.

The final part of the book, Part VIII, "Looking to the Future," projects into the future by consolidating recent progress in the area of *design for all* in HCI, and providing an account of critical research areas in the context of an emerging Information Society. The two chapters of this part, chapter 29, "Toward an Information Society for All: An International R&D Agenda" by C. Stephanidis et al., and chapter 30, "Toward an Information Society for All: HCI Challenges and R&D Recommendations" by C. Stephanidis et al., are reprints of two White Papers published in the *International Journal of Human–Computer Interaction* [Stephanidis, C., et al. (1999). Toward an information society for all: HCI challenges and R&D recommendations. *International Journal of Human–Computer Interaction, 11*(1), 1–28; Stephanidis, C., et al. (1998). Toward an information society for all: An international R&D agenda. *International Journal of Human–Computer Interaction, 10*(2), 107–134]. The two articles are the outcome of the first two meetings of the International Scientific Forum "Towards an Information Society for All," held in San Francisco, August 29, 1997, and in Heraklion, Crete, Greece, June 15–16, 1998, respectively.

—Constantine Stephanidis

Acknowledgments

This book would not have been possible without the tireless efforts of many people: authors of chapters, members of my research group at ICS-FORTH who assisted in the long and laborious editing process, and the production team of the publisher. To each and every contributor who helped making this book a reality I would like to convey my profound and sincere appreciation.

There are two people that I would like to mention by name. Pier Luigi Emiliani is the colleague and friend who over the years offered his faith, trust and unwithering support throughout the many and often difficult phases of research and development work that led to some of the results reported in this book. Gavriel Salvendy, the Editor of the LEA series under which this book is published, has been a source of inspiration and encouragement in carrying out this project to completion.

I would also like to thank the European Commission for partially funding some of the research and development work the results of which are reported in this book, as well as the cooperating partners in the following projects:

- IPSNI (R1066), partially funded by the RACE Programme of the European Commission (DG XIII). Partners of the IPSNI consortium are: IROE-CNR, Italy; KUL, Belgium; IRV, The Netherlands; CSELT, Italy; ICS-FORTH, Greece; Technical Research Centre of Finland, Finland.
- IPSNI-II (R 2009), partially funded by the RACE Programme of the European Commission (DG XIII). Partners of the IPSNI-II consortium are: IROE-CNR, Italy; DUMC, United Kingdom; IRV, The Netherlands; CSELT, Italy; KUL, Belgium; ICS-FORTH, Greece; VTT, Finland.
- GUIB (TP103), partially funded by the TIDE (Technology Initiative for the Disabled and Elderly People) Pilot Action Programme of the European Commission (DG XIII). The partners of the GUIB consortium are: IROE-CNR, Italy; F H Papenmeier GmbH & Co KG, Germany; IFI-University of Stuttgart, Germany; ICS-FORTH, Greece; RNIB, England; Institute of Telecommunications-TUB, Germany; Department of Computer Science-FUB, Germany; Vrije Universiteit Brussel, Belgium; VTT, Finland.

- GUIB-II (TP 215) project, partially funded by the TIDE Programme of the European Commission (DG XIII). Partners of the GUIB-II consortium are: IROE-CNR, Italy; ICS-FORTH, Greece; Vrije Universiteit Brussels, Belgium; Department of Computer Science-FUB, Germany; Institute of Telecommunications-TUB, Germany; IFI, University of Stuttgart, Germany; VTT, Finland; RNIB, England; F.H. Papenmeier Gmb & Co KG, Germany.
- TIDE - ACCESS (TP 1001) project, partially funded by the TIDE Programme of the European Commission (DG XIII). The partners of the ACCESS consortium are: CNR-IROE (Italy); ICS-FORTH (Greece); University of Hertforshire (United Kingdom); University of Athens (Greece); NAWH (Finland); VTT (Finland); Hereward College (United Kingdom); RNIB (United Kingdom); Seleco (Italy); MA Systems & Control (United Kingdom); PIKOMED (Finland).
- ACTS-AVANTI AC042 project, partially funded by the ACTS Programme of the European Commission (DG XIII). The partners of the AVANTI consortium are: ALCATEL Italia, Siette division (Italy); IROE-CNR (Italy); ICS-FORTH (Greece); GMD (Germany), VTT (Finland); University of Siena (Italy), MA Systems and Control (UK); ECG (Italy); MATHEMA (Italy); University of Linz (Austria); EUROGICIEL (France); TELECOM (Italy); TECO (Italy); ADR Study (Italy).

Finally, I would like to also mention the European Research Consortium for Informatics and Mathematics (ERCIM) Working Group "User Interfaces for All," and thank the colleagues - members for their participation and support.

Constantine Stephanidis
Heraklion, Crete, Greece
May 2000

List of Contributors

Demosthenes Akoumianakis
Institute of Computer Science
Foundation for Research and Techno-
logy - Hellas (FORTH)
Science and Technology Park of Crete
Heraklion, Crete, GR-71110 Greece
email: demosthe@ics.forth.gr

Margherita Antona
Institute of Computer Science
Foundation for Research and Techno-
logy - Hellas (FORTH)
Science and Technology Park of Crete
Heraklion, Crete, GR-71110 Greece
email: antona@ics.forth.gr

Albert Arnold
Delft University of Technology
Work and Organisation Psychology
Unit
De Vries van Heystplantsoen 2
Delft, 2628 RZ, The Netherlands
email: arnold@wtm.tudelft.nl

Liam Bannon
Interaction Design Centre, Foundation
Building
University of Limerick
Limerick, Ireland
email: Liam.Bannon@ul.ie

Len Bass
Software Engineering Institute
Carnegie Mellon University
PA 15213, Pittsburgh, USA
email: ljb@sei.cmu.edu

David Benyon
School of Computing
Napier University
42, Craiglockhart Ave.
EH14 1DJ, Edinburgh, UK
email: d.benyon@dcs.napier.ac.uk

Nigel Bevan
Serco Usability Services
4 Sandy Lane
Teddington
TW11 0DU, Middlesex, UK
email: nbevan@usability.serco.com

Judy Brewer
W3C, World Wide Web Consortium
c/o MIT Laboratory for Computer
Science
545 Technology Square
MA 02139, Cambridge, USA
email: jbrewer@w3.org

Alison Crerar
School of Computing
Napier University
Craiglockhart Campus
219 Colinton Road
EH14 1DJ, Edinburgh, UK
email: a.crerar@napier.ac.uk

Daniel Dardailler
W3C, World Wide Web Consortium
c/o INRIA
2004, Route des Lucioles - B.P. 93
06902, Sophia Antipolis Cedex,
France
email: danield@w3.org

Jan Ekberg
National R&D Centre for Welfare
& Health (STAKES)
Siltasaarenk 18.A
PB 220
FIN-00531, Helsinki, Finland
email: Jan.Ekberg@Stakes.Fi

Pier Luigi Emiliani
Institute of Research on Electro-
magnetic Waves "Nello Carrara"
National Research Council
Via Panciatichi, 64
I-50127 Firenze, Italy
email: ple@iroe.fi.cnr.it

Klaus-Peter Fähnrich
Fraunhofer IAO
Nobelstraße 12
D-70569, Stuttgart, Germany
email: Klaus-Peter.Faehnrich
@iao.fhg.de

Anthony Galetsas
European Commission
Rue de la Loi 200
B-1049 Brussels, Belgium
email: Antonis.GALETSAS@cec.eu.in

Dimitrios Grammenos
Institute of Computer Science
Foundation for Research and Techno-
logy - Hellas (FORTH)
Science and Technology Park of Crete
Heraklion, Crete, GR-71110 Greece
email: gramenos@ics.forth.gr

Seppo Haataja
Nokia Mobile Phones
P.O. Box 68
FIN-33721, Tampere, Finland
email: Seppo.Haataja@nokia.com

Kristina Höök
Swedish Institute of Computer Science
Box 1263
S-164 28 Kista, Sweden
email: kia@sics.se

Ilias Iakovidis
European Commission
Rue de la Loi 200

1049 Brussels, Belgium
email: Ilias.IAKOVIDIS@cec.eu.int

Julie Jacko
Department of Industrial Engineering
University of Wisconsin-Madison
1513 University Avenue
WI 53706 Madison, USA
email: jacko@engr.wisc.edu

Jan Jacobs
W3C, World Wide Web Consortium
c/o MIT Laboratory for Computer
Science
545 Technology Square
MA 02139, Cambridge, USA
email: ij@w3.org

Phill Jenkins
IBM Special Needs
11501 Burnet Road
TX 78758, Austin, USA
email: pjenkins@us.ibm.com

Victor Kaptelinin
Umeå University
Department of Informatics
S-901 87 Umeå, Sweden
email: Victor.Kaptelinin@
informatik.umu.se

Arthur Karshmer
Computer Science Department
New Mexico State University
PO Box 30001
NM 88003, Las Cruces, USA
email: arthur@cs.nmsu.edu

Judy Kay
Basser Dept. of Computer Science
University of Sydney
Madsen F09
2006 Sydney, Australia
email: judy@staff.cs.usyd.edu.au

Erkki Kemppainen
National R&D Centre for Welfare
& Health (STAKES)
P.O. BOX 220
FIN-00531, Helsinki, Finland
email: erkki.kemppainen@stakes.fi

Peter Korn
Sun Microsystems
901 San Antonio Road
MS UCUP02-103
CA 94043, Palo Alto, USA
email: peter.korn@sun.com

Shawn Lawton Henry
Optavia Corporation
613 Williamson Street
WI 53703-3543, Madison,USA
email: SLHenry@optavia.com

Aaron Marcus
Aaron Marcus & Associates
1144 65th Street
CA 94608-1053 Emeryville, USA
email: Aaron@AMandA.com

Mark Maybury
Information Technology Division
  (MS K308)
The MITRE Corporation
202 Burlington Road
MA 01730 Bedford, USA
email: maybury@mitre.org

Harry Murphy
Center on Disabilities
California State University, Northridge
18111 Nordhoff Street
CA 91330-8340, Northridge, USA
email: Harry.Murphy@csun.edu

Charles Oppermann
Microsoft Corporation
Active Accessibility
One Microsoft Way
WA 98052-6399, Redmond, USA
email: chuckop@microsoft.com

Alexandros Paramythis
Institute of Computer Science
Foundation for Research and Techno-
  logy - Hellas (FORTH)
Science and Technology Park of Crete
Heraklion, Crete, GR-71110 Greece
email: alpar@ics.forth.gr

Michael Pieper
German National Research Center for
  Information Technology

Institute for Applied Information
  Technology
Research Division on Human-
  Computer Interaction
Schloss Birlinghoven
D-53754 Sankt Augustin, Germany
email: michael.pieper@gmd.de

Gavriel Salvendy
Purdue University
1287 Grissom Hall
IN 47907-1287, West Lafayette, USA
email: salvendy@ecn.purdue.edu

Anthony Savidis
Institute of Computer Science
Foundation for Research and Techno-
  logy - Hellas (FORTH)
Science and Technology Park of Crete
Heraklion, Crete, GR-71110 Greece
email: as@ics.forth.gr

Michael Sfyrakis
Institute of Computer Science
Foundation for Research and Techno-
  logy - Hellas (FORTH)
Science and Technology Park of Crete
Heraklion, Crete, GR-71110 Greece
email: sfyrakis@ics.forth.gr

Christian Stary
Department for Business Information
  Systems
University of Linz
Freistaedterstrasse 315
A-4040 Linz, Austria
email: stary@ce.uni-linz.ac.at

Constantine Stephanidis
Institute of Computer Science (ICS)
Foundation for Research and Techno-
  logy - Hellas (FORTH)
Science and Technology Park of Crete
Heraklion, Crete, GR-71110 Greece
email: cs@ics.forth.gr

Hiroshi Tamura
Kyoto Institute of Technology
606-8585 Matsugasaki
Sakyoku, Kyoto, Japan
email: tamura@hisol.dj.kit.ac.jp

Manfred Tscheligi
Cure Centre for Usability Research
  & Engineering
Institute for Applied Computer Science
Dept. for Advanced Computer
  Engineering
University of Vienna
Lenaugasse 2/8
A-1080, Vienna, Austria
email: tscheligi@cure.at

Hirotada Ueda
3rd Research Department, R&D
  Laboratory
Hitachi Denshi Ltd.
32 Miyuki-cho, Kodaira-shi
187 Tokyo, Japan
email: hiro-u@po.iijnet.or.jp

Gregg Vanderheiden
Trace R&D Center
385 Mechanical Engineering Building
1513 University Avenue
WI 53705, Madison, USA
email:gv@trace.wisc.edu

Nikolaos Vernardakis
University of PatrasDepartment
  of Economics
GR-265 00 Rion-Patras, Greece
email: vernard@ics.forth.gr

Annika Waern
Swedish Institute of Computer Science
Box 1263
S-16429 Kista, Sweden
email: annika@sics.se

Willie Walker
Sun Microsystems Labs
Two Elizabeth Drive
MA 01824, Chelmsford, USA
email: william.walker@sun.com

Gerhard Weber
Fachhochschule HarzFB
Elektrotechnik/Informatik
Friedrichstr. 57–59
D-38855, Wernigerode, Germany
email: gweber@fh-harz.de

Simon Wilkinson
School of Computing
Napier University
Craiglockhart Campus
219 Colinton Road
EH14 1DJ, Edinburgh, UK
email: s.wilkinson@napier.ac.uk

Michael Wilson
Information Technology Department
CLRC Rutherford Appleton Laboratory
Chilton, DIDCOT
OX11 0QX, Oxon, UK
email: M.D.Wilson@rl.ac.uk

Terry Winograd
Department of Computer Science
Stanford University
Gates Computer Science 3B
CA 94305-9035, Stanford, USA
email: winograd@cs.stanford.edu

Juergen Ziegler
Fraunhofer IAO
Nobelstrasse 12
D-70569 Stuttgart, Germany
email: Juergen.Ziegler@iao.fhg.de

I

Introduction

1

User Interfaces for All:
New Perspectives into
Human–Computer Interaction

Constantine Stephanidis

This chapter introduces the notion of User Interfaces for All, *elaborates on the motivating rationale and examines its key implications on Human-Computer Interaction. The underlying vision of* User Interfaces for All *is to provide an approach for the development of computational environments catering for the broadest possible range of human abilities, skills, requirements and preferences. Consequently,* User Interfaces for All *should not be conceived as an effort to advance a single solution for everybody. Instead, it is a new perspective into Human-Computer Interaction, seeking to unfold and reveal challenges and insights, and to instrument appropriate solutions for alleviating the current obstacles to the access and use of advanced information technologies by the widest possible end-user population.*

INTRODUCTION

Human–computer interaction (HCI) is concerned with the design, implementation, and evaluation of interactive computer-based systems, as well as with the multidisciplinary study of various issues affecting this interaction. The aim of HCI is to ensure the safety, utility, effectiveness, efficiency, accessibility, and usability of such systems. The user interface is the part of an interactive system, application, or telematic service with which the user comes into contact cognitively, perceptually, and physically.

In recent years, HCI has attracted considerable attention by the academic and research communities, as well as by the information technology and telecommunications (IT&T) industry. An early argument in favor of HCI studies, from the perspective of system development, was the fact that the programming effort for the user interface, in many cases, exceeds 50% of the total programming effort required for the development of the entire system. Though this is generally true, today there are other more compelling reasons why HCI is becoming critical in the emerging information society.

The proliferation of computer-based systems and applications, and the widespread use of emerging telematic services, have introduced new dimensions to the issue of HCI. Human activities are increasingly becoming mediated by computers, thus replacing the initial view of computers as scientific devices with the percep-

3

tion of computers as tools for productivity enhancement. With the advent of the information society, the creation of new virtual spaces is likely to bring about yet another paradigm shift, whereby computers are no longer conceived as mere business tools, but as integrated environments, accessible by *anyone, anytime,* and *anywhere.* Therefore, in the information era, it is important to develop high-quality user interfaces, *accessible* and *usable* by a diverse user population with different abilities, skills, requirements, and preferences, in a variety of contexts of use, and through a variety of different technologies.

This chapter introduces the concept of *user interfaces for all* as a new perspective into HCI that aims at anticipating, analyzing, and tackling the challenges posed in the development of interactive applications and telematic services in the emerging information society. The chapter analyzes the ongoing paradigm shift, its driving factors and trends, and the challenging requirements it poses. The historical context of user interfaces for all is briefly described, and the implications on HCI are unfolded. Furthermore, the chapter provides an account of the new perspectives and insights that this concept brings about in the design, implementation, and evaluation of interactive software accessible and usable by *all* users in a variety of contexts of use.

THE PARADIGM SHIFT

The radical innovation in the IT&T sectors, the ever-growing demand for information access, and the diffusion of computers across industry sectors and application domains, are the driving forces of the ongoing paradigm shift toward an information society.

The Trends

From the early calculation-intensive tools that were prevalent in the 1970s, computer-based systems are progressively becoming tools for communication, collaboration, and social interaction among groups of people. From a specialist's device, the computer is being transformed into an information appliance for the citizen in the information society. It follows, therefore, that designers increasingly have to provide information artifacts to be used by diverse user groups, including people with different cultural, educational, training, and employment background, novice and experienced computer users, the very young and the elderly, and people with different types of disabilities (see chaps. 2 and 3, this volume).

Similarly, the context of use is changing. The "traditional" use of computers (i.e., scientific use by the specialist, business use for productivity enhancement) is increasingly being complemented by residential and nomadic use, thus penetrating a wider range of human activities in a broader variety of environments, such as the school, the home, the marketplace, and other civil and social contexts. As a result, information artifacts should embody the capability to interact with the user in all those contexts, independently of location, target machine, or run-time environment. Usability in such "nontraditional" usage contexts is likely to prove a harder target to meet than in the case of the workplace (Stephanidis & Akoumianakis, 1996).

Finally, technological proliferation contributes with an increased range of systems or devices to facilitate access to the community-wide pool of information resources (see, e.g., chap. 5, this volume). These devices include computers, standard telephones, cellular telephones with built-in displays, television sets, information kiosks, special information appliances, and various other "network-attachable" devices. Depending on the context of use, users may employ any of the aforementioned to review or browse, manipulate, and configure information artifacts, at any time.

HCI Challenges

The trends outlined previously bring about radical implications on the current and future HCI research focus and perspectives. Until recently, interaction in the context of interactive systems was considered as a dialogue between humans and computer-based embodiments of real-world analogies, conveyed predominantly in a visual manifestation. Design was largely conceived as the art of constructing this dialogue at different levels (i.e., semantic, syntactic, or lexical). Design knowledge, therefore, reflected the accumulated experience in performing design activities at various levels using suitable tools and techniques.

The aforementioned conceptions have changed significantly due to radical technological progress brought about to fulfil the increasing demand for information by an ever-growing information "client base." As a result, the scope and range of computer-mediated human activities have been expanded to facilitate more sophisticated business tasks of the "knowledge" worker, as well as a broad range of residential activities by nonexperts. Additionally, the emergence of novel application domains (e.g., World Wide Web) and advanced interaction technologies (e.g., mobile devices, wearable equipment) progressively shifts the focus of design to consider virtualities beyond the traditional desktop computer, and a variety of, until now, unconventional usage contexts. Consequently, user interactions with computer-embodied artifacts have become more complex to envision and design.

It is frequently argued that the hardest target to achieve en route to the information society is the design of new computer-embodied artifacts to facilitate the broad range of emerging activities (see chaps. 29 and 30, this volume; see also Winograd, 1997). For this to be attained, HCI should revisit some of the basic assumptions that have shaped recent work and progress. Some of these are briefly discussed next.

The "Average" Typical User.   One of the basic guidelines for HCI design is to study the user. Many recent methodological frames of reference introduce this guideline as a milestone for subsequent stages in the process. Additionally, there has been a wide variety of instruments for capturing requirements and studying users, such as questionnaires (e.g., *Usability Context Analysis;* Bevan & Macleod, 1994), checklists (e.g., *EVADIS;* Reiterer & Oppermann, 1995), formal and semiformal methods (e.g., *cognitive walkthrough;* Lewis & Polson, 1990; Lewis, Polson, Wharton, & Rieman, 1990), as well as various *ethnographically inspired approaches* (Simonsen & Kensing, 1997).

Notwithstanding the relative merits of each method or tool, it is important to mention that their usefulness should be carefully judged. This is not due to the validity of the techniques, but rather to the changing conceptions and contexts in studying users. More specifically, in the context of the emerging distributed and communication-intensive information society, users are not only the computer-literate, skilled, and able-bodied workers driven by performance-oriented motives, nor do users constitute a homogeneous mass of information-seeking actors with standard abilities, similar interests, and common preferences with regard to information access and use. Instead, our conception of users should accommodate all potential citizens, including the young and the elderly, residential users, as well as those with situational or permanent disability. Consequently, it becomes increasingly complex for designers to know the users of their products, and compelling to design for the broadest possible end-user population. This raises implications on design methodology and instruments, as well as on the technical and user-perceived qualities to be delivered.

The Business Environment.   Another assumption that has influenced HCI throughout its short history has been the context of use in which computer-based designs are encountered. Due to the unlimited business demand for information processing, the HCI community has progressively acquired a bias and a habitual tendency toward outcomes (i.e., theories, methods, and tools) that satisfy the business requirements and demonstrate performance improvements and productivity gains. However, since the early 1990s, analysts have been concerned with the increasing residential demand for information, which is now anticipated to be much higher than its business counterpart. Consequently, designers should progressively adapt their thinking to facilitate a shift from designing tools for productivity improvement, to designing computer-mediated environments of use. This leads to the next point, which addresses the need to extend the prevailing metaphors for interaction design to suit the changing requirements.

The Desktop Computer Embodiment.   The desktop embodiment of current interfaces is perhaps the most prominent innovation delivered by the user interface software industry, including the tool development sector. However, the diffusion of the Internet as an information highway and the proliferation of advanced interaction technologies (e.g., mobile devices, network attachable equipment) signify that many of the tasks to be performed by humans in the information age will no longer be bound to the visual desktop. New metaphors are likely to prevail in the design of the emerging virtual spaces, reflecting the broader type and range of computer-mediated human activities. Arguably, these metaphors should encapsulate an inherently social and communication-oriented character in order to provide the guiding principles and underlying theories for designing more natural and intuitive computer embodiments. Consequently, the challenge lies in finding powerful themes and design patterns to shape the construction of novel communication spaces. At the same time, it is more than likely that no single design perspective, analogy, or metaphor will be adequate for all potential users or computer-mediated human activities. Design will increasingly entail the articulation of diverse concepts, deeper knowledge, and more powerful representations to describe the broader range and scope of interaction patterns and phenomena.

USER INTERFACES FOR ALL—AN OVERVIEW

The term user interfaces for all denotes an effort to unfold and reveal the aforementioned challenges, as well as to provide insights and instrument appropriate solutions in the HCI field. The underlying vision of user interfaces for all is to offer an approach for developing computational environments that cater to the broadest possible range of human abilities, skills, requirements, and preferences. Consequently, user interfaces for all should not be conceived as an effort to advance a single solution for everybody, but rather, as a new perspective on HCI that alleviates the obstacles pertaining to *universal access* in the information society.

The roots of user interfaces for all can be traced in the notions of universal access and *design for all*. The term design for all (or *universal design*—the terms are used interchangeably) is not new. It is well known in several engineering disciplines, such as, for example, civil engineering and architecture, with many applications in interior design, building and road construction, and so on (Story, 1998). Though existing knowledge may be considered sufficient to address the accessibility of physical spaces, this is not the case with information society technologies, where universal design is still posing a major challenge. Universal access to computer-based applications and services implies more than direct access or access through add-on (assistive) technologies, because it emphasizes the principle that accessibility should be a design concern, as opposed to an afterthought. To this end, it is important that the needs of the broadest possible end-user population are taken into account in the early design phases of new products and services.

Several previous efforts addressed the accessibility of computer-based applications and services by disabled and elderly people through the a posteriori adaptation of interactive software. This amounts to a "reactive" approach, whereby assistive technology experts attempted to respond to contemporary technological developments by building accessibility features into interactive applications, as a result of specific user requirements. Such efforts to account for accessibility are, however, mainly governed by intuition and usually follow ad hoc procedures that may lead to suboptimal solutions with respect to user requirements (see chap. 6, this volume).

Although the reactive approach to accessibility may be the only viable solution in certain cases (Vanderheiden, 1998), it suffers from some serious shortcomings, especially when considering the radically changing technological environment, and, in particular, the emerging information society technologies. First, reactive approaches, based on a posteriori adaptations, though important to partially solve some of the accessibility problems of people with disabilities, are not viable in sectors of the industry characterized by rapid technological change. By the time a particular access problem has been addressed, technology has advanced to a point where the same or a similar problem re-occurs. The typical example that illustrates this state of affairs is the case of blind people's access to computers. Each generation of technology (e.g., DOS environment, windowing systems, and multimedia) caused a new "generation" of accessibility problems to blind users, addressed through dedicated techniques, such as text-to-speech translation for the DOS environment, off-screen models, and filtering for the windowing systems.

In some cases, adaptations may not be possible without loss of functionality. For example, in the early versions of windowing systems, it was impossible for the programmer to obtain access to certain window functions, such as window management. In subsequent versions, this shortcoming was addressed by the vendors of such products allowing certain adaptations on interaction objects on the screen.

Finally, adaptations are programming-intensive, which raises several considerations for the resulting products. Many of them bare a cost implication that amounts to the fact that adaptations are difficult to implement and maintain. Minor changes in product configuration, or the user interface, may result in substantial resources being invested to rebuild the accessibility features. The situation is further complicated by the lack of tools to facilitate "edit-evaluate-modify" development cycles (Stephanidis, Savidis, & Akoumianakis, 1995).

Due to the aforementioned shortcomings of the reactive approach, there have been proposals and claims for proactive strategies, resulting in generic solutions to the problem of access. One such proposal is user interfaces for all, which constitutes an attempt to apply, exemplify, and specify the principles of universal access and design for all in the context of HCI.

Proactive strategies entail a purposeful effort to build access features into a product, as early as possible (e.g., from its conception, to design and release). In the context of HCI, user interfaces for all advocates such a proactive paradigm for the development of systems accommodating the broadest possible end-user population (Stephanidis, 1995). In other words, user interfaces for all seeks to minimize the need for a posteriori adaptations and deliver products that can be adapted for use by the widest possible end-user population (adaptable user interfaces). This implies the provision of alternative interface manifestations depending on the abilities, requirements, and preferences of the target user groups.

Design for all is often criticized on the grounds of practicality and cost justification. In particular, there is a line of argumentation raising the concern that "many ideas that are supposed to be good for everybody aren't good for anybody" (Lewis & Rieman, 1994, Section 2.1, Paragraph 3). However, universal design in IT&T products should not be conceived as an effort to advance a single solution for everybody, but as a user-centered approach to providing products that can automatically address the possible range of human abilities, skills, requirements, and preferences. Another common argument is that universal design is too costly (in the short term) for the benefits it offers. Though the field lacks substantial data and comparative assessments as to the costs of designing for the broadest possible population, it has been argued that (in the medium to long term) the cost of inaccessible systems is comparatively much higher, and is likely to increase even more, given the current statistics classifying the demand for accessible products (Bergman & Johnson, 1995; Vanderheiden, 1990). What is really needed is economic feasibility in the long run, leading to versatility and economic efficiency (see chap. 28, this volume; see also Vernardakis, Stephanidis, & Akoumianakis, 1997).

The concept of user interfaces for all was originally introduced in 1995 (Stephanidis, 1995), following the results of several research initiatives in the context of collaborative project work cofunded by the European Commission. The ACCESS project (1994–1996; see Acknowledgments) was one of the first efforts to develop user interface development tools that would provide a vehicle toward user interfaces for all (ACCESS Technical Annex, 1993). ACCESS delivered a powerful methodology, the *unified user interface* development methodology, and a novel platform for designing and implementing user- and use-adapted interactions (see Part IV, this volume). The unified user interface methodology was demonstrated in specific applications in the field of disability in the context of the ACCESS project. Subsequently, it was applied by the AVANTI project (1995–1998; see Acknowledgments), which developed a unified Web browser inherently accessible by different categories of users with disabilities (see also chap. 25, this volume). Besides these demonstrations, the concepts addressed in the ACCESS project (e.g., development of nonvisual interaction metaphors, encapsulation of alternative dialogue patterns, platform abstraction, generation of user interface implementation through executable specifications rather than programming) proved to have wider significance. In particular, many of these concepts have been endorsed and adopted by recent accessibility initiatives, industrial organizations and consortia, and nonmarket institutions. In subsequent chapters of the volume, the reader will find the details about some of these developments.

IMPLICATIONS ON HCI

Universal access, and by implication user interfaces for all, introduce challenges that cover the broad spectrum of HCI work, including design, development, and evaluation. Some of these challenges are briefly addressed in an attempt to reveal new research issues that should guide ongoing and future efforts in this direction.

User Interfaces for All and HCI Design

In the relatively short history of HCI, several approaches have been developed to address the design of computer-based interactive artifacts. The paradigms of human factors evaluation (Salvendy, 1997), cognitive science (see chap. 8, this volume), and more recently, user-centered design (Norman & Draper, 1986) have been the most prominent. Human factors evaluation and cognitive science have been criticized with respect to the underlying scientific ground and the means (e.g., instruments, methods, and tools) they use for achieving the set objectives. The HCI literature reports many ongoing debates (e.g., Bødker, 1991; Carroll, 1991; Nardi, 1996; Suchman, 1987; Winograd & Flores, 1987), and offers an insight into the diversity that characterizes the field.

The realization of the shortcomings underpinning the previous two design approaches led many researchers to recognize that, on the one hand, traditional large-scale user testing is both costly and suboptimal, whereas, on the other hand, any attempt for a generalized model or theory for designing interactions between humans and machines is simply not feasible (see chap. 8, this volume). Instead, techniques are needed that focus on the requirements of end-users, and provide early feedback to design, so as to reduce the cost of design defects and to meet specific usability objectives. The term coined to characterize this approach was *user-centered design,* and the first comprehensive collection of articles on the topic appeared in Norman and Draper (1986). The normative perspective of user-centered design is to fulfill the need for usability by providing techniques that foster tight design-evaluation feedback loops, iterative prototyping, early design input, and end-user feedback. In subsequent years, and due to the compelling need to cost-justify usability throughout a product's life cycle, the field moved toward a variety of techniques, generally referred to as *inspection-based evaluation* (Nielsen, 1993), that, though inexpensive, are less formal in their conduct and deliverables. Following several success stories in the use and cost justification of these techniques, the consolidated experience gave rise to a generally applicable process model for constructing human-centered systems. Human-centered design is documented in a draft International Standards Organisation (ISO) document (ISO 13407, 1997), which provides a principled approach and guidelines for attaining usability.

In addition to the developments leading to user-centered design, there have been several proposals for remedying the shortcomings of human factors evaluation and information-processing psychology. These proposals stem from the social sciences, and aim to improve the experimental grounds for HCI design by emphasizing the need for analytical, prescriptive frameworks for studying users in specific contexts of use. Existing attempts toward the objective just discussed make suggestions for:

- Better exploitation of the existing knowledge and science base by utilizing the experimental ground of developmental approaches to psychology (e.g., cultural/historical/work psychology) and the social sciences (e.g., anthropology, sociology, humanities; see chap. 8, this volume).
- Broadening the range, or even extending the scope of information-processing psychology with concepts from developmental approaches to HCI, such as *activity theory* (Bødker, 1989, 1991; see also chap. 10, this volume), *language/action theory* (Winograd, 1988), *situated action models* (Suchman, 1987), and *distributed cognition* (Norman, 1991).

The normative perspective adopted in these efforts is that interactions between humans and information artifacts should be studied in specific *social contexts* and take account of the distinctive properties that characterize them. Despite this common commitment to the study of context, the aforementioned alternatives differ with regard to at least three dimensions, namely the unit of analysis in studying context, the categories offered to support a description of context, and the extent to which each approach treats actions as structured prior or during human activities (Nardi, 1996).

In the context of user interfaces for all, user-centered design, as well as the previously mentioned emerging approaches, have several contributions to make. First of all, the human-centered protocols and tight design-evaluation feedback loop of user-centered design provide a new insight into how interactive systems can be developed. Such an insight aims to replace the technocentric practices of the current paradigm with a human focus, which will help and guide designers to identify and attain accessibility, usability, and other *quality of use* targets (see chap. 18, this volume). To this effect, however, user-centered design needs to advance the current collection of tools so as to provide the means to study context and to complement existing artifact-oriented practices with analytical and process-oriented instruments.

Developmental approaches to HCI design, rooted in cultural psychology, the social sciences, and the humanities, hold the promise to provide contributions in this direction by offering richer tools for analytical design and prescriptive frameworks for studying varieties of context of use.

User Interfaces for All and User Interface Development

The second major implication of user interfaces for all is on user interface development. This entails several questions related to the sufficiency of both the currently available architectural models for guiding the construction of user interface software, as well as the overall engineering paradigm for user interface development.

Traditional Architectural Models for User Interface Software.   A challenging issue relates to a suitable reference model for user interface architectures facilitating design for all in HCI. Early work, following the appearance of graphical user interfaces, focused on window managers, event mechanisms, notification-based architectures, and toolkits of interaction objects. Such architectural models were quickly supported by mainstream tools, becoming directly encapsulated in the prevailing user interface software and technology. Today, all available user interface development tools support object hierarchies, event mechanisms, and callbacks as the basic implementation model.

In addition to these, there have been other proposals for architectural models of interactive software, which, however, have not gained the level of acceptance originally anticipated. The focus of these proposals has been to extract a reference model from concrete user interface architectures, in order to classify existing prototypes and to guide the construction of user interface software. Specifically, the Seeheim metamodel (ten Hagen, 1991) and its successor, the Arch metamodel (UIMS Developers Workshop, 1992), were proposed to foster the principle of separation between interactive code (i.e., dialogue control) and noninteractive code (i.e., functional core) of computer-based applications. These models influenced early research efforts in the area of user interface management systems (UIMSs) (Myers, 1995). There have also been alternative, implementation-oriented proposals, some of which have resulted from mainstream tools. For instance, the model view controller (MVC) (Goldberg, 1984) originates from the architectural abstractions underpinning the Smalltalk-80 programming model, whereas it has also influenced the Java's pluggable look-and-feel architecture. The PAC model (Coutaz, 1990) is similar to MVC, though not yet supported by any commercial tool.

These models fall short of addressing several of the requirements for user interfaces for all. First of all, they do not explicitly account for the notion of accessibility of an interactive application. Thus, they do not address issues such as multiple platform environments, toolkit integration, platform abstraction, and so on, which arise from the proliferation of novel interaction platforms and diversity in usage patterns. Second, these models offer no account of user interface adaptation (Dieterich, Malinowski, Kühme, & Schneider-Hufschmidt, 1993), which is a central theme in design for all in HCI. As a consequence, key decisions such as what and when to adapt are not addressed, whereas the components that are needed to drive adaptations at the level of the user interface are totally missing. Third, existing architectural models offer implementation-oriented views of user interface architectures; this limits the role of design by not addressing how design knowledge can be propagated to the development and implementation phases. As a result, the application of these models in current HCI practices leads to reimplementations (reactive approach) rather than instantiation of an alternative design (proactive approach).

Development Techniques.   Having presented the implications on the user interface software architecture, the next item addressed is whether or not prevailing development practices can facilitate the development of user interfaces for all. To assess this, we review some of the currently popular techniques for user interface development and consider their underlying focus. Some of these techniques have already been supported in commercially available user interface tools, others are embodied in tools that are currently available as public domain software, whereas yet another cluster is still in prototype versions. In considering these techniques, our objective is to assess their suitability for user interfaces for all. This means that we do not seek to compare them against any particular implementation approach but, instead, we wish to consider whether or not, and to what extent, the specific techniques considered make provisions for the fundamental requirements of user interfaces for all.

Currently, there are various interface development techniques and tools (Myers, 1995). For the purposes of our assessment, these are classified under six distinctive categories, namely presentation based, physical task based, demonstration based, model based, abstract objects and components, and declarative fourth-generation languages (4GLs). Presentation-based techniques include graphical construction tools, such as Visual Basic™ and TAE plus™.[1] Physical task-based techniques lead to a specific sequence of user actions and include TAG (Reisner, 1981) and UAN (Hartson, Siochi, & Hix, 1990). Demonstration-based techniques are similar to graphical construction methods, with the exception that they allow the interactive definition of a physical interface instance through an example, or a demonstration. Demonstration-based techniques have been embedded in systems such as Peridot (Myers, 1988), Pavlov (Wolber, 1996), and DemoII (Fischer, Busse, & Wolber, 1992). Model-based techniques represent a more recent effort based on the notion of generating interactive behaviors from suitable models about users, tasks, platforms, and the environment. A comprehensive retrospective account of model-based technology can be found in (Szekely, 1996). Examples of model-based tools include HUMANOID (Szekely, Luo, & Neches, 1992) and MASTERMIND (Szekely, Sukaviriya, Castells, Muthukumarasamy, & Salcher, 1995). Abstract objects/components are techniques that support alternative physical realizations through either object abstractions, such as meta-widgets (Blattner, Glinert, Jorge, & Ormsby, 1992), or component-ware technologies, such as Active X™ by Microsoft and JavaBeans™ by SunSoft. Finally, declarative 4GL methods are typically found in some UIMSs, such as SERPENT (Bass, Hardy, Little, & Seacord, 1990) and HOMER (Savidis & Stephanidis, 1998).

Because a key ingredient of user interfaces for all is the capability to encapsulate alternative interactive behaviors, it follows that development methods closer to the physical level of interaction (e.g., presentation based, physical task based and demonstration based) are less suited. On the other hand, techniques that focus on higher level dialogue properties, and offer mechanisms for articulating alternative interactive components, stand a better chance and could be considered as candidates for integrating unified user interface development facilities.

User Interfaces for All and Evaluation Cycles

In addition to the design and development implications identified earlier, user interfaces for all pose several challenges in terms of evaluation. Evaluation in this context entails both the formative element through which design feedback is provided to early prototypes, and the summative account of high-fidelity prototypes. The related challenge seems to have a dual substance: On the one hand, it addresses the instruments for evaluation, whereas on the other, it raises considerations on the process and the overall iterative software development cycle.

At the level of the instrument, we need to identify what is to be evaluated or measured before we can decide how to do it. Clearly, user interfaces for all expand the targets of evaluation beyond performance-oriented accounts of usability. But these new targets remain to be defined. In addition, user interfaces for all introduce a com-

---

[1]TAE Plus™: http://www.cen.com/tae/

pelling need to address the evaluation of adaptation-related behavior of user interfaces, which, in turn, can only be effectively accounted for when studied in context. The effectiveness of existing evaluation instruments in order to capture context and facilitate such an insight remains an open issue (see chap. 17, this volume).

At the level of the design process, user interfaces for all pose an equally important challenge. User-centered design and other peripheral or similar approaches foster a collection of guidelines for maintaining a multidisciplinary and user-involving perspective into systems development. However, these models do not specify how designers can cope with radically different user groups whose requirements are not known a priori. Most important, they do not specify how to design for a universal community. In particular, with the advent of the Internet and the emergence of a highly distributed and collaborative computing paradigm, it is difficult for designers to anticipate who the user may be. For example, Web sites today can in principle be accessed by anyone possessing an Internet connection and a modem, irrespective of gender, age, educational background, and level of expertise. However, it is unlikely that this can be practically obtained with currently available applications. Furthermore, prevailing process models have not much to offer to systems developers in terms of practical means and recommendations toward such an objective.

It follows that the evaluation challenge posed by user interfaces for all is equally important as the design and development challenges. Whether or not HCI, in its current form and focus, can provide the required tools to address this challenge remains a question. It is believed, however, that before such a target is met, the community should face the fundamental question of what constitutes representative quality attributes in the emerging information age, and how these can be practically managed throughout the system development. The view that distinguishes between functional and nonfunctional qualities seems to be a useful starting point.

CONCLUSIONS

In this chapter, user interfaces for all is introduced as a new perspective into HCI, emphasizing some of the implications on the design, implementation, and evaluation of interactive software accessible to and usable by all users in a variety of contexts of use. The scope of user interfaces for all is broad and complex, because it involves issues pertaining to context-oriented design, diverse user requirements, as well as adaptable and adaptive interactive behaviors. This complexity arises from the numerous dimensions that are involved and the multiplicity of contributions in each dimension. Accordingly, the fields of science that might provide useful insights toward user interfaces for all are many and diverse. This is not different from the recent history of HCI, in the course of which the field has grown to develop strong links with several disciplines such as, for example, psychology (e.g., information processing and developmental psychology), linguistics, and the social sciences (e.g., anthropology, sociology, etc.). With regard to technology, user interfaces for all links with recent advances in user interface software in terms of a variety of aspects including software engineering, advanced interaction platforms, tools for development, and so on.

This volume provides rich insights into the current state of the art, as well as into what remains a challenge for additional theoretical and applied research work.

REFERENCES

ACCESS Consortium (1993). *Development platform for unified access to enabling environments* (TIDE ACCESS TP-1001 project, Technical Annex). ACCESS Consortium (available from the prime contractor, IROE-CNR, Florence, Italy).

Bass, L., Hardy, E., Little, R., & Seacord, R. (1990). Incremental development of user interfaces. In G. Cockton (Ed), *Engineering for human–computer interaction* (pp. 155–173). Amsterdam: North-Holland, Elsevier Science.

Bergman, E., & Johnson, E. (1995). Towards accessible human–computer interaction. In J. Nielsen (Ed.), *Advances in human–computer interaction* (Vol. 5, pp. 87–113). Norwood, NJ: Ablex.

Bevan, N., & Macleod, M. (1994). Usability measurement in context. *Behaviour and Information Technology, 13*(1, 2), 132–145.

Blattner, M., Glinert, E., Jorge, J., & Ormsby, G. (1992). Metawidgets: Towards a theory of multimodal interface design. In *Proceedings of the COMPSAC '92 Conference* (pp. 115–120). New York: IEEE Computer Society Press.

Bødker, S. (1989). A human-activity approach to user interfaces. *Human–Computer Interaction, 4*(3), 151–196.

Bødker, S. (1991). *Through the interface: A human activity approach to user interface design*. Hillsdale, NJ: Lawrence Erlbaum Associates.

Carroll, J. (Ed.). (1991). *Designing interaction: Psychology at the human–computer interface*. New York: Cambridge University Press.

Coutaz, J. (1990). Architecture models for interactive software: Failures and trends. In G. Cockton (Ed.), *Engineering for human–computer interaction* (pp. 137–151). Amsterdam: North-Holland, Elsevier Science.

Dieterich, H., Malinowski, U., Kühme, T., & Schneider-Hufschmidt, M. (1993). State of the art in adaptive user interfaces. In M. Schneider-Hufschmidt, T. Kühme, & U. Malinowski (Eds.), *Adaptive user interfaces—Principles and practice* (pp. 13–48). Amsterdam: North-Holland, Elsevier Science.

Fischer, G., Busse, D., & Wolber, D. (1992). Adding rule-based reasoning to a demonstrational interface builder. In *Proceedings of ACM Symposium on User Interface Software and Technology* (pp. 89–97). New York: ACM Press.

Goldberg, A. (1984). *Smalltalk-80: The interactive programming environment*. Reading, MA: Addison-Wesley.

Hartson, H. R., Siochi, A., & Hix, D. (1990). The UAN: A user-oriented representation for direct manipulation interface design. *ACM Transactions on Information Systems, 8*(3), 181–203.

ISO 13407. (1997). *Human-centred design processes for interactive systems*. Geneva, Switzerland: International Standards Organisation.

Lewis, C., & Polson, D. (1990). Theory-based design of easily-learned interfaces. *Human–Computer Interaction, 5*(2–3), 191–220.

Lewis, C., Polson, D., Wharton, C., & Rieman, J. (1990). Testing a walkthrough methodology for theory-based design of walk-up-and-use interfaces. In *Proceedings of ACM*

Conference on Empowering People: Human Factors in Computing System (pp. 235–242). New York: ACM Press.

Lewis, C., & Rieman, J. (1994). *Task-centred user interface design: A practical introduction* [Online]. Available: http://www.syd.dit.csiro.au/hci/clewis/contents.html

Myers, B. (1988). *Creating user interfaces by demonstration*. Boston: Academic Press.

Myers, B. (1995). User interface software tools. *ACM Transactions on Computer–Human Interaction, 2*(1), 64–103.

Nardi, B. (1996). *Context and consciousness: Activity theory and human–computer interaction*. Cambridge, MA: MIT Press.

Nielsen, J. (1993). *Usability engineering*. Boston: Academic Press.

Norman, D. (1991). Cognitive artifacts. In J. Carroll (Ed.), *Designing interaction: Psychology at the human–computer interface* (pp. 17–38). New York: Cambridge University Press.

Norman, D., & Draper, W. (1986). *User-centered design: New perspectives on human–computer interaction*. Hillsdale, NJ: Lawrence Erlbaum Associates.

Reisner, P. (1981). Formal grammar and human factors design of an interactive graphics system. *IEEE Transactions on Software Engineering, 7*(2), 229–240.

Reiterer, H., & Oppermann, R. (1995). Standards and software-ergonomic evaluation. In Y. Anzai, K. Ogawa, & H. Mori (Eds.), *Symbiosis of human and artifact–Future computing and design for human–computer interaction*, Proceedings of the 6th International Conference on Human–Computer Interaction (vol. 20B, pp. 361–366). Amsterdam: North-Holland, Elsevier Science.

Salvendy, G. (Ed.). (1997). *Handbook of human factors and ergonomics*. New York: Wiley.

Savidis, A., & Stephanidis, C. (1998). The HOMER UIMS for dual user interface development: Fusing visual and non-visual interactions. *Interacting With Computers, 11*(2), 173–209.

Simonsen, J., & Kensing, F. (1997). Using ethnography in contextual design. *Communications of the ACM, 40*(7), 82–88.

Stephanidis, C. (1995). Towards user interfaces for all: Some critical issues. In Y. Anzai, K. Ogawa, & H. Mori (Eds.), *Symbiosis of human and artifact–Future computing and design for human–computer interaction*, Proceedings of the 6th International Conference on Human–Computer Interaction (vol. 1, pp. 137–142). Amsterdam: North-Holland, Elsevier Science.

Stephanidis, C., & Akoumianakis, D. (1996). Usability requirements for advanced information technology products. In A. Mital, H. Krueger, M. Menozzi & J. E. Fernandez, (Eds.), *Advance in occupational ergonomics and safety*, Proceedings of the 11th International Occupational Ergonomics and Safety Conference (pp. 145–149). Amsterdam: IOS Press.

Stephanidis, C., Savidis, A., & Akoumianakis, D. (1995). Tools for user interfaces for all. In I. Placencia-Porrero & R. Puig de la Bellacasa (Eds.), *The European context for assistive technology, Proceedings of 2nd TIDE Congress* (pp. 167–170). Amsterdam: IOS Press.

Story, M. F. (1998). Maximising usability: The principles of universal design. *Assistive Technology, 10*(1), 4–12.

Suchman, L. A. (1987). *Plans and situated actions: The problem of human machine communication*. Cambridge, England: Cambridge University Press.

Szekely, P. (1996). Retrospective and challenges for the model-based interface development. In F. Bodard & J. Vanderdonckt (Eds.), *Proceedings of 3rd Eurographics Workshop on the Design, Specification and Validation of Interactive Systems* (pp. 1–27). Belgium: Springer-Verlag.

Szekely, P., Luo, P., & Neches, R. (1992). Facilitating the exploration of interface design alternatives: The Humanoid model of interface design. In *Proceedings of the ACM Conference on Human Factors in Computing Systems* (pp. 152–166). New York: ACM Press.

Szekely, P., Sukaviriya, P., Castells, P., Muthukumarasamy, J., & Salcher, E. (1995). Declarative interface models for user interface construction tools: The mastermind approach. In L. Bass & C. Unger (Eds.), *Proceedings of the Conference on Engineering for Human–Computer Interaction* (pp. 120–150). London: Chapman & Hall.

ten Hagen, P. J. W. (1991). Critique of the Seeheim model. In D. A. Duce, M. R. Gomes, F. R. A. Hopgood, & J. R. Lee (Eds.), *User interface management and design, Proceedings of the workshop on UIMS* (pp. 3–6). Berlin: Springer-Verlag.

UIMS Developers Workshop. (1992). A metamodel for the run-time architecture of an interactive system. *SIGCHI Bulletin, 24*(1), 32–37.

Vanderheiden, G. (1990). Thirty-something (million): Should they be exceptions. *Human Factors, 32*(4), 383–396. [On-Line]. Available: http://www.trace.wisc.edu/text/univdesn/30_some/30_some.html.

Vanderheiden, G. (1998). Universal design and assistive technology in communication and information technologies: Alternatives or compliments? *Assistive Technology, 10*(1), 29–36.

Vernardakis, N., Stephanidis, C., & Akoumianakis, D. (1997). Transferring technology toward the European assistive technology industry: Mechanisms and implications. *Assistive Technology, 9*(1), 34–36.

Winograd, T. (1988). A language/action perspective on the design of co-operative work. *Human–Computer Interaction, 3*(1), 3–30.

Winograd, T. (1997). Interspace and an every-citizen interface to the national information infrastructure. In *More than screen deep: Toward every-citizen interfaces to the nation's information infrastructure* [Online]. Washington, DC: National Academy Press. Available: http://www.nap.edu/readingroom/books/screen

Winograd, T., & Flores, F. (1987). *Understanding computers and cognition*. Reading, MA: Addison-Wesley.

Wolber, D. (1996). Pavlov: Programming by stimulus-response demonstration. In *Proceedings of the ACM Conference on Human Factors in Computing Systems* (pp. 252–259). New York: ACM Press.

II

Dimensions

2 Individual Differences and Inclusive Design

David Benyon
Alison Crerar
Simon Wilkinson

Usability is widely recognized to be as important as functionality to the success of interactive computer systems. However, even where usability is considered seriously, the scope of thinking tends to be restricted to the so-called 'average' users in typical office situations. Our aim in this chapter is to summarize some of the key findings in the literature on individual differences, exploring their implications for human-computer interaction. A framework comprising the characteristics of accessibility, usability *and* acceptability *is presented to relate the salient factors to existing Human-Computer Interaction concepts. In doing this, the notion of 'user' is elaborated to reveal a much more complex and changing phenomenon than naive* novice / expert *or* frequent / intermittent *classifications suggest. In particular, we argue that we are all 'extra-ordinary users' when considered over time, though the term 'extra-ordinary users' is usually applied to disabled people (Edwards, 1994). We believe that the goal of* User Interfaces for All *is a timely and challenging concept which raises awareness of the* inadvertent exclusion *that too often results from nomothetic analysis and design. This chapter outlines a number of approaches that designers can take to accommodate individual differences, drawing on current research to illustrate some of the most promising directions being explored.*

This chapter on individual differences considers the important physical, cognitive, psychological, and social "dimensions" of human users, which impact on the nature and quality of their interactions with computer systems. We also emphasize the contexts in which human–computer interaction (HCI) takes place and other external factors that contribute to the environment and, therefore, influence both accessibility and usability. Within this framework, individuals are seen to be complex characters whose profiles are not fixed, but change over time, according to the circumstances, individuals' roles, and the systems facing them. All computing professionals have countless examples to cite of the frustration caused by such things as unfamiliar operating systems, unintuitive interfaces, badly written documentation, and incomprehensible error messages. Poor design of any artifact disables the best of us, and the more complicated the device, the greater the design challenge to make the full range of facilities accessible to the widest possible user base.

At a physical level, an increased awareness of this problem in society has led to better facilities for the physically less able: wider aisles and ramps for wheelchair users, larger handles for arthritics, speech input and output for the partially sighted, and so on. But computer systems have other dimensions, not shared by dedicated artifacts. Computers are general-purpose machines and using computer systems is, to a large extent, a cognitive activity. HCI involves information processing and is concerned with the acquisition, manipulation, and expression of abstract symbols that signify something else. Most other systems allow for some form of physical interaction, allowing the user to look inside and see how it works. The user of such physical artifacts can employ a broad range of strategies, which are unavailable to the computer user, who must judge the system purely by its external displays. This is the reason why individual differences in cognitive abilities, preferences, and learning styles are so important in HCI.

CATEGORIZING DIFFERENCES

There can be few less controversial observations than that people differ in a variety of ways. Two important questions arise from this observation. First, how do people differ? We might agree that people differ in physical characteristics, such as height, weight, and girth, but beyond that we enter a potential minefield. People have different personalities, but what is personality? People have different cognitive skills and preferences, but how should these be classified and measured? Second, for our pragmatic purposes of improving the usability of computer systems, we must ask which differences are pertinent to HCI, what the range of variation is in the target population, and whether these differences are stable over time (fixed) or are changing, or changeable. For example, people can differ in their handedness, but does this matter for mouse use? Hoffmann, Chang, and Yim (1997) found that the performance of left-handed individuals habitually using a mouse with their right hand was not significantly different from that of right-handed users, so it seems that left-handers are not disadvantaged if they conform to standard practice.

In the following sections, we consider some of the key personal differences under three headings: physiological, psychological, and sociocultural. The overview distinguishes between fixed, changing, and changeable characteristics because as we see in the section Accommodating Differences, they pose different demands on system designers.

Physiological

One of the most obvious biological classifications is by gender. Although there are evident physical differences between males and females, secondary gender characteristics are irrelevant to most computer use, and other differences, such as size, are more relevant to ergonomics than to interface design per se. Instead, a more detailed analysis of individual biological subsystems is more appropriate for informing computer system design. Because computer use is predominantly a cognitive activity, this section begins by highlighting a study that analyzed the effect of cerebral dominance in the brain on the performance of individuals using a hypermedia system. Subsequently, our discussion broadens to include the senses of vision and hearing and a discussion of disabilities affecting mobility and dexterity.

Cerebral Hemisphericity.   The human brain comprises two symmetrical hemispheres, each one controlling the motor and sensory capacities of the opposite side of the body. However, when it comes to supporting higher level cognitive functioning, hemispheric specialization is observed. For example, it has been found that the left hemisphere is dominant for language in most right-handed individuals, whereas the right hemisphere is dominant for visuo-spatial functions (Gardner, 1983). McCluskey (1997) used this neurological difference to investigate whether it influences creation of hypermedia stacks using HyperCard. As expected, left-brain-dominant subjects (right-handed individuals) included more text fields in their stacks than right-brain-dominant subjects. However, what was interesting was the lack of any significant difference in the quantity of graphics incorporated by each group, although left-brain-dominant subjects did include more cards in their stacks and used more buttons. McCluskey also found that both left- and right-brain-dominant subjects in general outperformed mixed-brain-dominant individuals (left-handed, right-eye dominant, or vice versa). He concluded that computer software should be designed to be nondiscriminatory and equally usable, regardless of the cerebral hemisphericity of its users. This could be achieved by providing a system with enough richness (in the case of cerebral hemisphericity—text and graphics) to permit all users to capitalize on their information-processing strengths (Liu & Reed, 1994).

Vision.   Color blindness is a common form of visual impairment that affects about 8% of Western males and 0.4% of Western females. The term *color-blind* covers a number of conditions; in fact only about 1 person in 30,000 perceives no color at all (only shades of gray). The most common effect of color blindness is the inability to distinguish between red and green colors, with other conditions confusing blue and yellow (Legge & Campbell, 1987).

There is a wide range of other visual impairments, ranging from short-sightedness (myopia) and long-sightedness (hyperopia), which can usually be accommodated by corrective lenses, through to more serious conditions, such as glaucoma, cataract formation, and total blindness. It is estimated that in Europe there are 1.1 million blind people and 11.5 million people with low vision (Gill, 1994). Changes in computer technology over the last decade have been of mixed benefit for visually impaired users. For those with poor but some vision, the proliferation of graphical user interfaces (GUIs) that use bitmapped displays has enabled the use of ancillary programs, which can magnify the font size, or magnify a screen area (e.g., UnWindows V1—Kline & Glinert, 1995). However, at the same time, the prominence of the GUI has created additional problems for those with no sight. Character-based operating systems, such as DOS, permitted the use of *screen readers,* small helper utilities that could verbally output all the words on a screen using appropriate hardware/software speech systems for output. But many of the newer operating systems do not process words and characters; instead the screen is treated as a large grid of addressable pixels. Although there are updated screen readers, specifically designed for GUI systems, the screen complexity favored by modern designs radically limits the effectiveness of these systems for blind users. For example, a relatively simple Internet Web page might contain a "title" frame at the top, a navigational "menu" frame down the side, and a main "content" frame occupying the rest of the screen. Although the screen might not be using anything but text, there is a problem in deciding which frame to start reading first. Relatedly, if a link from the menu frame is selected, how does one signify that this frame remains the same but that the content frame has altered? These questions are typical of the accessibility challenges facing user interface designers.

Hearing.   Most people with good hearing can perceive sounds in the range of 20 Hz to 20,000 Hz (20 kHz). There are different severities of deafness arising from a number of conditions, for example, congenital, infection, noise damage, or tinnitus. Hearing impairment is the largest single disability category in Europe, where there are approximately 1.1 million profoundly deaf people and 80 million others with mild to severe hearing loss (Gill, 1994). It is, therefore, important to consider how various degrees of hearing loss may affect an individual's interaction, when designing systems incorporating audio facilities. Youngson (1986) presented some effects of varying severities of hearing loss

- 25–30 db of loss results in no practical disability.
- 30–40 db of loss would make it difficult to hear at noisy parties, and would also require a television to be turned up more than the family would appreciate.
- 40–50 db of loss results in difficulty following normal conversation.
- 60 db of loss causes problems using the telephone.
- 90 db of loss means total deafness to speech.

Mobility/Dexterity.   In Europe there are 2.8 million wheelchair users (Gill, 1994). In the banking industry where automatic teller machines (ATMs) are in common use, most banks have introduced their latest generation machines at a wheelchair-friendly height. Ambulant users may have to stoop slightly, but this should not be too detrimental to their performance or comfort.

There are also estimated to be 1.1 million people who have impairments of manual dexterity (Gill, 1994). With the majority of computer systems involving keyboard input or the ability to click on a mouse button, this can create significant accessibility problems. For some users, alternative input devices/systems can facilitate access to the computer. For example, Noyes and Starr (1996) reported that some experiments employing voice recognition with disabled users have found recognition rates of up to 80%. This figure is lower than the recognition rates achievable by nondisabled users (in excess of 90%), because, as Noyes and Starr explained, environmental noise from medical equipment and the articulation difficulties of some disabled users (e.g., resulting from profound deafness, or a wide range of conditions such as cerebral palsy and multiple sclerosis) can make recognition of their speech less reliable.

Psychological

Unlike physiological differences, which can be observed and measured, psychological differences are more difficult both to categorize and quantify. As Boring (1923) argued, "intelligence" is what the intelligence tests test. A large number of psychological characteristics have been proposed in the literature. Fortunately most of them fall within three broad categories: *intelligence, cognitive style,* and *personality.* Interested readers are encouraged to explore the literature in the fields of psychology and especially computer-based learning, which has involved extensive research investigating individual differences and computer applications (e.g., *Journal of Educational Computing Research, British Journal of Educational Psychology, and Educational Technology Research and Development*).

Intelligence.   Three prominent models of intelligence have been proposed over the years. In the early years of intelligence testing, Spearman (1904) argued for a general form of intelligence, as suggested by the statistical similarity between the scores of individuals on a range of different intelligence tests. However, Thurstone (1938) was interested, not in the similarity between tests, but the differences. He argued for the existence of seven "primary mental abilities": perceptual speed, memory, verbal meaning, spatial ability, numerical ability, inductive reasoning, and verbal fluency. A similar model of multiple intelligences was proposed by Gardner (1983), and, as mentioned earlier, the idea of distinct intellectual capabilities is given credibility by findings showing cerebral localization of some of these proposed modules of intelligence. A third model combines aspects from the other two, proposing an overarching general intelligence with more specific semiindependent modules beneath it (Burt, 1940; Carroll, 1993; Cooley & Lohnes, 1976).

Of all the components of intelligence, spatial ability is the one that has been most frequently studied in connection with software use. Vicente, Hayes, and Williges (1987) found spatial ability to be positively associated with navigational efficiency in a hierarchical file system. Positive correlations have also been reported for hypertext navigation (Chen & Rada, 1996; Höök, Sjölinder, & Dählback, 1996; McGrath, 1992). Benyon (1993a) reported that, for subjects in a group with low spatial ability, the completion time for carrying out database tasks using a command interface was 35% longer than for a group of subjects with high spatial ability. However, when the user interface to the same database was changed to a menu system, the difference between the two groups narrowed to just 1%. This is an important result as it implies that it is not always necessary to create multiple interfaces, or adaptive interfaces to suit each category of user, if a solution can be found that accommodates all users without prejudice. This observation is also supported by the findings of Stanney and Salvendy (1992). In their study, they used three different interfaces: (a) a "challenge match" interface, which created a deliberate mismatch between the task and the user,[1] (b) a "compensatory match," which

---

[1]It has been suggested (e.g., Presland, 1994) that creating a deliberate mismatch between task and user can encourage more active cognitive processing of the stimulus material.

acknowledged various user deficiencies and so gave explicit user interface assistance, and (c) a "capitalization match," which ignored the deficiencies of the user by not requiring the implicated abilities. Stanney and Salvendy found that the performance times of high and low spatial groups were significantly different with the challenge match interface but not with the compensatory match or the capitalization match interfaces.

Cognitive Style.   Cognitive styles, like intelligence, are used to describe cognitive processing, but the two constructs differ in what they measure. Intelligence measures the capacity and speed of cognitive processing, whereas cognitive styles measure the form (or style) this cognition takes. A second fundamental difference concerns the nature of each construct's dimensions. For example, Messick (1976) described intelligence as unipolar and value directional (more intelligence being preferable to less), whereas cognitive styles are bipolar and value differentiated. The practical importance of this differentiation is that, for some tasks, it is beneficial to be at one end of a particular cognitive style dimension, whereas, for other tasks, it will be more advantageous to be at the opposite end. Furthermore, Messick stated that cognitive styles represent stable attitudes, preferences, or habitual strategies that determine an individual's typical modes of perceiving, remembering, thinking, and problem solving. He continued by stating that these styles influence almost all human activities involving cognition, including social activities. Riding and Cheema (1991) observed that as many as 30 different cognitive style labels can be found in the literature. Some of the more common styles are listed on the right of Fig. 2.1. It is interesting to note how pervasive cognitive styles are: They are used when acquiring new knowledge through perception, its subsequent storage in long-term memory, and finally its application in thought.

One cognitive style that has received much attention is *field articulation*. This measures an individual's predisposition to process a stimulus as a whole, or in parts. Persons tending toward processing the stimulus as a whole are termed field-dependent (they are dependent on the stimulus field), whereas those processing the stimulus in more focused parts are termed field-independent (they can process components independent of their surroundings). Research has shown that, in general, field-independent individuals take a more active approach to learning and can restructure information when required; conversely field-dependent persons are better at interpreting social cues. In a study of learning English as a second language, Liu and Reed (1994) found that field-dependent and field-independent subjects exhibited differences in the facilities they used within the educational hypertext provided. Field-dependent subjects selected more video contexts of words and sentence examples of words in use, whereas field-independent subjects selected more relationships of words. The videos and sentences were preferred by the field-dependent individuals because it suited their holistic approach to information processing, with each word being presented as part of a larger video sequence, or embedded in a larger sentence. Conversely, the field-independent subjects preferred word-based relationships, which are tightly focused in scope.

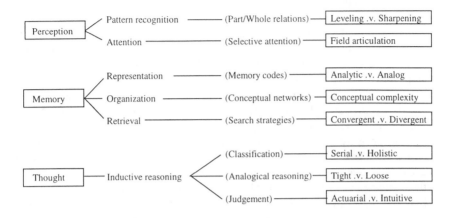

FIG. 2.1.   A model of cognitive styles and cognitive processes. From Miller (1987). Adapted by permission.

What is interesting to note is that subjects at both ends of the field articulation style adopted different learning strategies, but through the richness of the facilities provided by the hypermedia environment, they scored very similar marks. This study, together with those of Stanney and Salvendy (1992) and Benyon (1993a) cited earlier, demonstrate that, in some circumstances, it is possible to accommodate a range of individual differences using a single interface or system. It is important to note, however, that this is done by enriching the environment to cater to diverse preferences, not by forcing all users to interact in the same way.

Personality.   Personality is a complex psychological phenomenon, which may be thought of as complementary to cognitive styles: Whereas cognitive styles refer to *internal* processing behavior, personality refers to a range of characteristics that give rise to *external* observable behavior. Two personality models predominate: Eysenck's personality dimensions and the five-factor model (McCrae & John, 1992). Eysenck (1990) claimed that there are three important dimensions: psychoticism, extraversion, and neuroticism. The five-factor model, although similar, was developed through factor analysis of many traits and consists of five basic dimensions: extraversion, agreeableness, conscientiousness, neuroticism, and openness to experience.

Investigations into the affect of personality on computer system usability have taken two main approaches: The first seeks to understand more about how various facets of personality influence interaction with user interface components, whereas the second tries to actively influence, or stimulate various personality traits. The first approach is exemplified by Richter and Salvendy (1995). They found that, as expected, subjects with an *extrovert* personality were significantly quicker than *introverts* in performing a task using an "extrovert user interface."[2] However, although the introverts tended to perform better than the extroverts using the "introvert user interface," this difference was not statistically significant. The second approach is illustrated by Sproull, Subramani, Kiesler, Walker, and Waters (1996). This study used two interfaces: One presented an animated "talking-face," whereas the other presented a text-based interface. The system was a prototype career counseling service, which posed various questions. It was found that both male and female subjects were more aroused (less relaxed, less confident) and projected themselves in a more positive light when using the talking-face interface. Although the questions were identical between the two interfaces, it appears that subjects in the talking-face group attributed personality to the computer system and responded accordingly. This opens up interesting research and development avenues concerning software agents and whether they should have any humanlike personality traits. Indeed, the new field of affective computing is premised on the belief that HCI will be enhanced when computers can both recognize and synthesize emotion (Picard, 1997).

Sociocultural

In addition to their individual physiological and psychological characteristics, individuals operate within a larger sociocultural environment. Although all individuals share a great deal of physical commonality, differences of language, culture, and experience may be profound. This subsection mentions some of the sociological influences. A more detailed discussion concerning international and cultural differences can be found in chapter 3 of this volume.

Language.   Failure to adapt to linguistic differences can cause insurmountable accessibility problems. For example, a U.S.-English version of Microsoft Word might be perfectly usable to most British users, but not to Chinese or Japanese users. It is important to remember that, from a development perspective, the ease with which software can adapt can vary considerably. For example, European languages use very similar character sets and are written from left to right and top to bottom. However, other languages employ different character sets (e.g., Arabic and Russian), and different written orientations (e.g., Chinese), requiring radical interface redesign. However, language is not a simple skill that is either known or unknown. As Bourges-Waldegg and Scrivener (1998) reported in their cross-cultural study, even when a user knows the language used in an interface, the intended meaning may still be misunderstood. They related that one foreign subject who un-

---

[2]The "extrovert user interface" contained more words, a sense of humor, more sounds, and a quicker pace of screen changes compared with the "introvert user interface."

derstood English guessed that the "What's Cool" button in the Netscape Web browser had something to do with temperature!

Culture.   Whereas linguistic differences can cause *accessibility* problems, cultural differences can certainly result in *usability*[3] problems. For example, Fernandes (1995) discussed the problem of checkboxes. The Microsoft Windows 3.1 convention requires a user to "check" (place a diagonal cross) to select an option. However, as Fernandes highlighted, a cross in Switzerland and Korea is used to signify *not* wanting an item. Even within a culture, the cross symbol can still cause confusion. Some novice users of Excel have been tempted based on general Windows principles to select the cross icon to finish a formula (instead, this will actually cancel the operation). It is interesting to note that with the introduction of Windows 95, Microsoft changed the checkbox symbol to a tick mark. As the checkbox example highlights, although culture is important, it is not the entire story. Instead, Bourges-Waldegg and Scrivener (1998) posited that the key to effective interaction is whether the user can understand that *representation* R *means* M in *context* C. They illustrated this point by highlighting the fact that a Scots man who wears a kilt will not select the ladies' toilet just because the icon displays a skirt. The Scots man correctly interprets the symbolic meaning of the toilet sign despite his own attire.

Environmental.   In the field of HCI, it can be easy to fall into the trap of considering the user on one side of the equation and the computer system on the other and to forget about the larger environment in which both operate. For example, in a study of ATM usage by older users (61–81 years), Rogers, Gilbert, and Cabrera (1997) found that convenience, satisfaction, speed, and accuracy were all highly rated, but that safety (an environmental issue) was a concern raised by several subjects. As the researchers involved in this study suggested, a simple redesign of an ATM to include a "panic" button may be all that is required in order to reassure older users. In another example, one of the authors of this chapter experienced a *temporary* environmental effect. While using a south-facing ATM on a sunny day, it was very difficult to read the screen display; the author had to cup his hands to make the ATM messages readable. Although this usability problem resulted in minor annoyance, to an older or disabled user it could have represented an insurmountable barrier.

Changeability

Thus far, individual differences have been reviewed under three headings: physiological, psychological, and sociocultural. However, it is important to consider which characteristics are fixed and which are changing or changeable, because this will influence the approaches to *universal design,* or *design for all*. A few of the dimensions in which an individual can change are presented in Table 2.1. A distinc-

---

[3]We distinguish between the terms *accessibility, usability,* and *acceptability* in the *Terminology* subsection.

tion is made between those characteristics that are fixed and cannot be altered, for example gender, and those that have varying degrees of changeability. Although it is obvious that computer systems must accommodate the fixed characteristics of an individual, it is also often necessary to adapt to highly changeable user characteristics as well. For example, even though individuals have consistent handwriting and voice patterns, microchanges occur each time a word is pronounced, or a letter is written. Normally, we are unaware that these changes are taking place, as the human brain matches patterns very easily. However, a computer must constantly adapt if it is to successfully recognize the input. Psychological changeability is discussed first, followed by an overview of some of the changes brought on by aging (including physiological changes).

Psychological Changeability.   As Van der Veer, Tauber, Waern, and Van Muylwijk (1985) suggested, intelligence is one of the factors most resistant to change. This, as we have seen in the earlier discussion, interacts with cognitive style to shape the information processing of an individual. However, although cognitive styles represent habitual information-processing preferences, Witkin, Moore, Goodenough, and Cox (1977) claimed that some can be easily altered, just by asking an individual to think in a different way. Others (e.g., Miller, 1991; Pask, 1976) have suggested that some individuals may be naturally *versatile*, enabling them to choose between different cognitive styles and strategies to match the demands of the current task. Strategies are ways in which to deal with specific situations and tasks, can vary from time to time, and may be learned and developed (Riding & Cheema, 1991). Personal knowledge structures are the most easily changed psychological constructs: They can be changed in minutes just by personal observation, while reading a book, or by attending a lecture.

TABLE 2.1
Changeability of Individual Differences

	Fixed	Stable		Changeable
Physiological	Gender			
	Height		Weight	
	Eye Color			
Psychological	Cerebral Hemisphericity	Handedness	Perceptual Capabilities	Handwriting Speech
	Intelligence	Cognitive Styles	Strategies	Personal Knowledge
Sociological	Personality		Mood	Behavior
	Ethnicity	Nationality		
		Language		

A distinction is commonly made in the HCI literature between *novice* and *expert* users, the supposition often being that experts or "power users" are faster, more accurate, and more sophisticated in their use of a system. Interestingly, in the context of office-related tasks, Prümper, Zapf, Brodbeck, and Frese (1992) found that novices and experts did not make a significantly different number of mistakes, although, as expected, the experts recovered from their errors more quickly. However, a difference in the type of errors was observed: Experts were found to make fewer mistakes resulting from a lack of system knowledge, but instead experienced more functionality problems. In a separate study, Kalyuga, Chandler and Sweller (1998) found that, as subjects' expertise changed, so did the interface that was most efficient for them. Although the study investigated differences in expertise and the design of instructional materials, it should have implications for noneducational systems as well. The main finding was that novice subjects preferred diagrams that physically integrated text and graphics, whereas experts were more efficient with graphical diagrams with the text eliminated. The authors suggested that the reason for this can be explained in terms of the *split-attention effect* and the *redundancy effect*. The split-attention effect is caused by multiple sources of information that need to be combined in order to be understandable (e.g., topographical map plus key). The mental process of integration is cognitively demanding and can be helped by the physical integration of information in the stimulus material (placing text within the map/diagram). This was found by Kalyuga et al. to help novice subjects. However, when an individual gains expertise, the redundancy effect is more likely. This happens when redundant information is presented (e.g., diagram plus text describing the diagram), which causes unnecessary cognitive load. In this situation, the experts benefited more from a diagram that showed graphics only.

Like system knowledge, perceptions of various systems can change as a result of prior experience. For example, Wiedenbeck and Davis (1997) found that first-time users of a word processor judged the system by its intrinsic usability, whereas subsequent systems were compared with already used word processors. Analyzing users who were initially trained on a direct manipulation word processor and then given a command line version, Wiedenbeck and Davis found that subjects perceived the command line version to be harder to use and less useful than did users given this version initially. This study has important implications on the development of training material for a system, as far as the users' previous experience with other systems is concerned, as well as for facilitating likely transitions through system designs.

Age-related changes.   Both physiological and psychological faculties decline with age, but this process varies widely between individuals and can be affected by decade of birth, general health, nutrition, exercise, work, and social factors throughout life (Haigh, 1993).

Taking the eye first, there are a number of changes that occur with age. The lens tends to discolor, which reduces the amount of light that can enter. This means that older people require objects to be more brightly illuminated. On the other hand, increased opacity of the eye results in more scattering of light, which increases glare sensitivity in older individuals. The yellowing discoloration of the lens also changes blue-green and red-green color perception by filtering out incoming blue and violet light. Several of the negative effects of visual aging can be accommodated by careful workstation setup to ensure correct focal distances and to minimize glare. Alternative technologies can solve some of the other problems. For example, it is known that some people with low vision cannot read liquid crystal displays because of their low contrast (Gill, 1994).

Hearing loss due to aging (presbycusis) affects, in the United States, one third of elderly people aged 65 to 74 years and half of those aged 75 to 79 years (National Center for Health Statistics, 1985). High-frequency tone perception is normally the first to be affected by aging, followed later by loss of lower frequency perception. Men tend to be affected more than women. The practical effect of a loss in high-frequency perception can be difficulty in understanding children's and high-pitched female voices. At this point, it is worthwhile to highlight the importance of sound in certain safety-critical systems. For example, Prouix, Laroche, and Latour (1995) presented evidence that an average 55-year-old man with mild to severe hearing loss would not hear the emergency fire alarm in his apartment block. Although, as stated earlier, individuals deteriorate at different rates, a sensible design approach is to design for stereotypical hearing capabilities in specific age groups.

Mental abilities also change at different rates. For example, Haigh (1993) stated that tasks utilizing skills acquired over long periods of time, such as language, can be performed unimpaired well into later life; however, tasks requiring the use of logical or mathematical reasoning, rapid information processing, or spatial ability are markedly affected. Speed of information processing, complex tasks, working-memory capacity, and perception were also mentioned by Sharit and Czaja (1994). However, as Sharit and Czaja explained, although psychological capabilities inevitably decline, older persons are often able to compensate by using past experience with familiar tasks to adapt their strategies.

In a study of information retrieval using four different interfaces, Westerman, Davies, Glendon, Stammers, and Matthews (1995) found that in each condition the performance of older subjects was slower than that of younger subjects. However, an initially very large performance differential was dramatically reduced by the older subjects in later sessions. The authors suggested that the initial slowness was in forming a mental representation of a new system. Once such a representation was formed, the residual difference between young and old subjects was associated with differences in cognitive processing speed and motor skills due to age.

In a separate study, Mead, Spaulding, Sit, Meyer, and Walker (1997) investigated the effect of age on World Wide Web navigational strategies. Both young and old subject groups were Internet novices, but it was found that the older subjects were less efficient at searching. Specifically, the authors indicated that, to complete one of the tasks, the older users followed, on average, 20 links. As the Web site being used for the experiment consisted of 19 pages (no external Internet links), this raises some concerns. Mead et al. suggested that one reason for such a high number of links being used was the search strategy adopted by older users: When failing to locate information quickly, subjects were seen to try every link on a page. Evidence of older subjects having difficulty in remembering which pages they had already visited, and also difficulty in identifying the current page, was also reported.

RELATING INDIVIDUAL DIFFERENCES TO UNIVERSAL DESIGN: A FRAMEWORK

In the previous section we presented an overview of the human characteristics—physical, psychological, and social—that have a bearing on people's abilities to interact with technology. Physical characteristics are the easiest to comprehend and measure, whereas "unseen" characteristics, such as dyslexia, or cultural expectations, are harder to define and detect. Taking physical characteristics as a starting point then, it is evident that every interaction method (e.g., clicking, dragging, drawing, typing, touching, speaking), every specific interaction device (i.e., a particular piece of hardware with its unique design features), and every interface design (both genres such as GUI vs. command line, and individual instances of these) will suit some users well, will suit others less, and will be unusable by another portion of the user population.

Any computer system can be seen as consisting of three components: interaction method(s), interaction device(s), and interface design(s), together of course with the software functionality. It is clear that: (a) the combination of features available to the systems designer results in a potentially very large design space, and (b) that any limited choice of interaction devices and styles will necessarily disenfranchise significant numbers of potential users (as happened to blind users when GUIs appeared). It is, therefore, very important that as new technologies emerge, designers carefully consider the implications of their adoption for disabled users.

In the everyday world, "exclusive designs" are highly sought after; exclusivity is prized often in proportion to its costliness. In the current context *exclusive* is a pejorative term, because it implies the withholding from some people of facilities available to others. We call designs that needlessly exclude people *exclusive designs*. Whereas nonaccess to luxury goods brings no serious privation, nonaccess to information technology is fast becoming as debilitating and as divisive as physical immobility. In seeking to promote new paradigms, a definition, or redefinition, of terms is always beneficial, so that practitioners can share an understanding of key concepts. In this section we introduce a number of terms and concepts, some widely used already, others new, that may be helpful in moving toward *universal design*. We should stress that the term *universal design* does not mean *one* design, but design to be usable by *all*.

Terminology

The consideration of individual differences (which of them are important to HCI, the extent of their variation, and how we might accommodate them) inevitably forces us to try to classify the parameters of interest in a useful way. There are many possible ways of doing this, but one that helps to identify both the need and the potential cost is to look at human characteristics along two axes: *common* versus *uncommon* (i.e., whether a characteristic is frequently occurring or rare in the potential user population) and *fixed* versus *changing/changeable* (i.e., whether the characteristic is unalterable, such as congenital blindness; is changing, such as level of proficiency in the use of a system; or is changeable, such as an attitudinal prejudice that may perhaps respond to involvement in the design process).

An *inclusivity analysis* can then consider the solutions available, providing reasoned and costed recommendations, either for adoption of an *inclusive solution set* (i.e., hardware, software, training, etc., which together satisfy the—*universal*—design criteria), or, in some cases, for recommendation of an *exclusive solution set* (i.e., a mix of elements that acknowledges and details classes of users, of tasks, or of circumstances that cannot sensibly be supported). Though wholeheartedly endorsing the principle of universal design, we must accept that there will often be reasons (e.g., technical or financial) why total inclusion is not attainable. The important thing is that the process of considering universal accessibility (i.e., user interfaces for all) as an integral part of the specification and design process has two major ramifications: (a) *inadvertent exclusion* will be virtually eliminated, and (b) those commonly occurring user characteristics that are fixed and amenable to cost-effective solutions will be identified and ready solutions to these will become standard (as has begun to happen with customizable screen displays).

Inadvertent exclusion refers to the unintentional side effects of an information technology industry that is immature, subject to a phenomenal rate of change, and dominated by young, able-bodied, intelligent, male, computer-literate designers, who tend to design systems for users who are just like themselves. Though *usability* is now gaining in status as a software engineering concept, it is still too often dismissed as cosmetic tinkering, rather than as a major driver of the design process. Pushing the ambitions of usability further, toward universal design, calls for a more detailed analysis of where and why exclusion may occur. A useful framework for this task is to make a distinction between *accessibility, usability,* and *acceptability.*

Accessibility is a prerequisite for usability, and a high usability rating is essential to achieve widespread social acceptability. Accessibility refers to two things: (a) physical access to equipment (in sufficient quantity, in appropriate places, at convenient times), and (b) the operational suitability of both hardware and software for *any* potential user (even effortful participation qualifies as minimal accessibility). Clearly, vast numbers of people are excluded from computer use on one, or both of these accessibility criteria.

Once people are admitted to an interaction with a system, the issues of usability for different users, tasks, and environments come to the fore. Usability then refers to the quality of the interaction in terms of parameters, such as time taken to perform tasks, number of errors made, time to become a competent user, retention of performance for intermittent users, enjoyment in use, and so forth (Crerar & Benyon, 1998). Usability criteria should take into account the context of use and the profiles of intended users, but much usability testing is done by HCI experts, or is laboratory based, because there are well-understood principles by which a high proportion of general usability problems can be detected and removed. However, a system may be assessed as highly usable according to common usability evaluation criteria, but may still fail to be adopted, or to satisfy end-users.

Acceptability on the other hand cannot be gauged in laboratory conditions because it refers to fitness for purpose *in the context of use,* and also covers attitudinal factors, those elusive personal preferences that contribute to users "taking to" an artifact, or not. Table 2.2 gives a summary of the three terms discussed thus far along with typical factors that contribute to each.

The Multidimensional "User"

In the Accommodating Differences section, we provide some specific examples of how combinations of accessibility, usability, and acceptability problems can result in exclusion and discuss design options that can be considered to counteract them. Prior to that, there is a little more to be said about users, to enlarge the rather one-dimensional view of them that software engineering seems to take. We relate in the following discussion the individual differences criteria discussed in the Categorizing Differences section to notions of *roles, systems, tasks,* and *contexts,* which have been adapted from the familiar quartet of *constraints, users, tasks,* and *environment,* as understood in the HCI literature (Crerar & Benyon, 1998). Figure 2.2 illustrates the dynamic nature of users, both in terms of the range of things they do and of the changes they inevitably undergo as they move through life.

The ideas behind this conceptualization are that all users can be placed somewhere on a continuum between highly able (physically and mentally) and severely disabled.[4] Wherever they start off on this "conveyor belt," *all* users move, though at differing paces and in differing ways, toward the disabled end, simply as a function of the aging process. Moreover, during life, some individuals are subject to sudden incidents (such as a stroke) that may catapult them down the line, whereas others suffer transient problems (such as broken arm or bout of depression) that interfere temporarily with their physical and/or mental functioning. The model, then, is of changing requirements and of a diverse user base; design for elderly and disabled users is no longer of minority interest, it is clearly relevant to mainstream HCI. In the past, computers were the preserve of the relatively young; the older generation had not seen them at school, or at work. Today's computer-literate 20-year-olds will expect to be supported by technology for the rest of their lives and demographic predictions forecast that by the year 2020 almost 27% of the EU popula-

---

[4]For simplicity this is depicted as if unidimensional; of course it is, in fact, multidimensional, but here we are concerned with the general principle.

TABLE 2.2

Three Key Elements of Universal Design and Typical Factors That Should
Be Considered Under Each

Accessibility	Usability	Acceptability
Availability (times, places)	Input/output devices	aesthetics
Siting (ergonomics)	Interface design	convenience
Input/output devices	functionality	Cultural mores
Interface design	documentation/help/feedback	Social habits
cost		Perceived usefulness in context
		cost

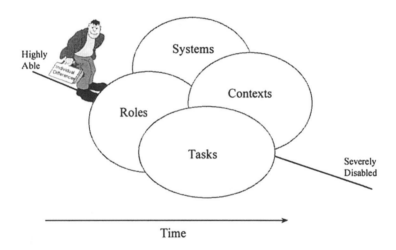

FIG. 2.2.   A multidimensional view of the user.

tion will be aged over 60 years. Moreover, by the year 2020, 20 million EU residents will be over the age of 80 (Ballabio & Moran, 1998). By then, computers will be truly ubiquitous and the "graying population" will be a very strong economic force indeed. The future social, economic, and political imperatives for universal design are plain to see.

Returning to Fig. 2.2, each user brings along personal "baggage," individual differences (physical, cognitive, psychological, social, cultural) that define his or her aptitudes, preferences, abilities, and attitudes. As we saw in the Categorizing Differences section, some of these individual differences impact more than others on computer use. In any one time slice, and over time, a user will be seen to play a number of different computer-mediated *roles,* involving a range of *tasks;* these activities will be performed using different *systems* (hardware and software) and will take place in a number of *contexts* (physical environments, projects, and interpersonal groupings). These factors are shown as overlapping ellipses in Fig. 2.2, to give the impression of interaction between them. To complete the picture we imagine our user, with his or her own personal baggage moving into this complex work and leisure environment, which is also inhabited by other users. HCI is not just about individuals using stand-alone machines, but is increasingly about cooperation between people, who may be geographically and/or temporally distant, yet who rely on technology to accomplish shared goals. The complex picture we have painted perhaps makes universal design seem even more elusive. In fact, we are merely recognizing that the model encompasses *all* users: that an able user today may be disabled tomorrow; that an expert UNIX programmer may be a desktop publishing novice; or, that a competent keyboardist with repetitive strain injury may find herself in the same boat, albeit temporarily, as a hemiplegic. These realizations demand the flexibility of universal design. We now consider some of the options that are available to system developers to accommodate the diverse needs outlined herein.

ACCOMMODATING DIFFERENCES

Taking the goal of universal design seriously means making "inclusivity criteria" an explicit component of any requirements specification, and making "inclusivity analysis" a mandatory part of design activities. In their work, Egan and Gomez (1985) suggested a three-stage approach to dealing with individual differences. First, it is necessary to "assay," or assess the extent of the differences. This involves considering what to measure and how to measure it. Once differences have been observed, the essential differences have to be isolated from confounding factors. Finally, when the important features have been isolated, it is then necessary to accommodate them.

As information technology pervades not just office life, but increasingly homes and public places, the need for much wider access becomes urgent. The response of private and public sector providers may be different, the former operating primarily on profitability, the latter perhaps also considering social responsibility and quality of life issues. In both cases, however, cost will be a key factor. Putting together the knowledge we have on individual differences and the conceptual framework provided in the previous section, a simple, generic decision tree for inclusivity analysis can be constructed, as shown in Fig. 2.3. Using this tree, we can see that, for example, a fixed, frequently occurring problem, such as color blindness, which has a cheap solution (avoidance of implicated color combinations, or customizable screen displays), should always be accommodated. Indeed, design guidelines for public access systems and the international standards they embody, are beginning to formalize exactly these principles (Gill, 1997).

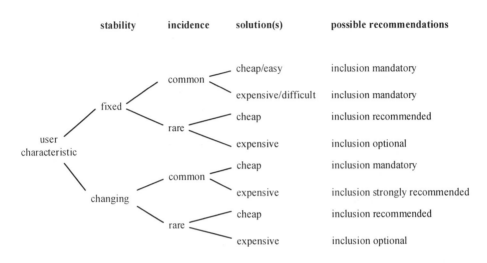

FIG. 2.3.   Generic decision tree for inclusivity analysis.

Of course, the effectiveness of any interaction between two systems (e.g., human and machine) depends on how well adapted the systems are, with respect to one another. Four broad types of facilitation can be identified, by recognizing two classes of system, each with two interaction mechanisms, as follows:

- Fixed systems can interact (a) by mutually compatible design or (b) by the use of an adaptor.
- Flexible systems can interact (a) by customization or (b) by automatic adaptation.

Systems with fixed characteristics may interact with other systems by design, or through the use of an intermediary system (an adaptor). For example, a "normal" user interacts with a standard PC using keyboard and mouse (as the designers intended), but a trackerball and a keyboard guard may be necessary for a user with cerebral palsy to gain access. The state of a system that has a more flexible design may be customized by the actions of another system, often the user, or changed automatically through incorporation of self-adaptive mechanisms ("adaptive systems").

Within this broad framework, we can analyze in more detail issues such as: which system makes a proposal for a change, which system takes the decision whether or not to adapt, how the change is executed, and how its effectiveness is evaluated (Dieterich, Malinowski, Kühme, & Schneider-Hufschmidt, 1993). Dieterich et al. identified several important classes of systems that can accommodate various degrees of adaptation and fit. For example, in "user-initiated self-adaptation," the user takes the initiative and then hands control to the system to decide on the nature of the change and to execute it, whereas in "computer-aided adaptation" the user takes the initiative, the system proposes a change, but it is the user who then takes the decision as to whether or not the change will be made. The system then implements the change.

Next, we elaborate on these four approaches to intersystem communication, highlighting their implications for universal design.

Fixed (Nonadaptive) Systems

In the realm of HCI, we are never dealing with two fixed systems that must be entirely compatible by design, because the human user is a sentient being. The intelligence and adaptability of humans makes the job of nonadaptive system designers so much easier, because they can rely on the users to compensate for and to circumvent so much poor design. The onus for adaptation falls squarely on the user. This places large burdens on "normal" users, as usability studies show; it also hinders uptake of technology because fixed designs so often inadvertently exclude individuals with special needs.

All nonadaptable systems require adaptation by the user to the system. The user must learn how to use the system and must invest effort in mastering the details. If the purpose of the interaction is important to the user, or the task is performed frequently, the user may decide that such an investment is indeed worthwhile. As we have seen previously, although some characteristics of people, such as cognitive style, are *resistant* to change, people are highly adaptable and can adopt strategies to deal with difficult systems, if they believe it to be worth the effort. However, for purposes that are only rarely encountered, or where the interaction is not deemed particularly important or enjoyable, the user will not take the time to learn a difficult system. For example, the basic tasks (the ones that are important) of "play" and "record" have been mastered by most users of home videocassette recorders. The range of other facilities offered such as timed recording, recording a program every Thursday for the next 3 weeks, or "rewind and playback" are rarely used.

Even the better designed nonadaptive systems will always exclude some users, or impoverish the interaction for others; moreover, the design will rarely satisfy purposes the design team did not foresee. For example, people with physical disabilities, such as arthritis, may be excluded from driving a beautifully designed car because the design presupposes that the user has two hands that can manipulate and grab in particular ways. The use of a particular natural language during interaction excludes those who cannot understand it. Successful interaction is facilitated by good design (with respect to some purpose). A chair that is well designed for its primary purpose of providing a comfortable and supportive seating position, is poorly designed for another purpose, such as helping one reach a high shelf, if it has wheels and a soft seat and is, therefore, unsuitable for standing on.

Systems that are adapted to interact with other systems through their inherent design will always be ill-adapted to interact with other systems, which are outside their scope. For example, a UK plug is well designed to fit into a UK socket, but it cannot fit a French or German socket; similarly, a particularly heavy person may break the aforementioned chair, a particularly large person may be unable to get into the chair, chairs with wheels do not stack. A fixed design embodies fixed models of interaction, which inevitably constrains the possible interactions.

Use of an Adaptor

Using an adaptor is another way of affording interaction. An adaptor is a third system, or a device, that mediates between two incompatible systems. For example, the electrical adaptor enables effective interaction to be accomplished between two ill-adapted, fixed design systems. The need for adaptors in order to enable interaction between systems is what gives rise to the network of agent–device interactions that characterize HCI. For example, a keyboard is needed to enter data and commands into a computer system, because the computer is generally ineffective at processing the signals that a human would usually use: speech or handwriting. A pointing device is needed because gestures have to be precisely communicated to the computer.

With bespoke software, the adaptor between user and system has typically been the systems analyst and the computer programmer. The analyst can understand the users' needs and translate these into system functions. Education and training are other examples of adaptors being used to mediate between a fixed design computer system and a human. More recently, we have seen speech recognition systems becoming available to act as adaptors between humans and computer systems; speech technology has wide application both for able-bodied and for physically disabled users. The use of an adaptor is appropriate when two systems cannot otherwise accommodate each other; this is the case when accessibility problems are alleviated by the choice of alternative input/output devices, or by communication via an alternative modality.

However, in HCI, we have another complicating dimension, that of continuous change in the human (by experience and by aging), and of discrete incidents (such as accidents) that present immediate requirements for accommodation. Hence, the fixed design will inevitably fail to meet some needs at some time, even through the use of an adaptor.

Customizable Systems

Customizable systems have built-in flexibility; they are designed to accommodate a wider range of interactions than systems with a single fixed design. For example, the comfort of a car driver is improved by the ability to alter some ergonomic parameters (seat and steering wheel positions) within limits. PCs' operating systems now come with a number of control panels that allow us to tailor screen resolution, colors used, sizes of fonts, menus and icons, date and time display formats, and much more. Indeed, the Microsoft Accessibility Properties Control Panel of the Microsoft Windows 95/98 operating system provides a range of useful functions designed to help disabled users. Example features include suppression of inadvertent repeated keystrokes, elimination of the need to press two keys simultaneously in control sequences, visual captions for auditory signals and speech, and automatic selection of high-contrast colors and fonts. In general, users can customize software by removing parts of the functionality, for example by selecting "short menus." In addition, shortcut command keys can be set up to make often-used functions easier and quicker to accomplish.

Many systems are customizable with respect to the purpose of use; large pieces of computer systems can be, or cannot be, installed, depending on the needs of the users. New functions can be added to systems by writing macros and these can significantly alter the capabilities of those systems. Customization at the physical level is most common; characteristics of mouse, keyboard, visual display unit, "desktop," and within-package functionality and layout are now routinely alterable.

The problem with customizable systems is that they rely on the users to configure them, thus forcing users to do something tangential to their main purpose. In driving the car, many of us will have found how frustrating it can be to have to spend time adjusting the seat, mirror, and so on before driving off. In using computer systems, the problem is magnified, because often new complexity is introduced by having to learn the tailoring mechanisms. Macro writing is one example of this. Users effectively have to become programmers in order to specify the system's behavior required. Indeed, frequently users require an adaptor to interface between themselves and the customizing system, which, in turn, involves learning yet another system.

Adaptive Systems

Adaptive systems have the ability to alter their state and/or behavior in response to an interaction with another system (Benyon, 1993b). Humans are clearly highly adaptive systems. When humans interact with other humans, they will happily accommodate all sorts of individual quirks and differences in knowledge, they will deal with errors, adapt to habits, and recognize cognitive characteristics, or emotional states, and they are able to adapt their behavior accordingly. We see this in people explaining things to others. In explaining how to get somewhere, if one method (perhaps a primarily verbal explanation) proves ineffective, another (perhaps a more visual explanation) is often adopted. Similarly, we see people change from a high-level explanation (dealing with general principles) to a more detailed one (offering specific examples) when necessary. Humans also benefit from meta-knowledge (knowing what they know). They also possess a good model of the typical person and how to interact with them. Notice, though, that we all have limits to our adaptive capability; we typically speak only one or two languages, have limited understanding of other cultures, and are knowledgeable about limited domains.

Humans are adaptive systems, but they can adapt only within the constraints imposed by their knowledge. This is true of any adaptive system. A thermostat is an adaptive system because it certainly interacts with another system (the ambient environment in this case). But its adaptive ability is hard-wired. Its model of the world consists of two pieces of thin metal that are suited to adapt to temperature changes, but to nothing else. This is an effective model for its purpose, but is hardly a sophisticated one. Between the human and the thermostat lie a host of systems that possess a more or less sophisticated model of the other systems and that can, therefore, adapt within the constraints of that model. This "notional world" (Dennett, 1989) is rarely a complete representation of the other system. It is a model that is effective for particular purposes.

Within HCI, one can identify two types of adaptive systems, both useful, but one requiring no sophistication on the part of the machine and the other requiring a great deal of sophistication. The first of these can be characterized as "automatic customization." For example, smart card technologies promise to automate the customization process and thereby enable users, wherever they may be, to access their own desktops, be presented with their own preferences, see displays in their own language, and have their disabilities automatically accommodated. In time, one might imagine such personal requirements to be held globally (centrally), accessible to any networked device, anywhere. The more sophisticated type of adaptation requires artificial intelligence, the ability to change with the user's changing cognitive, physical, and task requirements, to anticipate his needs, to understand her motivations, to support his activities as usefully as another human might. We are now in the realm of intelligent agents and intelligent tutoring systems. One of the problems that computers face in attempting to do this second type of adaptation is that it is difficult to infer high-level human characteristics, such as cognitive style, learning style, or personality, from the meager information that is available from the interaction.

Simple forms of automatic adaptation can be very effective for inclusive design. For example, the computer can recognize long commands or regularly used Internet addresses and fill in the details, thus saving users the effort of typing. However, the more elaborate applications of automatic adaptation seem unlikely to be a significant contributor to universal design solutions for several years to come.

Discussion

All systems are necessarily premised on a model of the other systems that they are designed to interact with; thus, all computer systems include, at least implicitly, some model of their users. These representations should include some specification of the range of variation that can be accommodated by the design. Systems also include a method by which that variation is accommodated—compatible design, the use of an adaptor, customization, or automatic adaptation. All nonadaptive systems suffer from the fact that they impose all the load of change onto the user. In some cases, this is highly desirable—for example, for security systems. In many circumstances, however, it is unacceptable. In a designed, nonadaptive system, the user may be unable to adapt or may not be sufficiently motivated to devote the resources necessary to adapt. In a customizable system, the user may be willing to adapt, but cannot understand the (designed, nonadaptive) customizing system. In other cases, there may be no adaptor available.

Fischer (1993) made the point that many systems need to be both adaptive and customizable. Frequently, an application needs *both* facilities to deal with the variety of support that users require.

SUMMARY AND CONCLUSIONS

In this chapter we have outlined the relationship between individual differences and HCI. Our aim has been to provide a perspective on the issue, rather than to discuss specific solutions. *The goal of designers must always be to maximize participation of the intended user groups within the resources available for the development of the system,* thus recognizing that total inclusion will sometimes be infeasible. A primary goal is that *inclusivity analysis* becomes the norm and that the implications of every design decision are understood and justified. We have offered an overview of the individual differences literature, picking out those aspects that seem particularly relevant to HCI. We then provided a conceptual framework for discussing inclusion and exclusion and embellished the notion of user to conjure up more of the richness and complexity of the individuals, the roles, and the tasks our computer systems have to support. Finally, we presented the broad options available to designers: nonadaptive (but well-designed) systems, use of an adaptor, customizable systems, and adaptive systems.

As in all design work, simplicity is advocated. When catering for special needs one should be certain that standard interfaces and interaction devices are not suitable before prescribing alternatives. Standard devices are generally cheaper, better supported by technical staff, and ubiquitous. By this token, there is a case for changing standard practice where current interfaces exclude a significant number of users, and a proposed alternative would accommodate current users and others equally well.

Finally, there is no getting away from the need for all complex systems to embody models of the other systems with which they interact. In all systems, this model is fixed at any given time. Adaptive systems depend crucially on the quality of their self-monitoring and of their internal representations of any cooperating systems. In searching for mechanisms to achieve a good fit between people and computers, a key question that designers must consider is how flexible can and should the representations of people be.

REFERENCES

Ballabio, E., & Moran, R. (1998). *Older people and people with disabilities in the information society: An RTD approach for the European Union.* Brussels: European Commission, Directorate-General XIII.

Benyon, D. R. (1993a). Accommodating individual differences through an adaptive user interface. In M. Schneider-Hufschmidt, T. Kühme, & U. Malinowski (Eds.), *Adaptive user interfaces: Principles and practice* (pp. 149–165). Amsterdam: North-Holland, Elsevier Science.

Benyon, D. R. (1993b). Adaptive systems; a solution to usability problems. *User Modelling and User Adapted Interaction, 3*(1), 1–22.

Boring, E. G. (1923). Intelligence as the tests test it. *New Republic, 35,* 35–36.

Bourges-Waldegg, P., & Scrivener, S. A. R. (1998). Meaning, the central issue in cross-cultural HCI design. *Interacting With Computers, 9,* 287–309.

Burt, C. (1940). *The factors of the mind.* London: University of London Press.

Carrol, J. M. (1993). *Human cognitive abilities: A survey of factor-analytic studies.* New York: Cambridge University Press.

Chen, C., & Rada, R. (1996) Interacting with hypertext: A meta-analysis of experimental studies. *Human–Computer Interaction, 11*(2), 125–156.

Cooley, W. W., & Lohnes, P. R. (1976). *Evaluation research in education.* New York: Irvington.

Crerar, A., & Benyon, D. R. (1998). Integrating usability into software development. In L. Trenner & J Bawa (Eds.), *The politics of usability—A practical guide to designing usable systems in industry* (pp. 49–60). Vienna: Springer Verlag.

Dennett, D. (1989). *The intentional stance.* Cambridge, MA: MIT Press.

Dieterich, H., Malinowski, U., Kühme, T., & Schneider-Hufschmidt, M. (1993). State of the art in adaptive user interfaces. In M. Schneider-Hufschmidt, T. Kühme, & U. Malinowski (Eds.), *Adaptive user interfaces: Principles and practice* (pp. 13–48). Amsterdam: North-Holland, Elsevier Science.

Egan, D. E., & Gomez, L. M. (1985). Assaying, isolating and accommodating individual differences in learning a complex skill. In R. Dillon (Ed.), *Individual differences in cognition* (pp. 173–217). New York: Academic Press.

Eysenck, H. J. (1990). Genetic and environmental contributions to individual differences: The three major dimensions of personality. *Journal of Personality, 58*(1), 245–261.

Fernandes, T. (1995). *Global interface design.* New York: Academic Press.

Fischer, G. (1993). Shared knowledge in cooperative problem-solving systems: Integrating adaptive and adaptable components. In M. Schneider-Hufschmidt, T. Kühme, & U. Malinowski (Eds.), *Adaptive user interfaces: Principles and practice* (pp. 49–68). Amsterdam: North-Holland, Elsevier Science.

Gardner, H. (1983). *Frames of mind: The theory of multiple intelligences.* New York: Basic Books.

Gill, J. (1994) *The forgotten millions.* Luxembourg: Office for Official Publications.

Gill, J. (1997). *Access prohibited? Information for designers of public access terminals.* London: Royal National Institute for the Blind.

Haigh, R. (1993). The aging process: A challenge for design. *Applied Ergonomics, 24*(1), 9–14.

Hoffmann, E. R., Chang, W. Y., & Yim, K. Y. (1997). Computer mouse operation: Is left-handed user disadvantaged? *Applied Ergonomics, 28*(4), 245–248.

Höök, K., Sjölinder, M., & Dählback, N. (1996). *Individual differences and navigation in hypermedia* (SICS Research Report), Swedish Institute of Computer Science, Kista, Sweden.

Kalyuga, S., Chandler, P., & Sweller, J. (1998). Levels of expertise and instructional design. *Human Factors, 40*(1), 1–17.

Kline, R. L., & Glinert, E. P. (1995). Improving GUI accessibility for people with low vision. In *Proceedings of the Human Factors in Computing Systems* (pp. 114–121). New York: ACM Press.

Legge, G. E., & Campbell, F. W. (1987). *Vision of color and pattern. Carolina Biology Reader No. 165* (J. J. Head, Series Ed.). Burlington, NC: Carolina Biological Supply Company.

Liu, M., & Reed, W. M. (1994). The relationship between the learning strategies and learning styles in a hypermedia environment. *Computers in Human Behavior, 10*(4), 419–434.

McCluskey, J. J. (1997). An exploratory study of the possible impact of cerebral hemisphericity on the performance of select linear, non-linear, and spatial computers tasks. *Journal of Educational Computing Research, 16*(3), 269–279.

McCrae, R. R., & John, O. P. (1992). An introduction to the five-factor model and its applications. *Journal of Personality, 60*(2), 175–215.

McGrath, D. (1992) Hypertext, CAI, paper, or program control: Do learners benefit from choices? *Journal of Research on Computing in Education, 24*(4), 513–532.

Mead, S. E., Spaulding, V. A., Sit, R. A., Meyer, B., & Walker, N. (1997). Effects of age and training on World Wide Web navigation strategies. In *Proceedings of the Human Factors and Ergonomics Society 41st Annual Meeting* (pp. 152–156). Santa Monica, CA: Human Factors and Ergonomics Society.

Messick, S. (1976). *Individuality in learning.* San Francisco: Jossey-Bass.

Miller, A. (1987). Cognitive styles: An integrated model. *Educational Psychology, 7*(4), 251–268.

Miller, A. (1991). Personality types, learning styles and educational goals. *Educational Psychology, 11*(3 & 4), 217–238.

National Center for Health Statistics. (1985). *Vital and health statistics* (Series 10, No. 160). Washington, DC: U.S. Department of Health and Human Services, Public Health Service.

Noyes, J., & Starr, A. (1996). Use of automatic speech recognition: Current and potential applications. *Computing & Control Engineering Journal, 7*(5), 203–208.

Pask, G. (1976). Styles and strategies of learning. *British Journal of Educational Psychology, 46*, 128–148.

Picard, R. (1997). *Affective computing.* Cambridge, MA: MIT Press.

Presland, J. (1994). Learning styles and CPD. *Educational Psychology in Practice, 10*(3), 179–184.

Prouix, G., Laroche, C., & Latour, J. C. (1995). Audibility problems with fire alarms in apartment buildings. In *Proceedings of the Human Factors and Ergonomics Society 39th Annual Meeting* (pp. 989–993). Santa Monica, CA: Human Factors and Ergonomics Society.

Prümper, J., Zapf, D., Brodbeck, F. C., & Frese, M. (1992). Some surprising differences between novice and expert errors in computerised office work. *Behavior & Information Technology, 11*(6), 319–328.

Richter, L. A., & Salvendy, G. (1995). Effect of personality and task strength on performance in computerized tasks. *Ergonomics, 38*(2), 281–291.

Riding, R., & Cheema, I. (1991). Cognitive styles—An overview and integration. *Educational Psychology, 11*(3 & 4), 193–215.

Rogers, W. A., Gilbert, D., K., & Cabrera, E. F. (1997). An analysis of automatic teller machine usage by older adults: A structured interview approach. *Applied Ergonomics, 28*(3), 173–180.

Sharit, J., & Czaja, S. J. (1994). Aging, computer-based task performance, and stress: Issues and challenges. *Ergonomics, 37*(4), 559–577.

Spearman, C. E. (1904). "General intelligence" objectively determined and measured. *American Journal of Psychology, 15,* 72–101.

Sproull, L., Subramani, M., Kiesler, S., Walker, J. H., & Waters, K. (1996). When the interface is a face. *Human–Computer Interaction, 11*(2), 97–124.

Stanney, K. M., & Salvendy, G. (1992). Diversity in field-articulation and its implications for human–computer interface design. In *Proceedings of the Human Factors Society 36th Annual Meeting* (pp. 902–906). Santa Monica, CA: Human Factors and Ergonomics Society.

Thurstone, L. L. (1938). *Primary mental abilities. Psychometirc Monographs, 1.* Chicago: University of Chicago Press.

Van der Veer, G. C., Tauber, M. J., Waern, Y., & Van Muylwijk, B. (1985). On the interaction between system and user characteristics. *Behavior and Information Technology, 4,* 284–308.

Vicente, K. J., Hayes, B. C., & Williges, R. C. (1987). Assaying and isolating individual difference in searching a hierarchical file system. *Human Factors, 29,* 349–359.

Westerman, S. J., Davies, D. R., Glendon, A. I., Stammers, R. B., & Matthews, G. (1995). Age and cognitive ability as predictors of computerized information retrieval. *Behavior & Information Technology, 14*(5), 313–326.

Wiedenbeck, S., & Davis, S. (1997). The influence of interaction style and experience on user perceptions of software packages. *International Journal of Human–Computer Studies, 46*(5) 563–588.

Witkin, H. A., Moore, C. A., Goodenough, D. R., & Cox, P. W. (1997). Field-dependent and field-independent cognitive styles and their educational implications. *Review of Educational Research, 47*(1), 1–64.

Youngson, R. (1986). *How to cope with tinnitus and hearing loss.* London: Sheldon Press.

3 International and Intercultural User Interfaces

Aaron Marcus

User interfaces for successful products enable users around the world to access complex data and functions. Solutions to global user interface design consist of partially universal and partially local solutions to the design of metaphors, mental models, navigation, appearance, and interaction. By managing the user's experience of familiar structures and processes, the user interface designer can achieve compelling forms that enable the user interface to be more usable and acceptable. The user will be more productive and satisfied with the product in many different locations globally.

The concept of *user interfaces for all* implies the availability of, and easy access to computer-based products among all peoples in all countries worldwide. Successful computer-based products developed for users in different countries and among different cultures consist of partially universal, or general solutions and partially unique, or local solutions to the design of user interfaces. By managing the user's experience with familiar structures and processes, the user's surprise at novel approaches, as well as the user's preferences and expectations, the user interface designer can achieve compelling forms that enable the user interface to be more usable and acceptable. Globalization of product distribution, as with other manufacturing sectors, requires a strategy and tactics for the design process that enables efficient product development, marketing, distribution, and maintenance. Globalization of user interface design, whose content and form is so much dependent on visible languages and effective communication, improves the likelihood that users will be more productive and satisfied with computer-based products, in many different locations globally.

Demographics, experience, education, and roles in organizations of work or leisure characterize users. Their individual needs, as well as their group roles define their tasks. User-centered, task-oriented design methods account for these aspects and facilitate the attainment of effective user interface design.

User interfaces conceptually consist of *metaphors, mental models, navigation, appearance,* and *interaction.* In the context of this chapter, these terms may be defined as follows (Marcus, 1995, 1998):

- *Metaphors*: essential concepts conveyed through words and images, or through acoustic, or tactile means. Metaphors concern both overarching concepts that characterize interaction, as well as individual items, like the "trashcan" standing for "deletion" within the "desktop" metaphor.
- *Mental models*: organization of data, functions, tasks, roles, and people in groups at work or play. The term, similar to, but distinct from cognitive models, task models, user models, and so on, is intended to convey the organization observed in the user interface itself, which is presumably learned and understood by users and which reflects the content to be conveyed, as well as the available user tasks.
- *Navigation*: movement through mental models, afforded by windows, menus, dialogue areas, control panels, and so on. The term implies process, as opposed to structure, that is, sequences of content potentially accessed by users, as opposed to the static structure of that content.
- *Appearance*: verbal, visual, acoustic, and tactile perceptual characteristics of displays. The term implies all aspects of visible, acoustic, and haptic languages (e.g., typography or color; musical timbre or cultural accent within a spoken language; and surface texture or resistance to force).
- *Interaction*: the means by which users communicate input to the system and the feedback supplied by the system. The term implies all aspects of command-control devices (e.g., keyboards, mice, joysticks, microphones), as well as sensory feedback (e.g., changes of state of virtual graphical buttons, auditory displays, and tactile surfaces).

For example, an application, its data, the graphical user interface (GUI) environment, and the underlying hardware, all contribute to the functional and presentational attributes of the user interface. An advanced English text editor, working within the Microsoft Windows 95 GUI environment, on a mouse- and keyboard-driven Intel Pentium processor-based PC, presents one set of characteristics. The LCD displays and buttons on the front panel of a French paper copier, or the colorful displays and fighter-pilot-like joysticks for a children's video game on a Japanese Sega game machine, present alternative characteristics.

This chapter discusses the development of user interfaces that are intended for users in many different countries with different cultures, languages, and groups, in the context of the emerging information society. The text presents a survey of important issues, as well as recommended steps in the development of user interfaces for an international and intercultural user population. With the rise of the World Wide Web and application-oriented Web sites, the challenge of designing good user interfaces, which are inherently accessible by users around the globe, becomes an immediate, practical matter, not only a theoretical issue. This topic is discussed from a user perspective, not a technology and code perspective. The chapter (a) introduces fundamental definitions of globalization in user interface design, (b) reviews globalization in the history of computer systems, and (c) demonstrates why globalization is vital to the success of computer-based communication products.

GLOBALIZATION

Definition of Globalization

Globalization refers to the worldwide production and consumption of products and includes issues at international, intercultural, and local scales. In an information-oriented society, globalization affects most computer-mediated communication, which, in turn, affects user interface design. The discussion that follows refers particularly to user interface design.

Internationalization issues refer to the geographic (in terms of location), political, and linguistic/typographic issues of nations, or groups of nations (see some examples in Table 3.1). An example of efforts to establish international standards for some parts of user interfaces is the International Standards Organisation's (ISO) draft human factors standards in Europe, for color legibility standards of cathode-ray tube (CRT) devices (ISO 9241-8, 1989). Another example is the legal requirement for bilingual English and French displays in Canada, or the quasi-legal denominations for currency, time, and physical measurements, which differ from country to country.

Intercultural issues refer to the religious, historical, linguistic, aesthetic, and other, more humanistic issues of particular groups or peoples, sometimes crossing

TABLE 3.1
Examples of Differing Displays for Currency, Time, and Physical Measurements

Measurement	U.S. Example	European Example	Asian Example
Currency	$1,234.00 (U.S. dollars)	DM I.234 (German marks)	¥1,234 (Japanese yen)
Time	8:00 PM, August 24, 1999	20:00, 24 August 1999 (England)	20:00, 1999.08.24, or Imperial Heisei 11or H11 (Japan)
	8:00 PM, 8/24/99	20:00, 24.08.99 (Germany, traditional)	
		20:00, 1999-08-24 (ISO 8601 Euro standard)	
Physical	3 lb, 14 oz	3.54 kg, 8.32 m (England)	3.54 kg, 8.32 m in Roman or Katakana characters (Japan)
	3' 10", 3 feet and 10 inches	3.54 kg. 8,32 m (Euro standard)	

national boundaries. Examples include calendars that acknowledge various religious time cycles, color/type/signs/terminology (see Table 3.2) reflecting various popular cultures, and organization of content in Web search criteria reflecting cultural preferences.

Localization refers to the issues of specific, small-scale communities, often with unified language and culture, and, usually, at a scale smaller than countries, or significant cross-national ethnic "regions." Examples include affinity groups (e.g., U.S. "twenty-somethings," or U.S. Saturn automobile owners), business or social organizations (e.g., German staff of DaimlerChrysler, or Japanese golf club members), and specific intranational groups (e.g., India's untouchables, or Japanese housewives). With the spread of Web access, "localization" may come to refer to groups of shared interests that may also be geographically dispersed.

History of Globalization in Computing

In the early years of computers, 1950–1980, globalization was not a significant issue. IBM and other early computer/software manufacturers produced English-based, U.S.-oriented products with text-based user interfaces for national and world markets. Large, expensive computers required an elite engineering-oriented professional group to build, maintain, and use such systems. This group consisted of either native English speakers or those with some fluency in English. Personal computers eventually emerged, but with textual user interfaces. In the 1980s, workstations and eventually low-cost personal computers and software brought dramatic increases in the worldwide population of users.

In the 1980s worldwide markets drove the development of multilingual editions of operating systems and software applications, which, by the late 1980s, had begun to acquire GUIs, such as those for Macintosh, Motif, OpenLOOK, and

TABLE 3.2
Examples of Differing Cultural References

Item	N. America/Europe Example	Middle-Eastern Example	Asian Example
Sacred colors	White, blue, gold, scarlet (Judeo-Christian)	Green, light blue (Islam)	Saffron yellow (Buddhism)
Reading direction	Left to right	Right to left	Top to bottom

Item	United States	France, Germany	Japan
Web search	Culture does not imply political discussions	Culture implies political discussions	Culture implies tea ceremony discussions
Sports references	Baseball, football, basketball	soccer	Sumo wrestling, baseball

Microsoft Windows. In addition, many countries outside of the United States became software suppliers for both local and international markets.

In the mid-1990s, with the rise of the World Wide Web and nearly instant global access and interaction via the Internet, determining the targets for global communication became immediately an issue. According to Alvarez, Kasday and Todd (1998), the European Information Technology Observatory's 1997 data stated that there were approximately 90.6 million Internet users globally, including 41.9 million in the United States. In addition, only 10% of people in the world are native speakers of English, yet 70% of Web sites are English only. It is likely that there will be strong interest among users outside North America to have more content in native languages.

Survey of the Literature

The existing literature on the subject of globalization for user interfaces, although relatively limited in size, is of great value to developers. Three primary references are the texts by DelGaldo and Nielsen (1996), Fernandes (1995), and Nielsen (1990).

The book edited by Nielsen, *Designing User Interfaces for International Use,* published in 1990, focuses on issues of language, translation, typography, and textual matters in user interfaces (including GUIs). Additionally, Ossner (1990) introduced semiotics-based analysis of icons and symbols and applied that knowledge to computer-based training (CBT).

Fernandes' book, *Global Interface Design,* published in 1995, is much more oriented to GUIs and provides chapters about topics such as language, visual elements, national formats, and symbols, together with design rules. The book is notable for providing chapters about culture, cultural aesthetics, and even business justification for globalization, a topic of significant importance to the managers of research and development organizations/groups, who must argue for the merits of considering globalization in product manufacturing budgets.

The book edited by DelGaldo and Nielsen, *International User Interfaces,* published in 1996, addresses topics covered in the previous books, but also adds consideration of many more cultural issues and their impact on user interface design. A notable contribution is that of Ito and Nakakoji (1996), which compares North American and Japanese user interfaces. They commented, for example, on the lack of familiarity of early Japanese users of word-processing software. This software used metaphorical references to typewriters, such as tab stops and margin settings, which were more familiar to Western users and more suited to 100-character alphanumeric keyboards and typesetting, rather than the 6,000 characters of Japanese-Chinese symbols used in Japan.

In the last few years, especially with the rise of the Web, more and more articles have appeared in conference proceedings and publications of professional organizations. Examples of these organizations in the United States include the Association for Computing Machinery's Special Interest Group on Computer–Human Interaction (ACM/SIGCHI), the International Conference in Human Computer Interaction (HCI International), the Usability Professionals Association (UPA), and the Society for Technical Communication (STC), all of which have had technical sessions, panels, and/or theme publications devoted to globalization and user interface design.

Currently, a casual Web search via AltaVista[1] returns 100,000 to 350,000 items that refer to user interfaces, globalization, internationalization, and/or localization. These large collections indicate more literature on these topics, although the quality of the entries and their relevance cannot be determined easily and efficiently.

Advantages and Disadvantages of Globalization

The business justification for globalization of user interfaces is complex. If the content (functions and data) is likely to be of value to target population outside of the original market, it is usually worthwhile to plan for international and intercultural factors in developing a product, so that it may be efficiently customized. Rarely can a product achieve global acceptance with a "one size fits all" solution. Developing a product for international, intercultural audiences usually involves more than merely a translation of verbal language. Visible (or otherwise attainable) language must also be revised, and other user interface characteristics may need to be altered.

Developing products that are ready for global use, although increasing initial development costs, gives rise to potential for increased international sales. However, for some countries, monolithic domestic markets may inhibit awareness of, and incentives for globalization. Because the United States has been in the past such a large producer and consumer of software, it is not surprising that some U.S. manufacturers have targeted only domestic needs. However, others in the United States have understood the increasingly valuable markets overseas. In order to develop multiple versions efficiently, languages, icons, and other components must be easily swappable. For example, in order to penetrate some markets, the local language is an absolute requirement (e.g., in France).

Some software products are initiated with international versions (but usually released in sequence because of limited development resources). Other products are "retrofitted" to suit the needs of a particular country, language, or culture, as needs or opportunities arise. In some cases, the later, ad-hoc solution may suffer because of the lack of original planning for globalization.

Several case studies demonstrating the value of localized products appear in Fernandes (1995).

Globalization Design Process

The general user interface development process, enhanced to address globalization issues, may be summarized as follows. This process is generally sequential

---

[1]URL: http://www.altavista.com

with partially overlapping steps, some of which are, however, partially or completely iterative. Additionally, the order in which these steps are performed within subsequent iterations may be modified to better suit the needs of the design process. For example, the evaluation step described in the following list may be carried out prior to, during, or after the design step:

- *Plan:* Define the challenges or opportunities for globalization; establish objectives and tactics; determine budget, schedule, tasks, development team, and other resources. Globalization must be specifically accounted for in each item of project planning; otherwise, cost overruns, delays in schedule, and lack of resources are likely to occur.
- *Research:* Investigate dimensions of global variables and techniques for all subsequent steps, for example, techniques for analysis, criteria for evaluation, media for documentation, and so on. In particular, identify items among data and functions that should be targets for change and identify sources of national/cultural/local reference. User-centered design theory emphasizes gathering information from a wide variety of users; globalization refines this approach by stressing the need to research adequately users' wants and needs according to a sufficiently varied spectrum of potential users, across specific dimensions of differentiation. In current practice, this variety is often insufficiently considered.
- *Analyze:* Examine results of research (e.g., challenges or opportunities in the prospective markets), refine criteria for success in solving problems or exploiting opportunities (write marketing or technical requirements), determine key usability criteria, and define the design brief, or primary statement of the design's goals. At this stage, globalization targets should be itemized.
- *Design:* Visualize alternative ways to satisfy criteria using alternative prototypes; based on prior or current evaluations, select the design that best satisfies criteria for both general good user interface design and globalization requirements; prepare documents that enable consistent, efficient, precise, accurate implementation.
- *Implement:* Build the design to complete the final product; for example, write code using appropriate tools. In theory, planning and research steps will have selected appropriate tools that make implementing global variations efficient.
- *Evaluate:* At any stage, review or test results in the marketplace against defined criteria for success; for example, conduct focus groups, test usability on specific functions, gather sales and user feedback. Identify and evaluate matches and mismatches, then revise the designs to strengthen effective matches and reduce harmful mismatches. Testing prototypes or final products with international, intercultural, or specific localized user groups is crucial to achieving globalized user interface designs.

- *Document:* Record development history, issues, and decisions in specifications, guidelines, and recommendation documents. As with other steps, specific sections or chapters of documents that treat globalization issues are required.

Critical Aspects for Globalization: General Guidelines

Beyond the user interface development process steps identified in the previous section, the following guidelines can assist developers in preparing a checklist for specific tasks. The recommendations that follow are grouped under user interface design terms referred to earlier:

1. *User demographics:*
   - Identify national and cultural target user populations and segments within those populations, then identify possible needs for differentiation of user interface components and the probable cost of delivering them.
   - Identify potential savings in development time through the reuse of user interface components, based on common attributes among user groups. For example, certain primary (or top-level) controls in a Web-based, data-retrieval application might be designed for specific user groups, so as to aid comprehension and to improve appeal. Lower level controls, on the other hand, might be more standardized, unvarying formslike elements.
2. *Technology:*
   - Determine the appropriate media for the appropriate target user categories.
   - Account for international differences to support platform, population, and software needs, including languages, scripts, fonts, colors, file formats, and so forth.
3. *Metaphors:*
   - Determine optimum minimum number of concepts, terms, and primary images to meet target user needs.
   - Check for hidden miscommunication and misunderstanding.
   - Adjust the appearance, orientation, and textual elements to account for national or cultural differences. For example, in relation to metaphors for operating systems, Chavan (1994) pointed out that Indians relate more easily to the concept of bookshelf, books or notebooks, chapters or sections, and pages, rather than the desktop, file folders, and files with multiple pages.
4. *Mental models:*
   - Determine optimum minimum varieties of content organization to meet target user needs.
5. *Navigation:*
   - Determine need for navigation variations to meet target user requirements, determine costs and benefits, and revise as feasible.

6. *Appearance:*
   - Determine optimum minimum variations of visual and verbal attributes. Visual attributes include layout, icons and symbols, typography, color, and general aesthetics. Verbal attributes include language, formats, and ordering sequences. For example, many Asian written languages, such as Chinese and Japanese, contain symbols with many small strokes. This factor seems to lead to an acceptance of higher visual density of marks in complex public information displays than is typical for Western countries.
7. *Interaction:*
   - Determine optimum minimum variations of input and feedback to meet target user requirements. For example, because of Web access-speed differences for users in countries with very slow access, it is usually important to provide text-only versions, without extensive graphics, as well as alternative text labels to avoid graphics that take considerable time to appear. As another example, some Japanese critics believe that office group-ware applications from Northern European countries match personal communication needs of Japanese users more closely than similar applications from the United States.

Specific Appearance Guidelines

Because of space limitations in this introductory chapter about a complex topic, complete, detailed guidelines cannot be provided for all of the user interface design terms listed in the previous section. Some detailed guidelines for one important topic, visual and verbal appearance, appear next. Further details can be found in DelGaldo and Nielsen (1996), Fernandes (1995), and Nielsen (1990).

1. *Layout and orientation:*
   - As appropriate, adjust the layout of menus, tables, dialogue boxes, and windows to account for the varying directions and size of text.
   - If dialogue areas use sentencelike structure with embedded data fields and/or controls, these areas will need special restructuring to account for language changes that significantly alter sentence format. For example, German sentences often have verbs at the end of sentences, whereas English and French place them in the middle.
   - As appropriate, change layout of imagery that implies or requires a specific reading direction. Left-to-right sequencing may be inappropriate or confusing for use with right-to-left reading scripts and languages.
   - Check for misleading arrangements of images that lead the viewer's eye in directions inconsistent with language reading directions.
   - For references to paper and printing, use appropriate printing formats and sizes. For example, the 8.5-by-11-inch standard office letterhead paper size in the United States is not typical in many other countries that use the

European A4 paper size of 210-by-297-mm with a square-root of two rectangular proportion.

2. *Icons and symbols:*
    - Avoid the use of text elements within icons and symbols to minimize the need for different versions to account for varying languages and scripts.
    - Adjust the appearance and orientation to account for national or cultural differences. For example, using a post letterbox as an icon for E-mail may require different images for different countries.
    - As a universal sign set reference, consider using as basic icon/symbol references the signs, or signs derived from them, that constitute the international signage set developed for international safety, mass transit, and communication (see American Institute of Graphic Arts [AIGA], 1981; Olgyay, 1995; Pierce, 1996).
    - Avoid puns and local, unique, charming references that will not transfer well from culture to culture. Keep in mind that many "universal" signs are covered by international trademark and copyright use, e.g., Mickey Mouse and the "Smiley" smiling face. In the United States, the familiar smiling face is not a protected sign, but it is in other countries.
    - Consider whether selection symbols, such as the X or check marks, convey the correct distinctions of selected and not-selected items. For example, some users may interpret an X as crossing out what is not desired rather than indicating what is to be selected.
    - Be aware that office equipment such as telephones, mailboxes, folders, and storage devices differ significantly from nation to nation.

3. *Typography:*
    - Use fonts available for a wide range of languages required for the target users.
    - Consider whether special font characters are required for currency, physical measurements, and so on.
    - Ensure appropriate decimal, ordinal, and currency number usage. Formats and positioning of special symbols vary from language to language.
    - Use appropriate typography and language for calendar, time zone, and telephone/fax references.

4. *Color:*
    - Follow perceptual guidelines for good color usage. For example, use warm colors for advancing elements and cool colors for receding elements; avoid requiring users to recall in short-term memory more than five plus or minus two different coded colors.
    - Respect national and cultural variations in colors, where feasible, for the target users.

5. *Aesthetics:*
    - Respect, where feasible, different aesthetic values among target users. For example, some cultures have significant attachment to wooded natural scenes, textures, patterns, and imagery (e.g., the Finnish and the Japanese), which might be viewed as exotic or inappropriate by other cultures.

- Consider specific culture-dependent attitudes. For example, Japanese viewers find disembodied body parts, such as eyes and mouths, unappealing in visual imagery.

6. *Language and Verbal Style:*
   - Consider which languages are appropriate for the target users, including the possibility of multiple languages within one country, for example, English and French within Canada.
   - Consider which dialects are appropriate within language groupings and check vocabulary carefully, for example, for British versus American terms in English, Mexican versus Spanish terms in Spanish, or Mainland China versus Taiwanese terms in Chinese.
   - Consider the impact of varying languages on the length and layout of text. For example, German, French, and English versions of text generally have increasingly shorter lengths.
   - Consider the different alphabetic sorting or ordering sequences for the varied languages and scripts that may be necessary and prepare variations that correspond to the alphabets. Note that different languages may place the same letters in different locations, for example, Å comes after A in French but after Z in Finnish.
   - Consider differences of hyphenation, insertion point location, and emphasis, that is, use of bold, italic, quotes, double quotes, brackets, and so on.
   - Use appropriate abbreviations for such typical items as dates, time, and physical measurements. Remember that different countries have different periods of time for "weekends" and the date on which the week begins.

A CASE STUDY

Planet SABRE

One example of user interface globalization is the design for Planet SABRE™, the graphical version of the SABRE™ Travel Information Network (STIN), one of the worlds largest private online networks. Planet SABRE is used exclusively by travel agents. The authors firm has worked closely with STIN marketing and engineering staff over a period of approximately 6 years in developing the user interface for Planet SABRE.

The SABRE system contains approximately 60 terabytes of data about airline flights, hotels, and automobile rental, and enables almost $2 billion of bookings annually. The system sustains up to 1 billion "hits" per day.

The Planet SABRE user interface development process emphasized achieving global solutions from the beginning of the project. For example, stated requirements mentioned allowing for the space needs of multiple languages for labels in windows and dialogue boxes.

Besides supporting multiple languages (English, Spanish, German, French, Italian, and Portuguese), the user interface design enables the switching of icons for primary application modules so that they would be more gender-, culture-, and nation-appropriate (see accompanying figures).

Figure 3.1 shows the initial screen of an earlier prototype for Planet SABRE, with icons representing the primary applications or modules within the system conveyed through the metaphor of objects on the surface of a planet, or in outer space. The postal box representing the electronic mail functions depicts an object that users in the United States would recognize immediately. However, users in many other countries would have significant difficulty in recognizing this object, because postal boxes come in very different physical forms.

Figure 3.2 shows a prototype version of the Customizer dialogue box, in which the user can change/select preferences; in particular, certain icons can be swapped, so that they appear in more recognizable images. This change could also be accomplished for other icons, such as the depiction of the passenger. Once changed in the Customizer, the icon's appearance is switched throughout the user interface wherever the icon appears.

At every major stage of prototyping, designs developed in the United States were taken to users in international locations for evaluation. The user feedback was relayed to the development team and affected later decisions about all aspects of the user interface design.

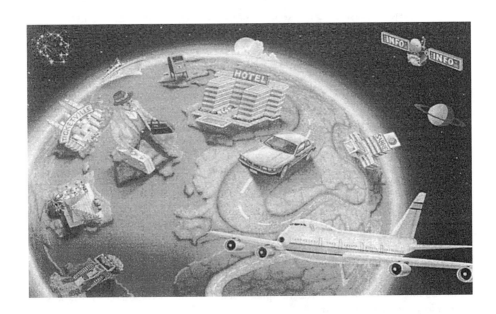

FIG. 3.1.   Example of Planet SABRE home screen showing typical icons for passenger, airline booking, hotel rental, car rental, and E-mail (post box).

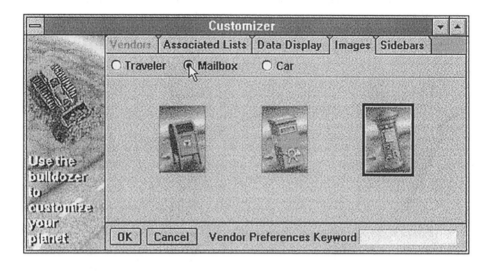

FIG. 3.2.   Example of dialogue box in the Customizer application, by means of which users can change the icons to become more culturally relevant.

CONCLUSIONS AND FUTURE RESEARCH ISSUES

The concept of user interfaces for all requires significant attention to globalization issues in the user interface development process. Progress in technology increases the number and kinds of functions, data, platforms, and users of computer-based communication media. The challenge of enabling more people and more kinds of people to use this content and these tools effectively will depend increasingly on global solutions. By recognizing the need for, and benefit to users of user interface designs intended for international and intercultural markets, developers will achieve greater success and increased profitability through the global distribution and acceptance of their products.

The recommendations provided in this chapter are an initial set of heuristics that will assist developers in achieving global solutions to their product development. (Additional resources and references on international standards and international issues are listed in the Appendix.) Design methodologies must support globalization throughout the development process.

In addition, it is likely that international and intercultural references will change rapidly, requiring frequent updating of designs. Future work on global user interface design may address the following issues:

1.   How might global user interfaces be designed to account for different kinds of intelligence? Gardner (1983) identified the following dimensions of intelli-

gence. These dimensions suggest users might have varying strengths of conceptual competence with regard to using user interfaces on an individual basis, but these might also vary internationally, or interculturally, due to influences of language, history, or other factors:

- Verbal/image comprehension.
- Word/image fluency.
- Numerical/graphical fluency.
- Spatial visualization.
- Associative memory.
- Perceptual speed.
- Reasoning.
- Interpersonal awareness.
- Self-awareness.

2. How might metaphors, mental models, and navigation be designed precisely for different cultures that differ by such dimensions as age, gender, national or regional group, or profession? The author posed this as a question to the user interface analysis/design community (Marcus, 1993). The topic is discussed broadly in DelGaldo and Nielsen (1996). Further, what means can be developed to enable these variations to be produced in a cost-effective manner using templates?

The taxonomic analyses of global issues for user interfaces, the theoretical basis for their component selection, the criteria for their evaluation, and their design methodology are still emerging in the user interface development field. Nevertheless, designers should be aware of the scope of the activity, know sources of insight, and incorporate professional techniques in their development process in order to improve the value and success of their international and intercultural products.

ACKNOWLEDGMENTS

Mr. Marcus thanks his staff at Aaron Marcus and Associates, Inc., as well as their client, SABRE Travel Information Network, and in particular, Mr. Michael Sites, Managing Director, Product Marketing, for their assistance in preparing this chapter. The author also acknowledges the books by DelGaldo and Nielsen (1996), Fernandes (1995), and Nielsen (1990) cited in the References section, which provided a basis for many points raised in this chapter. Finally, the author acknowledges Peter Siemlinger, Dipl.-Ing, Director, International Institute for Information Design, Vienna, Austria, Europe, and Prof. Andreas Schneider, Information Design Department, Tama Art University, Tokyo, Japan, for their advice about international time and space metrics.

REFERENCES

Alvarez, G. M., Kasday, L. R., & Todd, S. (1998). How we made the Web site international and accessible: A case study. In *Proceedings of the 4th Human Factors and the Web*

Conference [Online]. Available: http://www.research.att.com/conf/hfweb/proceedings/alvarez/index.html

American Institute of Graphic Arts. (1981). *Symbol signs*. Hastings House, NY: Visual Communication Books.

Chavan, A. L. (1994). *A design solution project on alternative interface for MS Windows*. Unpublished master's thesis, Guildhall University, London, England.

DelGaldo, E., & Nielsen, J. (Eds.). (1996). *International user interfaces*. New York: Wiley.

Fernandes, T. (1995). *Global interface design: A guide to designing international user interfaces*. Boston: AP Professional.

Gardner, H. (1983). *Frames of mind, the theory of multiple intelligences*. New York: Basic Books.

ISO 9241-8. (1989). *Computer display color*. Geneva, Switzerland: International Standards Organisation.

Ito, M., & Nakakoji, K. (1996). Impact of culture on user interface design. In E. M. del Galdo & J. Nielsen (Eds.), *International user interfaces* (pp. 105–126). New York: John Wiley & Sons.

Marcus, A. (1993). Human communication issues in advanced UIS. *Communications of the ACM, 36*(4), 101–109.

Marcus, A. (1995). Principles of effective visual communication for graphical user interface design. In R. Baecker, J. Grudin, W. Buxton, & S. Greenberg (Eds.), *Readings in human–computer interaction* (2nd ed., pp. 425–441). Palo Alto, CA: Morgan Kaufman.

Marcus, A. (1998). Metaphor design in user interfaces. *The Journal of Computer Documentation, ACM/SIGDOC, 22*(2), 43–57.

Nielsen, J. (Ed.). (1990). *Designing user interfaces for international use:* Vol. 13. *Advances in human factors/ergonomics*. Amsterdam: North-Holland, Elsevier Science.

Olgyay, N. (1995). *Safety symbols art*. New York: Van Nostrand Reinhold.

Pierce, T. (1996). *The international pictograms standard*. Cincinnati, OH: ST Publications. (For information: Design Pacifica, 725 NW Flanders Street, Portland, OR 97209, USA, E-mail: tpierce@designpacifica.com)

APPENDIX

This section lists organizations providing information about international standards and intercultural issues. Also listed are references that will help in further investigation of the issues discussed herein:

1. Organizations:
- American National Standards Institute (ANSI): This organization analyzes and publishes U.S. standards, including those for icons, color, terminology, user interfaces, and so on. Their contact data are the following:
   American National Standards Institute (ANSI)
   11 West 42nd Street, 13th Floor, New York, NY 10036, USA
   Tel: 212-642-2000
   E-mail: info@ansi.org
   URL: www.ansi.org
- China National Standards Organization: This organization analyzes products to be imported into China. Their contact data are the following:
   China Commission for Conformity of Electrical Equipment (CCEE)
   Secretariat

Postal address: 2 Shoudu Tiyuguan, NanLu, 100044, P.R. China
Office address: 1106, 11th floor, 2 Shoudu Tiyuguan, NanLu, Beijing,
P.R.China
Tel: +86-1-8320088, ext. 2659, Fax: +86-1-832-0825

- East–West Center: This organization, formerly funded by the U.S. Congress, is a center for technical and cultural interchange among Pacific Rim countries. Their research and publications cover culture and communication. Their contact data are the following:
  East–West Center
  1601 East-West Road, Honolulu, Hawaii 96848-1601, USA
  Tel: 808-944-7111, Fax: 808-944-7376
  E-mail: ewcinfo@ewc.hawaii.edu
  URL: http://www.ewc.hawaii.edu

- Information Technology Standards Commission of Japan (ITSCJ)
  Information Processing Society of Japan
  Kikai Shinko Building, No. 3-5-8 Shiba-Koen, Minato-ku, Tokyo 105,
  Japan
  Tel: +81-3-3431-2808, Fax: +81-3-3431-6493

- International Standards Organization (ISO): This organization analyzes and publishes world standards for all branches of industry and trade, including standards for icons, color, terminology, user interfaces, and so on. Their contact data are the following:
  International Standards Organization (ISO)
  Geneva, Switzerland
  Tel: +41-22-749-0111, Fax: +41-22-733-3430
  E-mail: central@iso.ch
  URL: http://www.iso.ch/
  Of special interest is ISO 8601's international time and date standards. Information about ISO's 8601 standards, and the particular one about time and date standards, may be found at the following URLs:
  http://www.aegis1.demon.co.uk/y2k/y2kiso.htm
  Http://www.roguewave.com/products/resources/exchange/iso 8601.html

- Japan National Standards Organization: This organization analyzes and publishes Japanese standards, including those for icons, color, terminology, user interfaces, and so on. Their contact data are the following:
  Japanese Industrial Standards Committee (JISC)
  Agency of Industrial Science and Technology
  Ministry of International Trade and Industry
  1-3-1, Kasumigaseki, Chiyoda-ku, Tokyo 100, Japan
  Tel: +81-3-3501-9295/6, Fax: +81-3-3580-1418

- World Wide Web Consortium: This organization provides information relevant to globalization issues, including accessibility. Two URLs of interest are:

http://www.w3.org/International, for information about internalization.
http://www.w3.org/WAI, for information about accessibility.

2. References:

- Eco, U. (1976). *A theory of semiotics*. Bloomington: Indiana University Press.
- Iwayama, M., Tokunaga, T., & Tanaka, H. (1990). A method of calculating the measure of salience in understanding metaphors. In T. Dietterich & W. Swartout (Eds.), *Proceedings of 8th National Conference on Artificial Intelligence* (pp. 298–303). Cambridge, MA: MIT Press.
- Lanham, R. A. (1991). *A handlist of rhetorical terms* (2nd ed.). Berkeley: University of California Press.
- Marcus, A. (1992). *Graphic design for electronic documents and user interfaces*. Reading, MA: Addison-Wesley.
- McLuhan, M. (1964). *Understanding media: The extensions of man*. New York: McGraw-Hill.
- Neale, D. C., & John, M. C. (1997). The role of metaphors in user interface design. In M. Helander, T. K. Landauer, & P. Prabhu (Eds.), *Handbook of human–computer interaction* (2nd ed., pp. 441–462). Amsterdam: North-Holland, Elsevier Science.

4

Intelligent User Interfaces for All

Mark T. Maybury

This chapter defines Intelligent User Interfaces, and describes a range of related issues, including multimedia input analysis, multimedia presentation generation, and the adoption of user, discourse and task models to personalize and enhance interaction. We contrast conventional commercial user interface architectures with those that have evolved from research on Intelligent User Interfaces over the past twenty years. We emphasize how Intelligent User Interfaces that reason about and exploit knowledge of the user, the task, and the communication process can mitigate application and domain complexity through such means as tailored presentation design and cooperative responses. Through reference to implemented systems, we exemplify ways in which intelligent interfaces tailor interaction for different user groups (e.g., novices versus experts; youth versus elderly) with differing capabilities and interests, from different access points (e.g., terminals, public information kiosks, portable digital assistants), and for a range of different applications and services (e.g., learning, traveling, shopping). Based on the examples, we conclude with a discussion of how Intelligent User Interface technologies can provide support in achieving interaction for all.

User interfaces bridge human and machine. *Intelligent user interfaces* (IUIs) specifically aim to enhance the flexibility, usability, and power of human–computer interaction for all users. In doing so, they exploit knowledge of users, tasks, tools, and content, as well as devices for supporting interaction within differing contexts of use. In a sense, knowledge that is explicitly represented in the system has the power to make the interface available to all by providing capabilities of understanding and adapting to individual users, the devices they use to communicate, and the environments and ways in which they interact.

IUIs focus on enhancing human–system and human–human interaction. This includes interfaces to people (e.g., computer-supported communication), interfaces to applications, and interfaces to information. According to Maybury and Wahlster (1998), IUIs are often distinguished from conventional interfaces by their *multimedia, multimodal,* and *multicodal* capabilities. In the context of our discussion, these terms are defined as follows:

- *Multimedia*—refers to the physical means via which information is input, output, and/or stored (e.g., interactive devices such as keyboard, mouse, displays; storage devices such as disk or CD-ROM).
- *Multimodal*—refers to the human perceptual processes associated with interaction such as vision, audition, taction.

- *Multicodal*—refers to the representations used to encode atomic elements, syntax, semantics, pragmatics, and related data structures (e.g., lexicons, grammars) associated with media and modalities.

Individual users may have different abilities or preferences in dealing with various media (e.g., keyboard vs. mouse input), modalities (e.g., visual vs. auditory), or codes (e.g., English vs. French).

There exists a large corpus of literature and associated techniques that address the development of learnable, usable, transparent interfaces in general (e.g., Baecker, Grudin, Buxton & Greenberg, 1995). This chapter focuses on intelligent and multimedia/multimodal/multicodal user interfaces (see also Maybury, 1993; Sullivan & Tyler, 1991). From the user's perspective, these interfaces are context sensitive, assist in tasks, adapt themselves appropriately in terms of time, place, and manner, and may:

- *Analyze* synchronous and asynchronous multimedia/multimodal input (e.g., spoken and written text, gesture, drawings) that might be imprecise, ambiguous, and/or partial.
- *Generate* (design, realize) coordinated, cohesive, and coherent multimedia/multimodal presentations.
- *Manage* the interaction (e.g., training, error recovery, task completion, tailoring interaction) by representing, reasoning, and exploiting *models* of the domain, task, user, media/mode, discourse, and environment.

Relatedly, when interacting with information spaces, there is a need to perform media content analysis and presentation (Maybury, 1997), which includes retrieval of text, audio, imagery, and/or combinations thereof. Finally, from the developer's perspective there is also the need to decrease the time, expense, and level of expertise necessary to construct successful systems, which has resulted in model-based user interface automated construction tools (Puerta & Eisenstein, 1999).

Individual users are different in many ways, not the least of which are their physical, perceptual, cognitive, and social characteristics. For example, some users have physical limitations (e.g., they may have poor dexterity and, therefore, be unable to type or use gestural interfaces well), perceptual limitations (e.g., poor vision or hearing, color blindness, difficulty in distinguishing shapes), or cognitive limitations (e.g., attention or memory limitations). In other cases, users may simply have different cognitive styles (e.g., learn and reason visually or logically—Kosslyn, 1983) or personality (e.g., introvert/extrovert, typical group roles), which may suggest the need for different interaction styles (e.g., the use of interface agents, collaborative filtering).

Intelligent interfaces that capture, represent, reason about, and exploit stereotypical and individual models of the user can use these characteristics as parameters for adaptation (e.g., voice vs. keyboard input, customization of presentations or interaction styles). Moreover, because distributed information services are already being delivered via a variety of end-user computing devices (e.g., desktop computers, public kiosks, or mobile handheld devices), consistency and cost considerations may necessitate the maintenance of a single source of content, from which multiple, tailored information interactions can be produced. Automated or semiautomated design of a range of presentations (Maybury, 1993) has already been demonstrated for menus, diagrams, graphics, (multilingual) text, and even entire interfaces. To clarify the possibilities of achieving these benefits for all users, this chapter contrasts conventional and intelligent interfaces. Second, the chapter describes an implemented system that represents an example of the possibilities offered by dynamic, tailored interaction in IUIs. Finally, it briefly discusses issues related to user-centered evaluation of intelligent interfaces.

CONVENTIONAL INTERFACES

Governments, industry and academia have increased their focus on the importance of the human–machine interface in the global information economy. More effective, efficient and natural human–computer or computer-mediated human–human interaction will require automated understanding and generation of multimedia, and will rely on precise information about the user, discourse, task, and context (Maybury, 1993).

Today's users are faced with a dizzying array of information sources and tools. Just as the amount of information available to users has grown exponentially, so too has the complexity of tools. For example, whereas early word processors had only a few dozen commands, word processors now contain several hundreds of commands. This inherent complexity of modern software creates usage problems for many (especially novice) users and the situation is further exacerbated for the physically, cognitively, or socially challenged, who may have to go through several levels of indirection (by using assistive technologies) in order to interact with a system. Additionally, a large number of users use only a small fraction of application functionality (Linton, Charron, & Joy, 1998). As a consequence of the aforementioned, and of a move toward smaller and mobile platforms, simplified, limited-feature applications have enjoyed resurgence in the marketplace.

In spite of the increasing need for new forms of communication between human and machine, the sophistication of user interfaces has remained relatively unchanged since the introduction of direct manipulation interfaces. As Fig. 4.1 illustrates, contemporary interfaces consist of three elements: presentation (e.g., windows, icons, menus), dialogue (e.g., selection or invocation via menus, dialogue boxes, keyboards, mice), and the application interface to the underlying application. The application could provide access to information, tools, or people (e.g., computer-mediated communication).

INTELLIGENT INTERFACES

Unfortunately, this conventional model does not support integrated multimedia input, generation of coordinated multimedia output, or tailoring the interaction to special needs related to the user, the task, or the specific application domain. In contrast, IUIs are human–machine interfaces that aim to improve the efficiency, effectiveness, and naturalness of human–machine interaction by representing, reasoning, and acting on models of the user, domain, task, discourse, and media (e.g., graphics, natural language, gesture) (Maybury & Wahlster, 1998). Figure 4.2 outlines their principal components, moving from analysis of input to generation of output. We describe these modules in turn.

Multimodal Input Processing and Analysis

Whereas traditional interfaces support sequential and unambiguous input from devices such as keyboard and conventional pointing devices (e.g., mouse, trackpad), intelligent multimodal interfaces relax these constraints, and typically incorporate a broader range of input devices (e.g., spoken language, eye and head tracking, three-dimensional gestures). For example, they support asynchronous, ambiguous, and inexact input by applying more sophisticated analysis techniques. These

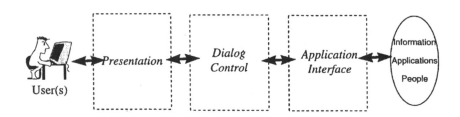

FIG. 4.1.   Conventional three-level user interface architecture.

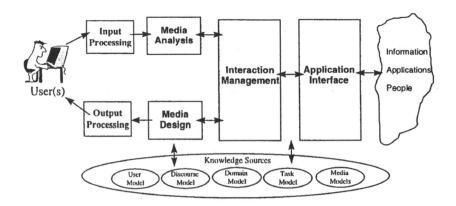

FIG. 4.2.   Intelligent user interface architecture.

systems allow the resolution of multimedia references, enabling the user, for example, to say "Put that there" while gesturing to a map, by correlating eye and hand gestures with the deictic expressions "that" and "there." Initial prototypes demonstrated this kind of multimodal reference resolution allowing strictly sequential input (Bolt, 1980) whereas more recent work relaxes this constraint to include synchronous input and ambiguous gesture (Koons, Sparrell, & Thorisson, 1993). In addition to providing flexibility, multimodal input can operate in a synergistic manner, by using different modalities as sources through which different constraints are derived for the interpretation of ambiguous input. For example, the TACTILUS subcomponent of XTRA (Wahlster, 1991), among other systems, resolves ambiguous gestures by processing associated linguistic input to select among multiple possible referents. TACTILUS also enables the user to choose between a pencil and a less exact "marker" (i.e., a point vs. a circle). Integrated input from multiple devices promises to simultaneously enhance communication efficiency, effectiveness (e.g., speed and accuracy), and naturalness. Intelligent interfaces can also detect and correct errors utilizing models of the media, user, discourse, and task.

Multimodal Output Generation

Whereas traditional interfaces draw on preprogrammed or "canned" presentations (e.g., windows, menus, dialogue boxes), automated interface and presentation generation concerns the ability of a system to select content, apportion that content to various media (e.g., typed or spoken language, graphics, gesture), and realize those media in an integrated and coordinated fashion. Figure 4.3 (Maybury, 1993) illustrates the key tasks associated with multimedia generation: managing the communication (i.e., reasoning about plans and intentions), selecting content to achieve given communicative goals, designing the presentation, allocating and coordinating information across media, realizing media, and laying them out. Figure 4.3 is intended to depict a cascaded, coconstraining set of processes that potentially share knowledge sources (e.g., multimedia lexica, grammars, semantics).

The principal advantage of automated output generation is that it can be tailored to the content that needs to be presented, as well as to the user, and to the context in which it appears. For example, if a system generates a map displaying locations of volcanic activity worldwide, and a user asks what the current most active location is, a presentation designed from an underlying formal representation can interpret the user's query and simply highlight the identified location on the displayed map. If the system knows the user is color-blind, it can decide, at run time, not to use color to encode the information to be presented. More fundamentally, if a system captures knowledge of human perceptual operations, it can exploit such knowledge during operation. For example, Casner (1991) presented a system that focuses on the automatic design of graphics that optimize human performance in information-processing tasks, by (a) designing them to enable users to substitute cognitively demanding "logical interfaces," with simpler and easier to learn "per-

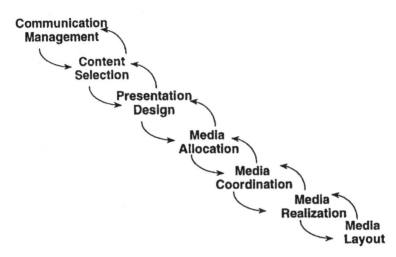

FIG. 4.3.   Generation tasks.

ceptual interfaces," and (b) streamlining information searching by providing users with custom, task-specific designs. In the former case, perceptual inferences, such as distance and size determinations, spatial collocation, and color comparison, yield more rapid results than cognitive operations (e.g., arithmetic comparison). In the latter case, group and ordering information, as well as color encoding, shading, and layout can support "preattentive and sometimes parallel visual search."

Research work has also addressed the automated layout of multimodal presentations (e.g., Graf, 1992), including chart, map, and arc/node diagram layout (e.g., Christensen, Marks, & Shieber, 1995). Multimedia generation applies not only to generic user interface applications, but also to the automated generation of instruction/documentation and explanation (Feiner & McKeown, 1991; Wahlster, André, Finkler, Profitlich, & Rist, 1993), support for interactive tutoring (Goodman, 1993), or the support for user creation of presentation graphics (Roth et al., 1996).

Key subareas of multimodal generation include text generation (Maybury, 1994), graphics generation (Mackinlay, 1986), and gesture generation. A fundamental challenge in this field is the generation of coordinated multimedia. Feiner and McKeown (1991) generated coordinated, three-dimensional explanations of devices, by extending rhetorical schemata originally developed for text generation. Other work involves the generation of multimedia presentations from communicative plans, to provide, for example, instructions or device descriptions (Wahlster et al., 1993), or geographical directions (Maybury, 1991). Researchers are now investigating the role of animated interface agents in multimedia generation (e.g., Nagao & Takeuchi, 1994).

Model-Based User Interfaces

Given the complexity of modern user interfaces, as well as the level of expertise and time required to build them, researchers have focused on the creation of environments that support and facilitate user interface design and development. User interface management systems (UIMS) and software development toolkits containing interactive components (such as windows, menus, and dialogue boxes) were originally designed to address this problem. Though UIMS foster design consistency and enhance programmer productivity via code reuse, they frequently mix interface code with application code. In contrast, ITS Tool of Wiecha, Bennett, Boies, Gould, and Green (1989), for example, separates applications into four layers: application actions, dialogue control, style rules (specifications of presentation and behavior), and style program layer (primitive toolkit objects composed by style rules). These roughly correspond to the application interface, interaction management, and media design and processing layers of Fig. 4.2. Models of each layer are developed, which support the declarative specification of applications. The authors illustrated the specification of a public kiosk application using such declarative models.

Model-based interface development environments, in general, seek to facilitate user interface design through automation, or provision of design assistance. They go beyond user interface toolkits by separating dialogue control from application code and decoupling presentation and style decisions from toolkit libraries. They are also distinct from UIMS in that they support finer grained design and usually provide more powerful tools to interface developers. Furthermore, in contrast to typical interface software repositories, these systems promise automated design critique, refinement, and implementation. Much research work in the area has focused on the definition of the separate functional components of an interface, in order to support declarative expression and modularization of interface functionality (Maybury & Wahlster, 1998). In this respect, model-based interface development practices and techniques can have significant impact in the development of automated input analysis and output generation modules.

Interaction Management

Context has always been recognized as critical to the effectiveness of interaction. Context comes in many forms, typically explicitly represented in models of the user, discourse, task, and situation, as shown in Fig. 4.2. Multiple investigations of discourse coherence, focusing, and reference have yielded an integrated theory of intention, attention, and the structure of discourse (Grosz & Sidner, 1986). Computational techniques derived from this theory enable systems to track, and react to, interactive dialogue. Because Judy Kay (chap. 14, this volume) addresses user modeling in detail, this chapter focuses on action-based models of communication, which promise new possibilities for interaction management. These efforts seek more principled models of communication, building on the philosophical view of language as action, and its subsequent development into speech acts theory (Searle, 1969).

Principled models of participant interaction are essential to enable intelligent communication behavior, such as negotiation, tailored explanation, and error detection and recovery among dialogue participants. The study of communication and action is necessarily an interdisciplinary endeavor, drawing on many scientific fields including sociology, philosophy, computational linguistics, human language processing, and human–computer interaction. For example, researchers have drawn on philosophical and linguistic notions of communication as an intentional and action-based endeavor (Austin, 1962; Wittgenstein, 1953), to formalize these as communication principals (Grice, 1975) and speech acts (Searle, 1969). Speech acts, which constitute a starting point for much work in communication and action, consist of three distinct but interrelated actions:

- *Locution*—the physical act of uttering.
- *Illocution*—conveying of the speaker's intent to the hearer (e.g., inform, request, warn, promise).
- *Perlocution*—the actions resulting from the successful performance of the illocution.

For example by uttering the sentence "it is raining outside" (locution), a speaker may inform a friend of the actual weather conditions (illocution), thus causing the hearer of the utterance to put on a raincoat (perlocution). Each speech act has associated propositional content (in the previous example, the fact that it is raining), and necessary and sufficient conditions for successful performance (e.g., the hearer has to be able to hear the locution in order to "uptake" it).

This formal basis of speech acts has made them very suitable for computational representation (e.g., as plan operators—Bruce, 1975) and use in real-world applications. A significant amount of computational research has subsequently investigated interpretation and reconstruction of the intentions of the speaker, as well as intent-based design and realization of output. The next section presents concrete examples that illustrate how communicative action-based interfaces can facilitate *universal access*.

EXAMPLE APPLICATIONS: DYNAMIC, TAILORED INTERACTION

The use of communicative actions has been explored in a number of application domains. These applications indicate their value in the creation of intelligent interfaces that can be readily adapted, in some cases online, to user populations with diverse abilities, skills, requirements, and preferences. We illustrate here one system that performs automatic generation of multimodal direction indications, namely the TEXPLAN system (Maybury, 1991). As Figs. 4.4 and 4.6 illustrate, TEXPLAN uses a high-level communication plan to select content, as well as to design integrated multimedia direction indications, including coordinated text, gesture, and graphical elements. Figure 4.4 illustrates the formal specification of a multimodal communicative action whose illocutionary purpose is to visually and linguistically identify an object. The specification in Fig. 4.4 consists of a plan operator from a library of plan operators, which specifies that in order for the speaker (in the specific case, the system) to identify a given entity to the hearer (HEADER), providing the entity is a cartographic one (CONSTRAINTS), the system must ensure the entity is visible to the user on the computer screen (PRECONDITIONS). Then the system must point to the object (indicate deictically) and assert its location (DECOMPOSITION), with the result that the user will know where it is (EFFECTS). The communicative action represented by such a plan is tailored, at execution time, to the content and the environment of the interaction.

In the TEXPLAN system, plan operators like the one in Fig. 4.4 are exploited by a traditional planner, to dynamically respond to communicative needs in the interface, at run time. For example, in response to a user query "How do I go from Wiesbaden to Frankfurt?", TEXPLAN uses a library of communicative action schemata (such as the one illustrated in Fig. 4.4) to plan the hierarchical communication decomposition. Such a decomposition is shown in Fig. 4.5, where a plan to explain a route between the objects #<Wiesbaden> and #<Frankfurt-am-Main> is specified, by highlighting and displaying the region around the departure city, highlighting and describing intervening segments, and finally identifying the destination city. This hierarchical communication plan (a tree structure) is linearized, and the resulting sequence of linguistic and visual primitive actions is generated (see Fig. 4.6).

A key benefit of providing an interface with communication knowledge, such as that just illustrated, to automatically plan multimedia direction indications, is the ability to utilize a small number of plans to communicate a lot of material, in many potential contexts. For example, the object-oriented application from which the aforementioned directions were generated includes thousands of German towns and tens of thousands of road segments. It would be very expensive to manually create directions for all cities and potential road segments connecting these. Given the importance of user-tailored output, it is also the case that, by introducing lexica and grammar of other languages, we can reuse the same communicative

NAME	Identify-location-linguistically-&-visually
HEADER	Identify(*S, H, entity*)
CONSTRAINTS	Cartographic-Entity?(*entity*)
PRECONDITIONS	Visible(*entity*)
EFFECTS	KNOW(*H*, Location(*entity*))
DECOMPOSITION	Indicate-Deictically(*S, H, entity*)
	Assert(*S, H*, Location(*entity*))

FIG. 4.4. Plan operator for graphical/textual display.

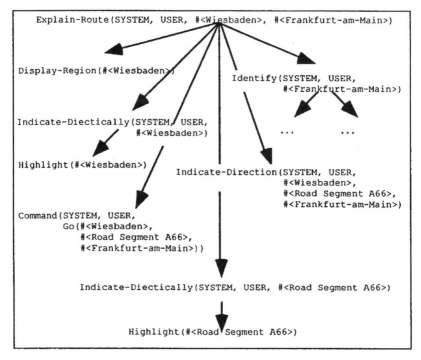

FIG. 4.5.   Hierarchical multimedia communication plan to give directions.

> *(Display map region around Wiesbaden) (highlight Wiesbaden)*
> From Wiesbaden take Autobahn A66 Northeast for thirty-one
> kilometers to Frankfurt-am-Main. *(highlight Autobahn A66) (indicate
> direction with blinking arrow) (highlight Frankfurt-am-Main)*
> Frankfurt-am-Main is located at 50.11° latitude and 8.66° longitude.

FIG. 4.6.   Realization of multimedia plan. Italicized expressions denote visual realizations.

content plans to generate multilingual presentations, thus dramatically decreasing
porting time via automated "localization" (the aforementioned system can generate directions in English and Italian).

Moreover, the presentation planner can deal with dynamic changes in available
media. For example, it can automatically substitute graphical demonstratives
(e.g., centering the display on the start and end points and highlighting the route)
with linguistic references, when this is appropriate in the current context (e.g., output must be directed to a text-only display). If the delivery devices (e.g., mobile

personal digital assistant), or user characteristics (e.g., a blind user), or the context of interaction (e.g., a hands- and eyes-busy driver) indicate that graphical communication elements are undesirable, the presentation planner can automatically construct a language-only message, by setting a high-level constraint specifying that graphical communication actions are undesired. This language could then be realized as speech. Similarly, if the user is elderly and known to have poor short-term memory and poor eyesight, an automated content planner like TEXPLAN can select highly memorable landmarks, lay out routes in smaller increments, and lay out text with larger fonts. Thus, explicit representation of communication knowledge provides the ability to extend the communicative capacity of interface elements to the broadest user population.

There are other important benefits arising from explicit reasoning about communicative actions, which are applicable not only to output generation, but also to the management of interactions. The integrated communication plan in Fig. 4.5, which underlies the presentation of direction indications, enables cross-modal referring expressions, error recovery and follow-up questions, and incremental replanning. For example, terms, clauses, and entire routes are associated with particular subtrees from Fig. 4.5 (e.g., the description of Frankfurt-am-Main). Thus, if a particular presentation confuses a user, each of these elements can be realized in alternative ways, or the motivation for using a specific communicative action can be expressed. For example, the system could tell the user what the higher level action of the current action in the plan hierarchy is, what desirable effects it will achieve, or the constraints that led the choice of this action over others. Moreover, it is possible to model additional types of interaction actions within a plan-based system, in order to manage discourse interactions (e.g., turn taking in dialogue, clarification subdialogues).

In related work, Paris (1987) analyzed expert and novice process explanations (e.g., adult vs. children encyclopedia entries), and designed explanation strategies that were then used to automatically generate explanations of the behavior of complex devices. By distinguishing structural from procedural explanations, and adopting domain-independent explanation schemas, Paris' TAILOR system is able to generate explanations that are customized to the individual user's knowledge (e.g., young vs. elderly, novice vs. expert). Moore (1989) developed techniques to respond to vague follow-up questions (e.g., "huh?" "why?"), arguing that systems should not rely on "one-shot" explanations and, instead, should exploit opportunities to react to the user.

Other investigations into lifelike characters (e.g., André, Rist, & Müller, 1999; Nagao & Takeuchi, 1994) aim to use communication actions to drive anthropomorphic interfaces that integrate speech, gesture, and facial displays. For example, Nagao and Takeuchi used facial expressions as a rich communication channel for conveying emotional and conversational signals in interface characters. The authors mapped several classes of system state (e.g., "thinking," "listening," "incredulity") onto 26 facial displays (e.g., "closing eyes," "eyebrow raising," "eyebrow lowering"). Their system integrates a speech dialogue system with facial animation and a plan recognition component. The recognition of system states (e.g., understanding, syntactic error, ambiguity) results in the generation of utterances coupled with matching facial displays. For example, if the speaker's request exceeds the systems knowledge, it displays a facial shrug and replies "I cannot answer such a question." A 32-subject ablation experiment comparing conversations with computer systems with and without facial displays revealed that users found conversations with facial displays smoother. Along the same lines, experiments with the Persona system (André et al., 1999) showed that users conceived a lifelike agent as helpful and entertaining, and tasks as less difficult if the agent was present.

Intelligent training systems also offer an important application area in which to explore intelligent communication. For example Aist and Mostow (1997) reported experiments in the systematic variation of a Reading Tutor's behavior, by adapting communication timing for individual children to help them learn to read. For example, the aforementioned system can dynamically adapt turn-taking rules that interrupt the students if they take a long time, self-interrupt if the student starts speaking, or provide a backchannel if the student pauses (e.g., "mm-hmm"). Aist and Mostow found that students are nearly twice as likely to continue reading after the tutor backchannels (54.5%) as they would be if the tutor did not backchannel even if he or she should (31.7%). This example illustrates the importance of interfaces that adapt to individual characteristics and preferences.

Finally, tedious tasks such as Web navigation or air travel planning can benefit from the communicative abilities of intelligent interfaces. For example, Rich and Sidner (1997) described an application independent toolkit called COLLAGEN (COLLaborative AGENt), based on the SharedPlan theory of discourse (Grosz & Sidner, 1990), to build a collaborative interface agent for air travel planning. The system automatically generates hierarchically structured interaction histories, based on the user's goals and intentions. This explicit interaction history is useful for the user's orientation and supports powerful manipulation of the application and discourse states (e.g., navigating, undoing, or replaying actions).

In summary, by creating interfaces whose behavior is driven by underlying principles, we can achieve interaction that is highly adaptable to diverse user populations and situations. Explicit representation of communicative principles, constraints, actions, alternatives, and effects, as well as planning mechanisms that exploit these to choose among alternatives, enables the interface to transform or "morph," thus presenting a suitable "face" to its user in order to facilitate interaction. The ability to tailor output to different users, interactive devices, and contexts of use, represents a significant step in achieving universal access, through real-time, user- and usage context-tailored interfaces.

USER-CENTERED EVALUATION

As systems increase in their complexity and in the knowledge required to operate them, it becomes important that we ensure user interfaces mitigate that complexity and empower users throughout the interactive experience. Human-centered analysis and design is even more important and complex when creating sophisticated intelligent interfaces, as those interfaces by their nature have more capability and flexibility. Because of the difficulty in distinguishing application from interface in a controlled evaluation, system developers frequently perform task-based, end-to-end systems evaluation. That is, evaluation is performed within a particular user task (e.g., transfer funds between a checking and savings account) for some controlled condition such as the type of user device (e.g., access from a PDA, public kiosk, desktop terminal), or the type of media employed (e.g., typed vs. speech vs. mixed mode input). If systems are architected in an open and component-based framework (e.g., Martin, Cheyer, & Moran, 1999), the preceding can be "plug and played" to evaluate particular hypotheses. Tools such as loggers, data analyst workbenches, and conventional evaluation instruments, such pre- and postinterviews, can then be used to discover task success and user satisfaction.

Current research at MITRE[1] is focusing on "instrumenting" the interface with mechanisms/facilities to collect and (automatically or semiautomatically) annotate corpora of human–machine interaction.[2] The aim is to establish a scientific methodology for the analysis and construction of models of interaction, which will enable the design of interfaces with predictable performance. "Instrumentation" can range from tracking application function invocations, to tracking interface widget-level events and generalizing these into meaningful application events, as well as controlling application scripts. In a related study (Linton et al., 1998), exposed application programming interfaces are used to track, at the client application, function usage by end-users, thus uncovering over time usage patterns of both individual users and groups of users.

CONCLUSIONS

Significant advances have been made in the theory and computational implementations of IUIs. Potential benefits include:

- More efficient interaction—enabling more rapid task completion with less work.
- More *effective* interaction—doing the right thing at the right time, tailoring the content and form of the interaction to the context of the user, task, dialogue.
- More *natural* interaction—supporting spoken, written, and gestural interaction, ideally as if interacting with a human interlocutor.

Perhaps the most significant benefit of IUIs over traditional interfaces is the ability of the former to tailor the interface to the needs of target user groups or individual users, both at design-time and at run-time. However, many challenges remain. For

---

[1]For more information, please refer to: http://www.mitre.org

[2]For more information, please refer to: http://www.mitre.org/research/logger

example, researchers are only now beginning to construct systems and instruments that allow for larger-scale usage data collection, and empirical evaluations that promise to uncover principals of communicative (linguistic and nonlinguistic) actions, as well as their relation to attention-, user-, and task-models. It is important that community-wide collaboration continues, and that corpus annotation and analysis, theory creation, and application development cross-pollinate, so that progress can proceed as rapidly as possible. This will also demand more explicit interdisciplinary collaboration as the collective expertise of, for example, epistemologists, computational linguists, cognitive psychologists, and human factors experts will be needed to make key advances in IUIs. Progress in our understanding of human–computer interaction, and human–human computer-mediated communication promises nothing less than fundamental advances in the way we interact with computers, or with one another via a computer.

REFERENCES

Aist, G., & Mostow, J. (1997). A time to be silent and a time to speak: Time-sensitive communicative actions in a reading tutor that listens. In *Working Notes of the AAAI-97 Fall Symposium on Communicative Actions in Humans and Machines* (pp. 1–5). Cambridge, MA: MIT Press.

André, E., Rist, T., & Müller, J. (1999). Guiding the user through dynamically generated hypermedia presentations with a life-like character. In M. T. Maybury (Ed.), *Proceedings of the 3rd International Conference on Intelligent User Interfaces* (pp. 21–28). New York: ACM Press.

Austin, J. L. (1962). *How to do things with words*. Cambridge, MA: Harvard University Press.

Baecker, R., Grudin, J., Buxton, W., & Greenberg, S. (1995). *Readings in human–computer interaction: Toward the year 2000* (2nd ed.). San Francisco: Morgan Kaufmann.

Bolt, R. A. (1980). "Put-that-there": Voice and gesture at the graphics interface. *ACM Computer Graphics, 14*(3), 262–270 (Quarterly Report of SIGGRAPH-ACM SIGGRAPH '80 Conference Proceedings, J. J. Thomas, Ed., Seattle).

Bruce, B. C. (1975). Generation as a social action. In B. Nash-Webber & R. Shank (Eds.), *Proceedings of Theoretical Issues on Natural Language Processing* (TINLAP-1); (pp. 64–67). Cambridge, MA: MIT.

Casner, S. M. (1991). A task-analytic approach to the automated design of information graphic presentations. *ACM Transactions on Graphics, 10*(2), 111–151.

Christensen, J., Marks, J., & Shieber, S. (1995). An empirical study of algorithms for point feature label placement. *ACM Transactions on Graphics, 14*(3), 203–232.

Feiner, S. K., & McKeown, K. R. (1991). Automating the generation of coordinated multimedia explanations. *IEEE Computer, 24*(10), 33–41.

Goodman, B. (1993). Multimedia explanations for intelligent training systems. In M. T. Maybury (Ed.), *Intelligent multimedia interfaces* (pp. 143–164). Menlo Park, CA: AAAI/MIT Press.

Graf, W. (1992). Constraint-based graphical layout of multimodal presentations. In T. Catarci, M. F. Costabile, & S. Levialdi (Eds.), *Proceedings of the Advanced Visual Interfaces Conference* (Vol. 36, pp. 365–385). Singapore: World Scientific Series in Computer Science.

Grice, H. P. (1975). Logic and conversation. In P. Cole & J. L. Morgan (Eds.), *Syntax and semantics: Vol. 3. Speech acts* (pp. 41–58). New York: Seminar Press.

Grosz, B. J., & Sidner, C. L. (1986). Attention, intentions, and the structure of discourse. *Computational Linguistics, 12*(3), 175–204.

Grosz, B. J., & Sidner, C. L. (1990). Plans for discourse. In P. R. Cohen, J. L. Morgan, & M. E. Pollack (Eds.), *Intentions and communication* (pp. 417–444). Cambridge, MA: MIT Press.

Koons, D. B., Sparrell, C. J., & Thorisson, K. R. (1993). Integrating simultaneous output from speech, gaze, and hand gestures. In M. T. Maybury (Ed.), *Intelligent multimedia interfaces* (pp. 257–276). Menlo Park, CA: AAAI/MIT Press.

Kosslyn, M. S. (1983). *Ghosts in the mind's machine: Creating and using images in the brain.* New York: Norton.

Linton, F., Charron, A., & Joy, D. (1998). OWL: A recommender system for organization-wide learning. In *Proceedings of the AAAI Workshop on Recommender Systems* (pp. 64–68). Menlo Park, CA: AAAI Press.

Mackinlay, J. D. (1986). Automating the design of graphical presentations of relational information. *ACM Transactions on Graphics, 5*(2), 110–141.

Martin, D., Cheyer, A., & Moran, D. (1999). The open agent architecture: A framework for building distributed software systems. *Applied Artificial Intelligence: An International Journal, 13*(1–2), 91–128.

Maybury, M. T. (1991). Planning multimedia explanations using communicative acts. In *Proceedings of the 9th National Conference on Artificial Intelligence* (pp. 61–66). Menlo Park, CA: AAAI/MIT Press.

Maybury, M. T. (Ed.). (1993). *Intelligent multimedia interfaces* [Online]. Menlo Park, CA: AAAI/MIT Press. Available: http://www.aaai.org/Press/Books/Maybury1/maybury1.html

Maybury, M. T. (1994). Automated explanation and natural language generation. In C. Sabourin (Ed.), *Computational text generation* (pp. 1–88). Montreal: Infolingua.

Maybury, M. T. (Ed.). (1997). *Intelligent multimedia information retrieval* [Online]. Menlo Park, CA: AAAI/MIT Press. Available: http://www.aaai.org/Press/Books/Maybury2/maybury2.html

Maybury, M. T., & Wahlster, W. (Eds.). (1998). *Readings in intelligent user interfaces.* San Francisco, CA: Morgan Kaufmann.

Moore, J. D. (1989). *A reactive approach to explanation in expert and advice-giving systems.* Unpublished doctoral dissertation, University of California at Los Angeles.

Nagao, K., & Takeuchi, A. (1994). Speech dialogue with facial displays: Multimodal human–computer conversation. In *Proceedings of the 32nd Annual Meeting of the Association for Computational Linguistics* (pp. 102–109). San Francisco: Morgan Kaufmann.

Paris, C. L. (1987). *The use of explicit user models in text generation: Tailoring to a user's level of expertise.* Unpublished doctoral dissertation, Columbia University, New York.

Puerta, A., & Eisenstein, J. (1999). Towards a general computational framework for model-based interface development systems. In *Proceedings of the International Conference on Intelligent User Interfaces* (pp. 171–180). New York: ACM Press.

Rich, C., & Sidner, C. (1997). Collaboration with an interface agent via direct manipulation and communication acts. In *Working Notes of AAAI '97 Fall Symposium on Communicative Actions in Humans and Machines* (pp. 130–131). Cambridge, MA: MIT Press.

Roth, S., Lucas, P., Senn, J., Gomberg, C., Burks, M., Stroffolino, P., Kolojejchick, J., & Dunmire, C. (1996). Visage: A user interface environment for exploring information. In *Proceedings of IEEE InfoVis '97* (pp. 3–12). New York: IEEE Press.

Searle, J. (1969). *Speech acts*. Cambridge, England: Cambridge University Press.

Sullivan, J. W., & Tyler, S. W. (Eds.). (1991). *Intelligent user interfaces* (Frontier Series). New York: ACM Press.

Wahlster, W. (1991). User and discourse models for multimodal communication. In J. Sullivan & S. Tyler (Eds.), *Intelligent user interfaces* (pp. 45–67). New York: ACM Press.

Wahlster, W., André, E., Finkler, W., Profitlich, H. J., & Rist, T. (1993). Plan-based integration of natural language and graphics generation. *Artificial Intelligence, 63*(1–2), 387–427.

Wiecha, C., Bennett, W., Boies, S., Gould, J., & Green, S. (1989). ITS: A tool for rapidly developing interactive applications. *ACM Transactions on Information Systems, 8*(3), 204–236.

Wittgenstein, L. (1953). *Philosophical investigations*. Oxford, England: Basil Blackwell.

5 Interaction Technologies: Beyond the Desktop

Len Bass

The manner in which users interact with computers is critically dependent upon the choice of input and output devices. This chapter outlines the design space formed by existing technologies in the area and discusses the implications that specific design decisions may have in the context of interaction that departs from the traditional desktop paradigm. Subsequently, the prospective enhancement of that design space is explored, with particular emphasis on a new interaction paradigm, based on the concept of "task dependent" computers.

During the 1960s, we interacted with computers through punch cards and paper, in the 1970s through online typewriters, in the 1980s through online video terminals, and in the 1990s through multimedia terminals. In all of these modes of interaction, the user is primarily intended to be sitting at a desk. In the 21st century, we will be moving beyond the desktop and interacting with computers in many different environments and through many different technologies. In this chapter, we discuss those technologies and the types of environments in which they will be useful. Initially, we discuss those items that are, or will soon be commercially available, but we also discuss innovative interaction technologies that are moving user interface interaction toward a new paradigm.

The desktop use of computers has prospered because there is a good match between the standard WIMP (window, icon, menus, and pointing) interface and the types of white-collar tasks that are typically done at the desktop. The problem of the desktop is both its lack of mobility and its assumptions about the physical abilities of its user. In the next decade, the computer will be used away from the desktop, in a wider variety of tasks. In particular, the computer will be used in an environment in which the users may be mobile, or using their hands for non-computer-related tasks. With the possibility of moving beyond the desktop, a new collection of tasks is amenable to computerization, as is a new collection of potential users. In this chapter, we focus on the technologies that make a computer usable while performing these tasks within a particular environment of use.

We take two different perspectives in this chapter. For the first portion, the perspective is that of designers who must construct a system, outlining the implications of particular choices they make on interaction. We explore the design space of input and output devices and discuss why one would make particular choices in particular situations. In the second portion of the chapter, the perspective is that of a researcher of input/output devices who wishes to extend the design space. We discuss an emerging paradigm of computer use, task-dependent computers, and give some examples of that paradigm. We discuss the critical role of input and output devices, and continue with the designer's perspective by discussing the motivations for the evolution, the environments of use, and the implications of the environment on the interaction technologies.

THE CRITICAL ROLE OF INPUT AND OUTPUT DEVICES

The manner in which users interact with their computer is critically dependent on the choice of input and output devices. This is such an obvious observation that sometimes system designers fail to take it explicitly into account. All interaction with the computer consists of issuing a command, providing data on which the command will operate, and, finally, providing a location for the output of the command to be placed; this holds even when the execution of the command is temporally disjoint from its invocation (e.g., a scheduled operation). The specification of those commands involves specifying the command, the source of data for the command, and the target for the output of the command. All must be specified using available input and output devices. Therefore, the choice of these devices will constrain the style of the interaction possible for the user.

Consider implementing a windows-style interface, without a pointing device such as a mouse. The windows style depends on the user being able to point to the source and destination of a command. Furthermore, the type of pointing device is important. The precision of pointing is not as important as the speed of pointing. Thus, a device such as a trackball is a possible pointing input device that admits to great precision. Not only does the trackball admit great precision, but also demands it. Thus, the use of a trackball is conditioned on the ability to very precisely specify a location on the screen, to serve as the current location. Because precise specification takes time, a lower precision, but inherently faster pointing device, such as the mouse, will allow a much smoother interaction with the computer.

The point of this discussion is that there is a coupling between the characteristics of an input or output device, and the style and manner in which a user can operate a computer. In this chapter, we discuss environmental and task constraints on the choice of input and output devices. It should be clear that any constraint on the devices may fundamentally alter the type of interaction that the user has with the computer. We also discuss computers in which the input and output portions are, essentially, hidden; the dominant function of these computers is to perform a single task, and to perform it very well. Now we turn to a discussion of environmental considerations in the choice of input and output devices.

ENVIRONMENTS OF USE

The forces driving the movement away from the desktop are partially technical and partially social. The technical forces are well known—the decreasing size and increasing power of all the components of a modern computer. The social forces are less well known. They include the increasing amount of time that a white-collar worker spends traveling, the extension of computer applications to the blue-collar worker, and the extension of computer applications to the entire population. The traveling white-collar worker is currently being served by a laptop computer that is logically an extension of the desktop. In this chapter, we are not concerned with laptop computers; we instead focus on devices that are appropriate for use by blue-collar workers, on the personal environment of an individual, and on users for whom standard desktops and laptops are not adequate.

Now, let us enumerate some factors of the environment of use that we can use to differentiate among interaction technologies. These are as follows:

- *Infrastructure required:* Some input devices (such as eye- or head-trackers) depend on calibration information that comes from the addition of some infrastructure. Input devices such as cameras, or other sensors used in ubiquitous computing environments, require infrastructure that must be established prior to its use by any application. Output devices, such as displays used in augmented reality, require objects to identify themselves. The establishment of the infrastructure needed to support such interaction devices represents an investment that will limit the environments in which they can be used.
- *Cleanliness of the environment:* The maintenance environment, for example, has grease and corrosive chemicals present. The input and output devices that are to be used in this environment must be resistant to these types of materials. A normal keyboard is not easily maintained because the keys are not protected from such substances.
- *Lighting conditions:* Certain lighting conditions cause problems for display screens. Bright sunlight, for example, washes out certain types of displays, rendering the contents of the screen unreadable.
- *Ambient noise:* Speech recognition is an appealing input technology when totally hands-free operation is required. The amount and type of ambient noise in the environment, however, will affect the accuracy of a speech recognition system.

These environmental conditions act as constraints to the choice of interaction technology in a rather obvious fashion. If a normal keyboard is not suitable because of the presence of grease, or corrosive chemicals, and free-text input is required as a condition of the task, then an alternative free-text input device must be used. The senses available to the user are also a variant on particular environmental conditions. The considerations involved when the user cannot see the display, for example, are much the same whether the cause of the inability to see is bright sunshine, or the blindness of the user. See Bass, Mann, Siewiorek, and Thompson (1997) for a more detailed discussion of interaction technologies suitable for wearable computers, taking into account the context of task to be accomplished and the environment in which the task will take place.

In the next section, we describe some of the task characteristics that will also affect the choice of interaction technologies.

TASK CHARACTERISTICS

Characteristics of the task to be performed that affect the interaction technology are as follows:

- *Degree of mobility:* If the user is moving while interacting with the computer system, then input devices that require stability of platform (such as a keyboard) are not suitable. Furthermore, for a mobile user, the output device must be visible and understandable during movement.
- *Number of hands available:* When performing a particular task, users may be using one or both of their hands for a purpose other than interacting with a computer. Additionally, some users do not have the use of both of their hands. Input technologies differ in the number of hands they require in order to operate.
- *Tasks that must be carried out in parallel:* It is possible that interaction with a computer occurs at the same time that another task is being performed by the user. For example, in the maintenance domain, a technician may be manipulating a tool while interacting with the computer. For this technician, the cognitive effort required to operate the computer may interfere with the repair task.
- *Structured input:* If a requirement exists for the input of free-form text, then clearly an input device that allows textual input is required. On the contrary, a large number of tasks (e.g., Web browsing) can be accomplished without any text input. These types of requirements will clearly have an impact on the choice of input device used in a given application.

These characteristics of the tasks to be performed act as constraints on the interaction technologies that can be used, much in the same way that the characteristics of the environment do. We now provide a brief survey of some of the possible input and output devices, and then we summarize this portion of the chapter.

Input Devices

In any computer application, a distinction exists between input intended to control the computer (commands) and input intended to be retained (data). The choice of appropriate input device will be heavily influenced by whether the device will be used primarily in the issuance of commands or in the input of data. In fact, the desktop has separate input devices for these two categories. The mouse is used primarily for commands, whereas the keyboard is used for the input of data. The types of applications that are primarily command oriented should not be discounted. Any data retrieval application (such as many of those based on the World Wide Web) is primarily command oriented. The commands are used to navigate to the desired data item(s), and then to specify how that data is to be displayed. In a data retrieval application, only search facilities require free-text input.

We now briefly survey some input devices in light of the distinction between command and free text input:

- *Speech:* In any discussion of alternative interaction technologies, speech recognition is always prominent. Speech recognition is appealing, in concept, because of the naturalness of speaking. People are trained from birth to communicate using speech. Speech requires no hands to operate. On the other hand, speech as a command-issuing mechanism is not particularly natural. Dialogues are somewhat stilted and constrained by the vocabulary of the speech recognizer. The utility of speech as a data entry mechanism depends on the type of data. If the data are truly free text—grammatical and with a rich context—then the error rates of speech recognizers are sufficiently low and data can be entered (and, subsequently, retrieved) very efficiently. If the data are of the "fill-in-the-form" variety—isolated words with no context—then speech recognizers generally require an error correction mechanism and error rates are high enough to make the added value of speech recognition questionable. Feedback from a speech recognizer can be visual, auditory, or nonexistent.
- *Chording keyboards:* Figure 5.1 shows a device called the "Twiddler," produced by HandyKey Corporation[1]. It is an example of a one-handed keyboard. Through "chording" (pressing combinations of keys simultaneously) all of the letters and numbers can be specified. Experienced users claim keying speeds of up to 50 words per minute with this device. A chording keyboard requires one hand in order to operate, and it is suitable for the input of free text. Feedback from a chording keyboard can be visual, auditory, or nonexistent. Chording keyboards are useful for command line interfaces, or textual data input, but they are not useful for controlling interactions and specifying commands, except through textual means.
- *Pointing devices:* Although the mouse is inherently a desktop device, a large number of alternative pointing devices exist that are suitable for two-dimensional pointing. Joysticks or joypads can be attached to either the body or a portable device, allowing arbitrary movement over a two-dimensional space. If the device is attached to the body, or on a separate fixed device, then it requires only one hand to operate. If it is attached on a

---

[1]For more information, see: http://www.handykey.com

mobile device, then the number of hands required depends on the size of the mobile device. For a large device, one hand is required to operate the pointer and another to hold the device steady. For a small device, one hand can both hold the device and operate the pointing device. Feedback from a pointing device is primarily visual—the cursor on a display—although auditory feedback mechanisms can be imagined. This type of device makes the input of free text very awkward, but it is suitable for navigation or for the input of two-dimensional data, such as locations on an image (e.g., a surface or a map). A pointing device must be used in conjunction with a selection device, such as a button, in order to specify that the current location is of interest. Pointing devices are useful for command-based input or for specifying locations, but are poor for free-text input.

- *Selection devices:* Any special key can be used as a selection device. Buttons are the most common selection devices, although other devices, such as a dial, can also be used for this purpose. Figure 5.2 shows the VuMan3 computer that has only a dial and a button for input (Bass, Kasabach et al., 1997). The dial is used to navigate around a collection of known locations on the output, and the buttons surrounding the dial are used to select the current location as being of interest. This device is used to navigate through a hierarchy, such as the World Wide Web, or through a questionnaire. Feedback is usually visual, to indicate the current location within the hierarchy. As with pointing devices, the input of free text is very awkward with a selection device. Selection devices are useful for command-based interfaces, but are poor when used to specify locations, or for the input of free text.

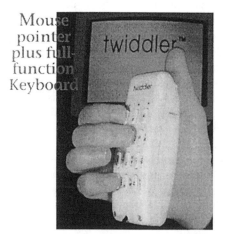

FIG. 5.1. Twiddler input device.

FIG. 5.2. VuMan3 computer.

- *Pen:* A pen is a device for the input of free text. It is used as a pointing, selection, and text input device. Text input requires the presence of text recognition software; the quite high quality of recognition and the low error rate of modern handwriting recognition systems make pen-based text input a viable solution. Pens are typically used in conjunction with flat-panel displays and require two hands to operate—one to hold the display and one to manipulate the pen (although as we see later, other types of pens are possible).
- *Gesturing (hands, head, eyes):* Gesturing includes input that is generated through movements or position of the head, the eyes, other portions of the body, or through objects being manipulated by the user. All uses of gesturing are for command input. For example, moving the hand in a particular manner can be mapped to a particular command. Eye position can be used to control the position of a cursor. Gesturing input devices usually require some infrastructure in order for the gestures to be recognized. Gestures can, in concept, be used as a pen replacement, but, in practice, they are used for command input and not for free-text input.

The choice of a particular input device depends on the environment, the task to be accomplished, and the capabilities of the user. It is also necessarily coupled to the choice of output device. As we indicated, feedback from input is usually required, and this feedback is presented through the output device. We return to this point when we discuss the construction of a system from input and output devices.

Output Devices

We discuss output devices based on the modality/human sense they apply to. Three of the senses are commonly considered for recognition of computer output: sight, sound, and touch. Of these, the most commonly used for computer output is sight. Details of these types of output devices follow:

- *Sight:* The visual devices designed for nondesktop computer usage are flat-panel displays and head-mounted displays. Flat-panel displays require either a hand to hold them, or a convenient place within the viewing space to mount them. Although, in concept, a magnifier could be placed in front of a flat-panel display and the physical size of the display made quite small, in practice these displays usually present an image where the apparent and the real sizes are identical. Head-mounted displays, on the other hand, rely on the distinction between real and apparent sizes. A very small display (liquid crystal in current technology) is reflected through the use of optics and made to appear much larger. Head-mounted displays can be monocular or binocular. Binocular displays are used for augmented reality applications where a computer image is imposed over the image the user would actually see without the display. Monocular displays are used as information displays, without any reference to the view the user would actually see. They are used, in essence, as mobile desktop displays. Because of the manner in which they are positioned, the user can look around them, or at them, much as bifocal glasses are used. Head-mounted displays suffer from a problem of social acceptance. Current displays are large and cumbersome, and give the wearer the appearance of a figure from a science fiction movie. These devices are becoming smaller and less obtrusive, however. Figure 5.3 illustrates a new eyeglasses-based head-mounted display, invented by Microoptics (Spitzer, Rensing, McClelland, & Aquilino, 1997), in which the display is embedded in a frame that makes it appear as if the user is wearing eyeglasses. In this device, the display is actually embedded into the left eyeglass. The glass used for mounting can then be ground to any prescription. The left temple contains a small liquid crystal display. This display is currently limited in resolution, but, in principle, the display can easily be improved to provide VGA quality (640 × 480 pixels). Both head-mounted and flat-panel displays are difficult to use in bright sunshine. Attempts to ameliorate the problems caused by glare lead to combinations of sunglasses and computer monitors. The glass in the display shown in Fig. 5.3 could be tinted to reduce glare. With a totally external device, some form of tinted glass is commonly used and this causes some problems with appearance. One consideration in the use of a head-mounted display is that of safety. If the normal view of the user is occluded, then users may not see obstructions,

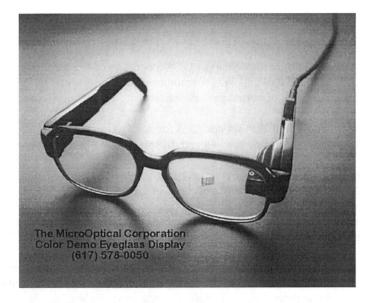

FIG. 5.3. MicroOptical Corporation eyeglass display.

or other safety hazards in their path. Clearly, this depends on the task(s) in which the user is engaged. A user walking down a street may have more requirements for full vision than a user squatting next to a piece of machinery (or less in the case of some repair environments). Motion is another consideration associated with the use of a head-mounted display. When purely textual information is displayed, the user can be moving, but when viewing images and attempting to understand them, the user must normally be still. The quality of the images, however, is sufficient for standard direct manipulation interfaces to be used, if they are appropriate to the task.

- *Audio:* Sound, both spoken and nonspoken, is also used as an output device. Speech can be used to replace a textual display, as well as for feedback for various forms of input. The use of sound to replace a textual display involves not only the presentation of the words, but also adjustment of the presented information because of differences between the temporal properties of sound versus the spatial properties of a display. Text shown on a display will remain there until it is removed, either by the system or by the user. Hence, the speed of understanding is at the control of the user. Audio output has a short duration and must be understood when it is spoken, or it must be repeated. Hence, the speed of understanding is at the control of the system. This leads to differences between carrying out dialogues with systems using visual output and with systems using audio output. Audio output is also sensitive to ambient noise. Nonspeech sound has been used both for alarms and for directional indications. Its use for direction has mainly addressed the needs of the blind. Using a combination of a route known by the computer and a positional location system, such as the Global Positioning System, sound is used to inform users that they have deviated from the planned route. By sending the sound to the right or left ear, as appropriate, users can be informed not only of the fact that they have deviated from the planned route, but also of the direction in which they should turn.

- *Tactile:* The use of tactile output for directional indicators has been suggested for drivers of automobiles. That is, the seat or steering wheel will inform the driver of the direction of the next turn. This suggestion has, as yet, not been implemented. The same is true of existing tactile technology that is currently used to present the output of computing systems to blind users in an attainable form, but has not yet found application in more mainstream systems.

SYSTEM DESIGN

It is in the design of the system that the considerations we have discussed so far come together. The style of the possible interactions is dependent on the choice of input and output devices; the choice of devices is, in turn, dependent on the constraints imposed by the environment and the task of the user.

Of the devices that we have discussed, the only inherent coupling between input and output devices is between the pen, as we have described it, and the flat-panel display. In a later section, we see that even this coupling is not necessary. In any case, using current—or soon to be available—technology, the system designer makes choices based on the task, the number of hands necessary to operate the input devices, the senses available from the user when operating the input device, the error rate of the device, and the extent to which training is required either to tailor the device to the user or to train the user to the device.

For example, suppose that in a maintenance application, the technician needs to navigate through manuals while performing some operation that requires the use of one hand. Then, the use of a pen as an input device for a system to support that technician is not appropriate, because a pen requires one hand to hold it and another to hold the display. The assumption in this example is that the technician is in an environment where there is no location to fix a display. For another example, suppose that the users of a particular system are assumed to be infrequent and naive with respect to computers. Then, using a chording keyboard as an input device is inappropriate, because chording keyboards require substantial training.

Table 5.1 gives, for each input device, the number of hands required to operate that device, the other senses that must be engaged while operating the device, the error rate for the device, and the training requirements. The term *error rate* is used here to refer to the recognition of user input by the computer. Thus, striking a key is always recognized correctly by the computer (even though the key may be struck in error), whereas words are not always recognized correctly by a speech recognition engine, even though the word may have been spoken correctly. The error rate is important because if a device has more than a small error percentage, then either the user must repetitively attempt to provide understandable input, or an alternative input mechanism must be provided. Small in this case means "very small," because a 1% error rate means one error in every 100 words and a chapter such as this has thousands of words. Again, the point is that the input device must be chosen to be appropriate to the task for which the system, including the particular device, is to be used.

TABLE 5.1
Characteristics of Devices in Terms of Important Usage Criteria

	Number of Hands	Other senses	Error Rate	Training Required
Speech	0	none	>0	high
Dial	1	none	0	low
Pointer	1	eyes	0	low
Chording keyboard	1	none	0	high
Keyboard	2	eyes	0	low
Pen	2	eyes	>0	high

COMPATIBILITY WITH DESKTOP SYSTEMS

Given that a system is to be mobile, to what extent should it be compatible with desktop systems? Three different types of compatibility need to be considered, namely: software compatibility, user interface compatibility, and data compatibility. Additionally, the environment of use remains of primary importance. We now briefly discuss each type of compatibility:

1.   Software compatibility is the ability of software to execute on multiple platforms; that is, the same software system will execute both on the desktop and on the mobile platform. This is, by and large, the situation with respect to desktops and laptops. Given the other types of compatibility that we discuss later, the motivations for this type of compatibility are economic. Software has only to be developed once—usually for the desktop—and it is then available for the mobile platform. Furthermore, development platforms are readily available, but include no special considerations because they are generally intended to be used for development for the desktop. This type of compatibility assumes that the distinctions between the desktop and the mobile platform are not fundamental, but merely those of size and weight.

Though the laptop may be merely a smaller desktop, in other types of mobile applications user or environmental characteristics may result in fundamentally different types of applications. Thus, for example, a spreadsheet application is not one that seems necessary to put on a mobile, nonlaptop platform. Furthermore, location-aware navigational assistance applications are most appropriate when the user is moving and, in many such situations, the burden of interacting with a laptop, using normal mechanisms, interferes with other tasks the user may be performing (such as driving).

Thus, the economic motivation for software compatibility must be weighed against the costs in terms of appropriateness for specific tasks or users.

2.   User interface compatibility is the use of the same "look and feel" on the mobile platform as on the desktop. The motivation for this type of compatibility is that it capitalizes on previous training/experience of the users, and ensures their comfort in interacting with the interface. The user is assumed to be a user of the desktop and, thus, the extension of the desktop metaphor to the mobile platform will simplify training and increase comfort for this user.

The assumption that the user is familiar with a desktop style interface is not necessarily true for users of mobile systems. Some unpublished data from the maintenance domain, for example, indicate that many technicians are not familiar with desktop systems. Certainly, familiarity with the desktop metaphor is growing, but many of the people reached by innovative technology are not computer literate.

Furthermore, the desktop metaphor is useful precisely because it bears some relationship to the user's familiar working environment. Once the user is mobile and using a device tailored to a particular task, it is unclear what relevance a desktop metaphor might have. User interfaces based on metaphors familiar to the user and pertinent to the task are more appropriate.

3.   Data compatibility refers to the ability to interchange data between a mobile device and a desktop. This is achieved via the use of accepted protocols and data formats. It also requires some mechanism for communicating data between the mobile device and the desktop. Data compatibility does not affect the user interface except for the requirement to provide an interface to control the transfer of data. This interface can be physical, such as a docking station, in which case the transfer of data is automatic.

NEW PARADIGM OF INTERACTION

Recently there has been interest in what is termed *task-dependent computers*. The terms *embodied interfaces* (Fishkin, Moran, & Harrison, 1999) or *information appliances* (Norman, 1998) have also been used to refer to the same concept. The actual concept is that a specific device is designed for a particular purpose, and it is so well designed for this purpose that its use is natural. In some senses, this type of interface has already arrived. The driver of an automobile, for example, is unaware of the myriad of computers being used in support of the driving task, such as those that manage the engine, the transmission, the cruise control, and even the power windows. Another example of a task-dependent computer is the digital camera. The user is not concerned with the underlying technology, but only with the fact the device can capture, preview, and transfer images.

The intent is to develop a special-purpose device, expressly designed for one set of tasks. Desktop computers are intended to be general-purpose devices and, consequently, must be able to support a myriad of different tasks. Task-dependent computers are intended to be special purpose and must support only one task. Therefore, they can be designed to be specific for that task and will be, as a result, easy to use.

We give some examples of devices that illustrate this type of embodiment. The first is called "Digital Ink" from Carnegie Mellon University (Kasabach, Pacione, Stivoric, Gemperle, & Siewiorek, 1998). Currently, a prototype version of Digital Ink is being constructed. Digital Ink is an input device (see Fig. 5.4) designed to use normal paper as its output device. That is, the user uses the pen to write or sketch on a piece of paper. Digital Ink contains real ink and so feedback as to the output is immediate and normal to the user. In addition to ink, Digital Ink also "remembers" the pen movements and sends those over a wireless communication link to any recipient. The following scenarios are all possible with Digital Ink:

1.   You are sitting in a field sketching a flower. On completion of the sketch, you wish to share it with a friend. You indicate "send" to the pen and the sketch is automatically sent because the pen has been recording your movements.

2.   You are sitting in a meeting taking notes on a piece of paper. In this case, when you indicate "send," the words are scanned using handwriting recognition and the scanned words are then transmitted to your desktop for inclusion in the meeting minutes.

3.   You are preparing a bid for a collection of equipment and you need a calculator. You write down a list of numbers together with a plus sign and a line under the numbers. The pen recognizes this as a request for computation, performs the addition, and displays the sum in the pen's LCD display.

For all of these scenarios, the user manipulates the pen in the manner that people have been using it for hundreds of years, and the system accommodates these types of interactions.

The next example of a task-dependent computer comes from the Xerox Palo Alto Research Center (Harrison, Fishkin, Gujar, Mochon, & Want, 1998). The task for which it is designed is locating names and addresses in a list. It is based on a Rolodex—a circular address or phone list that has been popular for decades. The physical device uses a hand-held computer, attached to which is a sensor that measures tilt angle. Interaction with the device takes place as follows: Tilting it forward causes the phone list to scroll forward and tilting it back causes it to scroll backward. Furthermore, the user interface displays pages that look like the pages on a Rolodex. Thus, knowledge of the Rolodex acts as a predictor for the user of how to use the computer-based Rolodex. On the other hand, the hand-held system is more sensitive to tilt than the physical Rolodex and, consequently, it is more difficult to ensure that it stops action on the correct entry. For this reason, side sensors were also added to the hand-held computer to enable the user to "squeeze" in order to slow down the speed of scrolling.

The growth of task-dependent computers is based on the underlying technology becoming "good enough" so that special-purpose devices can be designed and

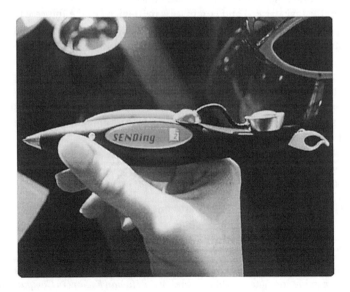

FIG. 5.4.   Digital Ink device.

constructed. The interaction technologies assumed by these devices are tightly coupled input and output devices, but limited in both aspects. Thus, the clear trend is to specialize, to sacrifice generality and to utilize the technology in the service of the user, rather than force the user to adapt to a few limited forms of interaction.

SUMMARY

It is undeniable that task-dependent computers and task-specific interaction techniques are emerging and will become more and more common. The types of interaction techniques used in particular applications and the types of devices available to implement these techniques will depend on the task being performed, the characteristics of the user, the environment of use, and the time frame for which the application is being envisioned.

For those applications being designed with current technology, we have presented a collection of device types and given some indications of the contexts for which they are appropriate. These devices are, by and large, components of the familiar desktop computer rearranged and configured independently. Even this configuration, as we have shown, leads to devices that are task- and user-specific in a manner that the desktop is not.

When looking to the future, however, we find devices in which input and output are integrated, again in a fashion uncommon for the desktop. Furthermore, by constructing these devices on a model of similar devices with which all users are familiar, their interfaces can be made intuitive and natural and their capabilities can be extended to realms that seem like science fiction.

In addition, the task-, user-, and context-aware devices of the future will introduce new capabilities in addressing the specific needs of the end-user population, which constantly increases in size and diversity. To take advantage of these opportunities, designers will have to acquire a better understanding of the design space of input and output devices, as well as of the implications that specific choices/combinations in that space have on interaction. Furthermore, by directing research efforts to the development of "targeted" interfaces for specific tasks and contexts of use, it is possible to constrain the intrinsic complexity in the development of user interfaces that are accessible and usable by all.

ACKNOWLEDGMENTS

This work is supported by the U.S. Department of Defense.

REFERENCES

Bass, L., Kasabach, C., Martin, R., Siewiorek, D., Smailagic, A., & Stivoric, J. (1997). The design of a wearable computer. In *Proceedings of the Human Factors in Computing Systems Conference* (pp. 139–146). New York: ACM Press.

Bass, L., Mann, S., Siewiorek, D., & Thompson, C. (1997). Issues in wearable computing: A CHI '97 workshop. *SIGCHI Bulletin*, 29(4). [Online]. Available: http://www.acm.org/sigchi/bulletin/1997.4/bass.html

Fishkin, K. P., Moran, T. P., & Harrison, B. L. (1999). Embodied user interfaces: Towards invisible user interfaces. In P. Dewan & S. Chatty (Eds.), *Proceedings of the IFIP Working Conference on Engineering for Human–Computer Interaction* (pp. 55–73). Norwalk, CT: Kluwer.

Harrison, B. L., Fishkin, K. P., Gujar, A., Mochon, C., & Want, R. (1998). Squeeze me, hold me, tile me! An exploration of manipulative user interfaces. In *Proceedings of the Human Factors in Computing Systems Conference* (pp. 17–24). New York: ACM Press.

Kasabach, C., Pacione, C., Stivoric, J., Gemperle, F., & Siewiorek, D. (1998). Digital ink: A familiar idea with technological might! In *Proceedings of the Human Factors in Computing Systems Conference* (pp. 175–176). New York: ACM Press.

Norman, D. (1998). *The invisible computer.* Cambridge, MA: MIT Press.

Spitzer, M. B., Rensing, N. M., McClelland, R., & Aquilino, P. (1997). Eyeglass-mounted displays for wearable computing. In L. Bass & A. Pentland (Eds.), *Proceedings of the 1st International Symposium on Wearable Computers* (pp. 48–51). Cambridge, MA: IEEE Press. Available: http://www.computer.org/conferen/proceed/8192/8192toc.htm

6 Special Needs and Enabling Technologies: An Evolving Approach to Accessibility

Pier Luigi Emiliani

Technological evolution is usually accompanied by considerable changes in the organization and functioning of our society, thus imposing new interaction and communication requirements for every citizen. For people with special needs, technological evolution could either improve their situation or could accumulate additional obstacles in their daily effort to efficiently and effectively function as members of our society. This chapter addresses the problems of people with disabilities in getting access to computer-based applications and telecommunication services, and provides an account of a decade of efforts in developing accessibility solutions through collaborative Research & Development projects funded by the European Commission. Moreover, the chapter presents an evolutionary path from "adaptation-based" towards "inclusive" approaches to accessibility, and indicates the changing paradigm in the approach to enabling technologies today.

The rapid evolution and fusion of information and telecommunications technologies has the potential to improve the quality of life of citizens, and to offer new opportunities for all citizens in society, including disabled and elderly people, if careful note is taken of their abilities, requirements, and preferences during the deployment of the corresponding equipment, services, and applications. A new technological environment emerges, characterized by multiplicity of information and redundant communication patterns (i.e., information can be coded and transmitted in different media). This environment, usually called the information society, enables potential access to new interpersonal communication telematic services and huge amounts of information that increasingly becomes available through different modalities and communication channels (e.g., access to textual, audio, and video resources through traditional and new interaction methods and techniques, such as speech, gestures, tactile, and kinaesthetic feedback). Furthermore, the development of signal-processing techniques, such as speech synthesis and image and video coding, enables the *transduction* of information between different modalities (e.g., from visual to acoustic), thus offering new interesting possibilities of exchanging information and interacting with the information and the environment. These developments have a great potential to contribute to the independent living and socioeconomic integration of all citizens in general, and disabled and elderly people in particular. At the same time, the opportunities resulting from these developments may also turn into further barriers, if appropriate measures are not taken in order to provide access to information and telematic services by the different sectors of the population (e.g., novice users of technology, disabled and elderly people).

This chapter provides a brief account of some research and development (R&D) efforts in Europe, which have explicitly addressed the issue of accessibility of computer-based applications and telematic services by disabled and elderly people. The review covers early R&D work in the area of accessibility, discusses its important contributions to the integration of people with disabilities in computer-based environments, and points out the emerging difficulties with existing adaptation-based approaches. The chapter also examines accessibility in the context of the emerging information society, addressing the shortcomings of the traditional approaches, and advocating the need for "inclusive" strategies in the design and development of new computer-based products and telematic services. References are included to examples mainly drawn from the direct experience of the author in the field of accessibility by different user groups, and in particular by blind people.

SPECIAL NEEDS

When considering particular requirements of the users, one usually refers to issues associated to obstacles resulting either from disability or aging. According to the international classification of impairments, disabilities and handicaps recommended by the World Health Organization (1989), *impairment* is defined as the loss of a function that entails a form of *disability,* whereas the *handicap* is a disadvantage that could be created by the environment, a gap between individual capabilities and environmental demand.

In case of impairment due to physical reasons, such as loss of vision, the consequent disability is the inability to see, whereas a potential handicap could be the computer screen. Such a handicap could be overcome by replacing the computer screen, which is a device or presentation medium for conveying visual information, with a speech output device. Therefore, impairment does not always lead to a handicapping situation, and a handicap can also depend on a person's personality, attitude, and opportunities. The role of accessible design and assistive technology is to reduce ability gaps and moderate the handicaps faced by people due to disability or old age. Table 6.1 provides a description of the main categories of users classified according to their impairment or disability.

Disabled and elderly people currently make up about the 20% of the market in the European Union, and this proportion will grow with the aging of the population to an estimated 25% by the year 2030 (Gill, 1996; Vanderheiden, 1990). Similar figures are found in the United States of America, and all other developed countries. There is not only a moral and legal obligation to provide access to computer-based applications and telematic services for this part of the population, but there is also a growing awareness among the industry that disabled and, in particular, elderly people can no longer be considered an insignificant minority. Instead they represent a growing market to which new services can be provided.

TABLE 6.1

Categories of Users Due to Their Impairment or Disability

Category	Description
Mobility Impaired	Reduced function of legs and feet means dependency on a wheelchair or other artificial aid to walking. In addition to people who are born with a disability, this group includes a very large number of users whose condition is caused by age or accidents.
Dexterity Impaired	Reduced function of arms and hands makes activities related to moving, turning, or pressing objects difficult or impossible.
Visually Impaired	Blindness implies a total to near-total loss of the ability to perceive visually presented information. Low vision implies an ability to utilize some aspects of visual perception, but with a great dependency on information received from other sources as well.
Hearing Impaired	Hearing impairment can affect the whole range or part of the auditory spectrum. The term deaf is used to describe people with profound hearing loss, whereas hard of hearing is used for those with mild to severe hearing loss.
Speech Impaired	Speech impairment may influence speech in a general way, or only certain aspects of it, such as fluency or voice volume.
Language Impaired	Language impairment may concern the difficulty to comprehend or express concepts in natural language, or may be associated with a more general intellectual impairment.

In this perspective, *accessibility* concerns the provision and maintenance of access by disabled and elderly people to encoded information and interpersonal communication, through appropriate interaction with computer-based applications and telematic services. In the present, information-rich and interaction-demanding communication environment, full access to different applications and services requires the ability of the user to perceive the provided information (the different media) as well as the ability to control the provided user interface (maintain a dialogue with the system).

Traditional approaches to accessibility, developed in the field of assistive technology, are based on the adaptation of existing systems and telematic services, so as to make them accessible to individual user groups, or on the development of "dedicated" solutions for specific tasks, or for particular user groups. These adaptations take place mainly at the level of the user interface, and aim at providing physical access to system functionalities, as well as transduction of information in appropriate modalities. For example, people with motor impairments require adaptations of interaction peripherals (input/output devices) in order to be able to physically interact with computers, whereas blind people require transduction of information to an appropriate modality (acoustic or tactile) in order to be able to explore displayed information.

Tables 6.2 and 6.3 analyze the relationships between different information types and impairments that may affect the user's perception and use of computer-based

TABLE 6.2

Accessibility of Information Types in Relation to Impairments Associated
With the User Sensory Channels

User Sensory (Input) Activities

Impairment		speech	sound	still picture	moving picture	text	graphics	moving graphics
Hearing	partial	−	−	+	+	+	+	+
	deaf	=	=	+	+	+	+	+
Visual	low vision	+	+	−	−	−	−	−
	blind	+	+	=	=	=	=	=
Language	partial	+	+	+	+	−	+	+
	total	+	+	+	+	=	+	+

Note.   + same use as able-bodied; − less use; = no use.

TABLE 6.3

Accessibility of Information Types in Relation to Impairments Associated
With the User Communication Channels

User Communication (Output) Activities

Impairment		speech	sound	still picture	moving picture	text	graphics	moving graphics
Speech	partial	−	+	+	+	+	+	+
	total	=	+	+	+	+	+	+
Dexterity	partial	+	+	+	−	+	+	+
	total	+	+	+	=	+	+	+
Language	partial	+	+	+	+	−	+	+
	total	+	+	+	+	=	+	+

Note.   + same use as able-bodied; − less use; = no use.

systems and telematic services and applications. Although this analysis treats each type of impairment in isolation, and divides the human activities into sensory (input) and communication (output) channels, rather than considering the different possible combinations of the human activities involved in an interaction session, it provides a preliminary understanding of the problem of accessibility of the provided media in relation to the physical abilities of the user.

Moreover, a list of adaptation solutions associated to the transduction possibilities of each information type to alternative media is provided in Table 6.4. Some of the listed adaptation solutions such as text-to-speech, text-to-Braille, and speech-to-text systems, are currently available in the market as stand-alone products, whereas most of the remaining ones concern possibilities that are currently underexploited (e.g., speech-to-signs, description of moving pictures, etc.).

With the progressive availability of highly computer-based applications and telematic services, to everyone and everywhere, *technology awareness* is becoming a dividing factor that classifies the users into experienced and inexperienced, or "able" and "disabled" with respect to the use of new technology. The dimensions of "disability" could, for example, include also the consideration of environmental conditions (e.g., hazardous environments) or work tasks (e.g., the use of telephone equipment while driving).

Special needs are, therefore, difficult to classify in the information society, because everyone has particular limitations and special requirements in interacting, communicating, working, and performing everyday activities. In this perspective, disability should no longer be considered as a situation faced by a particular small part of the population (usually defined as people with special needs). Rather, it

TABLE 6.4

Adaptation Possibilities Associated with the Different Information Types

Information type	Adaptation solutions
Speech	speech-to-text speech-to-symbolic and pictorial information speech-to-lip movement (animation) speech-to-signs (animation)
Sound	musical sounds to visual presentation (e.g., spectograms) environment sound to visual presentation
Still Pictures	tactile presentation of still pictures description of still pictures using sounds description of still pictures using text (speech and/or Braille) contrast/color/enlargement of still pictures
Moving Pictures	description of moving pictures using text (speech and/or Braille) description of moving pictures using sounds gesture-to-text
Text	text-to-speech text-to-Braille text-to-graphics/symbols text-to-sign (animation) text-to-lip movements (animation) text-to-gesture
Graphics	tactile representation of graphics description of graphics using text
Moving Graphics	description of moving graphics using text (speech and/or Braille)

concerns a large percentage of the citizens who want to fully utilize the provided services and actively participate in the everyday social activities.

ENABLING TECHNOLOGIES: A CHANGING PARADIGM

Having outlined the complexity involved in addressing access to computers by people with disabilities, this section presents the case of enabling technologies as depicted in a decade of evolutionary efforts. It should be noted that, though such an account is not intended to provide an exhaustive list of European R&D work, it does, however, illustrate the evolutionary thinking that more recent approaches such as *user interfaces for all,* have been based on. Additionally, the treatment of such efforts will inevitably reflect the author's personal experiences as project manager, without elaborating technical details that have been published elsewhere. These projects were funded by European Commission Programs,[1] have spanned across a decade, and have pursued an evolutionary path, starting from exploratory studies, to examining and developing adaptation solutions, to finally addressing the need for more general approaches to accessibility, following the principles concept of *design for all.* Their main contributions and interconnections are briefly outlined in Fig. 6.1, whereas their respective technical focus and achievements are summarized in Table 6.5.

Exploratory Studies

The RACE-IPSNI[2] project has investigated the possibilities offered by the multimedia communication network environment, and in particular B-ISDN (Broadband Integrated Services Digital Network), for the benefit of people with disabilities. Technological advances in this field include increased network bandwidth and reliability, as well as more powerful, more mobile, and less costly network terminals.

The starting point of the project was the consideration that increased bandwidth and reliability of the B-ISDN environment may offer new opportunities for the provision of multimedia information, which additionally can be manipulated by the end-user through innovative interaction techniques and styles. The utilization of network management techniques allows the application/service customization according to the end-user needs and abilities and the provision of special services, where appropriate. As a consequence, the introduction of B-ISDN applications and services may offer new opportunities for the socioeconomic integration and independent living of disabled and elderly people, including, but not limited to, distant learning, teleworking, teleshopping, sophisticated alarm systems, and so on.

---

[1]The European Commission Directorate General XIII Programs that have funded these projects are (in chronological order) RACE, TIDE, and ACTS.

[2]The IPSNI R1066 (Integration of People With Special Needs in IBC) project was partially funded by the RACE Program of the European Commission, and lasted 36 months (January 1, 1989 to December 31, 1991). For the list of consortium partners, see the Acknowledgments section.

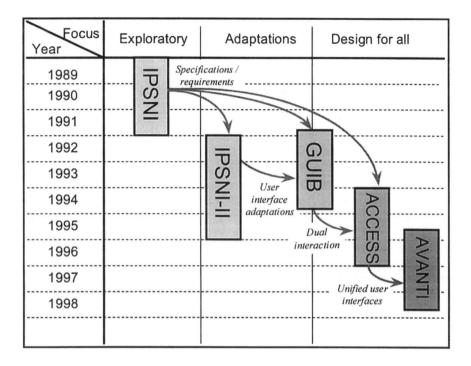

FIG. 6.1. Chronological sequence, focus, and some of the key outcomes of the projects reviewed.

In order to enable the accessibility of disabled people to the emerging telecommunications technology, the RACE-IPSNI project considered essential that the designers and/or providers of the services and terminal equipment take explicitly into account, at a very early stage of design, their interaction requirements. As an initial contribution, the project has addressed problems faced by people with special needs in accessing B-ISDN environments through an indepth analysis of interaction requirements, based on human factors issues and ergonomics criteria. Several barriers have been identified that prevent people with special needs from having access to information available through the network. The identified barriers are related to accessibility of the terminal, control of the anticipated services, and the perception of the service information.

In order to cope with these difficulties, different types of solutions have been proposed that address the specific user abilities and requirements at three different levels:

1. Adaptations within the user-to-terminal and the user-to-service interface, through the integration of additional input/output devices and the provision of appropriate interaction techniques, taking into account the abilities and requirements of the specific user group.

TABLE 6.5

Main Achievements of Some EC-funded Projects

Project	Main Achievements
IPSNI January 1, 1989 to December 31, 1991	• User requirements of people with disabilities for accessing telematic terminals, services, and applications (IPSNI Consortium, 1991a) • User-to-terminal and user-to-service interaction models (IPSNI Consortium, 1990) • Technical specifications for the development of adapted telematic terminals (IPSNI Consortium, 1991b)
IPSNI-II January 1, 1992 to December 31, 1995	• Demonstrators of adapted telematic terminals for the MS-Windows and the X-Windows platforms (IPSNI-II Consortium, 1993) • Prototypes of special purpose interaction peripherals, e.g., mouse emulators (IPSNI-II Consortium, 1992) • INTERACT: tool for developing user interfaces accessible by various user groups (prototype system supporting sighted, blind, low-vision, and motor-impaired user) (Stephanidis & Mitsopoulos, 1995) • Demonstrators of special-purpose services for people with disabilities (IPSNI-II consortium, 1994)
GUIB December 1, 1991 to May 31, 1993	• BRAILLEX 2D: transitory Braille display (developed and commercialized) (Papenmeier, 1999a) • WINDOTS: screen reader for transitory Braille displays for the MS-Windows platform (developed and commercialized) (Papenmeier, 1999b) • GUIB-ERL: formal language for the specification of appropriate interaction methods for blind users combining speech, sounds, and Braille output (Weber, Kochanek, Stephanidis, & Homatas, 1993)
GUIB-II June 1, 1993 to November 30, 1994	• Prototypes of speech-based screen readers for the MS-Windows (Weber et al., 1994) and the X-Windows platforms (Weber, Kochanek, Stephanidis, & Gogoulou, 1994) • Prototype system supporting spatialized three-dimensional sound (Crispien, Wurz, Weber, 1994) • CONFIG: screen reader configuration system (prototype system that facilitates customization of the nonvisual environment) (Stephanidis & Gogoulou, 1995) • COMONKIT: nonvisual toolkit (user interface toolkit for supporting nonvisual interaction) (Savidis & Stephanidis, 1995a) • Definition of dual user interfaces, and specification of tools for supporting their development (Savidis & Stephanidis, 1995a) • HOMER: user interface management system (prototype system for developing dual user interfaces accessible by both sighted and blind users) (Savidis & Stephanidis, 1998)

ACCESS
January 1, 1994 to
December 31, 1996

- Definition of Unified User Interfaces and specification of a new methodology for supporting their development (Savidis, Paramythis, Akoumianakis, & Stephanidis, 1997)
- A complete set of tools for developing Unified User Interfaces: USE-IT (Akoumianakis & Stephanidis, 1997a, 1997b), G-DISPEC (Savidis & Stephanidis, 1997b), HAWK (Savidis, Stergiou, & Stephanidis, 1997), MS-TOOL (Savidis, Vernardos, & Stephanidis, 1997), PIM (Savidis & Stephanidis, 1997a)
- I-GET: user interface management system (an integrated environment for designing and developing Unified User Interfaces) (Stephanidis & Savidis, 1997)
- Application of the ACCESS technology and tools in the development of a demonstrator multimedia system accessible by blind people (Petrie, Morley, McNally, O'Neill, & Majoe, 1997)
- ATIC: a novel architecture for developing interpersonal communication aids for people with disabilities (Kouroupetroglou, Viglas, Anagnostopoulos, Stamatis, & Pentaris, 1996)
- Application of the ACCESS technology and tools in the development of a demonstrator communication system accessible by motor and cognitive disabled people (ACCESS Consortium, 1996)
- Contribution to standardization activities regarding accessibility; proposal of a new ISO work item on accessibility (Stephanidis & Akoumianakis, 1997)

AVANTI
September 1, 1995
to August 31, 1998

- Prototype system of a user modeling component for supporting adaptable and adaptive interaction (Fink, Kobsa, & Nill, 1996, 1997)
- Prototype system of an adaptable and adaptive Web browser (Stephanidis, Paramythis et al., 1998)
- Demonstrator multimedia information systems with adaptable and adaptive information content and user interface (Bini & Emiliani, 1997; Bini, Ravaglia, & Rella, 1997)

105

2. Service adaptations through the augmentation of the services with additional components capable of providing redundant or transduced information.

3. Introduction of special services, only in those cases where the application of the two previously mentioned types of adaptation are not possible or effective.

Adaptations in Telecommunications

The RACE-IPSNI-II[3] project was built on the results of the RACE-IPSNI project, and demonstrated the technical feasibility of providing access to people with disabilities to multimedia services running over a broadband network. Adaptations of terminals and services were implemented and evaluated. In particular, two pairs of multimedia terminals (one UNIX/X-Windows based and one PC/MS-Windows based) were adapted according to the needs of the selected user groups.

Special emphasis was placed on the adaptation of the user interfaces, and for this purpose, a user interface design and construction tool named INTERACT was designed (Stephanidis & Mitsopoulos, 1995), which takes into account the interaction requirements of disabled users. INTERACT supports different interaction styles through the utilization of alternative interaction channels and media; the audio, visual, and tactile modalities can be selected by the user interface designer, taking into consideration the characteristics of the target user group and the scope of the particular application. For instance, in order to support the development of graphics-based applications accessible by blind users, INTERACT enhances the graphical objects with additional attributes (e.g., presentation in auditory or tactile form) and provides facilities for the exploration of the graphical objects of the various applications (e.g., audio-based navigation).

Furthermore, the RACE-IPSNI-II project has tested a number of special interaction peripherals and adaptation solutions (e.g., mouse emulators, screen keyboards, tactile and speech devices), and has developed special prototypes of services for disabled users, such as multimedia E-mail and computer-based interviewing (i.e., a non-real-time communication service based on structured multiple-choice answers).

The RACE-IPSNI-II project allowed an in-depth analysis of services and applications for the broadband telecommunications environment from the perspective of usability by disabled people, leading to the identification and testing of necessary adaptations and/or special solutions. This work led to the conclusion that, if emerging services, applications, and terminals were designed considering usability requirements of disabled users, many of their access problems would be automatically reduced with a negligible expense. As a minimum requirement, sufficient modularity and flexibility should be the basis of product implementa-

³The IPSNI-II (Access to B-ISDN Services and Applications by People With Special Needs) project was partially funded by the RACE-II Program of the European Commission, and lasted 48 months (January 1, 1992 to December 31, 1995). For the list of consortium partners, see the Acknowledgments section.

tion, in order to allow easy adaptability to the needs, capabilities, and requirements of an increasing number of users.

Adaptation of Graphical User Interfaces

The TIDE-GUIB[4] and TIDE-GUIB-II[5] projects aimed to identify and provide the technological means to ensure continued access by blind users to the same computer-based interactive applications used by sighted users. The starting point of the project was the consideration that graphical user interfaces (GUIs) can be thought of as totally inaccessible by blind users, due to the fact that they have been designed to exploit the visual capabilities of sighted users and do not support nonvisual interaction methods. On the other hand, multimedia user interfaces could potentially facilitate blind user interaction, provided that appropriate design allows for easy installation and handing of special input/output devices and supports nonvisual interaction methods, in addition, and in parallel, to the existing visual ones.

The short-term goal of the GUIB projects was to improve existing GUIs through adaptations. Specific developments were carried out through the implementation of appropriate demonstrators enabling access to MS-WINDOWS™ (PCs) and to interactive applications built on top of the X WINDOW SYSTEM (UNIX™-based workstations). The GUIB approach to interface adaptation for blind users was based on a transformation of the desktop metaphor to a nonvisual version combining Braille, speech, and nonspeech audio. Access to basic graphical interaction objects (e.g., windows, menus, buttons), utilization of the most important interaction methods, and extraction of internal information from the graphical environment were investigated. The system supports the specification of alternative output media for the various graphical interaction objects. The supported output media for nonvisual interaction include speech and nonspeech auditory cues, and Braille output. Input operations (e.g., exploration/selection of menu options, etc.) can be performed either by means of standard devices (keyboard or mouse) or through special devices (i.e., mouse substitutes, touch pad, and routing keys of Braille device). An important feature of the method is that the whole graphical screen is reproduced in a text-based form and simultaneously presented on a monochrome screen that can be explored by blind users by means of Braille and/or speech output. Additionally, sounds help navigation and provide spatial relationships between graphical objects. It is important to note that the text-based reproduction facilitates cooperation with sighted colleagues.

A tool was designed and implemented to facilitate the description of blind user interaction in a graphical environment and enable combinations of acoustic and

[4]The GUIB TP103 (Textual and Graphical User Interfaces for Blind People) project was partially funded by the TIDE Program of the European Commission, and lasted 18 months (December 1, 1991 to May 31, 1993). For the list of consortium partners, see the Acknowledgments section.

[5]The GUIB-II TP215 (Textual and Graphical User Interfaces for Blind People) project was partially funded by the TIDE Program of the European Commission, and lasted 18 months (June 1, 1993 to November 30, 1994). For the list of consortium partners, see the Acknowledgments section.

tactile media for presentation and access to graphical objects (Mynatt & Weber, 1994). Such a tool is mainly based on:

- Filtering of internal graphical data: Appropriate filters in MS Windows provide data about text appearing on the screen (fonts, size, and colors), mouse pointer and cursor positions, position and size of windows, icons, menus, and buttons. Additionally, information about more complex objects such as dialogue boxes and scrollbars is collected.
- Provision of a representation mechanism for synchronization of both static and dynamic filtered data.
- Definition of a formal language for the specification of appropriate interaction methods for the blind user, combining speech, sounds, and Braille output: The developed language (GUIB-ERL) (Weber, Kochanek, Stephanidis, & Homatas, 1993) is based on a formal language, called Event Response Language (ERL) (Hill, 1986), for modeling GUIs. GUIB-ERL supports easy handling of special input/output devices (Braille display, speech synthesizer, and sound generator) in addition to the handling of the conventional keyboard and pointing devices.
- Provision of a front-end module, the CONFIG screen reader configuration system (Stephanidis & Gogoulou, 1995) that hides the complexity of GUIB-ERL from the end-user, and enables easy and quick customizations of the screen reader in order to conform to user preferences and specific application peculiarities: Selection of devices and media, definition of key bindings, activation of monitors, and modification of the modeling of graphical objects constitute the basic functionality of the front-end module.

The GUIB project also investigated a variety of issues related to user interaction in a graphical environment, particularly for blind users. For example, the project investigated different input methods that can be used instead of the mouse. It also studied the problem of how blind users can efficiently locate the cursor on the screen, and examined issues related to combining spatially localized sounds (both speech and nonspeech) and tactile information in order to present available information. Finally, the project addressed the design and implementation of real-world metaphors in a nonvisual form and the development of an optimal method to present graphical information from within applications.

ACCESSIBILITY IN THE EMERGING INFORMATION SOCIETY

This section examines how R&D efforts, such as those described earlier, match to the new interaction and communication requirements imposed by the emerging information society. More specifically, this section briefly examines the effectiveness of the traditional approaches to accessibility in providing adaptation solutions to the new range of products and services, as well as their capacity to cope with the rapid acceleration of the technology changes and address recent developments on some new accessibility strategies.

In the context of the information society, citizens are required to carry out some of their daily activities (e.g., for work and leisure) as well as to interact and communicate with each other through the use of network attachable devices, and telematic applications and services. The ability of the citizens to efficiently use the new telecommunication products is, therefore, becoming critical for their successful integration in this evolving environment. This entails not only the physical and sensory capacity of users, but also their experience in using new technologies.

The projects described earlier in this chapter have mainly addressed accessibility of computer-based applications and telematic services to disabled and elderly people through an a posteriori adaptation of interactive software and representation of information. This amounts to a reactive approach, whereby assistive technology experts attempted to react to contemporary technological developments by building accessibility features into interactive applications, as a result of specific user requirements. Such efforts to account for accessibility are, however, mainly governed by intuition and usually follow ad hoc procedures, which need to be proven in practice; often, they may lead to suboptimal solutions with respect to user requirements.

Despite the value and usefulness of the adaptations-oriented approach and the accumulated body of knowledge, the current state of the art clearly neglects aspects of accessibility, which become promptly relevant and important in the context of the emerging information society (Müller, Wharton, McIver, & Kaux, 1997; Stephanidis, Salvendy et al., 1998, 1999). The inadequacy of the reactive approach stems directly from the shortcomings of adaptations. In particular, with rapid technological change, each new wave of technology introduces a new generation of access problems. By the time suitable adaptations have been devised, the technology is substituted by a subsequent generation, thus introducing the same problem all over again. It follows, therefore, that adaptations-oriented solutions not only fail to guarantee adequate interaction quality to the users, but also lag behind at least one technological cluster.

Recently, the concept of user interfaces for all (Stephanidis, 1995) has been proposed, applying the concept of design for all in the domain of human–computer interaction (HCI), as the vehicle to efficiently and effectively address the numerous and diverse accessibility problems, adopting a *proactive* approach. The underlying principle of such approach is to ensure accessibility at design time and to meet the individual needs, abilities, and preferences of the user population at large, including disabled and elderly people.

A first step toward a proactive approach to accessibility was carried out in the already mentioned GUIB and GUIB-II projects. The goal of these efforts was the development of innovative user interface software technology aiming to guarantee access to future computer-based interactive applications by blind users. In particular, these projects conceived, designed, and implemented a User Interface Management System (the HOMER UIMS; Savidis & Stephanidis, 1998) as a tool for the efficient and modular development of user interfaces that are concurrently accessible by both blind and sighted users. As a result, the concept of *dual user interfaces*

(Savidis & Stephanidis, 1995b) has been proposed and defined as an appropriate basis for "integrating" blind and sighted users in the same working environment.

The TIDE-ACCESS[6] project goes one step forward toward the development of new technological solutions for supporting the concept of user interfaces for all. The developed solutions aimed to provide universal accessibility to computer-based applications by facilitating the development of user interfaces automatically adaptable to individual user abilities, skills, requirements, and preferences. More recently, the developed technologies were applied in the context of the ACTS-AVANTI[7] project, which developed a Web browser inherently accessible by sighted, blind, and motor-impaired users. Detailed information on the outcomes of these projects and the developed technology are provided in Part VI of this volume.

DISCUSSION AND CONCLUSIONS

This chapter has aimed to highlight a perspective on accessibility, namely that of disabled and elderly people, which has been traditionally underserved by the mainstream information technology and telecommunications industries. As a result, access to computer-based applications and telematic services to people that deviate from the so-called "average user" had been traditionally granted through the employment of adaptation solutions and assistive technology aids.

In this context, the chapter has briefly outlined recent progress in some of the fields that have been addressed by collaborative R&D projects in the context of the European Commission, and provided the experience gained in developing accessibility solutions based on the *adaptations* approach. Such experiences indicate that although adaptations may be the only viable solution in the short term, they would probably not suffice in the longer term as an adequate and economically feasible approach to accessibility. The latter is also supported by the rapid changes in the information and telecommunications technology industries, which bring to light a new paradigm for communication, interaction, work, and collaboration among the citizens of the emerging information society.

The chapter argues that, in order to accommodate the new situation, there is a compelling need for "inclusive" approaches to the design of user interfaces for the new range of products and services that go beyond addressing the requirements of the average user. The latter was considered until recently as the design target by the mainstream industry, whereas the new perspective addresses equally the distinctive characteristics of *all* users and of the different environments of use.

---

[6]The ACCESS TP1001 (Development Platform for Unified ACCESS to Enabling Environments) project was partially funded by the TIDE Program of the European Commission, and lasted 36 months (January 1, 1994 to December 31, 1996). For the list of consortium partners, see the Acknowledgments section.

[7]The AVANTI AC042 (Adaptable and Adaptive Interaction in Multimedia Telecommunications Applications) project was partially funded by the ACTS Program of the European Commission, and lasted 36 months (September 1, 1995 to August 31, 1998). For the list of consortium partners, see the Acknowledgments section.

The main conclusions are twofold: First, a considerable body of knowledge concerning user characteristics and requirements, which has been progressively accumulated through the work on adaptations, constitutes a fundamental basis for ensuring that requirements of disabled and elderly people are systematically and consistently taken into account during the design of user interfaces. Second, technical solutions provided for alleviating specific access problems (e.g., three-dimensional auditory direct manipulation, scanning) can be exploited by more general, proactive approaches to accessibility and be integrated as "compulsory" design alternatives in the development of mainstream products and services. To this end, the recent developments in the domain of HCI, concerning the development of methodologies and tools for the design and implementation of user interfaces following the principles of design for all and universal access, provide a vehicle for bringing accessibility from the assistive technology field to the mainstream industry.

REFERENCES

ACCESS Consortium. (1996). *Report on the implementation of the demonstrator interpersonal communication aid* (TIDE TP-1001, ACCESS Project, Deliverable 2.6). Athens, Greece: University of Athens.

Akoumianakis, D., & Stephanidis, C. (1997a). Knowledge-based support for user-adapted interaction design. *Expert Systems With Applications, 12*(2), 225–245.

Akoumianakis, D., & Stephanidis, C. (1997b). Supporting user adapted interface design: The USE-IT system. *Interacting With Computers, 9*(1), 73–104.

Bini, A., & Emiliani, P. L. (1997). Information about mobility issues: The ACTS AVANTI project. In *Proceedings of the 4th European Conference for the Advancement of Assistive Technology* (pp. 85–88). Amsterdam: IOS Press.

Bini, A., Ravaglia, R., & Rella, L. (1997). Adapted interactions for multimedia based telecommunications applications. In *Proceedings of 3rd International Conference on Networking Entities* (paper IH6, pp. IH6.1–IH6.4). Italy: CNR.

Crispien, K., Wurz, W., & Weber, G. (1994). Using spatial audio for the enhanced presentation of synthesised speech within screen readers for blind computer users. In W. L. Zagler, G. Busby, & R. R. Wagner (Eds.), *Proceedings of 4th International Conference on Computers for Handicapped Persons* (pp. 25–31). Vienna: Springer Verlag.

Fink, J., Kobsa, A., & Nill, A. (1996). User-oriented adaptivity and adaptability in the AVANTI project. In *Proceedings of Designing for the Web: Empirical Studies* [Online]. Available: http://fit.gmd.de/publications/ebk-publications/Fikonill.pdf

Fink, J., Kobsa, A., & Nill, A. (1997). Adaptable and adaptive information access for all users, including the disabled and the elderly. In A. Jameson, C. Paris, & C. Tasso (Eds.), *Proceedings of the 6th International Conference on User Modeling* (pp. 171–173). Vienna: Springer Verlag.

Gill, J. (1996). *Telecommunications: The missing links for people with disabilities* (COST 219, European Commission, Directorate General XIII, Telecommunications, Information Market and Exploration of Research). Brussels: European Commission.

Hill, R. D. (1986). Supporting concurrency, communication and synchronization in human–computer interaction—the Sassafras UIMS. *ACM Transactions on Graphics, 5*(3), 179–210.

IPSNI Consortium. (1990, June). *WP 2.2: Potential solutions for accessibility of IBC terminals for PSN users* (RACE R1066 IPSNI Deliverable, CEC-DG XIII). Brussels: RACE Central Office.

IPSNI Consortium. (1991a, December). *WP 4: A guide for engineers to people with disabilities and access to telecommunications* (RACE R1066 IPSNI Deliverable, CEC-DG XIII). Brussels: RACE Central Office.

IPSNI Consortium. (1991b, December). *WP 4: IBC terminal functional specifications* (RACE R1066 IPSNI Deliverable, CEC-DG XIII). Brussels: RACE Central Office.

IPSNI-II Consortium. (1992, December). *WP 1: Proposed terminal access solutions* (RACE R2009 IPSNI II Deliverable, CEC-DG XIII). Brussels: RACE Central Office.

IPSNI-II Consortium. (1993, June). *WP 2: Hardware and software specifications of adapted terminal emulators for PSN* (RACE R2009 IPSNI II Deliverable, CEC-DG XIII). Brussels: RACE Central Office.

IPSNI-II Consortium. (1994, December). *WP 3: Identification of special services for PSN users* (RACE R2009 IPSNI-II Deliverable, CEC-DG XIII). Brussels: RACE Central Office.

Kouroupetroglou, G., Viglas, C., Anagnostopoulos, A., Stamatis, C., & Pentaris, F. (1996). A novel software architecture for computer-based interpersonal communication aids. In J. Klaus, E. Auff, W. Kremser, & W. Zagler (Eds.), *Interdisciplinary aspects on computers helping people with special needs, Proceedings of 5th International Conference on Computers Helping People with Special Needs* (pp. 715–720). Wien, Austria: Oldenberg.

Müller, M. J., Wharton, C., McIver, W. J., & Kaux, L. (1997). Toward an HCI research and practice agenda based on human needs and social responsibility. In *Proceedings of the ACM Conference on Human Factors in Computing Systems* (pp. 155–161). New York: ACM Press.

Mynatt, E. D., & Weber, G. (1994). Nonvisual presentation of graphical user interfaces: contrasting two approaches. In *Proceedings of the ACM Conference on Human Factors in Computing Systems* (pp. 166–172). New York: ACM Press.

Papenmeier, F. H. (1999a). *BRAILLEX 2D screen Braille display, The two-dimensional Braille display for personal computers* [Online]. Available: http://www.papenmeier.de/reha/products/2de.htm

Papenmeier, F. H. (1999b). *WinDOTS 2, The ultimate Braille access software for Microsoft® Window™ 3.x and Windows™ 95* [Online]. Available: http://www.papenmeier.de/reha/products/windotse.htm

Petrie, H., Morley, S., McNally, P., O'Neill, A-M., & Majoe, D. (1997). Initial design and evaluation of an interface to hypermedia systems for blind users. In *Proceedings of Hypertext '97* (pp. 48–56). New York: ACM Press.

Savidis, A., Paramythis, A., Akoumianakis, D., & Stephanidis, C. (1997). Designing user-adapted interfaces: The unified design method for transformable interactions. In *Proceedings of the ACM Conference on Designing Interactive Systems: Processes, Methods and Techniques* (pp. 323–334). New York: ACM Press.

Savidis, A., & Stephanidis, C. (1995a). Building non-visual interaction through the development of the rooms metaphor. In *Companion Proceedings of the ACM Conference on Human Factors in Computing Systems* (pp. 244–245). New York: ACM Press.

Savidis, A., & Stephanidis, C. (1995b). Developing dual user interfaces for integrating blind and sighted users: The HOMER UIMS. In *Proceedings of the ACM Conference on Human Factors in Computing Systems* (pp. 106–113). New York: ACM Press.

Savidis, A., & Stephanidis, C. (1997a). Agent classes for managing dialogue control specification complexity: A declarative language framework. In *Proceedings of HCI International '97* (pp. 461–464). Amsterdam: Elsevier Science.

Savidis, A., & Stephanidis, C. (1997b). Unifying and merging toolkits: A multi-purpose toolkit integration engine. In *Proceedings of HCI International '97* (pp. 457–460). Amsterdam: Elsevier Science.

Savidis, A., & Stephanidis, C. (1998). The HOMER UIMS for dual user interface development: Fusing visual and non-visual interactions. *Interacting With Computers, 11,* 173–209.

Savidis, A., Stergiou, A., & Stephanidis, C. (1997). Generic containers for metaphor fusion in non-visual interaction: The HAWK Interface toolkit. In *Proceedings of the Interfaces '97 Conference* (pp. 194–196). EC2 & Development (ISBN 2-910085-21X).

Savidis, A., Vernardos, G., & Stephanidis, C. (1997). Embedding scanning techniques accessible to motor-impaired users in the WINDOWS object library. In *Proceedings of HCI International '97* (pp. 429–432). Amsterdam: Elsevier Science.

Stephanidis, C. (1995). Towards user interfaces for all: Some critical issues. In *Proceedings of HCI International '95, Panel Session "User Interfaces for All—Everybody, Everywhere, and Anytime"* (pp. 137–142). Amsterdam: Elsevier Science.

Stephanidis, C., & Akoumianakis, D. (1997, January). *Report on issues in standardisation* (TIDE TP1001 ACCESS Project, Deliverable D.4.2.3).

Stephanidis, C., & Gogoulou, R. (1995). Enhancing non-visual interaction in a graphical environment through a screen reader configuration system. In *Proceedings of RESNA '95 Conference* (pp. 467–469). Arlington, VA: RESNA Press.

Stephanidis, C., & Mitsopoulos, Y. (1995). INTERACT: An interface builder facilitating access to users with disabilities. In *Proceedings of HCI International '95* (pp. 923–928). Amsterdam: Elsevier Science.

Stephanidis, C., Paramythis, A., Sfyrakis, M., Stergiou, A., Maou, N., Leventis, A., Paparoulis, G., & Karagiannidis, C. (1998). Adaptable and adaptive user interfaces for disabled users in the AVANTI project. In S. Trigila, A. Mullery, M. Campolargo, H. Vanderstraeten, & M. Mampaey (Eds.), *Proceedings of the 5th International Conference on Intelligence in Services and Networks, "Technology for Ubiquitous Telecommunication Services"* (pp. 153–166). Berlin: Springer-Verlag.

Stephanidis, C., & Savidis, A. (1997, January). *Final report on the I-GET tool* (TIDE TP1001 ACCESS Project, Deliverable D.1.4.2).

Stephanidis, C. (Ed.), Salvendy, G., Akoumianakis, D., Arnold, A., Bevan, N., Dardallier, D., Emiliani, P. L., Iakovidis, I., Jenkins, P., Karshmer, A., Korn, P., Marcus, A., Murphy, H., Oppermann, C., Stary, C., Tamura, H., Tscheligi, M., Ueda, H., Weber, G., & Ziegler, J. (1999). Toward an information society for all: HCI challenges and R&D recommendations. *International Journal of Human–Computer Interaction, 11*(1), 1–28.

Stephanidis, C. (Ed.), Salvendy, G., Akoumianakis, D., Bevan, N., Brewer, J., Emiliani, P. L., Galetsas, A., Haataja, S., Iakovidis, I., Jacko, J., Jenkins, P., Karshmer, A., Korn, P., Marcus, A., Murphy, H., Stary, C., Vanderheiden, G., Weber, G., & Ziegler, J. (1998). Toward an information society for all: An international R&D agenda. *International Journal of Human–Computer Interaction, 10*(2), 107–134.

Vanderheiden, G. C. (1990). Thirty-something (million): Should they be exceptions? [On-line]. *Human Factors, 32,* 383–396. Available: http://www.trace.wisc.edu/text/univdesn/30_some/30_some.html

Weber, G., Kochanek, D., Stephanidis, C., & Gogoulou, R. (1994, November). *Report on the XWindows demonstrator* (TIDE-GUIB II Deliverable No 15.2, GUIB-II Consortium). Germany: Universtiy of Stuttgart.

Weber, G., Kochanek, D., Stephanidis, C., & Homatas, G. (1993). Access by blind people to interaction objects in MS Windows. In *Proceedings of the 2nd European Conference on the Advancement of Rehabilitation Technology* (Section 2.2). Välingby: The Swedish Handicap Institute.

World Health Organization. (1989). *Monitoring of the strategy for health for all by the year 2000: Part 1. The situation in the European region.* Copenhagen: Regional Office for Europe.

7 Everyone Interfaces

Gregg C. Vanderheiden
Shawn Lawton Henry

"Everyone interfaces" allow people with disabilities, those with low or no technological skills or inclinations, and those with literacy and other language barriers, to effectively access and use these systems. This chapter discusses the possibility and practicality of "Everyone interfaces", and provides examples which demonstrate how a single system can be made to work (at different times) in hands-free, vision-free, or hearing-free mode. Furthermore, the chapter introduces strategies and techniques that facilitate building "Everyone Interfaces".

TECHNOLOGIES BECOMING UBIQUITOUS

We are rapidly incorporating emerging technologies into our daily lives, much in the way that electricity is currently incorporated. At one time, electricity was available at a couple of points in only some homes, and electrical appliances were special devices. Today, electricity is inherent in almost every device we use, from our computers, to our telephones, cars, and wristwatches. In the future, information technologies and computer systems will be similarly integrated. We will not use specific isolated appliances for accessing and using information. Rather, these appliances will be integrated into our environments and our lives.

Information access points and appliances will be built into the walls, incorporated into our working environments, carried and even worn by us, and used as an integral part of most of our daily activities. Computer systems will control refrigerators, camcorders, and the temperature, humidity, and lighting in houses. As technology is being incorporated into household products, we progressively find out that the programming ability required to run a VCR is rapidly becoming necessary to perform tasks as basic as setting the living room temperature, or starting the coffeepot. As information technologies are being woven into the fabric of education, business, and daily life, greater attention is being focused on whether most

115

people will be able to access and use these systems. This includes not only people with disabilities, but also people who find that the information technologies are too complicated, or require high literacy or other skills. There is risk that individuals who have worked out strategies for education, work, and daily living may find they are no longer able to function adequately or independently, because of new technologies.

Everyone Interfaces

In order to address the issue of accessibility and ensure the broadest possible access to the next-generation systems and products, there has been increased attention to the development of *everyone interfaces*. Such interfaces allow people with disabilities, those with low or no technology skills or inclinations, and those with literacy and other language barriers, to effectively access and use these systems. As our society ages, the majority of us will encounter disabilities. Thus many who consider people with disabilities to be "them" will find that "them" is "us," and the term *temporarily able-bodied* takes on a slightly different meaning.

Figure 7.1 shows a series of pie charts depicting the percentage of people with functional limitations and severe functional limitations as a function of age. These start at younger people with functional limitations in the 5% range and severe functional limitations in the 1% range, and moving up to age 75+ where approximately 72% have functional limitations and about 45% have severe functional limitations.

RANGES OF USERS AND ENVIRONMENTS

Range of Users

Everyone interfaces must meet the needs of a wide range of users, from novices to power users. The systems of the future will need to be usable and understandable by individuals who, today, avoid technologies or use them only when they must. Technologies will have to be operable by people who have difficulty figuring out household products. They will also need to address the issues of individuals with literacy problems, as well as individuals with physical, sensory, and cognitive disabilities. Those with disabilities account for between 15% and 20% of the overall population, and close to 50% of the population who are elderly.

Everyone interfaces must be perceived, understood, and operated by a person who:

- Cannot see very well, or at all.
- Cannot hear very well, or at all.
- Cannot read very well, or at all.
- Cannot move their head or arms very well, or at all.
- Cannot speak very well, or at all.
- Cannot feel with their fingers very well, or at all.
- Are short, are tall, use a wheelchair, and so on.

Functional Limitation as a Function of Age

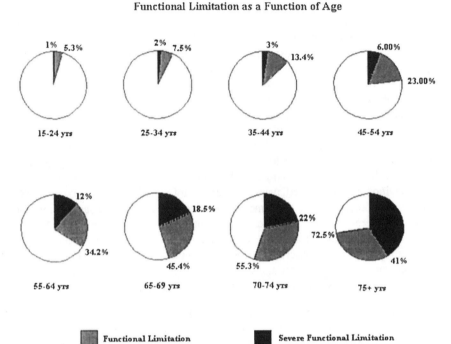

FIG. 7.1.   A series of pie charts depicting the percentage of people with functional limitations and severe functional limitations as a function of age. From Bureau of the Census, Series P-70, 8; SIPP (1984).

- Cannot remember well.
- Have difficulty learning or figuring things out.
- Have little or no technological inclination or ability.
- Have any combination of these difficulties (e.g., are deaf-blind, or have reduced visual, hearing, physical, and cognitive abilities—which occurs in many older individuals).

At the same time, these interfaces need to be efficient for experienced and power users. Interestingly, it turns out that many of the individuals who have disabilities, such as blindness, are some of the best power users as well, as long as the interfaces stay within their sensory capabilities.

Range of Environments and Mobility

The information appliances of tomorrow, which might be referred to as tele/trans/info/comm (telecommunications/transaction/information/communication) devices, will be operating in a wide range of environments. Miniaturization, advances in wireless communication, increasing power, and thin-client architectures are all quickly breaking our need to be tied to a workstation, or to carry a large device with us, if we want to have access to our computing, communication, and information services and functions.

As a result, we will need interfaces that can be used not only when sitting at a desk, but also while driving a car, sitting in an easy chair, sitting in a library, participating in a meeting, walking down the street, sitting on the beach, walking through a noisy shopping mall, taking a shower, or relaxing in a bathtub. The interfaces will also need to be usable in hostile environments; when camping or hiking, in factories, or in shopping malls during holiday time.

In addition, many of us will need to access our information appliances in very different environments within the same day—and perhaps within the same communication or interaction activity. Users will want to be able to transition from one environment to another, from one device to another (e.g., workstation to hand-held), and from one mode to another (visual to voice), in the midst of a task. These different environments will put constraints on the type of physical and sensory input and output techniques that will be appropriate in each case. For example:

- It is difficult to use a keyboard when walking.
- It is difficult and dangerous to use visual displays when driving a car.
- Speech input and output, which work great in a car, may not be usable in a shared environment, in a noisy mall, in the midst of a meeting, or while in the library.

Systems designed to work across these environments will, therefore, need to have flexible interface options in order to work in the different environments, even for a single individual, in a single day, and for a single task.

Interaction techniques, however, must operate essentially the same from a conceptual point of view, even though they may be quite different (such as visual vs. aural). Users will not want to master completely different interface paradigms in order to operate their devices in different environments. Continuity will be necessary in the metaphor and the "look and feel" of interfaces, even though the devices may be operating entirely visually at one point (e.g., in a meeting), or entirely aurally at another (while driving a car).

Juxtaposition Between User Characteristics and Environment Characteristics

It is interesting to note that the issues involved in providing access to people with disabilities will, in large part, be addressed if we simply address the issues raised by the range of environments discussed earlier. For example:

- When we create interfaces that will work well in noisy environments such as prop airplanes, construction sites, or shopping malls at holiday time, or for people who have to be listening to something else while they use their device, we will have created interfaces that work well for people who cannot hear well or at all.
- When we create interfaces that will work well for people who are driving a car or doing something else where it is not safe to look at the device, we will have created interfaces that can be used by people who cannot see.
- When we develop very small pocket and wearable devices where it is hard to use a full-size keyboard or even a large number of keys, we will have developed techniques that can be used by individuals with some types of physical disabilities.
- When we create interfaces that can be used by people doing something that occupies their hands, we will have systems that can be used by people who do not have use of their hands.
- When we create interfaces for individuals who are very tired, under a lot of stress, under the influence of drugs (legal or illegal), or in the midst of a traumatic event or emergency, and have little ability to concentrate or deal with complexity, we will have interfaces that can be used by people who have naturally reduced abilities to concentrate or deal with complexity.

Thus two seemingly different objectives—designing systems that work in a wide range of environments and designing systems for a wide range of user characteristics—have similar solutions. If we design a system that is truly universal and mobile, we will have created a system that is accessible to almost anyone with a physical or sensory disability. We will also have gone a long way toward creating a system that is usable by a large percentage of the population who currently find systems aversive or difficult to learn. Thus, although there may be a few specific residual disability access issues that need to be covered, the bulk of the issues are addressed automatically through the process of developing interfaces that can be used in a wide range of environments and situations.

SOLUTIONS

Possibility and Practicality of Everyone Interfaces

No Single Interface Technique Will Work.   Creating an everyone interface sounds wonderful, but it can sound unobtainable. Trying to design to a single least common denominator interface clearly does not work. If we use only those abilities or input techniques that everyone has and can use in any environment, we would have to rule out all visual, auditory, and tactile interfaces.

Even the Most Flexible Systems Will Be Inaccessible to Some.   In addition, no matter how flexible the interface is, there will always be someone with a combination of two or three severe disabilities that, in combination, will render the interface unusable. There are also applications such as telepresence (e.g., a cultural tour of the museums and orchestras of Europe) that cannot be made fully accessible to people who are blind or deaf. Some aspects can be made accessible, and the whole application can be made partially accessible to both user groups, but neither group would be able to have full access to all the information presented, because of the very nature of the information.

However, It Is Possible to Design Most Products So That They Are Usable by Most People.   It is possible to create products that are: (a) directly usable by the vast majority of people with and without disabilities, and (b) usable by even more people, through the use of assistive technologies.

Three examples, namely a touchscreen kiosk, a quick-time movie, and a reference design for a cell phone, demonstrate how a single system can be made to work (at different times) in hands-free, vision-free, or hearing-free mode.

Example 1: A Touchscreen Kiosk

The first example is a touchscreen kiosk interface that is currently used in over 30 locations, including the Mall of America and the international airport in Minneapolis, Minnesota. The techniques employed in rendering the kiosk accessible are also being incorporated into some voting systems in the United States. The touchscreen kiosk interface includes a set of features developed at the University of Wisconsin, called the EZ Access package. The EZ Access features add flexibility to the user interface for those who would ordinarily have difficulty using, or would be unable to use, a touchscreen kiosk. This flexibility is added without changing the way that the kiosk looks or behaves to users without disabilities. With the EZ Access features in place, the kiosk can now be used by individuals:

- Who have difficulty reading.
- Who cannot read at all.
- Who have low vision.
- Who are completely blind.
- Who are hard of hearing.
- Who are deaf.
- Who cannot speak.
- Who have physical disabilities.
- Who are completely paralyzed.
- Who are deaf-blind.

Moreover, the techniques can be implemented on a modern multimedia kiosk by adding only a single switch and incorporating the EZ Access features into the standard interface software for the computer. Once the EZ Access features are built into a company's standard user interface software for kiosks, implementing the techniques on subsequent kiosk designs is simple and straightforward, and adds less than 1% to the product's cost. The kiosk demonstrates the feasibility of very flexible interfaces implemented on public commercial information systems (see Fig. 7.2).

FIG. 7.2.   Curtis Chong, President of the Computer Science Division of the National Federation of the Blind, using Mall of America, Knight-Ridder Newspaper's Jobs kiosk at the Mall of America. Professor Gregg Vanderheiden looks on.

Example 2: Modality-Independent QuickTime Movies

QuickTime movies on the Web are being captioned and described in order to make them accessible to, and viewable by, people who cannot listen to them (because they are deaf, because they cannot turn up the volume in the environment they are in, or because the environment is too noisy), as well as people who cannot see them (because they are blind, or because their vision is otherwise occupied). These movies take advantage of QuickTime's ability to have multiple time-synchronized audio and text tracks. What would be thought of as closed captions on a television show are stored in a text track as a part of the QuickTime data structure. Users who cannot hear or listen to the sound track can turn on the text track and have the "captions" of the audio track displayed immediately below the QuickTime movie as it plays. Similarly, an alternate audio track can be pulled up that adds a verbal description of what is visually happening on screen, so that someone who cannot see the image can "view" the QuickTime movie (see Fig. 7.3).

FIG. 7.3.   Screen shot from a captioned and described Quicktime movie prepared by the CPB/WGBH National Center on Accessible Media in Boston. For actual samples, go to: http://www.wgbh.org/ncam

It is also possible for a user to use the QuickTime built-in search command to search for any occurrence of a particular word in the movie and jump to that instant in the movie. These movies can also be searched by intelligent software agents that can pull a movie or clips out of a movie in response to a user's request.

When full-length movies are prepared in this way, they will be accessible in audiovisual mode (standard viewing format), as well as viewable in audio-only or video-only format. This will allow them to be "viewed" in a wide variety of fashions. A person viewing a movie in standard format could (if they have to get up) switch to audio-only mode while they give Jimmy a drink of water, or go pick up milk at the store. They can also switch to all video mode with the sound turned off if their spouse decides to go to sleep while they finish watching the movie, the in-laws call in the middle of the game, or the vacuum cleaner wipes out the audio. Newer technologies, which have emerged, such as SMIL (Synchronized Multimedia Integration Language) and SAMI (Synchronized Accessible Media Interchange) will facilitate the addition of synchronized, alternative and overlay audio and video tracks and features, in ways not available previously (such as allowing the author to automatically pause the action, when captions are showing, to allow the viewer to read captions that accompany critical visual information).

Example 3: Reference Design for Cell Phone

The third example is a reference design rather than an existing product. However, it is useful in demonstrating cross-disability access on a small portable device, as well as the usefulness of the access features to all disabilities. The phones are described in somewhat more detail to provide an example of exactly how this can be accomplished across disabilities on a single product.

Figure 7.4 shows sketches of two phones. One is a standard-size cell phone and another is a miniature phone. In both cases the phones are designed to be cross-disability accessible and can be used by individuals with low vision, blindness, hearing impairments, deafness, and physical and cognitive disabilities. The same features also make the phones useable by individuals, who do not have disabilities, while they are driving, while they are sitting in a meeting, when they have their gloves on, and when they are distracted, as well as in other times or circumstances where they are operating with constraints on their physical or sensory abilities. Some key features and capabilities of the phone are reviewed here.[1]

For individuals who are blind, the phone has a number of tactile and auditory features that provide full access and use of the phone. A plastic ridge is provided just above and below the standard 3×4 button-dialing block, which is found on all touch-tone telephones. The presence of these two ridges along with a tactically distinctive 5 key makes it very easy for individuals who are blind to locate any of the standard dialing keys, as well as to separate them from the function keys. All of

---

[1]A more complete description of these reference designs can be found at: http://trace.wisc.edu/docs/phones/

the keys on the keyboard are adjacent to a tactile landmark and most are adjacent to two or three (including the two ridges, the sides, and the 5 key). On the larger phone, the menu scrolling keys are also tactually distinct and provide tactile shape cues as to their function. These features not only allow better use by individuals who are blind, but also facilitate use by all users who are dialing the phone while driving or when their eyes are otherwise occupied.

To provide nonvisual access to the display and other keys, the phone incorporates the EZ Access Quick-Help feature (as well as other EZ Access features). Pressing and releasing the green diamond-shaped key located on the top of the large phone (and the side or top of the small phone) causes the contents of the display and any other visual status indicators to be read to the individual. Holding the EZ button down while pressing any of the other keys causes them to be named (but not activated). Pressing them again causes help text or description for the keys to be read out. This provides access by individuals who are blind, people who cannot read, and people with cognitive disabilities, as well as providing help functions for each of the keys. This method also allows users who are blind to maneuver and access all of the items in the menus and has the advantage that menu items, which are currently abbreviated (especially on the smaller phone's display) and sometimes hard to decipher (for all users), would be spoken in full.

FIG. 7.4.   A miniature phone and a standard-size cell phone.

For individuals with low vision, the contents of the keys are also presented on the LCD screen in large print. Thus, even the small print on the buttons and keys can be easily seen by those with low vision (or those who do not happen to have their reading glasses with them).

For individuals who are hard of hearing, the phone has an extended volume range. In addition to allowing the phone to be used by individuals who have hearing disabilities, this extended volume output allows the phone to be used in a speakerphone fashion by all users. The phone is T-coil compatible for those who have T-coils in their hearing aids and a stereo phone jack is provided to allow direct audio or electromagnetic coupling to hearing aids, cochlear implants, headsets, and other audio enhancement devices.

For individuals who are deaf, the phone provides a voice-over TTY capability. Information received in baudot is displayed on the screen. Messages can also be typed using the keyboard on the phone; a disambiguation routine is available for those instances where a key sequence could generate more than one word (e.g., 228 could be either "cat" or "bat"). Thus, the phone can be used by individuals who are deaf to receive calls, as well as by anyone to make a call to someone who is deaf and uses a TTY. The phone also allows for voice carryover, so that a person who is deaf but can speak, could speak into the phone and have the return message spelled out on the screen. The message coming back to them could be from an individual using a TTY, from someone who is speaking through a TTY relay service, or, in the future, through an automated voice to TTY or digital text service within the phone network. The phone also supports digital text messaging in non-TTY formats.

The same audio jack that is used to connect headsets can also be used to directly connect this phone to a full-size TTY. This would allow individuals to use their cellular phone in connection with a full-size keyboard for additional speed and convenience. The phone has a built-in vibrator for silent ringing and also displays the line status on the LCD display. These features are useful to individuals who are hard of hearing or deaf. They also allow full use of the phone in a meeting or other situation when the individual can see and manipulate their phone but would not be able to hold it up to their ear or carry on a conversation. Thus, someone with a standard voice telephone would be able to call and carry on a conversation with an individual in a meeting who was using this phone in a completely nonaudio fashion.

For individuals who have physical disabilities, the phone has a Touch-and-Confirm and a Delayed-Confirm mode of operation. Both of these modes allow an individual to operate the phone even if they frequently bump or activate other unwanted keys on the way to each target key. With the Touch-and-Confirm mode the individual tries to hit the desired key. If they hit any wrong keys, these keys are ignored. When they finally hit the desired key they confirm it by pressing the green diamond button.

With the delay activation mode, individuals attempt to hit the desired key. When the desired key is pressed, they simply hold their finger on the key and it is dialed. Any other keys bumped along in the process are ignored. In both modes the current or last-activated key is shown in large print on the display and optionally spoken. These functions are useful for individuals with physical disabilities and also for individuals wearing gloves, in situations where they are being jostled, or in any other situation where it is difficult for people to accurately hit individual keys on the phone. The phone also has an optional mounting bracket that allows for custom mounting to car dashboards, wheelchairs, desks, or other locations.

For individuals with cognitive disabilities, the phone has a number of features, some of which have been described earlier. The voice output feature, the help feature, and the ability to have all the menus read aloud were all discussed previously. In addition, the phone has the ability to be programmed to operate in a number of "restricted keypad" modes. In one mode, pressing any key on the large phone (or opening the small phone) will cause it to automatically dial a given number. This number could be a person's home, or it can be a service that is designed specifically to provide assistance to individuals with cognitive disabilities or reduced cognitive abilities due to aging. Simply pressing any button on the phone (or opening the small phone) would put them in instant contact with someone who would (via caller-ID) be able to identify them and have a file called up on their computer with full information about the individual so that they could be of assistance.

Other modes of operation that can be invoked include one that simply provides a series of names that the individual can step through. The names are displayed on the screen and read aloud when the desired name is reached (heard). The individual simply presses the appropriate button and that number is dialed for them. These features are useful for individuals with cognitive disabilities. They can also be locked into a phone provided by a parent to their child or to others (e.g., babysitters, etc.) so that the latter are able to use the phone only with a restricted series of numbers. A side benefit is that the phone is very easy to use in this mode, so no instruction is needed for its operation. Another benefit of this phone is that it be used even in an emergency situation, when individuals are confused or have difficulty in operating an unfamiliar piece of equipment while under high stress.

The cost for adding all of the aforementioned features is quite low as most of them can be implemented directly through modification of the software in the phones. The most significant change would be an enhancement of the signal-processing circuitry to allow it to do the speech output and the baudot decoding. With the signal-processing circuitry already in the phones, however, this would not be substantial. And the rapidly dropping cost and size of memory would allow for the room needed for the extra software and canned speech for the function announcements.

STRATEGIES AND TECHNIQUES

Future Interfaces

In order to be able to design everyone interfaces, we need to develop interface strategies and techniques that facilitate building interfaces that are: straightforward and easy to learn, flexible and adaptable, and modality independent.

Straightforward and Easy to Learn.   Interfaces must be simple, straightforward, and easy to learn, so that as much of the population as possible is able to use them, and so that all users can master new functions and capabilities easily.

Flexible and Adaptable.   We will need interfaces that take advantage of fine motor movements and three-dimensional gestures when a user's situation and abilities allow. At the same time, the interfaces should be operable using speech, keyboard, or other input techniques as necessitated by the user's environment, activities, and motor or other constraints.

Modality Independent.   The interfaces will need to allow the user to choose the presentation modalities that are appropriate to their environment, situation, or individual characteristics. Systems will need to let users obtain information visually at some times and auditorially at others; on high-resolution displays when they are available, and on smaller, low-resolution displays, when that is all that is at hand.

Enhanced interface architectures are needed to achieve everyone interfaces. These new interfaces must support multiple modalities. In addition, users will need to seamlessly, coherently, and intuitively switch between modalities. They will need to switch between input and control techniques as well as display formats, all based on compatibility with changing environments.

Simple Set of Alternate Selection Techniques

Although people have quite different abilities and, therefore, different interaction needs, it is possible to create a simple set of alternate selection techniques that can accommodate a very wide range of physical and sensory abilities, and environmental situations.

A suggested selection of operating modes might be:

- Standard mode: the way the device should most effectively behave for individuals who have no restrictions on their abilities, due to task, environment, or disability.
- A list mode: where the user can call up a list of all information and action items, and use this list to select items for presentation or action. This mode should not require vision to operate. It could be operated using an analog transducer to allow the individual to move up and down within the list, or using a keyboard or arrow keys, combined with a confirm button. This mode can be used by individuals who are unable to see or look at a device.
- External list mode: that makes the aforementioned list available externally through a hardware or software port and accepts selections through the same port. This mode can be used by individuals who are unable to see and hear the display, and, therefore, must access it from an external auxiliary interface. This would include artificial intelligent agents, which are unable to process visual or auditory information that is not also available in text form.

- Select and confirm mode: that allows individuals to obtain information about items without activating them (i.e., a separate confirm action is used to activate items after they are selected or, alternatively, items can be activated by holding down a button for a longer period of time). This mode can be used by individuals with reading difficulties, low vision, or physical movement problems, as well as by individuals in unstable environments or whose movements are awkward due to heavy clothing or other factors.
- Auto-step scanning mode: that presents the individual items in groups or sequentially for the user to select. This mode can be used by individuals with severe movement limitations, or movement and visual constraints, when direct selection techniques are not usable.
- Direct text control techniques: including keyboard or speech input.

The purpose of a text-based auxiliary interface port is to allow external hardware or software to query the system and receive a list of information and action objects available, as well as to make a selection from among the available actions. This port would be used in conjunction with the "external list mode" described previously. The text-based auxiliary interface port can be either a software connection point or a hardware connection point (e.g., an infrared port).

The port might, for example, be used to connect an external dynamic Braille display for viewing and controlling devices such as a kiosk, PDA, or a tele/trans/info/comm device. This port can also allow intelligent agents or other devices to have (text-based) access to the information and functions in the system.

Uni-List-Based Architecture

By maintaining an updated list of all information and actions or commands currently available to the user, it is relatively easy to provide great flexibility in the techniques that can be used to operate the device or system.

For example, in a three-dimensional (3-D) virtual shopping mall, a database is used to generate the image seen by the user, and to react to user movements or choices of objects in the view. The database could be constructed to provide a list of all the objects in view, as well as information about any interactive objects presented to the user at any point in time. By including verbal (e.g., textual) information about the various objects and items, it is possible for individuals to navigate and use this 3-D virtual shopping system in a wide variety of ways, including purely verbal ones. In addition:

- Individuals who are unable to see the screen (because they are driving their car, because their eyes are otherwise occupied, or because they are blind) can have the information and choices presented verbally (or via Braille). They can then select items from the list in order to act on them, in much the same way that an individual may reach down and pick up or "click on" the object in the environment.

- Individuals with movement disabilities can have a highlighter or sprite step around to the objects, or they could indicate the approximate location and have the items in that location highlighted individually to select the desired item. Other methods for disambiguating the user input could also be used.
- Individuals who are unable to read can touch, or select any printed text presented and have it read aloud to them.
- Individuals with low vision (or who do not have their glasses with them) can use the system in the same way as a fully sighted individual. When they are unable to see well enough to identify the objects, they can switch into a mode that lets them touch the objects (without activating them) and have them named or described.
- Individuals who are deaf-blind could use the device in the same fashion as an individual who is blind. Instead of having the information spoken, however, it could be sent to the individual's dynamic Braille display.

Modality-Independent or Modality-Parallel Source Material

Key to achieving everyone interfaces is the provision of all basic information in either a *modality-independent* or a *modality-parallel* (flex-modal) form.

Modality-independent formats store information in a form that is not tied to any particular mode of presentation. An example of a modality-independent format is ASCII text, which is not inherently visual, auditory, or tactile. ASCII text is easily presented visually on a visual display or a printer. It is easily presented auditorily through a voice synthesizer. And it is easily presented tactually through a dynamic Braille display or a Braille printer.

Some types of information, however, are inherently difficult to present in a modality-independent form. For example, a photograph is by nature easiest to present in visual form. Music is easiest to present in auditory form. The types of information that cannot be effectively presented in a modality-independent format can be handled by providing modality-parallel formats.

Modality-parallel formats store information in packages with the information formatted for multiple modalities. An example of information provided in modality-parallel formats is audio speech, accompanied by a transcript of the speech in electronic text format. Some packages would include more formats than others. For example, a movie might include:

- A video track.
- An audio track.
- A description of the audio track in electronic text format ("captions").
- A description of the video track in audio format.
- A description of the video track in electronic text format.

As noted previously this would allow access to all the information by users who are deaf or hard of hearing, or blind or with low vision, as well as allowing access by search engines, intelligent agents, and movie players with "jump to scene," "jump to phrase," or "jump to event" features.

If information is not accessible as it is received, it is generally difficult or impossible to make it accessible to the user at presentation time. For example, if available and reserved airplane seats are shown on a picture only through color-coding, it is very difficult to make this information available to someone who is blind. However, if the information is also available in text, it will be accessible not only to people who are blind or have low vision, but also to artificial agent software.

If multiple modalities are available, users can request that information be presented visually, auditorially, or tactually according to their needs, preferences, and environmental situation.

Roles From Source to Delivery

Each system component plays a role in supporting the multiple modality approach to everyone interfaces, including those handling information production, storage, serving, transportation, translation, delivery, presentation, control, and interaction.

Production.   Information production technologies should allow for the creation of modality-parallel packages. Sophisticated production technologies would both facilitate and encourage the creation of usable alternative modality formats. Examples of usable modality-parallel packages include:

- Audio tracks with text description. Preferably, the text description would be available both separate from and time-synchronized with the audio track.
- Graphic image files with a text description of the important information within the graphic. Preferably, the text would include both a functional and an aesthetic description.
- Video or movie files with time-synchronized text translation or description of the audio component, as well as time-synchronized text description of the video content. Preferably, the description of video content would be provided in both auditory and text format, and would, where feasible, include both functional and aesthetic description.

Storage.   Technologies and formats providing a storage function should support the linked storage of alternate modality formats in modality-parallel packages.

Technologies and formats providing a transport function should transport modality-parallel packages without distorting or removing any components of the package. Transport technologies and formats should support user-directed routing of specified formats to translators, so that they automatically show up in alternate formats, or in a combination package that contains both the original format and the translated format of the information.

An example of technology not supporting modality-parallel package transportation was a data compression technique for movies, which inadvertently stripped out all captioning. Movies arrived at the destination point in excellent condition, except that captions had been lost in the pipeline.

A new and very powerful development is inline transmission services, which provides the ability to have E-mail turned into voice mail, voice mail into E-mail, faxes into E-mail or voice mail, and the like. The use of these services will increase dramatically as optical character recognition, page layout recognition, and voice recognition technologies improve.

Serving.   Technologies providing a serving function should support the serving of information either in *flex-modal* form, or in *select-modal* form:

- In flex-modal form, the server sends the entire modality-parallel package. After the information arrives, the user chooses the presentation modality or modalities. An example of flex-modal serving would be a closed captioned, closed descriptive movie with a full script text track. All modalities are sent, and users at the destination choose whether to view the movie with, or without the captions. Someone who can both see and hear may want to watch the movie in traditional format. If they have to watch it in a very noisy room, or if they have hearing problems, they could turn the captions on. Individuals driving a car, or whose vision is otherwise occupied, might want to turn on the video description component, as might an individual who is blind.
- With select-modal form, users specify the modality in which they want the information served. An example of select-modal form would be a server that sends a graphic, a text, or an audible version of the information on request. Only the selected modality or modalities are served, rather than the entire package.

Select-modal serving has many benefits; for example, users may want to save time by having only selected portions of material downloaded. A reporter researching presidential speeches may want to take the time to download only the transcripts, or the caption tracks from audio or movie files of the presidents' speeches. These can be downloaded very quickly and searched for keywords or phrases. The reporter might then download only the specific, much larger, auditory or movie files that are of interest. Once the files are downloaded, the reporter could again use the caption tracks in the movies to quickly jump to the particular portions of interest within the speeches. Individuals with disabilities may also opt to download only those components they will use. For example, a person who is blind may download only the audio tracks and save themselves time and money if there is no one with them who will view the video track. Individuals who are deaf may wish to download the video track, the caption track, and/or a second video track that shows a sign language interpreter signing the material of the first video track.

Select-modal serving also applies to information that is generated on demand, rather than retrieved from storage. For example, the airline reservation system does not store available and reserved seats in picture format. Rather, at the time of inquiry, it generates the airplane picture with the available and reserved seats indicated. With select-modal serving, individuals inquiring about seating could request either the visual presentation, or a nonvisual presentation of the information, as best meets their current needs and circumstances. An example where this would be useful is an individual who is driving to the airport and, having missed one flight, is frantically trying to find a seat on the next plane. This person would want flight availability information presented nonvisually and quickly.

Presentation and Control.   Technologies that present information to the user for any purpose (directly or indirectly) should be able to present the information in multiple presentation modalities. Information should be presented so that it can be perceived and controlled by individuals with any type of disability (including visual, hearing, physical, and cognitive disabilities of all degrees and combinations).

Technologies that present information, sometimes referred to as viewers, include computer-based programs, such as Internet Explorer, Netscape, and AOL browsers, as well as kiosks, television set-top boxes, hand-held personal data assistants (PDA), and next-generation home information appliances. Also included are touchtone phones and other audio-based systems that might be used by someone who is driving a car to access diverse information services. These viewers should support select-modal presentation of information. For example, they should provide a means for displaying captions and playing alternate audio tracks.

CONCLUSIONS

The interface strategies and techniques presented herein to facilitate everyone interfaces have additional benefits. For example, ASCII text, a prime modality-independent form, is easily searchable and easily handled by most technologies. The provision of select-modal information serving would benefit Internet usage by decreasing the time needed to download the information, minimizing unnecessary packet traffic on the Internet.

Using modality-independent data storage and serving as discussed previously has a number of advantages besides supporting future mobile systems and providing disability access. Because it allows access via a number of modalities, it also allows information to be made available through a variety of channels. The same information or service can be accessed via a graphic Web browser or a telephone. Different resolution displays can be easily supported. Even very small, low-resolution displays can be used. In fact, the problems posed by small low-resolution displays resemble low-vision issues. Low-bandwidth systems can also take advantage of the text-only access that would be available in such a system. Those with higher bandwidth would not have to be limited to this format but could take advantage of the full graphic interface that their displays and bandwidth would support. As a result, information and service providers could use a common information or service server to handle inquiries from a wide variety of people and agents, using devices with a wide range of speed and display technologies. Also, as technologies evolve, the same serving structure could be used across different technologies.

As described earlier, designing systems that work in a wide range of environments overlaps significantly with designing systems for a wide range of user abilities. A similar parallelism of design objectives and solutions also exists between issues of facilitating access by users with various abilities and access by intelligent agents. Intelligent agent software faces similar barriers accessing audiovisual information as an individual who is blind or deaf. Providing text descriptions of graphic and auditory information, which provides access through readers such as Braille displays, also allows that information to be indexed and found using standard text-based search engines. For example, providing alternative text ("ALT" tags) for images on Web pages allows intelligent agents to search and index the information provided in the graphic. Text descriptions also allow access to the graphical information by people who are blind and access Web pages through nonvisual browsers.

Approaching Everyone Interfaces

There are many examples where interfaces that were previously thought to be unusable by individuals with a particular disability were later made easily accessible. The difference was simply the presence or absence of an idea. The challenge, therefore, is to discover and develop approaches that can make next-generation interfaces accessible to, and usable by, greater numbers of individuals and easier for all to use. The strategies and techniques presented herein help us to design everyone interfaces to next-generation technologies.

Everyone interfaces address the equity issues of providing access to people with a wide range of abilities and inclinations, as well as access in a wide range of environments and situations. They provide flexibility in the types of physical and sensory input and output techniques used, and allow access by people with different data requirements, operating under conditions in which particular data types are not useful.

Because of the parallelism between the objectives and solutions in designing disability-accessible interfaces and environment-independent interfaces, one of the best ways to explore environment-independent, mobile interface strategies may be the exploration of past and developing strategies for providing cross-disability access to computers, information systems, and other technologies. Though many of the aforementioned examples are Internet or media based, the same basic strategies can be and are being applied to everything from computers to coffeepots, from PDAs to microwave ovens, from vending machines to watches. All these devices, and those that we encounter in our environments, can and should be designed to work better for all of us, now and as we age.[2]

---

[2]For more information on designing a more usable world (including a design tool and references to work internationally), see: http://trace.wisc.edu/world

Design

8 Theory and Practice From Cognitive Science

Michael Wilson

This chapter discusses the cognitive science approach to Human-Computer Interaction, and in particular how some cognitive science models can be used to facilitate the goal of User Interfaces for All. Two major approaches to modeling Human-Computer Interaction are described, showing the benefits they offer, but also the limitations in their use by system developers. Two tools which overcome these usability problems, at the cost of constraining the cognitive phenomena they can capture, are also described. Finally, recent developments in computing technology and cognitive science are outlined, which could work synergistically to create future computer interfaces that are suitable for the broadest possible end-user population.

Cognitive science is concerned with the understanding of mental life, and the expression of that understanding in the form of theories and models. It develops frameworks and models of human behavior, which can be used to provide insights into human–computer interaction (HCI). Usually, these insights are either adopted into requirements capture or evaluation methods, or lead to user interface design ideas that require engineering into future computing interfaces (such as those discussed in chap. 4, this volume). However, cognitive science approaches are difficult to directly embed into tools, which are usable by general designers, because background knowledge is required to select the appropriate models and to interpret results.

The models produced by cognitive science are designed to capture the generalities of a population and are poor at addressing individual differences, or small groups of specialized users, because they require statistically valid data to be collected in order to justify them. However, the general architectures provided by cognitive science have given rise to much of the work on individual differences (see chap. 2) or user modeling (see chap. 14) discussed elsewhere in this volume.

This chapter does not review the whole canon of work in cognitive science, nor even that subset related to HCI (for a general review of the contribution of cognitive science and human information-processing psychology to HCI see, e.g., Barnard, 1995). The next section discusses the cognitive science approach in order to lay a foundation for the later descriptions of how some of its models can be used within the *user interfaces for all* enterprise. The third section describes two major approaches to modeling HCI (TAL and Interacting Cognitive Subsystems [ICS]), showing the benefits they offer, but also the limitations in their usability. The fourth section outlines two tools that overcome these problems of usability for system developers, but at the cost of constraining the phenomena they can address (GOMS and KADS). Finally, recent developments in computing technology and cognitive science are outlined, which could work synergistically to create future computer interfaces capable of being used by the broadest possible end-user population, following the user interfaces for all principles.

THE COGNITIVE SCIENCE APPROACH

Methodology in Cognitive Science

The big questions about the structure of matter and the universe were moved beyond the arena of ecclesiastical and philosophical debate to be the subject of logical reasoning, supported by empirical results, in 1688, when Isaac Newton published his *Principia*. The big questions about the functioning of the human mind were not addressed by a similarly rigorous scientific method until 1879, when Wilhelm Wundt established the first laboratory of psychology at Leipzig.

Wundt reacted against the introspective method of the time so extremely that he performed thousands of trials on users in each experiment, to ensure that his empirical results were replicable (Miller, 1966). Because of this rigor, he could address only local episodes of behavior, which were amenable to such repeated measurement in detailed experiments, whereas more general phenomena remained the province of the introspectionists.

The tension between the desire to answer broad important questions about mental life, and the need to use arguments that draw upon methodologies that are sound, is a dialectic that drives the progress of psychology. Indeed, this conflict has been present throughout the history of ideas, and has led to swings between rationalism and romanticism. However, this chapter explores only the recent developments in cognitive science, and the application of the results to HCI.

The paradigm that dominated psychology through the first half of this century was *behaviorism,* which accepted that humans learn and act in the world as a result of stimuli that they encounter, and that a linking of those actions to the stimuli is all that can be validly argued. Behaviorism maintained a strong defense of scientific methods in psychology against introspectionism. However, by the 1960s, behaviorism was felt to have been too restrictive in the range of questions that it could address, and cognitive or information-processing psychology had taken over as the dominant paradigm. Cognitive psychology permitted arguments to postulate mental processes and intermediate representations between stimuli and actions, on which those processes could operate. It also introduced methodologies to interpret the performance of humans in the execution of various tasks, with respect to the time taken and the number of errors made, in terms of these postulated processes and representations.

The methodology, initially accepted as valid by cognitive psychology, was restrictive, as illustrated by the interpretation, in terms of processes and representations that could be proposed as a result of the time that people take to make decisions in reaction to stimuli (Sternberg, 1969). This logic proposes that variables whose effects are additive affect different processing stages, but two variables whose effects are interactive affect the same processing stage. For example, variables can be assigned a role in perceptual processing when they interact with marker variables whose effects can be assumed to be in perceptual processes, but not with variables whose effects are on response selection or action execution. Analyses based on the distributions of reaction times have been used to develop models of the stages of human information processing as boxes linked by arrows.

Although not as much as behavioral psychology, this methodology is still limited to interpreting overall task performance and does not permit observations of the intermediate stages of processes. Methodological developments in cognitive psychology were, therefore, directed at gaining information as to the intermediate states of processes, by looking at secondary actions performed during the performance of the primary task, which would be effected by it. The most extreme methodology acceptable in cognitive psychology permits retrospective answers to questions about intermediate cognitive states during previous task performance as data suitable for validating theory. Various procedures for collecting verbal protocols, while tasks are being performed, have been suggested (e.g., Ericsson & Simon, 1984), which are accepted by the discipline with intermediate weight as validating evidence. However, their production may interfere with the performance of the primary task, and they may be rationalizations of behavior rather than accounts of intermediate processing.

This methodology is extreme for cognitive psychology in its closeness to introspection, but is clearly rigorous compared to the methodology of the deconstruction of cultural artifacts accepted in many humanities, where texts, paintings, films, and so forth, can be deconstructed in terms of any of the major cultural distinctions (class conflict, feminism, racism, environmentalism, self/society or ego/superego, dominance/submission, etc.). The extreme reaction against these overly free analytic methods in social science and the humanities has been to the equally extreme *ethnomethodology,* where talk, or artifacts may merely be collected and presented. Any form of analysis or even juxtaposition is prohibited because it merely adds the interpretation of the analyst to those of the creator and reader/viewer, which amounts to creating the original artifact, as well as the analysis underpinning it. To balance protocol analysis as rigorous deconstruction, a rigorous form of ethnomethodology, termed *situated action,* has been introduced into cognitive science (e.g., Norman, 1993). In the case of HCI, Suchman (1987) and Winograd and Flores (1987) argued that to develop better interfaces, we must focus on how people use them, rather than on how people think, or what computers can do. Although its origins are not within cognitive psychology, this approach has many similarities to the perception-centered proposals of Gibson (1979). Such a proposal argues that cognitive psychology should consider how humans perceive *affordances* in objects (e.g., a dial, a button) to act on them in particular ways (e.g., turn the dial, push the button), and that more direct links can be made between perception and action, than can be facilitated by a single centralized representation. It is possible to view this as a cautious step back, toward the constraints behaviorism imposed, in order to curb the freedom to theorize about intermediate representations.

When the postulation of mental processes and representations was first accepted, theory was required to be strictly tied to experimental data, and theories that were not strongly predictive of experimentally testable predictions were given little weight. If a prediction that a theory made was not experimentally supported, then the theory should be rejected, following the strict arguments of Popper (1992). Such strictures were necessary in order to prevent the abuse of the freedom, which was allowed by postulating mental processes and representations. These restrictions gradually mellowed, so that a layering arose in theoretical proposals between models as instances of theories, within general frameworks (e.g., Morton, Barnard, Hammond, & Long, 1979). Models make predictions that are strictly experimentally testable. Theories can give rise to sets of models, so if the predictions of one model failed, other predictions that arose from the same theory could still be held. Integrative frameworks provide a higher level structure, which guides the integration of theories. Because integrative frameworks can combine theories accounting for data from different experimental paradigms, they are not in themselves testable by any single experimental paradigm. Rather, the predictions arising from a set of integrated theories are tested for their ability to achieve a purpose, such as the engineering goals of HCI. However, there are always questions raised about the validity of integrating theories derived from different paradigms into a single framework. The cost of this increasing freedom to theorize is an uncertainty as to methodological soundness, in both testability and the generalizability of results.

The trend toward further softening the methodological demands on theory led to the development of cognitive science from cognitive psychology in the mid-1970s. Answers to broad questions concerning mental life require broad theories encompassing many aspects of that mental life, beyond performance on limited tasks. These are hard to motivate purely within the tasks addressed by experimental psychology. It was necessary to introduce theoretical findings from linguistics into theories, and to build theories whose complexity went far beyond the experimentally testable predictions arising from "box and arrow" models. To build such theories, the modeling techniques used in artificial intelligence were also combined with methods from linguistics and cognitive psychology to form the discipline of cognitive science.

The merge of cognitive psychology and artificial intelligence introduces sound representational and computational modeling techniques to express cognitive psychology theories. Psychology has determined many empirical regularities and small-scope theories to account for them, but it has been argued that to move forward it needs to search for a single unified theory of cognition as an alternative to integrative frameworks (Newell, 1990). Although artificial intelligence may provide the mechanism for this, it also has the potential to permit theorizing that is not sufficiently grounded in evidence for the purpose of psychology as a science.

The merge of cognitive psychology and linguistics can also cause problems in determining a sound methodology in cognitive science, arising from the diverging methods the two disciplines use to generalize results to populations. The core methodology of cognitive psychology is to develop hypotheses that are tested by experiments on small samples of subjects. Results of experiments are subjected to statistical analyses. The power of this methodology in generalizing from the particular experiment to the population relies on statistics to show that differences are significant for the population sampled. The sample can be taken from the population as a whole, or from subpopulations that are shown to differ on a secondary measure, correlated or interacting with the investigated phenomena. Thereby, individual differences among the population can be determined (see chap. 2, this volume). In contrast, the methodology commonly adopted in linguistics is either to identify a phenomenon in the real world by a single example as a proof of existence, by relying on the judgment of a single individual as a native speaker of a language, or to abstract away from the performance of individuals to determine a generalization of a phenomenon to a universe of speakers—often represented as a grammar that captures the general competence of a population in a language, independently from its use in performance. These differing methodologies do not fit well together and lead to conflicts within cognitive science, particularly when considering variations between individuals in language-related capacities and abilities, which are the noise to the linguist's general competence signal, yet can be the signal to the psychologist.

The distinctions drawn in this section between paradigm-specific theories and integrative ones, between integrative theories and frameworks, between evaluation of theory for scientific validity or utility, and the relationships between experimentation on samples and variations in the population, are all drawn on in the description of the results of cognitive science modeling in the section Applicable Cognitive Science Theory. The next section provides a brief mapping from cognitive science to the engineering discipline of HCI.

The HCI Problem in Cognitive Science Terms

The objective of cognitive science is to understand mental life, and to express that understanding in the form of theories and models. In contrast, the objective of HCI comes from an engineering tradition, and is to design artifacts that facilitate and improve users' performance of tasks (e.g., computer use at work), or attainment of emotional states (e.g., computer use as entertainment). In order to perform design, an understanding of the users' performance of tasks is insightful, whereas specific tools that guide or evaluate designs are most immediately useful. To achieve these engineering objectives, it is necessary to (a) identify those results of the cognitive science enterprise that are applicable in principle, (b) transform these into a form that can be directly applied, and (c) hopefully reduce them to methods that can be adopted and executed by designers, without the need of considerable education in the disciplines that gave rise to them.

As described previously, cognitive science promotes the use of explicit representations of knowledge and the reasoning processes of humans. The problem of HCI can be characterized in cognitive science terms as follows: Given that a user wishes to perform a task in some domain of application using a computer system as a tool, and communicate with the latter by establishing some form of dialogue, such a dialogue should be constructed in a way that maximizes effectiveness and efficiency of task performance, and thereby minimizes the complexity of the communication between the user and the computer system. Users hold cognitive representations of the domain, task, computer system, and dialogue, which they reason over to take actions, and perform the task. In order to maximize the efficiency of task performance, we need to distribute processing between the user and system to match the abilities of each—sometimes called *distributed cognition*. In order to simplify the dialogue, we need to ensure that the system side is as compatible with the user's representations as possible, and as internally consistent as possible, so that users can identify the compatibility. In order to design computer systems with these properties, we need to model the domain, task, dialogue, and user, and the system itself. We can use these models in three ways: First, in the development process we can embody these models to establish requirements, guide design decisions, or establish evaluation methods to measure compatibility between the system's and user's representations; second, we can embody these models in documentation and training material to bring the user as close to the system as possible; or third, we can embody these models in the system itself, allowing it to adapt to the user at run time. Of course, the initial scenario is too simple, because users' representations change over time through learning and forgetting. A task is not a disembodied goal, but is linked to higher motivational and emotional states. Users perform the task in an environmental context, where they may need to communicate with other humans who are involved in the performance of the task, through computer systems or by other means. Equally, single users perform many different tasks in different domains, whereas multiple users perform the same task. To address this last point in the spirit of user interfaces for all, it is necessary for the models to capture the variation between users (and within individual users as they change over time). Models should also be applied in any of the three ways mentioned previously, so that task performance can be as effective and efficient as possible for each individual user, rather than a stereotyped or average user.

"For the most part, useful theory is impossible, because the behavior of human–computer systems is chaotic or worse, highly complex, dependent on many unpredictable variables, or just too hard to understand" (Laundauer, 1996). In the sense intended by Laundauer, a single useful theory that can dictate system design characteristics to produce better designs than are possible through human skill and emulation is certainly not available, and probably impossible. However, as with other aspects of cognitive science, a large number of different models have been developed from different experimental paradigms, which address aspects of the solution. Although these models cannot produce better designs, they can often provide insights into the options or trade-offs involved in design decisions, or provide constraints on those design options.

APPLICABLE COGNITIVE SCIENCE THEORY

Three examples of applicable cognitive science theory are described in this section. The first example is paradigm specific, and theoretically weak, but very important in accounting for user variations in HCI, which have consequences for design. The second moves to the theoretically stronger approach of a model (TAL), developed within an integrative theory that captures changes during learning. Finally, an integrative framework (ICS) is described, which can address the widest aspects of variation in HCI from a cognitive perspective, but is consequently underspecified in too many areas to be usable by system developers. This trade-off between the detailed representation necessary to provide predictive power, and the usability of a modeling technique, is then further taken up in the next section, User Interface Development Tools and Methods as Cognitive Science in Practice, which describes two techniques, intended as tools for system developers.

Variation in Users' Domain Language Across the Population

The exhortation, mentioned in the previous section, for design to be compatible with users' previous experience is well founded (Barnard, Hammond, Morton, Long, & Clark, 1981). To achieve such a compatibility in the terminology about the domain to be adopted, it appears straightforward to follow a linguistic competence approach in creating a dictionary of the terms typical of the specific domain, and use this dictionary in the system dialogue. However, considering the high degree of variation in vocabulary usage, compatibility can be impossible to achieve, and trying to produce it only leads to frustration for designers. An elegant series of studies (Furnas, Gomez, Landauer, & Dumais, 1982; Gomez & Lochbaum, 1984; Landauer, 1987) have been undertaken on a wide range of problem domains, including text editing, the contents of yellow pages directories, cooking, and goods "wanted/for sale" services. These studies show that people use a variety of descriptions when referring to the same items. Furthermore, they show that such a variability can be quantified, and that the likelihood of any two people using the same term to refer spontaneously to the same concept ranges from 0.07 to 0.18. Therefore, any designer has a very low chance of choosing any single keyword that more than 20% of the population are likely to use if they can use only one. Statistical simulations based on these data were used to explore the probable success of alternative access schemes, which showed that the probability of a successful match could be raised to 75% to 80% where the system accepts many terms for target information (Furnas, Gomez, Landauer, & Dumais, 1983).

These studies, involving considerable empirical evaluation and statistical modeling, have brought about a reformulation of the design objective from selecting the name most compatible with the expectations of a population of users, to designing an efficient aliasing system, to capture the variation in the population's language use.

Variation in Users' Mental Models During Learning

Users construct mental models of the system as they learn about it. These models become richer during learning, and, therefore, users require different support as learning progresses. In order to develop interfaces suitable for the broadest possible end-user population, at whatever stage of learning about a system users are, we need to understand, and model the learning process, applying the cognitive science method.

Halasz and Moran (1983) conducted an experiment in which half of a group were taught an explicit conceptual model of a calculator's stack, and the other half were not given a model but were taught only how to perform the necessary tasks. No difference was found between the groups as far as the accuracy in solving routine or even complex problems was concerned. However, when it came to solving problems that required some level of invention—that is, where the users had to invent new methods—the group that was taught the model got more problems right. In particular, this group was better able to perform novel tasks that required more cognitively intense problem solving. It was argued that the model helped users to construct a better problem space in which to carry out the problem-solving process necessary for creative solutions.

This example shows how different styles of teaching can produce different mental representations in users, which in turn result in different performance. A study by Wilson, Barnard, and MacLean (1990) concerning users learning to use an office software suite, showed similar distinctions, due, not to the type of training, but to variations between users. In this study, users were trained initially on the core, or habitable subset of functions in the office suite, which all users require to perform basic tasks. They were subsequently trained on more advanced functions. Performance tests on these and further untrained functions were given to users in order to assess their skills and the transfer of learning, along with various off-line measures of their knowledge of the system, including questions based on screen images, and questionnaires to assess verbalizable knowledge (either triggered *affordances* or conversational language). This study investigated a general view of users learning a mapping from a task to the actions required to perform it (the *task-action mapping* after Payne & Green, 1986), which is characterizable as a general model of skill acquisition (after Fitts, 1964). This model divides learning into three phases. In the first, users acquire sufficient fragments of knowledge about a system to support the performance of some tasks. In the second phase, the knowledge recruited to perform tasks is compiled into procedures. In the third phase, users draw on these compiled procedures, and exhibit a level of performance that is considered expertlike. A further distinction concerns the nature of the types of knowledge in the first and third phases. It is assumed (after Anderson,

1983) that the type of knowledge, which accumulates in the first phase, is accessible to the processes of verbalization, whereas the compiled procedures drawn on in the third phase would not be possible to articulate. During these changes in the representation of knowledge, there is a hypothesized parallel reduction in the time necessary to perform the task, and in the number of errors during performance. This change in performance time is called the *power law of practice,* which states that the time to perform a task decreases as a power law function of the number of times the task has been performed. It has recently been argued that this law applies not only to the domain of motor skills, to which it was originally applied (Snoddy, 1926), but to the full range of human tasks (Newell & Rosenbloom, 1981), including perceptual tasks such as target detection (Neisser, Novick, & Lazar, 1963), and purely cognitive tasks such as supplying justifications for geometric proofs (Neves & Anderson, 1981). Consequently, it is often assumed that as experience with a computer system increases, the general task performance time and error count reduces.

The progressive teaching of more sophisticated commands, used in the Wilson et al. (1990) study, is similar to one advocated by Carroll and his colleagues under the banner of *training wheels* (e.g., Catrambone & Carroll, 1987). This approach can be integrated with the general view of learning outlined earlier through the *chunking hypothesis,* and three assumptions that support it (after Laird, Newell, & Rosenbloom, 1987):

- *The Chunking Hypothesis:* A human acquires and organizes knowledge of the environment by forming and storing expressions, called chunks, that are structured collections of the chunks existing at the time of learning.
- *Performance Assumption:* The performance program of any system is coded in terms of high-level chunks, with the time to process a chunk being less than the time to process its constituent chunks.
- *Learning Assumption:* Chunks are learned at a constant rate on average from the relevant patterns of stimuli and responses that occur in the specific environments experienced.
- *Task Structure Assumption:* The probability of recurrence of an environment pattern decreases as the pattern size increases.

This hypothesis suggests that the component chunks of command sequences are progressively clumped together into chunks until a whole command sequence is represented as a single chunk. The fewer chunks required to perform a command sequence, the less time the sequence takes to perform. Consequently, commands learned at an early stage have no or few relevant constituent chunks, and require the development of a procedure for the sequence from its minimal parts. The representation for commands learned later may include chunks that were developed for commands already learned. However, they still require the development of some structures that were not previously defined, and the appropriate recruitment of those that are. This process may be easier if the new commands conform to an already learned characterization of commands—hence the often cited guideline for consistency in the user interface, and compatibility with previous computer interfaces, and natural-language structures (e.g., Barnard et al., 1981). In contrast to the chunks of performance sequences, such characterizations may take the form of high-level rules governing the system command structure and operation that users have abstracted.

The results of the Wilson et al. (1990) study showed that there are both users and tasks for which performance remains poor, and others for which it improves. The major reductions in time are due not to the speeding up of proceduralized tasks, but instead to the ability of some users to identify local errors, and correct them locally, rather than having to go back to the start of a major sequence and reattempt it completely. It was clear, from verbalizable knowledge, that users had developed general characterizations of the command structure, and indeed they often overgeneralized these rules to apply to the few exceptional commands where they did not, which accounted for a large proportion of the performance errors (e.g., they asserted the need to select menu items for all actions, including the default action of typing into a text editor when no selection was required, and they asserted the need to terminate all menu sequences with a "done" item, which was not always the case).

The individual differences between subjects on this study show the same effects as those in the Halasz and Moran (1983) study summarized earlier. Subjects who developed clear verbalizable rules about the system, which included the exceptions, performed best on most tasks because they improved their error recovery times. This was the dominant change observed during learning in this study. It is reasonable to classify the users who systematically derived generalization rules and consequently performed better in these tests as employing a *systematic* rather than a *heuristic* learning style (Bariff & Lusk, 1977). Those who performed worse and did not abstract rules may be characterized as employing a *heuristic* learning style, although in this study no explicit measuring was performed of the users' learning styles.

A general result supported in the aforementioned studies is that the easiest task action mappings to learn are those that are consistent, compatible, interactive, and meaningful. A task-action mapping is consistent if it shares structural properties, such as syntax, with other task-action mappings at the interface. It is compatible if it shares properties with other task-action mappings from previously experienced interfaces, or natural language. It is interactive if the task environment perceptually cues the actions that can be performed on it, and the correct mappings; in the language of Gibson (1979), if the object has an affordance for the action. Finally, it is meaningful if the actions can be generated from the semantics of the task, or, as in the previously described studies, if the users can extract generalization rules about the interface to facilitate understanding rather than merely learning by rote.

The chunking hypothesis and the associated view of learning have been used as one of the main components in the design of the SOAR problem-solving architecture. SOAR is a problem space theory of human cognitive architecture proposed by Laird et al. (1987) as an integrative theory. In contrast to the specific experimental studies and limited theories described so far, SOAR was used by Howes and Young (1996) to construct an integrated model of learning and performance at the user interface (TAL).

TAL models a user who starts with some knowledge of the primitive interface actions, for example how to use a keyboard and a mouse, but without knowledge of how to combine these actions into sequences for achieving new goals. Given a task, TAL interacts with a device simulation and an instructor. If it has a rule that determines actions to be performed, those actions are performed. Otherwise, the system requests instructions from the human instructor. The instructor responds by giving some instruction, which TAL interprets and uses to determine the next action to be performed. In this process, TAL learns chunks that encode an interpretation of the instructions. Chunks are rules that consist of a left-hand side of conditions and a right-hand side of actions. Chunks learned in one task may be transferred to many similar tasks. The more similar the task semantics are, the greater the possibility of transferring rules is. For example, a chunk from a task to open a file whose condition names a file, and whose action is to move to the file menu, could be transferred to all other tasks on that file including closing or saving it, and to creating a new file. A significant effect of transfer in TAL's behavior is that, if transfer is successful, then less instruction is required to learn a new set of task-action mappings.

TAL does not attempt to be the complete and correct model whose existence Landauer (1996) denied (see subsection The HCI Problem in Cognitive Science Terms). Indeed, it is still only in its infancy. It has been used to simulate interaction with both display- and keyboard-based user interfaces, and has produced similar outcomes with many empirical results with respect to consistency, compatibility, interactivity, and meaningfulness. It does not contain a model of reportability, so the observed differences between verbalizable and nonverbalizable procedural knowledge cannot be captured. Neither does it model error recovery, so many of the phenomena reported previously are not captured. Equally, it does not yet implement alternative learning strategies, so the individual differences in learning cannot be duplicated. However, Hegarty, Just, and Morrison (1988) produced a SOAR simulation to account for the results of a study about individual differences in mechanical ability, which could be incorporated into TAL because it has been carried out within the same integrative theory—SOAR.

A widely used test of mechanical ability is the Bennett Mechanical Comprehension Test (Bennett, 1969), on which a large body of data has been collected for the population as a whole, and its predictive power and the variations in scores across the population are well documented. Such psychometric tests allow a quick paper-and-pencil assessment of a subject, in order to predict their mechanical ability through correlation with previous scores, with a known reliability for a wide range of mechanical tasks. However, there is no explicit or implicit attempt to explain the cognitive processes behind such scores; the tests are a tool designed to serve a specific function. The Bennett test contains items pertaining to many aspects of mechanics including fluid and thermal dynamics, levers, gears, and pulleys. Hegarty et al. (1988) studied subjects who undertook the pulley questions from the Bennett test, and proposed an explanation and SOAR simulation of the cognitive process that contribute to mechanical ability. The test questions are mostly of the form where two pulley systems are presented, and the subject must decide which one requires more force to lift a weight. Some pulley systems differed only in a single attribute, whereas others differed in more than one attribute, allowing a measure of how subjects combine information from different attributes. Similar problems, in which the variation of both relevant and irrelevant attributes allowed a determination of whether subjects could distinguish the attributes relevant to the mechanical function, were also posed. A subset of the 43 subjects taking the test gave verbal protocols of the task; these verbal protocols were used to derive the rules applied by the subjects to answer the test questions, and the preference orderings for the use of rules when they yield different results. A second experiment investigated the range of ways in which subjects addressed problems where both relevant and irrelevant attributes of a problem varied. The range of individual differences was characterized by low-scoring subjects using rules based on visible components of the pulley systems. These rules are qualitative, the attributes on which they are based can be either relevant or irrelevant, and subjects have no clear preference among rules, so that their responses appear inconsistent with any particular rule. High-scoring subjects, on the other hand, have rules that are quantitative and take configurational properties of the system into account. They prefer rules based on attributes that are highly correlated with mechanical advantage. Three abilities accounted for the individual differences in performance: (a) the ability to discriminate relevant from irrelevant attributes, (b) the consistency of rule use, and (c) the ability to quantitatively combine information about two attributes within a single rule. A simulation model was used to specify mechanisms that can account for these three sources of individual differences. The model suggested that the process of applying rules is similar for high-scoring and low-scoring subjects, but the content of the rules changes with increases in mechanical ability. A subsequent study (Hegarty & Just, 1993) tested eye movements and comprehension of descriptions of pulley systems involving text and diagrams. The results of this study were consistent with the view that the reader's choice of modality depends on the cognitive effort involved in each modality. The construction of mental models from the text emerged as being the less effortful choice for high-ability subjects, whereas the construction of mental models from a diagram was the less effortful choice for low-ability subjects (modality variations are discussed in more detail in the next section of this chapter).

These results show that high-scoring subjects are flexible problem solvers who can use either qualitative or quantitative mental models, depending on the demands of the problem. For example, some subjects did not resort to using a quantitative model unless the problem required it. If the relative effort required to answer the test question could be determined by comparing a single attribute, then just that attribute was evaluated and compared. This invocation of quantitative reasoning, when qualitative reasoning is insufficient, is generally consistent with the difference between expert performance on a task and the performance of novices who apply only qualitative reasoning (De Kleer, 1985; DiSessa, 1983).

As TAL develops, there is no reason why reportability effects, error recovery, individual differences, and many other phenomena cannot be accounted for. TAL is not the only model of the acquisition of task-action mappings. CE+ (Polson & Lewis, 1990) models the exploratory acquisition of task-action mappings. In this system, when the available knowledge is insufficient to achieve a goal, problem solving is performed. Similar approaches have been developed to capture compatibility of designs, the most notable being Display-based Task-Action Grammar (D-TAG) (Howes & Payne, 1990), which has its origins in the competence grammars of linguistics. However, TAL is computationally implemented, whereas, the components of CE+, although implemented in isolation, have not been integrated into a single running simulation. Perhaps most important, TAL implements a single learning system in SOAR's chunking, which operates in both plan-based tasks and those tasks where the cognition is strongly situated in the environment, due to the perceptual cues or affordances it offers.

These examples of experimental studies, localized theories, and the use of an integrative theory to host different models show how the elements provided by cognitive science can come together. However, these results are not yet of direct practical use in designing user interfaces for all, because they require cognitive science skills to both apply and interpret them. Despite this, they provide an understanding of how the learning process is modeled, and can move on to address individual differences. This, in turn, provides a basis for user interfaces that *support all potential users throughout system learning and use.*

Modality Modeling and Design Choices

SOAR provides an integrative theory within which models of experimental data derived from different paradigms can be constructed. However, SOAR primarily represents information at a propositional level, which is appropriate for reasoning or planning tasks, but is less compatible with the problems and variations encountered in more perceptual tasks involving sensory modalities (e.g., voice, vision), and the performance of action with a range of devices. Such modality issues must be modeled in order to design interfaces for those with accessibility problems due to sensory or action impairments, to overcome interaction problems that may arise only for users without impairments in rapidly changing environments (such as flying planes, or managing nuclear reactors), and to investigate new interface devices that may support novel interaction methods (such as virtual or augmented reality).

ICS represent the human information-processing system as a highly parallel organization with a modular structure. The ICS architecture (Barnard, 1985) contains a set of functionally distinct subsystems, each with equivalent capabilities, yet each specialized to deal with a different class of representation. The ICS architecture can be considered as an integrative framework (in contrast to SOAR, which is an integrative theory) in which some components are further specified as theories, or indeed models, whereas others are outlined only at the framework level, awaiting elaboration from other theories.

ICS exchange representations of information directly, with no role for a *central processor* or *limited-capacity working memory*. Acting together, nine component subsystems deal with incoming sensory information, structural regularities in that information, the meanings that can be abstracted from it, and the creation of instructions for the body to respond and act both externally, in the real world, and internally, in terms of physiological effects. Figure 8.1 outlines the overall architecture, whereas Table 8.1 lists the nature of the mental representations that each subsystem processes. The subsystems are classed as peripheral if they exchange information with the world via the senses of the body, and as central if they only exchange information with other subsystems.

Each subsystem has the same internal structure, even though they all address different classes of information. They each receive a representation at an input array, which is *copied* to an Image Record, while simultaneously being operated on by a number of *transformation processes*. The Image Record acts as a local memory for the subsystem. Any representation ever received is stored in the Image Re-

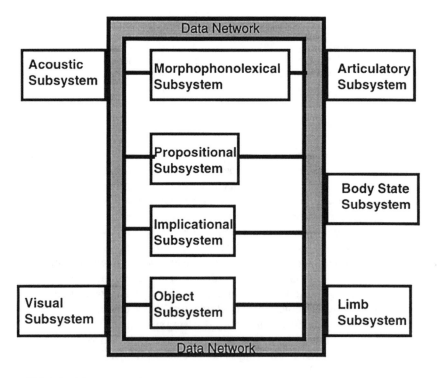

FIG. 8.1. The systematic organization of ICS.

TABLE 8.1

The Subsystems Within ICS and the Type of Information With Which They Deal

Peripheral Subsystems

A) Sensory

1) Acoustic (AC)	Sound frequency (pitch), timbre intensity, etc.Subjectively "what we hear in the world."
2) Visual (VIS)	Light wavelength (hue), brightness over visual space, etc.Subjectively "what we see in the world" as patterns, shapes, and colors.
3) Body States (BS)	Type of stimulation (e.g., cutaneous pressure, temperature, olfactory muscle tension) its location, intensity, etc.Subjectively, bodily sensations of pressure, pain, positions of parts of the body, as well as tastes and smells, etc.

B) Effector

4) Articulatory (ART)	Force, target positions, and timing of articulatory musculature (e.g., place of articulation).Subjectively, our experience of subvocal speech output.
5) Limb (LIM)	Force, target positions and timing of skeletal muscles.Subjectively, "mental" physical movement.

Central Subsystems

C) Structural

6) Morphophono-lexical (MPL)	An abstract structural description of entities and relationships in sound space. Dominated by speech forms, where it conveys a surface structure description of the identity of words, their status, order, and the form of boundaries between them.Subjectively, what we "hear in our head," our "mental voice."
7) Object	An abstract structural description of entities and relationships in visual space, conveying the attributes and identity of structurally integrated visual objects, their relative positions, and dynamic characteristics.Subjectively, our "visual imagery."

D) Meaning

8) Propositional	A description of entities and relationships in semantic space conveying attributes and identities of underlying referents and the nature of the relationships among them.Subjectively, specific semantic relationships—"knowing that."
9) Implicational (IMPLIC)	An abstract description of human existential space, abstracted over both sensory and propositional input, and conveying ideational and affective content: schematic models of experience.Subjectively, "senses" of knowing (e.g., "familiarity") or of affect (e.g., apprehension, desire).

cord, and in the long term, any commonalties and regularities of the representations in a subsystem can be abstracted from the stock of past experience. The transformation processes are the key to the function of the overall organization. In normal operation, these processes transform the information represented on the input array into a different representation, for use by another subsystem. For example, the VIS subsystem contains a VIS → OBJ transformation process that transforms the visual information into a more abstract object representation. These transformation processes are independent and work in parallel, with the consequence that a subsystem can produce multiple simultaneous outputs.

Subsystems do not contain transformation processes from their own to all other representations, with the consequence that chains of subsystem operation may be required to produce the final representation. For example, to transform information from a visual to a propositional representation requires the intermediate use of the object subsystem, because a direct transformation is not available. A consequence of this is that cognition is a result of a series of subsystems acting in a chain or configuration as information flows through the system. The more complex configurations can include cyclical exchanges of information between pairs and even triplets of central subsystems (e.g., PROP → IMPLIC & IMPLIC → PROP, and OBJ → MPL, MPL → PROP & PROP → OBJ). Having built up an understanding of the resources needed to perform a particular task using a particular interface, it is then possible to reason about the suitability of the interface by, for example, looking at the conflicts that arise in the use of the identified cognitive resources. ICS provide a framework for answering such questions as "how many information channels can be used simultaneously?" using concepts such as stability of configurations, oscillation between configurations competing for resources, and the requirements of blending different data streams.

As TAL and CE+ represent both the user and the system, so ICS can be represented in a formal notation along with a system. Modal Action Logic (MAL) has been used on several occasions to reason about gestural interaction, or about interaction using novel input and display devices (Duce, 1995; Duce, Barnard, Duce, & May, 1995) in an approach termed *syndetic modeling* (Faconti & Duce, 1996). Such a formal modeling approach requires a detailed understanding of the notations used, and takes considerable effort to perform. Clearly, this is not required in every interface design. However, to understand the properties of novel interface devices, which may provide support for users with individual needs due to some impairment in one or more sensory modality, they can be efficient as well as effective in pursuing the goal of user interfaces for all.

USER INTERFACE DEVELOPMENT TOOLS AND METHODS AS COGNITIVE SCIENCE IN PRACTICE

The previous section has described an integrative theory (SOAR) instantiated with a model (TAL) that draws on experimental data collected from various paradigms to model learning, and an integrative framework (ICS) that can be used in conjunction with formal system models (as syndetic modeling) to model device level interactions. These approaches have potential to provide insights into HCI, which could contribute to the development of user interfaces for all, but these approaches also require considerable craft skill in cognitive science to use them; in both cases the techniques are models of the user's task performance per se, and they need models to be populated to a fine granularity, thus requiring considerable effort.

There are very few user interface development tools or methods, which have been derived from cognitive science, that do not suffer from these shortcomings. Available tools model the task to be performed using constructs derived from human information-processing models, rather than presenting models of the human processing explicitly. They also assume the simplest cases of static knowledge of an expert, rather than addressing the dynamic knowledge of learning. However, these constraints allow them to be specified without cognitive science skill, and to be specified at either coarse or fine granularities, making them adequate for use by system developers when the benefits of modeling outweigh the costs (Bias & Mayhew, 1994).

Two methods are briefly described here (GOMS and KADS), illustrating how they capture a range of tasks—user interfaces for all tasks—and how they capture the variation in the performance of those tasks by users as selection rules to choose between methods, or as alternative strategies. GOMS models a task as a set of goals, operators, methods, and selection rules for user interface design evaluation, whereas KADS models reasoning tasks in terms of domain knowledge, inference rules, task structures, and strategies for requirements engineering.

GOMS—Engineering Task Performance

The GOMS model was developed by Card, Moran, and Newell (1983) based on previous work in the domain of human problem solving by Newell and Simon (1972). GOMS has been designed as a cognitive engineering model. Therefore, its primary purpose is to be used in the design of user interfaces rather than as a foundation of scientific theories. The model is used to outline the cognitive performance of a person by decomposing a problem into hierarchical goals and goal stacks. The basic GOMS model is best at making qualitative predictions about differences between tasks in which users make no or few errors. By associating times, or time distributions with each operator, GOMS models are able to make total performance time or statistical predictions. Depending on the granularity of analysis, several variations of GOMS models can be explored to make quantitative predictions, for example, from unit-task to keystroke level.

The four basic elements of a GOMS model are goals, operators, methods, and selection rules. A goal is a symbolic representation of a state of affairs to be achieved, which determines a set of possible methods to be used to achieve it. Operators are elementary perceptual, motor, or cognitive acts whose execution is necessary to change any aspect of the user's strategy or the task environment. Methods describe procedures for achieving goals in terms of operators, or other goals. Operators that are often used together are grouped into methods, in the way that the chunking mechanism in SOAR would. However, GOMS assumes all such expert chunks have been constructed, and includes no learning mechanism, thus being applicable only to expert performance. The fourth element is the rules to select between alternative methods to perform goals—(selection rules).

GOMS has received much more empirical testing than any other analytic model of HCI tasks (Gugerty, 1993). Models have been developed for a wide variety of applications including simple text editors (Card et al., 1983), spreadsheets (J. R. Olsen & Nilsen, 1988), and hypertext applications (Carmel, Crawford, & Chen, 1992). The strongest validation of its use has been on an interface for telephone inquiry operators where cutting task time by seconds mounts up to savings of millions of dollars over a year (Gray, John, & Attwood, 1993; Gray, John, Stuart, Lawrence, & Attwood, 1990). However, this is primarily a sensorimotor task performed by highly skilled operators, for whom the model is intended to perform best.

GOMS analyses allow flexibility for modeling variations of task performance with respect to the times chosen for operators, which will effect only the quantitative predictions of performance time, and in the selection rules. Selection rules provide the control structure to the model in terms of if–then rules. In an example where different methods are available for cursor control, a selection rule may state:

IF the desired position and the current position are both on the screen
THEN the arrow key method would be used.
IF the desired position is on a different screen than the current position
THEN the search command method would be used.

The original presentation of GOMS (Card et al., 1983) stated that expert method selection would always be made on the basis of the fastest method, and this reasoning should be used in constructing selection rules. Although this is a useful guideline for developing engineering models, it has been empirically shown to be false (MacLean, Barnard, & Wilson, 1985), and users show some inertia against changing methods from one, recently used, to another, even if the second is faster. Clearly, the data on learning, reported previously, and the training wheels approach to learning computer systems, suggest that for nonexperts, many methods may not be within the habitable subset of commands, and may be selected only very rarely, even when they are considerably more efficient.

In order to make the GOMS approach more usable for practical applications, Kieras (1988) developed the Natural GOMS Language (NGOMSL), which allows the modeler to describe user–computer interaction in a specification language similar to computer-programming languages. Cognitive complexity theory (CCT) represents another extension of the basic GOMS model (Kieras & Polson, 1985). With respect to CCT, the CE+ proposal mentioned earlier (Polson & Lewis, 1990) is a further advance to address "walk up and use" interfaces (e.g., public information kiosks). John (1988) extended the approach to the analysis of parallel activities. The GOMS approach and its followers have dominated the research on models in HCI.

However, to date, such an approach has had little influence on the practice of user interface design. The technique is still very dependent on the skill of the analyst, and although Kieras' work has addressed learning time predictions, expert–novice differences, and estimates of mental work load from the number of items in working memory at any time, the approach is not tuned to address individual differences (J. R. Olsen & G. M. Olsen, 1990).

KADS—Engineering Task Specification

GOMS provides a tool to predict interaction time, and some aspects of complexity for a clearly specified task, once a system has been designed. Cognitive science has also produced task models that can guide the developer at an earlier stage of system development, when a problem is being analyzed. KADS was developed as a Knowledge Acquisition and Design System for knowledge-based systems with a firm theoretical foundation (see Schreiber, Wielings, & Breuker, 1993), as well as a clearly stated method to be followed by developers (see Tansley & Hayball, 1993). KADS provides a library of task models that can be instantiated to become descriptions of future systems and their domains.

As a complete method, KADS provides models of the system, the context within which it will work, the users, cooperation with other agents, the system's organizational environment, and the details of the documentation required for the complete development process from analysis through design to implementation (see Tansley et al., 1993). The current description covers only the central library of expertise models that can be applied widely to guide requirements engineering.

The problem-solving components of tasks are investigated and described by building up an Expertise Model from its four layers: the *domain layer,* which describes the static factual knowledge about the domain; the *inference layer,* which defines the inference steps that the system can make; the *task layer,* which defines the basic problem-solving tasks; and the *strategy layer,* which defines how tasks are constructed, modified, or chosen. Although an instantiated expertise model is an implementation-independent representation of the problem-solving capability of a prospective system, it is more concrete than an uninstantiated model would be. KADS provides a library of uninstantiated models for the inference and task layer called Generic Task Models, which can be used to motivate the analysis (see Breuker & Van de Velde, 1994). Each of the four layers of the Expertise Model are outlined here.

The domain layer defines static domain knowledge as an ontology, consisting of structures of domain concepts and relationships between concepts. This knowledge is termed "static," because it describes a domain while being neutral as to how it is used for inference purposes.

The inference layer describes basic inferential capability, in terms of *inference types, domain roles,* and *inference structures* linking these with tasks at the task layer. It identifies the inferences that can be made for selected tasks over the static knowledge in the domain layer. Inference types are descriptions of the way in which domain concepts, structures, or relations can be used to make inferences. For example, the "classify" inference type takes an object with its attributes, and derives the class of the object. Inference types direct the way in which static domain knowledge may be used, and provide "handles" for the control of inference by the next, task layer. Domain roles define the functions that domain structures may perform. For example, in a diagnosis task in the medical domain, HIV may be either a hypothesis to be verified or a solution resulting from a reasoning process. These are two different roles for a single domain concept within a problem-solving process. The third component in the layer is the inference structure, which is a network of inference types and domain roles constraining a reasoning process by explicitly describing which inferences can be made, and implicitly defining which cannot be made. Inference structures are depicted in a graphical form where domain roles are represented by rectangles and inference types by ellipses (e.g., Fig. 8.2).

The third layer in the Expertise Model is the task layer, which describes how the individual inferences within the inference layer may be sequenced in order to satisfy the required problem-solving goals. The representation used is a task structure that can be defined statically (with a fixed control structure) or dynamically (e.g., as a result of planning at the inference layer). Task structures are typically simple sequences of inferences wrapped in some conventional procedural control structures, such as selection (IF … THEN … ELSE) and repetition (e.g., FOR, WHILE, and REPEAT). However, they may include more complex control structures such as parallelism, and temporal dependency.

The fourth layer of the Expertise Model is the strategy layer, which provides the strategic knowledge to select, sequence, plan, or repair the corresponding task structures. The following types of strategic knowledge may be described: goal selection, task structure selection, goal sequencing, task structure configuration, mode of system operation, inference control, and repair.

KADS allows analysts to approach problem-solving tasks with poorly expressed algorithms in the same way as data-intensive tasks or detailed algorithm

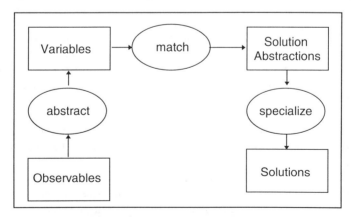

FIG. 8.2.   Inference structure for heuristic classification.

implementation may be approached. However, the major contribution of KADS addresses the identification of problem-solving tasks and the reuse of task-level descriptions through Generic Task Models (GTM). KADS provides a library of GTMs, which are models of problem-solving tasks not tied to a particular domain. GTMs are used to initiate and drive the analysis process during Expertise Model development. For each task, the model describes it at the task and inference layers. Both the domain and strategic layers are strongly domain dependent and, therefore, little generic information can be given for them.

The library of GTMs is presented as a hierarchy. The top node represents tasks; the next layer has three entries: (a) systems analysis, which deals with an examination of the elements or structure of some entity, (b) system modification, which deals with tasks that update or change some entity (often after a process of analysis and synthesis, but that modify the entity after finding a solution), and (c) system synthesis, which deals with tasks that build up an entity from constituent parts. The full hierarchy of GTMs is presented in Table 8.2.

For each GTM, the input and output descriptions are presented (see Fig. 8.3) to aid the analyst in selecting the appropriate GTM for a system by considering the knowledge based tasks in the Process Model in terms of their inputs and outputs. GTMs are used as a starting point for developing Expertise Models. The selection of the GTM and the analysis of static knowledge are the first two stages of the expertise analysis, and are performed in parallel. Once the GTM is selected, the specialization of its terminology to that of the problem domain collected in the analysis of static knowledge can begin. This population is then continued as the

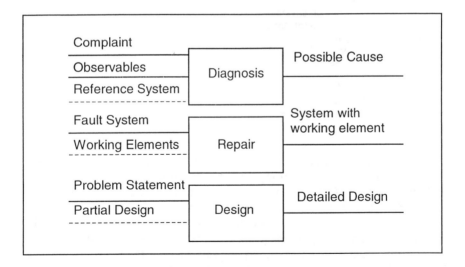

FIG. 8.3.   Input/output descriptions of three task models (input on left, output on right; solid lines necessary, dashed lines optional).

TABLE 8.2

The Hierarchy of Generic Task Models in the KADS Task Library

- SYSTEM ANALYSIS
    - Identification
        Diagnosis
            Single Model Diagnosis
                Systematic Diagnosis
                Localization
                Causal Tracing
            Multiple Model Diagnosis
                Mixed Mode Diagnosis
        Verification
        Correlation
            Assessment
        Monitoring
        Classification
            Simple Classification
            Heuristic Classification
            Systematic Refinement
    - **Prediction**
        Predication of Behavior
        Prediction of Values
- SYSTEM MODIFICATION
    - Repair
    - Remedy
    - Control
    - Maintenance
- SYSTEM SYNTHESIS
    - **Design**
        Hierarchical Design
        Incremental Design
    - Configuration
        Simple Configuration
        Incremental Configuration
    - **Planning**
    - **Scheduling**
    - **Modeling**

model develops through step-by-step refinement, eliciting knowledge from documents, task simulations, and through structured interviews.

Both GOMS and KADS allow tasks to be modeled at different granularities: in the case of GOMS, at the unit-task level, or lower levels down to the keystroke; in the case of KADS, at the task-model level, or lower levels down to complete domain knowledge ontologies. Equally, both GOMS and KADS offer a mechanism to choose between user strategies, through selection rules in GOMS or strategic knowledge rules in KADS. Both are derived from cognitive science models of human information processing, but are abstracted away from the processing itself, and provide representation languages, or libraries embodying them. In both cases the methods are accessible to system developers for use in designing systems for a range of tasks and users.

CONCLUSIONS

This chapter has tried to convey the spirit of the cognitive science enterprise, and to show how HCI can benefit from the results of freedom in modeling behavior by using cognitive representations and processes. The use of integrative theories and frameworks to capture local theories and experimental results of cognitive science has been illustrated with TAL and ICS. The limitations in the accessibility of these theories and frameworks to system developers are overcome in methods that hide the underlying cognitive processing, and model task representations, as illustrated by GOMS and KADS. However, these accessible tools are limited in the range of phenomena they can address, as a result of their simplifying assumptions.

Variations in the way different users learn, and the different knowledge and skills that they start with, must be addressed within the user interfaces for all approach. A major problem with developing tools based on cognitive science theory is that it requires detailed descriptions of knowledge, which are usually created too late to influence the technology because of the need for experimental data from those technologies once they are created to tune models—"Human computer interaction ... will not sustain approaches that are too low level, too limited in scope, too late and too difficult to apply in real design ..." (Carroll & Campbell, 1986). This paradox has no clear solution, although cognitive science clearly has much to offer to the user interfaces for all approach.

Perhaps the greatest influence of cognitive science on user interfaces, and on the user interfaces for all approach in the future, has, paradoxically, not been mentioned here, but in other chapters in this volume. The cognitive science method produces models and representations of mental life that are normally implemented on computers. The human is the only generally accepted example of intelligent mental life we have. Therefore, these models are the best detailed models of intelligent mental life we have that are amenable to computer programming. Cognitive science models of human reasoning with prototypes were used as the basis of most of the early work on user modeling (described in chap. 14, this volume). Cognitive science models of human communication comprehension and generation were the basis of the architectures of intelligent multimedia systems (described in chap. 4, this volume).

The aforementioned systems, which adopt cognitive science models as the inspiration for representations and architectures, have already been developed. If the cognitive science method is too detailed and slow to refine interface design once innovation has happened, the important question to answer is: *What will be the next innovation, and what does cognitive science offer to guide it?*

The 1998 user interface, like the 1984 one, is a two-dimensional graphical representation of the desktop controlled by direct manipulation. Recent advances in several core software technologies have made possible a new type of human–computer interface: the conversational character. Conversational characters are autonomous, anthropomorphic, animated figures (Badler, Phillips, & Webber, 1993) that have the ability to communicate through multiple modalities, including spoken language, facial expressions (Pelachaud, Badler, & Steadman, 1996; Sproull, Subramani, Kiesler, Walker, & Waters, 1996), and gestures. Conversational characters inhabit three-dimensional virtual worlds along with avatars of humans, to facilitate human–human (Benford, Bowers, Fahlen, Greenhalgh, & Snowdon, 1995), as well as human–computer interaction. Unlike textual natural-language interfaces, conversational characters have the ability to perceive and produce the verbal and nonverbal signals that identify discourse structure and regulate the flow of information between interlocutors. Such signals include intentional patterns, gestures, back-channel feedback signals, and turn-taking protocols. These capabilities enable them to engage in complex interactions with human users via natural speech, rather than complex command languages, menus, or graphical manipulations. For conversational characters to maintain realistic embodiments, they require rich representations of emotions, goals, plans, affordances, learning, and cultural and other individual differences, which must be adequately conveyed in conversation in order to maintain a sense of presence and engagement.

The fully anthropomorphic interface envisioned in Arthur C. Clark's 2001 as HAL 9000 is still a long way off (see Stork, 1996). However, cognitive science has many models that provide insights into how to construct such complex "beasts." Much of the research in the last 10 years in cognitive science has been directed at integrating emotional aspects of mental life with the cognitive aspects (Oatley & Johnson-Laird, 1987; Ortony, Clore, & Collins, 1988). A second advance has been to address discourse and dialogue rather than syntax. If you agree that conversational characters in three-dimensional multiuser virtual worlds are the way of the future, and this chapter has helped you to understand the basis of cognitive science models, then cognitive science research may be the place to look further for the insights that will provide user interfaces for all.

ACKNOWLEDGMENTS

The research work reported in this chapter was partly funded by CEC Esprit IV Grant 20597 to the CHAMELEON project, partly by CEC HCM Grant CHRX-CT93-0085 to the ERCIM Computer Graphics Network, and partly by CEC TMR Grant FMRX-CT97-01333 to the TACIT consortium.

REFERENCES

Anderson, J. R. (1983). *The architecture of cognition*. Cambridge, MA: Harvard University Press.

Badler, N. I., Phillips, C. B., & Webber, B. L. (1993). *Simulating humans: Computer graphics animation and control*. Oxford, England: Oxford University Press.

Bariff, M. L., & Lusk, E. J. (1977). Cognitive and personality tests for the design of management information systems. *Management Science, 23*, 820–829.

Barnard, P. J. (1985). Interacting cognitive subsystems: A psycholinguistic approach to short term memory. In A. Ellis (Ed.), *Progress in the psychology of language* (Vol. 2, pp. 197–258). London: Lawrence Erlbaum Associates.

Barnard, P. J. (1995). The contributions of applied cognitive psychology to the study of human computer interaction. In R. Becker, J. Grudin, W. Buxton, & S. Greenberg (Eds.), *Readings in human computer interaction: Towards the Year 2000* (pp. 640–658). San Francisco: Morgan Kaufmann.

Barnard, P. J., Hammond, N., Morton, J., Long, J., & Clark, I. (1981). Consistency and compatibility in human computer dialogue. *International Journal of Man–Machine Studies, 15,* 87–134.

Benford, S., Bowers, J., Fahlen, L. E., Greenhalgh, C., & Snowdon, D. (1995). User embodiment in collaborative virtual environments. In *Proceedings of the Human Factors in Computing Systems Conference* (pp. 242–249). New York: ACM Press.

Bennett, C. K. (1969). *Bennett Mechanical Comprehension Test.* New York: The Psychological Corporation.

Bias, R. G., & Mayhew, D. J. (Eds.). (1994). *Cost-justifying usability.* New York: Academic Press.

Breuker, J., & Van de Velde, W. (1994). *CommonKADS library for expertise modelling.* Amsterdam: IOS Press.

Card, S. K., Moran, T. P., & Newell, A. (1983). *The psychology of human–computer interaction.* Hillsdale, NJ: Lawrence Erlbaum Associates.

Carmel, E., Crawford, S., & Chen, H. (1992). Browsing in Hypertext: A cognitive study. *IEEE Transactions on Systems, Man and Cybernetics, 22*(5), 865–884.

Carroll, J. M., & Campbell, R. L. (1986). Softening up hard science: Reply to Newell and Card. *Human–Computer Interaction, 2*(3), 227–250.

Catrambone, R., & Carroll, J. M. (1987). Learning a word processor system with training wheels and guided exploration. In J. M. Carroll, & P. P. Tanner (Eds.), *Proceedings of the Human Factors in Computing Systems and Graphics Interface* (pp. 169–174). New York: ACM Press.

De Kleer, J. (1985). How circuits work. In D. Bobrow (Ed.), *Qualitative reasoning about physical systems.* Cambridge, MA: MIT Press.

DiSessa, A. A. (1983). Phenomenology and the evolution of intuition. In D. Gentner & A. L. Stevens (Eds.), *Mental models.* Hillsdale, NJ: Lawrence Erlbaum Associates.

Duke, D. J. (1995). Reasoning about gestural interaction. *Computer Graphics Forum, 14*(3), C-55–C-66.

Duke, D. J., Barnard, P. J., Duce, D. A., & May, J. (1995). Systematic development of the human interface. In *Proceedings of the 2nd Asia-Pacific Software Engineering Conference* (pp. 313–321). Brisbane, Australia: IEEE Computer Society Press.

Ericsson, K. A., & Simon, H. A. (1984). *Protocol analysis: Verbal reports as data.* Cambridge, MA: MIT Press.

Faconti, G. P., & Duke, D. J. (1996). Device models. In F. Bodart & J. Vanderdonckt (Eds.), *Proceedings of Design, Specification and Verification of Interactive Systems '96* (pp. 73–91). Vienna: Springer-Verlag.

Fitts, P. M. (1964). Perceptual-motor skill learning. In A.W. Melton (Ed.), *Categories of human learning.* New York: Academic Press.

Furnas, G. W., Gomez, L. M., Landauer, T. K., & Dumais, S. M. (1982). Statistical semantics: How can a computer use what people name things to guess what people mean when they name things? In *Proceedings of Human Factors in Computer Systems* (pp. 251–253). New York: ACM Press.

Furnas, G. W., Gomez, L. M., Landauer, T. K., & Dumais, S. M. (1983). Statistical semantics: Analysis of the potential performance of key-word information systems. *The Bell System Technical Journal, 62*(6), 1753–1906.

Gibson, J. J. (1979). *The ecological approach to visual perception.* Boston: Houghton Mifflin.

Gomez, L. M., & Lochbaum, C. C. (1984). People can retrieve more objects with enriched keyword vocabularies. But is there a performance cost? In B. Shackel (Ed.), *Human computer interaction—Interact '84* (pp. 257–261). Amsterdam: North-Holland, Elsevier Science.

Gray, W., John, B., & Attwood, M. (1993). Project Ernestine: A validation of GOMS for prediction and explanation of real-world task performance. *Human–Computer Interaction, 8*(3), 237–309.

Gray, W, John, B., Stuart, R., Lawrence, D., & Attwood, M. (1990). GOMS meets the phone company: Analytic modelling applied to real world problems. In D. Diaper, D. Gilmore, G. Cockton, & B. Shackel (Eds.), *Proceedings of INTERACT '90* (pp. 29–34). Amsterdam: North-Holland, Elsevier Science.

Gugerty, L. (1993). The use of analytic models in human–computer interface design. *International Journal of Man–Machine Studies, 38*, 625–660.

Halasz, F., & Moran, T. P. (1983). Mental models and problem solving using a calculator. In *Proceedings of the Human Factors in Computing Systems* (pp. 212–216). New York: ACM Press.

Hegarty, M., & Just, M. A. (1993). Constructing mental models from text and diagrams. *Journal of Memory and Language, 32*, 717–742.

Hegarty, M., Just, M. A., & Morrison, I. R. (1988). Mental models of mechanical systems: Individual differences in quantitative and qualitative reasoning. *Cognitive Psychology, 20*, 191–236.

Howes, A., & Payne, S. J. (1990). Display-based competence: Towards user models for display driven interfaces. *International Journal of Man–Machine Studies, 33*, 637–655.

Howes, A., & Young, R. M. (1996). Learning consistent, interactive, and meaningful task-action mappings: A computational model. *Cognitive Science, 20*(3) 301–356.

John, B. (1988). *Contributions to engineering models of human computer interaction.* Unpublished doctoral dissertation, Carnegie Mellon University, Pittsburgh, PA.

Kieras, D. E. (1988). Towards a practical GOMS model methodology for user interface design. In M. Helander (Ed.), *Handbook of human computer interaction* (pp. 135–158). Amsterdam: North-Holland, Elsevier Science.

Kieras, D. E., & Polson, P. G. (1985). An approach to the formal analysis of user complexity. *International Journal of Man–Machine Studies, 22*, 365–394.

Laird, J. E., Newell A., & Rosenbloom, P. S. (1987). SOAR: An architecture for general intelligence. *Artificial Intelligence, 33*, 1–64.

Landauer, T. K. (1987). Relations between cognitive psychology and computer systems design. In J. M. Carroll (Ed.), *Interacting thought: Cognitive aspects of human–computer interaction* (pp. 1–25). Cambridge, MA: MIT Press.

Landauer, T. K. (1996). Let's get real: A position paper on the role of cognitive psychology in the design of humanly useful and usable systems. In R. Becker, J. Grudin, W. Buxton, & S. Greenberg (Eds.), *Readings in human computer interaction: Towards the year 2000.* San Francisco: Morgan Kaufmann.

MacLean, A., Barnard, P. J., & Wilson, M. D. (1985). Evaluating the human interface of a data entry system: User choice and performance measures yield different trade-off functions. In P. Johnson & S. Cook (Eds.), *People and computers: Designing the interface* (pp. 172–185). Cambridge, England: Cambridge University Press.

Miller, G. A. (1966). *Psychology: The science of mental life.* London: Pelican Books.

Morton, J., Barnard, P. J., Hammond, N. V., & Long, J. (1979). Interacting with the computer: A framework. In E. J. Boutmy & A. Danthine (Eds.), *Proceedings of Teleinformatics '79* (pp. 201–208). Amsterdam: North-Holland, Elsevier Science.

Neisser, U., Novick, R., & Lazar, R. (1963). Searching for ten targets simultaneously. *Perceptual and Motor Skills, 17,* 427–432.

Neves, D. M., & Anderson, J. R. (1981). Knowledge compilation: Mechanisms for the automatisation of cognitive skills. In J. R. Anderson (Ed.), *Cognitive skills and their acquisition* (pp. 57–84). Hillsdale, NJ: Lawrence Erlbaum Associates.

Newell, A. (1990). *Unified theories of cognition.* Cambridge, MA: Harvard University Press.

Newell, A., & Rosenbloom, P. S. (1981). Mechanisms of skill acquisition and the law of practice. In J. R. Anderson (Ed.), *Cognitive skills and their acquisition* (pp. 1–56). Hillsdale, NJ: Lawrence Erlbaum Associates.

Newell, A., & Simon, H. (1972). *Human problem solving.* Englewood Cliffs, NJ: Prentice-Hall.

Norman, D. A. (1993). An Introduction to the special issue on "situated action". *Cognitive Science, 17*(1), 1–7.

Oatley, K., & Johnson-Laird, P. N. (1987). Towards a cognitive theory of emotions. *Cognition and Emotion, 1,* 29–50.

Olsen, J. R., & Nilsen, E. (1988). Analysis of the cognition involved in spreadsheet software interaction. *Human–Computer Interaction, 3,* 309–350.

Olsen, J. R., & Olsen, G. M. (1990). The growth of cognitive modelling in human computer interaction since GOMS. *Human–Computer Interaction, 5,* 221–265.

Ortony, A., Clore, G. L., & Collins, A. (1988). *The cognitive structure of emotions.* Cambridge, England: Cambridge University Press.

Payne, S., & Green, T. (1986). Task-action grammars. A model of the mental representation of task languages. *Human–Computer Interaction, 2,* 93–133.

Pelachaud, C., Badler, N. I., & Steadman, M. (1996). Generating spatial expressions for speech. *Cognitive Science, 20*(1), 1–46.

Polson, P. G., & Lewis, C. H. (1990). Theory based design for easily learned interfaces. *Human–Computer Interaction, 5,* 191–220.

Popper, K. R. (1992). *The logic of scientific discovery.* London: Routledge.

Schreiber, G., Wielings, B., & Breuker, J. (1993). *KADS: A principled approach to knowledge based systems development.* London: Academic Press.

Snoddy, G. S. (1926). Learning and stability. *Journal of Applied Psychology, 10,* 1–36.

Sproull, L., Subramani, M., Kiesler, S., Walker, J. H., & Waters, K. (1996). When the interface is a face. *Human–Computer Interaction, 11*(2), 97–124.

Sternberg, S. (1969). On the discovery of processing stages: Some extensions to Donder's method. *Acta Psychologica, 30,* 276–315.

Stork, D. G. (1996). *HAL's legacy: 2001's computer as dream and reality.* Cambridge, MA: MIT Press.

Suchman, L. (1987). *Plans and situated action: The problem of human–machine communication.* Cambridge, England: Cambridge University Press.

Tansley, D. S. W., & Hayball, C. C. (1993). *Knowledge based systems analysis and design: A KADS developer's handbook.* London: Prentice-Hall.

Wilson, M. D., Barnard, P. J., & MacLean, A. (1990). An investigation of the learning of a computer system. In P. Falzon (Ed.), *Cognitive ergonomics: Understanding, learning and designing human computer interaction* (pp. 151–172). London: Academic Press.

Winograd, T., & Flores, F. (1987). *Understanding computers and cognition: A new foundation for design.* Reading, MA: Addison-Wesley.

9 From Programming Environments to Environments for Designing

Terry Winograd

As the field of programming has matured over the years, attention has shifted from the program to the programmer, i.e., from the logical and computational structure of algorithms to the cognitive structures of the people who produce them. Innovations such as interactive programming environments, object-oriented programming, and visual programming have not been driven by considerations of algorithmic efficiency, or formal program verification, but by the ongoing drive to increase the programmer's effectiveness in understanding, generating, and modifying code. This chapter aims to broaden the view still further: from programming to design.

In this chapter, our intention is to highlight a field in transition—from programming-intensive to design-oriented practice. Such a transition is especially relevant for human–computer interaction (HCI) and *user interfaces for all,* as the latter proclaim a paradigm shift toward designing for the broadest possible end-user population.

The first part of the chapter portrays the evolving field of software design, and its relationships to the traditions of programming, analysis, and design that have served the computer field throughout its history.

The second part relates the concerns that are highlighted in software design to traditional approaches in programming and programming environments. It draws analogies between four specific aspects of current programming environments and four corresponding aspects of environments for software design. As with most analogies, the results are intended to be suggestive, not rigorous. Starting with the cognitive aspects of well-understood programming tools, we can get insights into the demands for an environment that will support the ongoing development of computer software in a rapidly changing industry.

SOFTWARE DESIGN

Technological Maturity

In his widely read "Software Design Manifesto" a few years ago, Mitchell Kapor (1991) bemoaned the fact that:

> Despite the enormous outward success of personal computers, the daily experience of using computers far too often is still fraught with difficulty, pain, and barriers for most people.... The lack of usability of software and poor design of programs is the secret shame of the industry. (p. 62)

These are strong words to throw into the face of a multibillion-dollar industry that by all standard measures must be doing things right. But Kapor is highly respected as the founder of Lotus Software and the designer of Lotus 123, the "killer app" that gave a major impetus to the whole microcomputer industry. His concerns have stirred a resonance among many people who work with software. Although the unprecedented power of computing systems makes them highly useful, there is a big gap between what we see in most products today and what could be done to make them really usable. This becomes more pressing as we begin to reach beyond the current applications to new audiences and new ways of taking advantage of computation in people's lives.

In the last few decades the computing profession has matured from its early days, when ingenuity was required at every turn to make programs work at all. Today we are part of a major industry in which the expectations for successful programming constantly move upward. Software designers today have the opportunity (and necessity) of moving to a broader view of what they need to achieve, because of the tremendous successes of computing.

We are now entering a new phase of computer product development, which can be understood as a step in a history of technological maturity that has been repeated for many new technologies, such as the radio, the automobile, and the telephone:

Phase 1: Technology-Driven.   In the first phase, a new technology is difficult to employ, its benefits are not yet obvious, and its appeal is mainly to those who are fascinated with it for its own sake—the "early adopters." We find clubs of enthusiasts who love to share stories about how they fought the difficulties and overcame them. The general public is seen as not having sufficient understanding or merit to really use the new inventions. Ham radio is a good example of a technology that was adopted by a small but dedicated group in the technology-driven stage. In the same vein, the legends of Silicon Valley include many stories of the early computer days and the brave pioneers who tackled the Altair, or the Osborne.

Phase 2: Productivity-Driven.   In the second phase, the economic benefits of using a technology are developed to the point where people in industry and business will adopt it for practical uses. The measure is in the bottom line—not whether the technology is fascinating or easy to use, but whether it can be shown (or at least believed) to produce greater efficiency, productivity, and profits. The use of radios for truck and taxi dispatch, police, and military communication falls into this class, as do most of the major microcomputer applications sold today. Spreadsheets, word processors, databases, desktop publishing, and a host of other applications have been sold as tools to increase the productivity and competitiveness of companies that buy and use them. Design considerations are measured primarily in the realm of cost-effectiveness. If better design can speed up use, cut training time, or add to efficiency in any other such way, then it is important. If it cannot produce a measurable difference in one of these dimensions, then it is a "frill."

Phase 3: Appeal-Driven.   A maturing technology may never enter this phase. For a technology that does, it reaches a wide audience of "discretionary users" who choose the technology because it satisfies some need or urge. The emphasis is not on measurable cost–benefit analyses, but on whether it is likable, beautiful, satisfying, or exciting. The market attractiveness of a product rests on a mixture of its functionality, its emotional appeal, fashion trends, and individual engagement. CB radio and cellular phones for personal use are examples of radio technology moved into this third product phase. Computer games have been there since the beginning, and an increasing portion of computer use is shifting to the consumer end of the spectrum. The huge new markets of the future—the successors to the productivity markets conquered by IBM and Microsoft in the past—will be in this new consumer arena, responsive to different dimensions of human need.

Software design that focuses on the user, not on the mechanisms, is moving to center stage.

The Movement Toward Software Design

"Software design" has become a slogan for the emerging shift of perspective, away from what the computer does, toward the experience of the people who use it. From this perspective, the task of those who create new software is to design the interaction, not to design the program. Although the difference may be subtle (good applications programmers have always paid some attention to designing the interaction), many people are feeling the need to bring it to explicit attention, by creating new professional identities and new affiliations.

Some notable recent examples are as follows:

- In 1993, a professional organization named The Association for Software Design (ASD) was founded with the mission to "transform and elevate the status and quality of software design as an activity." It already has chapters in several locations around the United States and is initiating a program of educational activities.
- In 1994, a publication named *interactions* was founded by the Association for Computing Machinery, in conjunction with the ASD and SIGCHI, the

ACM special interest group on computer–human interaction. In the inaugural issue, the editors stated, "We seem to have moved well beyond the idea that making a computer 'useful' is simply to design a good interface between 'man and machine.' Our ideas have evolved to the point where the richness of human experience comes to the foreground and computing sits in the background in the service of these experiences" (Rheinfrank & Hefley, 1994, p. 88).

- The entire first 1994 issue of the journal *Human–Computer Interaction* (the primary academic journal in the field) was devoted to a dialogue around an article by John Seely Brown and Paul Duguid on the role of context in design (Brown & Duguid, 1994). The editor said "We can look at the development of the field of human–computer interaction as an evolution of what we in the field of HCI consider to be the significant aspects of context for computer-based artifacts" (Moran, 1994, p. 2).
- At the CHI '94 SIGCHI conference, Mitchell Kapor gave the keynote, in which he argued for the primacy of design as an approach to HCI. The conference offered an unprecedented number of papers and sessions devoted to design issues. The 1995 conference instituted a new section called "Design Briefings," which are "specifically intended to provide increased exposure to user interface design and to practical user interface work. They involve the presentation of notable designs and a discussion of how these designs came to be."
- In August 1995 the first Symposium on Designing Interactive Systems was sponsored by ACM SIGCHI, IEEE, and ASD. The call for participation said, "The time is ripe to address designing as a coherent activity—technical, cognitive, social, organizational, and cultural. The goal is to come to a better understanding of how designing works in practice and how we can improve it."

What Is Software Design?

It is evident that software design is coming into prominence. But that should give us a moment's pause. Just what is software design? How does it differ from programming, software engineering, software architecture, human factors, or any of the other labels that have been applied to the activities around creating computers and the programs that people interact with? How does it relate to other fields that call themselves design, such as industrial design, graphic design, urban design, and even fashion design? It is easy to make a new label. The real work is in creating a change in perspective that gives new directions and ideas.

The education of computer professionals has generally concentrated on understanding the nature of computational devices and the engineering that makes them behave as the builder intends. The focus is on the things being designed—the devices and programs and the parts that go to make them up. The goal is to fully implement a specified functionality in a manner that is robust, reliable, and efficient.

When a software engineer says that a piece of software "works," he or she typically means that it is robust, reliable, and meets its functional specification. These concerns are indeed important. A designer ignores them at the risk of disaster.

But this inward-looking perspective with its emphasis on function and construction is one-sided. In order to design software that really works we need to move from a constructor's eye view to a designer's eye view, a view that takes the system, the users, and the situation of use all together as a starting point. When a designer says that something works (e.g., a book cover layout or a design for a housing complex) the sense is much broader—it works for people in a context of values and needs, to produce quality results and a satisfying experience. The key to this shift of perspective is in turning our attention to the larger context in which the object of design resides.

Traditional software engineering has dealt with context in an operational sense, relating a program to the operating systems, networks, programming interfaces, and the like that will surround its operation. Software engineering techniques are geared to expand the possibilities for a program to be modified, ported to other systems, extended to new functionalities, and adapted by users over a lifetime of use. But the focus is always on the mechanisms, not the human situations in which they will be embedded.

The perspective of software design shifts from the "outside-looking-in" focus on mechanisms to an "inside-looking-out" focus on people and their situations: how people experience software, what they do with it, and the larger situation in which they encounter it (see Winograd, 1996, for a number of current perspectives).

ENVIRONMENTS FOR SOFTWARE DESIGN

The development of programming environments was an important step forward in software engineering. A good environment can embody and facilitate the principles and practices that make programming more productive. It brings the programmer's activities into focus along with the activity of the program being produced.

In an analogous way, we can better understand the user-oriented view of software design by looking at what might constitute a "software design environment." In a traditional programming environment, the objects of interest are programs, and the programmer's tools are designed to operate on various representations of those programs. The software design environment is concerned with designing the interactions, and works with a broader array of representations, including different kinds of conceptual models, mock-ups, scenarios, storyboards, and prototypes. The design methods reach outside of the workstation to include the setting and the thinking of the people who will use the software. With recent technological advances (e.g., ubiquitous access, wearable equipment, virtual and augmented realities), the setting of use becomes even more complex to prescribe, and embed into design. Designers will increasingly be called upon to anticipate variety in the context of use and design artifacts that comply to the "anyone-anywhere-anytime" access principle.

The activities of a software designer include the traditional activities of software engineering and programming, such as specification writing, coding, and debugging, along with the user—and experience-focused design activities we emphasize in this chapter. In giving traditional programming concerns short shrift in the sections that follow, we are not implying that they are superfluous or that environments to facilitate them are unnecessary. The emphasis here is on developing our understanding of the additional activities that go on around and through them.

In order to highlight the software design perspective, we look at four current topics of focus in programming environments and lay out the analogous issues in software design point of view, as outlined in Table 9.1 and explained in the following sections.

Of course, as with all analogies, there is not a perfect fit, but the parallels can help elucidate the motivations and criteria for new design environments.

Interactive Programming Versus Responsive Prototyping Media

Modern interactive programming environments emphasize quick turnaround—the ability of a programmer to try something out, see what it does, make changes, and try again in a tightly coupled cycle. In this activity, the nature of the programming language and environment makes a large difference—perhaps as large as the difference between sculpting in clay and sculpting in stone. The ability to quickly shape and reshape requires a capacity for turning an unarticulated idea into a working object quickly enough to be able to change it, listen to it, even throw it out and go on to another.

This kind of "reflective conversation with the materials" (see Schon, 1983, for an excellent analysis of the nature of design activities, in which he introduced this term) is a key to effective design and is even more important for the interaction-intensive programs that dominate today's software world. Both the interface and underlying functionality of the application are incrementally designed through interaction with the intended users. Both users and designers need to be able to experience what the program will be like and what can be done with it, even before it is programmed.

TABLE 9.1

In Expanding From Programming Environments to Environments for Design, There Are Suggestive Correspondences Between Current Techniques and What Is Needed for User-Oriented Software Design

Programming Environments	Environments for Software Design
Interactive programming	Responsive prototyping media
Specifications	User conceptual models
Reusable code	Design languages
Interactive debugging	Participatory design

Abstract representations, such as written descriptions, flow charts, and object class hierarchies cannot provide a grounded understanding. In the past few years, a number of techniques have been developed for initiating a dialogue with the user (and with designers) before writing program code, through mock-ups, storyboards, scenarios, and prototypes (Laurel, 1990; Müller & Kuhn, 1993).

In classical engineering practice, a prototype has been a kind of laboratory test, taking the concept for a device and demonstrating that a simplified version of it could be made to work. In current design practice, prototyping is primarily a vehicle for exploration and communication. Prototypes not only give feedback to the designers, but also serve as an essential medium for information, interaction, integration, and collaboration. The emphasis is on quickly providing an artifact that can be a concrete vehicle for letting the users (and the designers) see both possibilities and problems with the proposed design.

The key element is not the accuracy or thoroughness of the prototype, but the communicative role it plays, both in the designer's interaction with the materials and the user's interaction with the designer. A traditional programming environment emphasizes getting the prototype to do the right things; a design perspective emphasizes getting it to communicate.

Traditional programming environments emphasize visual constructions; a design environment should enable the articulation of design concepts in abstract and alternative concrete manifestations (e.g., tactile, auditory, visual) and combinations thereof. A software design environment needs to support a variety of prototyping levels, suitable for different projects and different phases of a project, each making use of different tools.

Rough Hand Sketches and Scenarios.   The initial step in presenting a concept to a user is to get something that has enough of the general "look" to suggest the functionality and interaction to those who feel it and talk about it. This requires little in the way of technology. Poster board and marker pens or narratives may be all that is needed, and the relevant skills lie in being able to quickly sketch a rough vision, not a polished piece of art. By working with a sequence of sketches, a designer can explore a large space of possibilities for a program—not just its look, but also its functionality. The sketches serve as a communication vehicle for letting users envision what a piece of software might do that they never thought of, and for being able to give insight into how it will actually work in their situation.

Low-Fidelity Prototypes (Wizard of Oz).   Moving beyond static sketches, a number of techniques have been developed for giving the user a sense of the dynamics of a program without having to build a functional version of it. The simplest techniques can be implemented with paper technologies such as Post-its® and transparent overlays, manipulated by a human "machine surrogate" who can pop up and pull down menus, select by gesture, switch window contents, and so on. Even a rough attempt at duplicating the dynamics of the program being designed can give a surprising amount of new insight into what will work and what will falter.

The fact that these prototypes do not feel like a real product is not a problem. In fact, in many design settings it is often important to make sure that prototypes at various levels have a feeling of "roughness"—even to the point of using scanned pencil sketches in place of more polished bitmap art. A user faced with something that has the feel of a rough sketch is more likely to respond with substantive suggestions. A designer is more likely to see strikingly different possibilities. A highly polished prototype—even if it is only a first attempt at the functionality and interface structure—fosters a sense of finality that tends to inspire suggestions for minor improvements and further visual niceties.

Programmed Facades.   "Potemkin village" prototypes can be built on the computer using prototyping tools such as Hypercard, Supercard, Macromind Director, and Toolbook. An interface produced in these languages can present a facade that appears on the surface to be a real program, and that may mimic some illustrative aspects of the functioning of the intended program. But this facade is often supported only by an illusion. The underlying logic of the prototype may duplicate only a tiny fraction of what is intended for the finished design, and it may work for only a carefully selected set of possible interaction sequences.

Even though the programmers know that much is missing underneath, the effect in communicating to others can be tremendous. They get a "feel" for the program that is impossible to get from looking at static screens, and they will be able to see many of its flaws and its new possibilities. In fact, designers have sometimes found that by showing this kind of prototype to users or managers, they create false expectations—it looks so good that it seems like the real thing should be only a short step away. Usually there is a lot more to be designed!

Prototype-Oriented Languages.   There is no clear dividing line between facade-building languages and full-fledged programming languages that are designed to support the prototyping process (often at the expense of traditional computer language concerns, such as execution speed and economy of storage). Environments such as Hypercard, Smalltalk, and Visual Basic include interface builders that make it especially easy to design the screens that people will see and to attach working code to the visual elements. Although these languages are sometimes thought of as the basis for throwaway demonstration prototypes, it has often turned out that for the specific intended use, a program written in one of these languages will be adequate for the job, and does not need to be reprogrammed into a "real" programming language. In deciding how much programming effort should go into a prototype, it is important for the designer to look at these trade-offs and see whether it should be thought of as a throwaway, or be written with the expectation that it could be the basis for the final implementation.

The full design environment is a mix-and-match of all of these prototyping levels. Some projects lend themselves more to one of them. Some projects will best use a mix of all. The goal is to use the level(s) that will best facilitate the two primary interactions—designer with design and user with designer.

Much effort has gone into the design of prototyping systems that can increase the software designer's fluidity of iterative design. This same kind of "creativity acceleration" is produced in other design disciplines in what Michael Schrage (1993) called a "culture of prototyping." He pointed out that different organizations develop and use prototypes in different ways. In some cases a prototype is a rough sketch to be passed around for quick comment and change. In others it is a carefully crafted selling aid, designed to get approval from a manager, customer, or committee that will decide on the future of a project. Some media (like the clay models used in automotive design) lead to resource-intensive prototypes in which a highly finished look leads to assessments of quality. Others, such as carved foam, lead to rough-cut prototypes whose visual and material qualities suggest their provisional status and openness to being changed.

As the field of software design develops, we too are developing not one but many cultures of prototyping, and design environments to facilitate them. We are learning which of them is most appropriate to a given organization and task.

Specifications Versus User Conceptual Models

A key element of many software engineering methodologies is the creation of abstract specifications. These characterize the desired system at a higher level than the operational code, and they can, therefore, be more easily understood, described, and manipulated. There are of course many controversies about the values and limitations of different specification formalisms and methodologies, but we do not address those here.

The analogy we want to draw to software design is with the "conceptual model" (Norman, 1986) or "virtuality" that lies behind the interface seen and manipulated by the user. One of the key differences between software and most other kinds of artifacts that people design is the freedom of the designer to produce a world of objects, properties, and actions that exist entirely within the created domain. The comprehensible but arbitrary consistency of a virtuality is most immediately evident in computer games, which gain tremendous appeal through the ability of the player to engage in the virtual world in earnest, exploring the vast reaches of space, fighting off the villains, finding the treasures, or whatever the designer creates. But there is also a world created in a desktop interface, a spreadsheet, or an information network. We are familiar today with the virtuality of the graphical user interface (GUI) with its windows, icons, folders, and the like. Although these are loosely grounded in analogies with objects in our everyday physical world, they exist in a unique world of their own, with its special "physics" and potentials for action by the user.

The literature on interface design uses a number of terms for the world created by the software, such as conceptual model, cognitive model, user data model, user's model, interface metaphor, user illusion, virtuality, and ontology. What they all share is the recognition that the designer and user are engaged in creating a world, not in simply bringing to the computer what existed outside of it.

In early computer program development, the virtual world was usually a side effect of the implementation. Users of UNIX, for example, work in a world of files, directories, and links (symbolic and direct) because those were elements of the underlying system implementation. This direct mapping onto the implementation model works well for certain kinds of applications and certain kinds of users, but in general the way of dividing up the world that works best for the computer is not the same as the one that will work well for human understanding and acting.

In the development of the Xerox STAR in the early 1970s, designers began to directly confront the question of building a clear and understandable conceptual model (Johnson et al., 1989). Although the STAR did not have the commercial success of its later derivatives, it was the original model for consistent integration of now-familiar mechanisms for windows, icons, dialogue boxes, drawings, and onscreen formatted text. Its interface innovations have been the basis for a whole generation of systems, including the Macintosh, Microsoft Windows, and Motif.

Rather than deciding what the system would do and then figuring out how to produce interfaces, the developers engaged psychologists and designers from the beginning in an extensive set of storyboards, mockups, prototypes and user tests to see what would work, and how. In doing this, they recognized that the most important thing to design properly was the users' conceptual model and that everything else should be subordinated to making that clear, obvious, and substantial. Users could manipulate documents by moving and acting on the icons that appeared on the screen. But, of course, the icon is not the document. The interface could just as well have used pinwheels or little text fragments, and could have let the user operate on them with different physical devices, commands, and visual effects. The key part of the design was the creation of a coherent and consistent world, or "virtuality," with an understandable underlying structure or model.

Tools for designing virtualities have often been based on object-oriented models, in which the object classes reflect the user's perspective rather than being driven by implementation concerns. In fact, object-oriented programming began with the simulation language SIMULA, which started from the standpoint of representing and simulating real-world objects. With later developments, such as Smalltalk and its descendants, designers realized that many of the objects they were creating did not reflect the existence of things outside the computer, but had a life of their own in the virtual world with which a person interacted. It is notable that current methodologies talk about "object-oriented design" rather than "object-oriented programming." This is not to say that the only methods for conceptual design are object oriented, but they have in common a concern with defining and describing the objects, properties, and operations that the user interacts with, rather than the algorithms or representations used by the computer.

Reusable Code Versus Design Languages

One of the major efforts in software engineering today is to find better tools for reusability. Object-oriented software, component software, linked libraries, and many other mechanisms are being explored to enable significant elements of a program to be rearranged and reused in others. In the design world, this kind of borrowing has always been standard, and is technically easier. The concept for a widget such as a tool bar, or an interaction style like a multiple selection can be copied from one application to another without technical difficulty.

A significant part of the larger design environment is the collection of design elements that have been previously used and are standard in a software culture. It does not take deep analysis to see that all of the current GUIs draw on a basic vocabulary and interaction style that was pioneered in the STAR and then the Macintosh. In fact the great success of the Macintosh can be attributed to a large degree to the efforts of the early Apple "evangelists" to encourage applications developers to use a common design vocabulary. They facilitated this by publishing explicit guidelines (Apple Computer, 1987), by providing tools for all of the standard elements (menus, dialogue boxes, window management, etc.) and by working directly with developers. They convinced the developers that they would gain more from promoting the popularity of the Mac platform by making it seem easy to use through uniformity, rather than through having minor differences ("improvements") unique to their interface.

The Role of Design Languages. The Macintosh was the first open platform to publicize a "design language" to use in designing software interaction (the STAR had a carefully articulated design language, but all of the applications were developed by Xerox). A number of design theorists have pointed out how the use of consistent and understandable language by the designer makes it possible to communicate functionality to users in a natural and unintrusive way (Rheinfrank, Hartman, & Wasserman, 1992).

Whenever people construct objects, they draw on a background of shared design language in their community and culture. Even something as apparently simple as a door is built to communicate to the user through convention. A door with a flat plate near shoulder level says "Push me!" One with a round knob says "Turn here," and one with a fixed graspable handle says "Pull." Although these messages are related to the underlying ability to perform the acts, they are also a matter of convention and learning, as every tourist finds out in trying to deal with everyday objects in an unfamiliar culture. We learn such languages from our everyday experience, and when a designer defies them, the result is confusion (Norman, 1988).

Design languages can be more or less natural, more or less "intuitive" (comprehensible to the user on the basis of previous expectations). As a simple example, a slider that moves horizontally can be used to control a dimmer on a light. If it were wired to make the light brighter when moved to the left and dimmer when moved to the right, it would confuse most users from a European culture. There are even some conceptual mappings and metaphors such as "up is more" that cut across all kinds of phenomena and often even across languages and cultures (Lakoff & Johnson, 1980). The designer of an artifact for interaction needs to harness these general cognitive resources and languages to the specifics of the particular interface at hand.

Just as a modern programming environment provides the programmer with a base language and libraries of common program elements, the design environment is populated by the collection of design languages on which the designer can draw in creating something new. This is in spite of all the lawsuits and concerns about "look and feel" infringement. The designer needs to be well versed in all of the common design languages and elements that users will encounter, either to employ them, or to avoid them if economic and legal concerns require that.

Genres/Styles.   It is important to recognize that there is not a best design language for interacting with computers, just as there is not a best kind of building, or a best kind of literature. Every piece of software conveys a "genre," with its own language of expectations and interpretations. We are familiar with genres in literature (the Greek tragedy, the Victorian novel, the pulp romance) and architecture (the Greek temple, the Gothic cathedral, the postmodern office complex). The concept is equally applicable in computer software with the spreadsheet, the video game, and the word processor. The designer working within a background of experience in these genres can effectively use the expectations that go with them to situate the user in previous experiences and to move beyond them (Brown & Duguid, 1994). The power of genres is clear if we try to imagine a spreadsheet with a joystick and lifelike explosive sound effects whenever a formula is entered into a cell, or a word processor that requires the user to assemble words by chasing letters around on the screen.

Of course, the genre with its language can never be taken as the boundary of what can be designed. Just as poets (and even technical writers) will creatively bend language to new purposes, the creative designer will mix, distort, and at times completely violate language conventions for a desired effect. KidPix, a drawing program, comes close to the purported counterexamples cited earlier, bringing video game design language elements (such as wacky sound effects and cartoon icons) into a drawing program. For the intended audience of young children, the mix is quite appealing. In effect, KidPix has introduced a new language that is now being duplicated in other products.

Just as in the cases mentioned previously, software designers will most certainly need to integrate new design languages and/or augment existing ones to derive representations and artifacts that meet diverse or evolving requirements and contexts of use. Such a requirement raises additional challenges for environments for software design, which should, at least partially, provide (computational) support for design language integration and augmentation. To address the level and scope of design language integration/augmentation, it is perhaps appropriate to distinguish between the tacit and embodied manifestations of a design language. The tacit manifestation usually entails the historical underpinnings of the design language, the range and scope of phenomena that can be represented, and so on. On the other hand, embodied manifestations of a design language are typically provided in the form of tools, such as, for example, collections of reusable software libraries, referred to as toolkits for user interface development. Out of the two, the tacit manifestation of design languages is more difficult to capture and encode into software design environments. Embodied design language manifestations, such as toolkits, are easier, though extremely demanding from a software architectural point of view. Nevertheless, the capability of software design environment to support this requirement is likely to be a critical quality attribute, as no single design language is likely to prevail for all types and range of emerging computer-mediated human activities, thus necessitating both design language integration and augmentation.

Interactive Debugging Versus Participatory Design

All approaches to software engineering require a form of testing. Some methodologies call for carefully developed test suites and rigorous testing of components before and after integration into a larger program. As with the other issues addressed earlier, testing takes on a more complex and broader meaning for programs that do not just calculate a result from a few inputs, but that enter into a dynamic (and unpredictable) sequence of interactions with users. There are standard practices in the software industry for testing in use—alpha test, beta test, usability laboratories, and the like (Nielsen, 1993).

An environment for software design includes the tools for testing both the technical tools (e.g., the observational technology of usability labs) and the social tools (the people and practices required to identify, recruit, and interact with testers at different levels and points in the design process). Going further, the process of interaction does not follow the classical "generate and test" where the designer develops a working program and then sends it off to users to test. The dialogue with users can begin with the first sketch of an idea and continue through all the stages in which the functionality as well as the interface are determined. The "debugging" starts with the ideas, not with the code. The environment for this dialogue goes well beyond the workstations and files of the traditional programming environment (Müller & Kuhn, 1993; Schuler & Namioka, 1993).

Expanding the Debugging Environment.   There are limits to what can be learned about software while working in the programming office or the software-testing laboratory. Some aspects, such as the speed and convenience of different interface mechanisms, can be tested to a high degree of accuracy. But others, often much more important, do not show up unless the user is in the natural context in which the system will be employed. What happens when the phone rings in the middle of doing an activity with the system? What if the person at the next desk is using a different word processor and you need to share a document? A designer who creates a system that works in idealized conditions may end up blaming (and alienating) the user when those conditions do not hold in the chaotic realities of his or her life. A designer who can understand and anticipate the chaotic realities can produce a new level of usability.

To get people (both designers and users) to think about these interactions early in the design process, when they can most easily be taken into account, it is often important to interact in the actual setting where the final product will be situated. Rheinfrank et al. (1992) described the insights for copier design that came from extensive field visits to see where and how the copiers were really used, and by whom. Much of the success of the Quicken program for personal finance is attributed to an explicit "follow the user home" policy, in which the designers worked with people who purchased early copies to see what actually happened when they tried to install, use, and integrate them into their everyday practices.

In the area of office software, this problem has been addressed through a method called "contextual design" (Holtzblatt & Jones, 1993), in which the designer enters into the situated context of the user to learn about possibilities. An environment for the prototyping and conceptual design tools discussed previously needs to extend beyond the walls of the software organization to engage users in the process.

Use in Organizations.   When we think of a piece of software on a personal computer, we tend to visualize the user as a person—an individual sitting in front of the machine. On the other hand, when we think of a traditional mainframe-based system, such as an airline reservation system, inventory control system, or payroll system, there is not a single prototypical individual user, but an "organizational user" composed of many people with different roles and functions. This distinction between personal and mainframe software is blurring in today's age of distributed client-server software, interconnected information networks, and groupware. The design of a computer application, regardless of its specific details, is intertwined with the design of the organizational interactions that surround its use.

Over the past few years, a number of approaches have emerged to looking at how the design of computer systems interacts with the design of organizations and their activities. Conferences, journals, and books have appeared on computer-supported cooperative work (CSCW), groupware, and organizational computing (Grudin, 1991). Of course, since the beginning of computing, people have used computers in group and organizational settings. The shift lies in asking the designer to focus on the way that design will affect people in those settings.

One interesting indicator is in the change of terminology used by information systems professionals. In earlier days they talked about the importance of "systems analysis"—getting a model of the organizational system before designing the information structures for it. Today we hear more in the management and information technology magazines about "business process reengineering." The structures and practices of the business are not taken as a fixed environment to be analyzed and fit into, but as a domain of potential change and new design (see Winograd & Flores, 1987).

With this shift there has been an increasing interest in what can be offered by those who bring to software, systems, and design, lessons from systematic studies of the nature and structure of work, independently of its technological augmentations. Anthropologists, ethnographers, sociologists, organizational theorists, and others have become a part of the interdisciplinary teams that approach the design of systems from a "work-oriented" perspective (Greenbaum & Kyng, 1991).

Coevolution of Practices, Tools, and Social Systems.   The final extension in scope of the dialogue with users is that the design cycle does not start and end with a product. The overall environment of computer use is a constant coevolution in which new tools lead to new practices and ways of doing business, which in turn create problems and possibilities for technical innovation. We tend to think of the designers and programmers in the development laboratory as the primary part of the design environment, but in this larger picture, the people in customer support are also central participants in the dialogue. Many companies are beginning to integrate this critical source of feedback into the design cycle explicitly, some even to the degree of requiring that system designers spend a significant amount of time in a help-desk role, to see the consequences of their design in actual use (Adler & Winograd, 1993).

The Designers' Organizational Environments.   Many of the design activities outlined in the preceding sections have been advocated for over a decade and yet are far from standard practice in the industry. The picture is oversimplified in its implicit notion of a designer or design team, working in concert to produce software. In fact the software organization contains many disjoint parts concerned with software design, from marketing to interface design and development, to customer support, training, and documentation. The coordination of these often far-flung groups with diverse interests and responsibilities makes it a very different matter to put into real practice the theoretical practices that go into our design environment (Poltrock & Grudin, 1994). In many cases, the most significant elements in creating a productive environment for software design are the organizational structures and changes that need to be made in order to support the communication and flow of activities that constitute software design.

CONCLUSION

The environment for the designer goes well beyond the traditional bounds of programming environments. In fact the descriptions herein may feel to many readers like an extension without bounds—opening up the concerns of the designer to so many issues and methods that nothing can ever get done. Of course, not every environment or every piece of software requires explicit attention to all the dimensions of design. A project to port an E-mail interface from one window system to another may require careful attention to design languages, but it can take for granted most of the initial analysis of the setting and patterns of use. An attempt to create a totally novel kind of application may require situated observation of what people do and how their lives might be changed by a new technology, but it may be the kind of application that does not require careful study of how the new software will modify practices of a group.

The point in taking a broad view here is to prompt awareness—awareness on the part of software designers of the issues they may need to think about, and awareness on the part of those who create environments (computational, physical, and social) for those designers of the objects and methods they need to support. As with all tools, there is no magic—the environment does not produce the result. But a comprehensive and thoughtfully constructed environment can facilitate the human creativity that is always at the core of design.

ACKNOWLEDGMENTS

This chapter was adapted from "Environments for Software Design" by T. Winograd, 1995, *Communications of the ACM, 38*(6), pp. 65–74. Copyright 1995 by ACM, Inc. Adapted with permission of the author.

REFERENCES

Adler, P., & Winograd, T. (Eds.). (1993). *Usability: Turning technologies into tools.* New York: Oxford University Press.

Apple Computer. (1987). *Human interface guidelines: The Apple desktop interface.* Reading, MA: Addison-Wesley.

Brown, J., & Duguid, P. (1994). Borderline issues: Social and material aspects of design. *Human–Computer Interaction, 9*(1), 3–36.

Greenbaum, J., & Kyng, M. (1991). *Design at work: Cooperative design of computer systems.* Hillsdale, NJ: Lawrence Erlbaum Associates.

Grudin, J. (Ed.). (1991). Special issue on collaborative computing. *Communications of the ACM, 34*(12).

Holtzblatt, K., & Jones, S. (1993). Contextual inquiry: A participatory technique for system design. In D. Schuler & A. Namioka (Eds.), *Participatory design: Principles and practices* (pp. 177–210). Hillsdale, NJ: Lawrence Erlbaum Associates.

Johnson, J., Roberts, T., Verplank, W., Smith, D.C., Irby, C., Beard, M., & Mackey, K. (1989). Xerox Star, a retrospective. *IEEE Computer, 22*(9), 11–29.

Kapor, M. (1991). A software design manifesto: Time for a change. *Dr. Dobb's Journal, 16*(1), 62–67.

Lakoff, G., & Johnson, M. (1980). *Metaphors we live by.* Chicago: University of Chicago Press.

Laurel, B. (1990). *The art of human–computer interaction.* Reading, MA: Addison-Wesley.

Moran, T. (1994). Introduction to the special issue on context in design. *Human–Computer Interaction, 9*(1), 1–2.

Müller, M., & Kuhn, S. (Eds.). (1993). Special issue on participatory design. *Communications of the ACM, 36*(6).

Nielsen, J. (1993). *Usability engineering.* Boston: Academic Press.

Norman, D. (1986). Cognitive engineering. In D. Norman & W. S. Draper (Eds.), *User-centered system design: New perspectives on human–computer interaction* (pp. 31–62). Hillsdale, NJ: Lawrence Erlbaum Associates.

Norman, D. (1988). *The design of everyday things.* New York: Basic Books.

Poltrock, S., & Grudin, J. (1994). Organizational obstacles to interface design and development: Two participant-observer studies. *ACM Transactions on Computer–Human Interaction, 1*(1), 52–80.

Rheinfrank, J., Hartman, W., & Wasserman, A. (1992). Design for usability: Crafting a strategy for the design of a new generation of Xerox copiers. In P. Adler & T. Winograd

(Eds.), *Usability: Turning technologies into tools* (pp. 15–40). New York: Oxford University Press.

Rheinfrank, J., & Hefley, W. (1994). Reflections. *Interactions, 1*(1), 88.

Schon, D. (1983). *The reflective practitioner.* New York: Basic Books.

Schrage, M. (1993). The culture(s) of prototyping. *Design Management Journal, 4*(1), 55–56.

Schuler, D., & Namioka, A., (Eds.). (1993). *Participatory design: Principles and practices.* Hillsdale, NJ: Lawrence Erlbaum Associates.

Winograd, T. (Ed.). (1996). *Bringing designing to software.* Reading, MA: Addison-Wesley.

Winograd, T., & Flores, F. (1987). *Understanding computers and cognition: A new foundation for design.* Reading, MA: Addison-Wesley.

10 From Human–Computer Interaction to Computer-Mediated Activity

Liam J. Bannon
Victor Kaptelinin

In this chapter the authors outline how activity theory - an approach originating from Russian cultural-historical psychology - might be used in the understanding and development of technological artifacts that meet the needs of various categories of users. A brief introduction to some of the key ideas of activity theory and their historical development is followed by a discussion of the potential strengths and limitations of this approach as a theoretical foundation for the understanding and design of computer-based artifacts. It is shown that the focus of the approach on issues such as development, social context, and mediation of purposeful human activities radically transforms the object of study in Human-Computer Interaction. Activity theory requires that technology is integrated into the context of meaningful human activities, and provides concrete conceptual tools for capturing key aspects of the context. The implications of the activity theory perspective presented in this chapter for designing User Interfaces for All are discussed, and a useful bibliography for further reading is included.

The concerns of this book are to increase awareness of, and possible solutions to, the difficulties we have today in the use of the ubiquitous information technology that plays an increasingly large role in all our lives—at work, at home, and in our leisure activities. The development of an information society, as heralded by many national and transnational bodies, will involve increased acceptance and use of a myriad of information appliances in our everyday lives. Yet, there are a number of serious concerns about the kinds of technologies that we are developing, in terms of their accessibility to many sections of the population. Although the need for human factors input into the design of artifacts is generally recognized, often this input is reduced to a very minimal role (e.g., discussing with industrial designers the layout of controls on devices).

The need for a strong and continued emphasis on usability issues throughout the design process, from initial conception, through to testing of early prototypes, is still not universally recognized, or, if recognized, is not always acted on in design practice. Likewise, the need for involvement of those people who will ultimately use the devices in the whole design process is still not accepted by many design groups. As noted in chapter 1 of this volume, we need to move from a *reactive* stance—evaluating systems after they are deployed—into a *proactive* stance, where both the users themselves and human factors experts play a significant role in the whole design process.

Over the years, the importance of these issues has grown. When technologies were used only by few specialized "operators," then special training could be given to them in order to learn how to use the complex equipment. Nowadays, computers permeate every aspect of our lives, and are used on a discretionary basis by millions of people all over the world with different professional competencies, different cultural backgrounds, and different individual capabilities. Therefore, the onus of fitting hardware and software to the varieties of human users and use situations is on the designers of the technologies. This contrasts with the situation that has existed for many years, where people had to learn to conform to the machine's demands, rather than the other way around.

The evolution of the human–computer interaction (HCI) field in the early 1980s, and its rapid growth in size and importance over the intervening years, attests to the increased awareness by industry that in order to achieve the purported advantages of the information society, further effort must be put into making their technologies more "user friendly."[1] Despite the growth of interest in HCI issues, it is still the case that most HCI work operates on the implicit assumption that the intended user population consists of relatively well-off, well-educated, reasonably fit Northern European or North American men in the age range 15–45.[2] The field of ergonomics has always had a broader brief, especially in Europe, and large-scale ergonomic studies have been done in such areas as health and education, but the resources available for this work have almost always been minimal in comparison to that available for military human factors and ergonomics work. Again, this has meant that our knowledge of the general population's range of capacities and competencies is seriously incomplete, in comparison to the specialized capacities of military personnel. More narrowly, it is true that, within the HCI field, this human factors problem has not been as noticeable, but it is still the case that most funded HCI work does not take into account the needs of those who have extraordinary capabilities (cf. Edwards, 1995). In his keynote address to the InterCHI '93 conference, Alan Newell noted the important point that a good user interface is one that should be usable by *all* sectors of the user community—that is, we should not partition the design space into a space for "normal" users and a separate design space for those categorized as non-normal "extraordinary" users. All of us, to some degree, have "special needs"—in certain tasks, in certain situations, at certain times. The increased interest in "usability issues" by the computer industry, and the increased awareness concerning cultural differences among user populations (e.g., the fact that colors, such as red

---

[1]Despite the popularity of this term, especially in advertising literature on technology, we are not that happy with it, as good design of, for example, information appliances, has more to do with understanding the use context and making a tool fit for use than it does to having the device appear "friendly." We would not describe many useful artifacts in everyday life as being "user friendly," so why should we use the term for computer-based artifacts?

[2]In earlier days in the human factors community, due to the fact that so much human factors work was done for the military, a dissenting human factor group used to have special sessions at human factors conferences entitled "not everyone is a 29-year-old air force pilot"—to emphasize this point!

and green, do not have a universal cultural interpretation of "danger/stop" and "OK/go") are steps in the right direction—toward the design of *user interfaces for all*. However, much work still needs to be done.

In our own HCI work over the years, we have attempted at both a theoretical and a practical level to improve the accessibility, usability, and utility of technology for people. We have emphasized the importance of viewing the computer as a medium through which people interact, and not simply as a calculator, or even a tool (Bannon, 1986). We have emphasized the fact that people are attempting to accomplish an activity through computers, and not simply "using the computer" as an end in itself. Thus, the issue is not improving instruction for computer users, but making more effective tools and media that help people in different walks of life to accomplish their goals. Therefore, the problems people have with computers are seen not as a lack of "computer knowledge," but as a failure of designers to understand the nature of the work and the work setting. We prefer to speak of "computer-mediated activity" rather than "human–computer interaction" for the same reason (Kaptelinin, 1996a). Our work has contributed in the shift from a system-centered to a user-centered design process (Norman & Draper, 1986). We have also emphasized the importance of participatory design practices, as a way of ensuring that the designs we develop truly meet the needs of people (Bannon, 1990). We also highlight the importance of studying use as a prelude to design (Bannon & Bødker, 1991). We study use throughout the design cycle, through developing mock-ups and scenarios of future use that allow people to experience the future use situation, and then again, in developing early prototypes of systems that can be tested, so that the results of these tests can be fed back into the design process in order to improve the system (Bannon, 1996; Grønbaek, Grudin, Bødker, & Bannon, 1993).

In this chapter, our focus is on introducing the conceptual framework that we both have found helpful in developing our understanding of HCI issues, namely activity theory. This framework shifts attention away from the interface per se, and focuses on computer-mediated activity. We believe that this shift in focus is extremely important if we are to develop truly useful and usable systems that support people in their everyday activities. The framework emphasizes the concept of mediation in all human activities, and its strongly historical approach provides us with a powerful tool for viewing the computer system as yet another, albeit much more powerful and flexible mediational device that is used by people to accomplish certain goals. Although the conceptual framework can be at times obscure, it provides a useful conceptual tool for understanding such issues as user goals, mediational means, work context or environment, and collective human activities (Kaptelinin, 1996b; Kuutti & Kaptelinin, 1997). We show how some of these issues tend to be neglected in the traditional information-processing account of human cognition, dominant in the HCI tradition to date. As we show, recently introduced techniques for practical systems development, such as the use of scenarios, rapid prototyping, and so on, can be seen to fit very well with activity theoretical considerations.

ACTIVITY THEORY AS AN ALTERNATIVE TO THE COGNITIVE
SCIENCE PERSPECTIVE

Donald Norman, one of the pioneers in the field of cognitive science, wrote a prescient article that pointed to the gaps existing in the then fledgling new discipline (Norman, 1980). He was particularly concerned about the basic building block in this approach, which he referred to as the model human-information processor:

> The problem seemed to be in the lack of consideration of other aspects of human behavior, of interaction with other people and with the environment, of the influence of the history of the person, or even the culture, and of the lack of consideration of the special problems and issues confronting an animate organism that must survive as both an individual and as a species. (p. 2)

Many researchers now agree that the information-processing model of the human is too limited, and too fractionated, neglecting important aspects of human activity. The social nature of human learning is downplayed. Questions of motivation in the performance of tasks are not considered sufficiently. The way people perform activities in everyday life is not studied adequately, with the assumption that lab studies can be generalized readily to many work situations. A focus on representational formalisms predominates. The intent is often to understand human activities in order to substitute with computing procedures, rather than to support people with better computer tools. The underlying model of the user seems at times patronizing and misguided—naive users, idiot-proof system design, and so on.

A number of challenges to the standard rationalistic mainstream cognitive science theoretical framework have emerged. Some theorists focus on the individualistic nature of much cognitivist theorizing (e.g., Velichkovsky & Zinchenko, 1982), arguing for greater attention to the setting in which cognition takes place and how it is shaped by this setting (Lave, 1988). Some have argued for a radically different epistemology for the discipline, eschewing the Cartesian model for a hermeneutical interpretation. Such a radical critique has been popularized within the computing fraternity by the work of Winograd and Flores (1987) starting out from the work of Heidegger, Maturana, and others. Another long-standing critique of Cartesianism comes from the dialectical materialist tradition within Soviet thought, developed from ideas about activity present in the work of Hegel, then elaborated by Marx and applied within a psychological framework by psychologists such as Vygotsky (1978) and Leont'ev (1978). This cultural-historical or activity approach is the one that we mention here, as it has a number of interesting features within our present context.

Perhaps the term *activity theory* is somewhat of a misnomer for the perspective. Although the concept of activity is central to this approach, there is not some monolithic theoretical superstructure that is accepted as defining the theory. Although elaborated most completely in the domain of psychology, the concept is not restricted to this domain. For many years, activity theory was the leading theoretical orientation in Russian psychology, and a large number of studies were conducted within this framework. Nowadays, activity theory is not an exclusively Russian approach. Recent developments in activity theory are associated with a larger research community, which also include researchers from Finland, Germany, Denmark, the United States, and other countries. There are also attempts to expand the coverage of activity theory beyond a purely psychological realm toward more general socially and organizationally oriented problems in understanding the dynamics of work activities, like the "Developmental Work Research" developed in Finland by Yrjo Engeström and his coworkers (e.g., Engeström, 1990). The theory is difficult to comprehend without a background in German philosophy and Soviet thought, and the reader is forewarned that the following comments barely scratch the surface of the theoretical concepts that underlie this perspective. For further accounts of this approach, see Engeström (1987), Kozulin (1986), Kuutti (1991a, 1991b), Raeithel (1992), Wertsch (1981, 1998) and Leont'ev (1978). Backhurst (1988) referred to the cultural-historical tradition as "communitarian" in distinction to the "individualistic" Cartesian tradition, and noted three key aspects of this approach:

1. Activity (social forms of material activity) explains the nature and origin of human consciousness. "We become human through labor" (Leont'ev, 1978).
2. The higher mental functions are social in nature and origin—in other words, to use a phrase coined by Cole et al. (Vygotsky, 1978), mind is "in society."
3. The higher mental functions are internalized forms of social activity (Backhurst, 1988).

What is of interest in this approach is a more theoretical framing of certain issues, which are difficult to conceptualize within traditional information-processing accounts of human behavior. For example, the problem of context, which has become increasingly recognized as a crucial issue, is built into the very basis of the theory, in terms of activities. The conceptual framework of activity theory can be presented as a set of underlying principles. We focus on the following basic principles of the approach: object-orientedness, internalization/externalization, tool mediation, hierarchical structure of activity, and development.

Object-Orientedness. The principle of object-orientedness describes the specific activity theory point of view on the nature of objects with which human agents interact. On the one hand, as mentioned previously, activity theory is based on materialistic Marxist philosophy, and it assumes that human beings live in an objective reality that determines and shapes the nature of subjective phenomena. This basic assumption makes it possible to seek for an objective account of subjective phenomena. Psychology, according to Leont'ev, can be (and should be) no less a thorough, rigorous science than natural sciences are. On the other hand, Leont'ev clearly understood that the concept of object in psychology cannot be limited to physical, chemical, biological, and so on, properties of things. Socially determined properties of things, especially those of artifacts, and the very involvement of things in human activity, are also objective properties, which can be studied with objective methods. So, the principle of object-orientedness states that human beings live in a reality that is objective in a broad sense; the things that constitute this reality not only have the properties that are considered objective according to natural sciences, but socially/culturally defined properties as well.

Internalization/Externalization.   Because human interaction with reality is the subject matter of several disciplines, and because activity theory was originally developed as a psychological approach, the problem emerges of how to define the specifically psychological perspective on activity and how activity theory is related to other psychological approaches. Activity theory differentiates between internal and external activities. The traditional notion of mental phenomena (e.g., attention, consciousness, imagination) corresponds to internal activities. External activities, on the other hand, are physically observable movements and actions—such as grasping a tool, counting on one's fingers, moving objects in the world, speaking or writing, and so on. Activity theory emphasizes that internal activities cannot be understood if they are analyzed separately, in isolation from external activities, because there are mutual transformation between these two kinds of activities: internalization and externalization. It is the general context of activity (which includes both external and internal components) that determines when and why external activities become internal and vice versa.

Internalization, that is, transformation of external activities into internal ones, provides a possibility for human beings to try potential interactions with reality without performing actual manipulation with real objects. In some cases, external components can be omitted in order to make an action more efficient (e.g., in the case of calculations). In other cases, internalization helps to identify the optimal way of action before performing this action externally. Externalization, that is, transformation of internal activities into external ones, is often necessary when an internalized action needs to be "repaired," or when collaboration between several agents requires their activities to be performed externally in order to be coordinated.

Tool Mediation.   The activity theory emphasis on social factors, and on the interaction between agents and their environments, explains why the principle of tool mediation plays a central role within the approach. First of all, tools shape the way human beings interact with reality. According to the aforementioned principle of internalization/externalization, shaping external activities finally results in shaping internal ones. Second, tools usually reflect the experience of other people who tried to solve similar problems before and invented the tool or modified the tool to make it more efficient. This experience is accumulated in the structural properties of tools (shape, material, etc.), as well as in the knowledge of how the tool should be used. Tools are created and transformed during the development of the activity itself, and carry with them a particular culture—historical remains from that development. So, the use of tools is a way toward the accumulation and transmission of social knowledge. It influences the nature not only of external behavior but also of mental functioning of individuals. According to Vygotsky, there are two categories of tools: technical and psychological. Technical tools are intended to manipulate physical objects (e.g., a hammer), whereas psychological tools are used by human beings to influence other people or themselves (e.g., the multiplication table, a calendar, or an advertisement).

On the one hand, tools expand our possibilities to manipulate and transform different objects, but on the other hand the object is perceived and manipulated, not "as such," but within the limitations set by the tool. Thus, mediating tools have both an enabling and a limiting function. Activity theory also directs attention toward the contexts of use. The tools are never used in a vacuum. Instead, tools and all the conceptions related to them have been shaped by the social and cultural context where the use is taking place.

Hierarchical Structure of Activity.   Leont'ev's version of activity theory is often associated with a three-level scheme describing the hierarchical structure of activity. The central level (or, rather, group of levels) is that of actions. Actions are oriented toward goals, which are the objects of actions. Usually, goals are functionally subordinated to other goals, which may be subordinated to still other goals, and so forth. Moving up the hierarchy of goals we finally reach a top-level goal that is not subordinated to any other goal. This top-level goal, which in activity theory is designated as "motive," is the object of a whole activity. Basically, motives correspond to human needs. They are the objects that motivate human activities, whereas goals are the objects human activities are directed at. Moving down the hierarchy of actions we eventually cross the border between conscious and automatic processes. The latter, which individuals are not aware of, are "operations" closely linked to the actual conditions. For instance, when driving a car people often notice and follow road signs without thinking about them. Operations do not have their own goals; they rather provide an adjustment of actions to current situations. According to activity theory terminology, they are operations. Therefore, activities, which are driven by motives, are performed through certain actions that are directed at goals and that, in turn, are implemented through certain operations.

Development.   Activity theory requires that human interaction with reality should be analyzed in the context of development. Of course, activity theory is not the only psychological theory that considers development as a major research topic. However, in activity theory development is not only an object of study, but also a general research methodology. Activity theory sees all practices as the result of a certain historical development under certain conditions and continuously reforming and developing processes. According to the philosophy of dialectical materialism and, more specifically, to dialectical logic developed by Evald Ilyenkov (see Backhurst, 1991), any system can be understood only through analysis of its developmental transformations. That is why the basic research method in activity theory is not a traditional laboratory experiment, but the so-called "formative experiment," which combines active participation with monitoring of the developmental changes of the object of study.

The activity theory perspective on human development is based on Lev Vygotsky's ideas about development as being both individual and social. In particular, the notion of the *zone of proximal development,* proposed by Vygotsky (1978), is one of the most central notions within the conceptual framework of activity theory. According to Vygotsky, it is important to take into consideration not only the actual level of development, determined by individual problem solving, but also the level of potential development, determined by problem solving *in collaboration with other people.* Assessment of the distance between these two levels, that is, the zone of proximal development, has important implications for creating optimal conditions for learning, including learning with computer-based artifacts, and learning how to use such artifacts.

These basic principles of activity theory should be considered as an integrated system, because they are associated with various aspects of the whole activity. A systematic application of any of these principles makes it necessary eventually to engage with all the others. For instance, an analysis of the mechanisms underlying the social determination of the human mind should take into consideration tool appropriation, internalization of social knowledge, and transformations of the structure of activity resulting from learning and development. After this general presentation of the framework, we now investigate its use in the field of HCI. (Please see the Appendix for additional sources on activity theory and the design of computer systems, activity theory in general, and other related papers.)

ACTIVITY THEORY AND HCI

Can we apply the alternative activity theory framework discussed earlier to the field of HCI? The recent history of HCI research based on this approach suggests that the answer to this question is affirmative. First of all, it should be noted that, in Russia, activity theory has been used for many years as a conceptual foundation for theoretical and practical work in the area of human factors and ergonomics (Munipov, 1983; Zinchenko & Munipov, 1989). Perhaps the first attempt to introduce activity theory to the field of HCI was made by Susanne Bødker about 10 years ago (Bødker, 1989, 1991). Since that time, there has been a growing HCI tradition of activity-theory-based HCI studies in Europe, especially in Scandinavia (e.g., Bannon, 1990; Bannon & Bødker, 1991; Christiansen, 1996; Draper, 1993; Engeström & Escalante, 1996; Kaptelinin, 1994, 1996a; Kuutti, 1991a, 1991b; Raeithel, 1992), as well as in North America (e.g., Bellamy, 1996; Blumenthal, 1995; Cohen, Candland, & Lee, 1995; Nardi, 1992, 1993, 1996) and Australia (e.g., Bourke, Verenikina, & Gould, 1993).

Let us note some of the features of the theory that have made some researchers consider it as a promising conceptual framework for HCI. Fundamental to this approach is the *mediation of activity through tools or artifacts.* As noted earlier, such tools may include symbolic signs, language, as well as physical instruments. Tools are developed by a community over time, they can be viewed as crystallized knowledge, and they are changed through use. Activities are accomplished through actions, which are in turn implemented as a series of subconscious operations corresponding to the context in which the action is being performed. Through learning, we transform conscious actions into operations. However, if conditions change, then the flow of operations can be broken and they can again reappear consciously as actions. In "normal" use situations, our handling of artifacts is done through operations, and is not conscious to us. Thus, Bødker noted that an artifact works well in our activity, if it allows us to focus our attention on the real object, and badly, if it does not; therefore, we should talk about human operation of a computer application rather than of HCI.

This approach focuses on the character of the operations performed, and the conditions under which they are activated. Through design we will change operations and their conditions, and in order to understand this, we have to allow the user to try out the new artifact in the work process, as we cannot predict in advance what the future operations that accomplish an action will be. Here we have a firmer basis for arguing why some form of envisioning or prototyping is required in design. The role of practice within groups in the theory makes it possible to deal with HCI not just concerning an individual user, but focusing on groups who share a practice. Design of artifacts is a process in which we determine and create the conditions that turn an object into an artifact of use. The future use situation is the origin for design, and we design with this in mind. Use, as a process of learning, is a prerequisite to design. Through use, new needs arise, either as a result of changing conditions in human activities, or as a recognition of problems with the present artifacts. To design an artifact means much more than designing the object used by human beings in a specific kind of activity. As the use of artifacts is part of social activity, we design new conditions for collective activity, for example, a new division of labor, and other new ways of coordination, control, and communication. Design of educational support is important too, because the artifact is to be integrated into an existing practice.

According to activity theory, the computer is just another tool that mediates the interaction of human beings with their environment. The only way to come to an adequate understanding of HCI is to reconstruct the overall activity of computer use. As Kuutti (1992) argued, activity provides a "minimal meaningful context" for HCI. The questions that arise when computer use is considered from the point of view of activity theory are: What is the hierarchical level of HCI within the structure of activity? Does computer use correspond to the level of particular activities, to the level of actions, or to the level of operations? Which tools, other than computerized ones, are available to the user? What is the structure of social interactions surrounding computer use? What are the objectives of computer use by the user, and how are they related to the objectives of other people and the group or organization as a whole?

Another general idea directly relevant to the field of HCI is that of development. The importance of analyzing computer use within a developmental context is relevant to both the individual level and the group/organizational level. The assimilation of new technologies causes the emergence of new tasks (the so-called "task-artifact cycle"; see Carroll, Kellogg, & Rosson, 1991). A possible way to cope with unpredictable structural changes on a user activity is to support users in customizing the system according to their current needs (Henderson & Kyng, 1991). Yet this is not a universal solution because users often need substantial assistance even in formulating their own needs. So, a conceptual analysis of the basic factors and regularities of organizational development is needed to predict this development and to provide an efficient use of information technologies.

The development of individual expertise is also an important factor that is not adequately addressed by the cognitive approach. Cognitive models of skill acquisition, based on ideas of procedural knowledge compilation or chunking, have troubles with accounting for the qualitative changes that cognitive skills undergo in the process of development (Kaptelinin, 1993). Yet these very transformations can be studied and predicted from the standpoint of Bernstein's (1967) theory, which is usually closely associated with activity theory.

The tool mediation perspective suggests a structure for HCI that is radically different from the information-processing loop. The components of the structure should not only be the user and the computer, but also the object the user is operating on through the computer application, and the other people with whom the user is communicating (Bødker, 1991). The tool mediation perspective means that there are actually two interfaces that should be considered in any study of computer use: the human–computer interface and the computer–environment interface. The interface in the traditional sense is not only a border separating two entities, but also a link that provides the integration of a computer tool into the structure of human activity. The mechanisms underlying this integration can be understood from the point of view of activity theory as the formation of a functional organ. This means that computer applications are the extensions of some natural (precomputer) human abilities. One of the most important functions of computer tools in the structure of human activity is the extension of the cognitive structure referred to within activity theory as the "internal plane of actions" (see Kaptelinin, 1996b).

One fundamental difficulty related to building up a theory of HCI is the changing nature of the subject matter of the study. In contrast to physical laws, the laws of HCI are not necessarily invariant over time. When the current methods, styles, standards, and so on, of HCI are used, the results are inevitably obsolete soon after they are formulated. Activity theory puts HCI into the context of basic, invariant principles underlying human activity, so it provides a better chance for creating a theoretical framework that has a predictive potential.

Papers discussing the activity theory perspective in HCI have for the most part dealt mainly with theoretical arguments as to why this approach might be of benefit to the HCI field. However, there are a growing number of applications of activity theory to actual design and evaluation of computer technologies. The first examples of such applications have been *retrospective analyses,* that is, attempts to look at already finished projects and, with the benefits of a hindsight, see if activity theory could be useful if it were applied to those cases (e.g., Bødker, 1991). Second, a number of *conceptual tools* have been developed to support various phases of system development, from identifying potential types of computer applications (Kuutti, 1992), to capturing the context of computer use (Kaptelinin & Nardi, 1997), to structuring the process of cooperative design (Bødker & Christiansen, 1997). Third, activity theory has been used as a conceptual framework in projects aimed at *understanding the large-scale social context* in which specific information technologies are being developed and used (Engeström, 1990; Engeström & Escalante, 1996; Norros & Hukki, 1995). Finally, there are also cases of a direct use of activity theory in very concrete *design projects*. For instance, Blumenthal (1995) reported a combined application of several approaches, namely industrial design, traditional HCI, and activity theory, in design of the BodyWise watch that was actually an interactive device, a component part of complex fitness equipment. It was shown that activity theory could help to identify problems related to human use of technology, which could hardly be revealed by more traditional approaches. The basic conclusion of the study was that "activity theory is most useful in determining just what components and functionality the system should support" (Blumenthal, 1995).

In our view, there are good reasons to expect even more tangible results from activity theory in the coming years. We believe that a new model of HCI will replace the information-processing loop underlying the cognitive approach. This model will identify and present in a thorough way the most important aspects of computer use by individuals and groups or organizations. We hope this model will provide various parties involved in the study and design of HCI with a framework that can make their mutual understanding and cooperation more efficient.

Also, activity theory can make an important impact on the further development of design support tools. The design of a new interactive system involves the design of a new activity—individual or organizational. However, even the perfect design of an ideal activity does not guarantee the success of a system. The transformation of an activity from an initial to a target state can be difficult, and even painful. Activity theory may be used as a basis for the development of a representational framework that would help designers to capture current practice, as well as to build predictive models of activity dynamics. Such conceptual tools would enable designers to achieve appropriate design solutions, especially during the early phases of design.

USER INTERFACES FOR ALL: AN ACTIVITY THEORY PERSPECTIVE

The main ideas underlying activity theory have direct implications for the central theme of this book, that is, development of computer technologies that can be considered useful and usable by a wide variety of potential users. First of all, the conceptual approach presented previously shifts the focus of inquiry from low-level processes of human interaction with technology, toward predominantly purposeful human activities and the role of computers in mediating such activities. In other words, the object of design should be not only *user interfaces,* but *meaningful activities,* which are mediated by computers. This offers a more powerful unit of analysis for the study of context and a richer means for understanding why and how diversity can be accommodated. The main advantage of this approach is that it provides a general conceptual framework that can be instrumental in directing and coordinating concrete projects. In particular, activity theory can help to formulate research questions and set priorities for development.

As emphasized earlier, human activities are shaped by a number of interrelated factors. Activities are hierarchically structured and mediated, they have internal and external components that transform into each other, and they develop and take place in a social context. Taking all these aspects of human activities into account might help to approach the issue of user interfaces for all in a systematic way. Next we formulate several tentative conclusions based on the discussion in the previous section in this chapter.

Let us begin with some ideas that might seem rather trivial, but are often being overlooked by system developers. In many cases, developments in HCI have been stimulated more by technological possibilities than by real needs of users. In order for the ideas and approaches in this book to make a difference, they should be applied to situations where improving the quality of user interfaces helps to resolve a real-life problem. For instance, meeting specific interface requirements of people over 65 is more important in the case of Internet banking (especially if switching to the Internet is associated with closing down regular bank branches in scarcely populated areas) than in the case of arcade-style computer games. It would be an illusion to expect that the problem of user interfaces for all is just a technical one. Because human activities are shaped by various factors, technical issues are inseparable from social and psychological ones. It would be reasonable to consider increased accessibility of information technologies as a result of concerted efforts aiming at the development of better technologies, efficient user learning, and creating appropriate social environments for use. Perhaps helping users over a certain age to communicate with each other about problems and solutions can in some cases be more successful than a sophisticated individualized adaptable interface (see Bannon, 1986).

Also, activity theory can make a more concrete contribution to theoretical and practical work in the area by providing specific ideas and tools. As mentioned in the introduction of this chapter, it is important for designers to understand the needs and requirements of specific users to move away from the preoccupation of many approaches in the field of HCI with well-educated young White men. There are several examples of checklists and other practical techniques based on activity theory that have been developed to support practitioners in focusing on the most relevant aspects of the context in which people use technology (Bødker & Christiansen, 1997; Kaptelinin & Nardi, 1997).

Finally, the notions of functional organs and the zone of proximal development seem to be very promising for further work on extending accessibility of computer technology. Human activities have enormous plasticity, mostly because there is a huge range of technical and psychological tools existing in the culture. Understanding the mechanisms of how people create functional organs on the basis of various artifacts appears to be an important factor for further progress in the area of assistive technology (Kaptelinin, 1996b; Kuutti & Kaptelinin, 1997). Also, perhaps the most serious barriers precluding some categories of people from using computer technologies more successfully are cultural ones. Computer use often brings with it its own culture, which might look alien to some people. Part of the problem is caused by the computer culture itself, which definitely needs to become more open. However, it is not possible to put all the blame on the traditional attitude of developers. Computer technologies are inevitably associated with changes in human interaction with the world. Changing the way people interact and accomplish activities in the world is not easy. The problem here is not simply that features of the interface are inscrutable, but more fundamentally, the whole nature of the activity may have changed as a result of the technological possibilities. According to activity theory, the best way to help such users would be to create for them the zone of proximal development. In other words, such users need special support to get involved in new activities that utilize the new tools to accomplish some meaningful result for them, even if, in the beginning, they have to heavily rely on others people's help.

CONCLUSIONS

What we hope this sketch has demonstrated is that the activity theoretical approach provides an interesting framework for discussing human cognition and action. In particular, we find that the approach allows one to discuss not just individuals, but also relationships between people and settings in accomplishing work activities. The approach can help in extending the boundaries of traditional HCI to elaborate a more nuanced and extensive framework for understanding the varieties of human experience, cultural practices, and competencies. Our focus is on what people *do* with computer systems, not with how the interface looks. Emphasis is on the use situation, as we have noted. The idea of a collective practice is an important one when we think of HCI implementations. Likewise, the realization of the historical co-determination of both work settings and artifacts affects how we think about (re)design. The framework seems to "hold in" some of the complexity of real work situations when we start to think about developing computer tools for people, something that is missing in the individualistic cognitive science accounts. Finally, we would like to emphasize that activity theory is not a "silver bullet" that will automatically solve all the theoretical problems of HCI. This theory has serious limitations, too. Let us consider some of them.

First of all, as it was already emphasized, activity theory was mainly developed as a psychological theory of individual activity. This is an important limitation, because the current meaning of the term *user* includes not only individuals, but also groups and organizations. Many researchers agree that activity theory can be applied to supraindividual units, such as groups and organizations. However, the specific conceptual system necessary for the analysis of social systems is still under discussion. In the former USSR, the opportunity to study social phenomena were limited for political reasons. Probably, the only relevant idea developed by the Soviet proponents of activity theory, was the notion of "collective agent." This concept is less elaborated compared to the aspects of activity theory related to individual agents, and it is still to be clarified to what extent the conceptual apparatus of activity theory is applicable to collective agents. Important developments toward the extension of activity theory to the level of social processes were made by Western researchers (Cole, 1984; Engeström, 1987; Raithel, 1992), but this problem seems not to be solved yet.

Second, compared to the cultural-historical approach developed by Vygotsky (1978), activity theory adopted a more narrow view of culture. Activity theory was oriented to practical needs of the society, it was greatly influenced by the example of natural science, and it always tended to interpret reality in formal schemes (see Zinchenko, 1992, 1996). Although the phenomena of culture, values, motivation, emotions, human personality, and personal meaning are embraced by the conceptual system of activity theory, the theory does not aim at giving a comprehensive description of all these phenomena. It captures only some of their aspects—those related to the rational understanding of human interaction with the world. This feature of activity theory can be considered as a benefit, because it is similar to the way many system developers think; but it might also be viewed as a disadvantage, because activity theory cannot completely substitute anthropology.

Third, the tool mediation perspective, which is considered as the most important advantage of activity theory, can also impose some limitations on its potential application. In virtual realities the border between a tool and reality is rather unclear: Information technology can provide the user not only with representations of objects of reality, but also with a sort of reality as such, which does not obviously represent anything else and is intended to be just one more environment the individual can interact with. Virtual realities present a problem to activity theory, a problem that probably cannot be solved without enriching its basic principles with new ideas from either the cultural-historical tradition, or other related approaches.

Finally, in the field of HCI, compared, for instance, to the field of education, activity theory is not yet operationalized enough. There are very few methods and techniques that can be directly utilized to solve specific problems. So, it would be unrealistic to expect immediate results from accepting activity theory as an approach guiding theoretical research or practical efforts. The aforementioned limitations of activity theory are not inevitable. It is a developing approach, and probably one of its strengths is its potential for integration with other conceptual systems. We hope this chapter has given some insight into how this approach may contribute to a better conceptual framework for HCI that will ensure our future computer-mediated activities will be accessible to all.

ACKNOWLEDGMENTS

We would especially like to note the contribution of Kari Kuutti, our co-researcher for many years on issues in activity theory, and co-author of tutorial notes on activity theory that form the basis for this chapter.

REFERENCES

Backhurst, D. (1988). Activity, consciousness, and communication. *The Quarterly Newsletter of the Laboratory of Comparative Human Cognition, 10*(2), 31–39.

Backhurst, D. (1991). *Consciousness and revolution in Soviet philosophy: From the Bolsheviks to Evald Ilyenkov.* Cambridge, England: Cambridge University Press.

Bannon, L. (1986). Helping users help each other. In D. Norman & W. Draper (Eds.), *User centered system design: New perspectives on human–computer interaction* (pp. 399–410). Hillsdale, NJ: Lawrence Erlbaum Associates.

Bannon, L. (1996). Use, design, and evaluation: Steps towards an integration. In D. Shapiro, M. Tauber, & R. Traunmueller (Eds.), *The design of computer-supported cooperative work and groupware systems.* (Series: *Human Factors in Information Systems,* Vol. 12, pp. 423–444). Amsterdam: North-Holland, Elsevier Science.

Bannon, L. J. (1990). A pilgrims progress: From cognitive science to cooperative design. *AI and Society, 4*(4), 259–275

Bannon L., & Bødker, S. (1991). Beyond the interface: Encountering artifacts in use. In J. Carroll (Ed.), *Designing interaction: Psychology at the human–computer interface* (pp. 227–253). Cambridge, England: Cambridge University Press.

Bellamy, R. (1996). Designing educational technology: Computer-mediated change. In B. Nardi (Ed.), *Context and consciousness: Activity theory and human–computer interaction* (pp. 123–146). Cambridge, MA: MIT Press.

Bernstein, N. (1967). *The Co-ordination and regulation of movements.* Oxford, England: Pergamon.

Blumenthal, B. (1995). Industrial design and activity theory: A new direction for designing computer-based artifacts. In B. Blumenthal, Y. Gornostaev, & K. Unger (Eds.), *Selected papers: Human–computer interaction, Proceedings of the 5th East–West International Conference on Human–Computer Interaction* (pp. 1–16). Berlin: Springer-Verlag.

Bødker, S. (1989). A human activity approach to user interfaces. *Human–Computer Interaction, 4*(3), 151–196.

Bødker, S. (1991). *Through the interface: A human activity approach to user interface design.* Hillsdale, NJ: Lawrence Erlbaum Associates.

Bødker, S., & Christiansen, E. (1997). Scenarios as springboards in design. In G. Bowker, L. Gasser, S. L. Star, & W. Turner (Eds.), *Social science research, technical systems, and cooperative work* (pp. 217–234). Hillsdale, NJ: Lawrence Erlbaum Associates.

Bourke, I., Verenikina, I., & Gould, E. (1993). Interacting with proprietary software users: An application for activity theory? In *Proceedings of the East–West International Conference on Human–Computer Interaction* (vol. 1, pp. 219–226). Moscow: ICSTI.

Carroll, J. M., Kellogg, W. A., & Rosson, M. B. (1991). The task-artifact cycle. In J. Carroll (Ed.), *Designing interaction: Psychology at the human–computer interface* (pp. 74–102). Cambridge, England: Cambridge University Press.

Christiansen, E. (1996). Tamed by a rose: Computers as tools in human activity. In B. Nardi (Ed.), *Context and consciousness: Activity theory and human–computer interaction* (pp. 175–198). Cambridge, MA: MIT Press.

Cohen, A., Candland, K., & Lee, E. (1995). The effect of a teacher-designed assessment tool on instructors' cognitive activity. In *Proceedings of IFIP TC13, 5th International Conference on Human–Computer Interaction*. London: Chapman & Hall.

Cole, M. (1984). The zone of proximal development: where culture and cognition create each other. In J. Wertsch (Ed.), *Culture, communication, and cognition: Vygotskian perspective* (pp. 146–161). Cambridge, England: Cambridge University Press.

Draper, W. (1992). Activity theory: The new direction for HCI? *International Journal of Man–Machine Studies, 37*(6), 811–821.

Edwards, A. D. (Ed.). (1995). *Extra-ordinary human–computer interaction: Interfaces for users with disabilities* (Series on human–computer interaction, No 7). Cambridge, England: Cambridge University Press.

Engeström, Y. (1987). *Learning by expanding: An activity-theoretical approach to developmental research.* Helsinki, Finland: Orienta-Konsultit Oy.

Engeström, Y. (1990). *Learning, working, and imagining: Twelve studies in activity theory.* Helsinki, Finland: Orienta-Konsultit Oy.

Engeström, Y., & Escalante, V. (1996). Mundane tool or object of affection? The rise and fall of the postal buddy. In B. Nardi (Ed.), *Context and consciousness: Activity theory and human–computer interaction* (pp. 325–375). Cambridge, MA: MIT Press.

Grønbaek, K., Grudin, J., Bødker, S., & Bannon, L. (1993). Improving conditions for cooperative system design—shifting from product to process focus. In D. Schuler & A. Namioka (Eds.), *Participatory design: Principles and practices* (pp. 79–97). Hillsdale, NJ: Lawrence Erlbaum Associates.

Henderson, D., & Kyng, M. (1991). There's no place like home: Continuing design in use. In J. Greenbaum & M. Kyng (Eds.), *Design at work: Cooperative design of computer systems* (pp. 219–240). Hillsdale, NJ: Lawrence Erlbaum Associates.

Kaptelinin, V. (1993). Item recognition in menu selection: the effect of practice. In *Adjunct Proceedings of the Conference on Human Factors in Computing Systems* (pp. 183–184). New York: ACM Press.

Kaptelinin, V. (1994). Activity theory: Implications for human computer interaction. In M. D. Brouwer-Janse & T. L. Harrington (Eds.), *Human machine communication for educational systems design* (pp. 5–16.). Berlin: Springer-Verlag.

Kaptelinin, V. (1996a). Computer-mediated activity: Functional organs in social and developmental contexts. In B. Nardi (Ed.), *Context and consciousness: Activity theory and human–computer interaction* (pp. 45–68). Cambridge, MA: MIT Press.

Kaptelinin, V. (1996b). Distribution of cognition between minds and artifacts: Augmentation or mediation? *AI and Society, 10,* 15–25.

Kaptelinin, V., & Nardi, B. (1997). *The activity checklist: A tool for representing the "space" of context* (Research Report No. ISSN 1401-4572). Umeå, Sweden: Umeå University, Department of Informatics.

Kozulin, A. (1986). The concept of activity in Soviet psychology. *American Psychologist, 41*(3), 264–274.

Kuutti, K. (1991a). Activity theory and its applications in information systems research and design. In H. E. Nissen, H. K. Klein, & R. Hirschheim (Eds.), *Information systems research arena of the 90's* (pp. 529–549). Amsterdam: North-Holland, Elsevier Science.

Kuutti, K. (1991b). The concept of activity as a basic unit for CSCW research. In *Proceedings of the 2nd European Conference on Computer Supported Cooperative Work (ECSCW '91)* (pp. 249–264). Amsterdam: Kluwer Academic.

Kuutti, K. (1992). HCI research debate and activity theory position. In *Proceedings of the East–West International Conference on Human–Computer Interaction* (pp. 13–22). Moscow: ICSTI.

Kuutti, K., & Kaptelinin, V. (1997). Rethinking cognitive tools: From augmentation to mediation. In *Proceedings of the 2nd International Conference on Cognitive Technology "Humanizing the Information Age"* (pp. 31–32). Los Alamitos, CA: IEEE Computer Society.

Lave, J. (1988). *Cognition in practice.* Cambridge, England: Cambridge University Press.

Leont'ev, A. N. (1978). *Activity. Consciousness. Personality.* Englewood Cliffs, NJ: Prentice-Hall.

Munipov, V. M. (1983). Vklad A. N. Leontjeva v razvitie inzhenernoj psikhologii i ergonomiki [Leont'ev's contribution to engineering psychology and ergonomics]. In A. V. Zaporozhets, V. P. Zinchenko, & O. V. Ovchinnikova (Eds.), *Leont'ev and Contemporary Psychology* (pp. 88–96). Moscow: Izdatelstvo MGU.

Nardi, B. (1992). Studying context: A comparison of activity theory, situated action models, and distributed cognition. In *Proceedings of the East–West International Conference on Human–Computer Interaction* (pp. 352–359). Moscow: ICSTI.

Nardi, B. (1993). *A small matter of programming.* Cambridge, MA: MIT Press.

Nardi, B. (Ed.) (1996). *Context and consciousness: Activity theory and human–computer interaction.* Cambridge, MA: MIT Press.

Norman, D. (1980). Twelve issues for cognitive science. *Cognitive Science, 4,* 1–32.

Norman, D., & Draper, W. (Eds.). (1986). *User centered system design: New perspectives on human–computer interaction.* Hillsdale, NJ: Lawrence Erlbaum Associates.

Norros, L., & Hukki, K. (1995). Contextual analysis of the operator's on-line interpretation of process dynamics. In *Proceedings of the 5th European Conference on Cognitive Science Approaches to Process Control* (pp. 182–195). Espoo: Technical Research Center of Finland.

Raeithel, A. (1992). Activity theory as a foundation for design. In C. Floyd, H. Züllighoven, R. Budde, & R. Keil-Slawik (Eds.), *Software development and reality construction* (pp. 391–415). Berlin: Springer-Verlag.

Velichkovsky, B., & Zinchenko, V. (1982). New perspectives on cognitive psychology. In L. Cohen, J. Los, H. Pfeiffer, & K-P. Podewski (Eds.), *Logic, methodology and philosophy of science VI.* Amsterdam: North-Holland, Elsevier Science.

Vygotsky, L. S. (1978). *Mind in society—The development of higher psychological processes* (M. Cole, V. John-Steiner, S. Scribner, & E. Souberman, Eds.). Cambridge, MA: Harvard University Press.

Wertsch, J. (1981). The concept of activity in Soviet psychology: An introduction. In J. Wertsch (Ed.), *The concept of activity in Soviet psychology.* Armonk, NY: M. E. Sharpe.

Wertsch, J. (1998). *Mind as action.* New York: Oxford University Press.

Winograd, T., & Flores, F. (1987). *Understanding computers and cognition: A new foundation for design.* Reading, MA: Addison-Wesley.

Zinchenko, V. (1992). Activity theory: Retrospect and prospect. In *Proceedings of the East–West International Conference on Human Computer Interaction* (pp. 1–5). Moscow: ICSTI.

Zinchenko, V. (1996). Developing activity theory: The zone of proximal development and beyond. In B. Nardi (Ed.), *Context and consciousness: Activity theory and human–computer interaction* (pp. 283–324). Cambridge, MA: MIT Press.

Zinchenko, V., & Munipov, V. (1989). *Fundamentals of ergonomics.* Moscow: Progress Publishers.

APPENDIX

Selected Bibliography

I. Activity Theory and the Design of Computer Systems

Bødker, S. (1993a). Historical analysis and conflicting perspectives: Contextualizing HCI. In L. Bass, J. Gornostaev, & C. Unger (Eds.), *Proceedings of the East–West International Conference on Human–Computer Interaction* (pp. 1–10). Moscow: ICSTI.

Bødker, S. (1993b). Re-framing research in human–computer interaction from the point-of-view of activity theory? *The Journal of Psychology [in Russian], 14*(4), 71–81.

Favorin, M., & Kuutti, K. (1996). Support learning at work by making work visible through information technology. *Machine-mediated learning, 5*(2), 109–118.

Holland, D., & Reeves, J. (1996). Activity theory and the view from somewhere: Team perspectives on the intellectual work of programming. In B. Nardi (Ed.), *Context and consciousness: Activity theory and human–computer interaction* (pp. 257–282. Cambridge, MA: MIT Press.

Imaz, M., & Benyon, D. (1996). Cognition in the workplace: Integrating experientialism into activity theory. In T. R. G. Green, J. J. Cañas, & C. Warren (Eds.), *Proceedings of the 8th European Conference on Cognitive Ergonomics*. Granada, Spain: University of Granada Press.

Kaptelinin, V. (1992a). Human computer interaction in context: The activity theory perspective. In *Proceedings of the East–West International Conference on Human–Computer Interaction* (pp. 7–13). Moscow: ICSTI.

Kaptelinin, V. (1992b). Integration of computer tools into the structure of human activity: Implications for cognitive ergonomics. In *Proceedings of the 6th European Conference on Cognitive Ergonomics* (pp. 285–294). Rome: CUD.

Kaptelinin, V., Kuutti, K., & Bannon, L. (1995). Activity theory: Basic concepts and applications. In B. Blumenthal, J. Gornostaev, & C. Unger (Eds.), Human-computer interaction, EWHCI'95, Selected Papers; Lecture Notes in computer science 1015 (pp. 189–201). Berlin: Springer.

Kuutti, K. (1993). Notes on systems supporting "organizational context"—An activity theory viewpoint. In L. Bannon & K. Schmidt (Eds.), *Issues of supporting organization context in CSCW systems* (pp. 101–118). Lancaster, England: Lancaster University Press.

Kuutti, K., & Arvonen, T. (1992). Identifying CSCW applications by means of activity theory concepts: A case example. In *Proceedings of the 3rd European Conference on Computer Supported Cooperative Work* (pp. 233–240). New York: ACM Press.

Kuutti, K., & Bannon, L. (1993). Searching for unity among diversity: Exploring the interface concept. In *Proceedings of the Conference on Human Factors in Computing Systems* (pp. 263–268). New York: ACM Press.

Kuutti, K., & Favorin, M. (1993). Tools for research-simulating learning. In B. Z. Barta, J. Eccleston, & R. Hambusch (Eds.), *Computer-mediated education of information technology professionals and advanced end-users*. Amsterdam: North-Holland.

Kuutti, K., & Virkkunen, J. (1995). Organizational memory and learning network organization: The case of Finnish labour protection inspectors. In *Proceedings of the 28th An-*

nual *Hawaii International Conference on System Sciences-1995* (pp. 313–322). Los Alamitos, CA: IEEE Computer Press.

Nardi, B. (1994). Studying task-specificity: How we could have done it right the first time with activity theory. In *Proceedings of the East–West International Conference on Human–Computer Interaction.* Moscow: ICSTI.

Nardi, B. (1996). Some reflections on the application of activity. IN B. Nardi (Ed.), *Context and consciousness: Activity theory and human–computer interaction* (pp. 235–246). Cambridge, MA: MIT Press.

Rogalski, J. (1994). Analyzing distributed cooperation in dynamic environment management: The "distributed crew" in automatized cockpits. In *Proceedings of the 7th European Conference on Cognitive Ergonomics* (pp. 187–199). GMD, Sankt Augustin: GMD = Studien Nr. 233, Germany.

Saarelma, O. (1993). *New work—New tools. Information system development in the context of development of work activity* (CEC ESPRIT Project No. 6225, COMIC Report No. Oulu-2-2). Oulu, Finland: University of Oulu.

Sjöberg, C. (1994). *Voices in design: Argumentation in participatory development* (Linkoping Studies in Science and Technology, Thesis No. 436). Linköping, Sweden: Linköping University.

Tikhomirov, O. K. (1981). The psychological consequences of computerization. In J. Wertsch (Ed.), *The concept of activity in Soviet psychology.* Armonk, NY: M. E. Sharpe.

Timpka, T., & the MDA Group. (1993). *A methodology for the definition of clinical support system requirements* (CEC AIM Project No. A2005 "DILEMMA" deliverable D04). Linköping, Sweden: Linköping University.

II. Activity Theory in General

Lektorsky, V. A. (1984). *Subject, object, cognition.* Moscow: Progress Publishers.

Leont'ev, A. N. (1981). *Problems of the Development of Mind.* Moscow: Progress Publishers.

III. Other Related Papers

Bannon, L. (1991). From human factors to human actors: The role of psychology and human–computer interaction studies in system design. In J. Greenbaum & M. Kyng (Eds.), *Design at work: Cooperative design of computer systems* (pp. 25–44). Hillsdale, NJ: Lawrence Erlbaum Associates.

Brusilovsky, P., Burmistrov, I., & Kaptelinin, V. (1993). Structuring the field of HCI: An empirical study of experts' representations. In *Proceedings of the East–West International Conference on Human–Computer Interaction* (vol. 3, pp. 18–28). Moscow: ICSTI.

Cypher, A. (1986). The structure of user's activities. In D. Norman & S. Draper (Eds.), *User centered system design: New perspectives on human–computer interaction* (pp. 243–264). Hillsdale, NJ: Lawrence Erlbaum Associates.

Ehn, P. (1988). *Work-oriented design of computer artifacts.* Stockholm: Arbetslivscentrum.

Grudin, J. (1990). The computer reaches out: The historical continuity of interface design. In *Proceedings of the Conference on Human Factors in Computing Systems* (pp. 261–268). New York: ACM Press.

Monk, A., Nardi, B., Gilbert, N., Mantei, M., & McCarthy, J. (1993). Mixing oil and water? Ethnography versus experimental psychology in the study of computer-mediated communication. In *Proceedings of the Conference on Human Factors in Computing Systems* (pp. 3–6). New York: ACM Press.

Norman, D. (1991). Cognitive artifacts. In J. Carroll (Ed.), *Designing interaction: Psychology at the human–computer interface* (pp. 17–38). Cambridge, England: Cambridge University Press.

Wood, C. (1992). *A cultural-cognitive approach to collaborative writing* (Cognitive Science Research Report No. 242). Sussex, England: University of Sussex, School of Cognitive and Computing Sciences.

11 Sociological Issues in HCI Design

Michael Pieper

A precondition for an Information Society for All is that interaction environments are personalized. This chapter argues that personalization should not be confused with the traditional software ergonomic demand of individualizable user-interfaces, which, for example, is being promoted by Part 10 "dialogue requirements" of the most comprehensive user interface design standard (ISO 9241, 1995). Instead, personalities, as opposed to individuals, need to be conceived as a social phenomenon. Individualization in Human-Computer Interaction design is, up to now, largely based on psychological explanation, focussing on individual strivings, which prevents any analysis of social structure. In contrast to this, the social dynamics of interacting personalities impacting on stabilizing (virtual) communities in the Information Society will be described by a sociological paradigm which on the one hand is general enough to have a sound theoretical foundation, and on the other hand is precise enough to generate recommendations for (socio-) technical manifestations of 'sociable Human-Computer Interaction'.

INTRODUCTION: SOCIABLE HUMAN–COMPUTER INTERACTION

Virtual communities are rapidly becoming an important part of the information society. The degree of social immersion, that is, the awareness of being an integral part of a comprehensive social context, is becoming an ever more important factor for the success and the subjective satisfaction of end-users of networked telematic systems:

> Typically a user is only aware of a portion of the social network to which he or she belongs. By instantiating the larger community, the user can discover connections to people and information that otherwise lay hidden over the horizon. (Kautz, Selman, & Shah, 1997, p. 65)

In the virtual social environment of public cyberspace, technology is thus expected to increasingly push toward "sociability."

From the perspective of an information society *for all,* the social dimension of virtual communities is particularly relevant. In the future, Western industrial societies will be demographically overaged; that is, elderly and aging citizens will be one of the most important social groups. Up to now they usually have been considered as passive receivers of information. Future information environments, however, involve a shift toward the conscious reception and meaningful expression of information, which may create the risk of excluding this part of the population from important parts of the economic, social, and cultural life of the society they live in, if appropriate measures are not taken to ensure that they contribute to the described process. The same argument holds for people with disabilities.

Often, disabled and elderly people are bound to a tendency toward social isolation in everyday life, due to mobility restrictions caused either by sensory impairments (e.g., low vision or hearing) that reduce spatial orientation, or by functional restrictions in spatial movement abilities. Empirical sociology has proved that spatial distance correlates with social distance and isolation (Dodd & Nehnevajsa, 1954). Moreover, expressive and receptive problems may be accompanied by slow cognitive recollection due to aging, which puts conventional face-to-face interaction in the social environment under the load of delayed or extended reaction time. This, far too often, leads to further marginalization or even stigmatization of disabled and elderly people. Such an experience on their side may lead to feelings of inferiority—an attitude that may further restrict the range of behavioral approaches needed to preserve social competence.

Networked telematic systems offer opportunities to reinforce "virtual mobility." They provide "extension in space," thus allowing one to overcome the effects of mobility restrictions caused by aging or disability. Systems can make end-users "mobile," thus providing access to information while "on the spot," regardless of one's physical location, or, alternatively, they can make the information mobile so that users get access to it from a constant place.

Moreover, network-mediated interaction allows keeping disability- and aging-related delays out of the dynamics of social interaction. This can help in maintaining an equalized social environment, and avoiding the intentional or unintentional stigmatization of disabled and elderly people as "strenuous" interaction partners. New system functions may be developed in order to enhance this social dimension of the information society, as, for example, information environments that notify volunteers about disabled and/or elderly people requiring some service. Similarly, virtual meetings may be set up for the elderly who are interested in discussing social topics, or may simply arrange appointments between people who want to overcome (a disability-related) isolation.

Besides special needs interface adaptations to meet *information presentation* requirements at the input/output level, compensating for functional disabilities and/or aging-related perceptual disorders, much research and development work is still to be carried out with regard to the *information content* adaptation. This means that human–computer interaction (HCI) design increasingly needs to push toward "sociability." To meet these ends, information needs to be presented and stored at a high level of abstraction, to enable structured transformation and presentation in different forms, both at the semantic (i.e., content) and the lexical (i.e., presentation) levels. To this effect, disabled and able-bodied people should form an integral part in the construction of information environments, as they indirectly determine the content and presentation of information, through their abilities, requirements, and preferences.

This brings about the need for a sociological perspective to interaction design, to unfold the social, in addition to the informational, insight of systems. Currently adopted design methodologies do not adequately address the requirements of such a social insight. They are largely based on psychological explanation and common standards for *individualizing* user interfaces (e.g., Part 10 of ISO 9241, 1995), instead of conceiving the end-user as a social actor, who should accordingly be supported in using the computer as a medium to connect to others through what may be called a "sociable user interface." In the context of virtual communities, HCI design should progressively advance to facilitate communication and exchange of information; to this effect, new social metaphors need to guide and inform HCI design practices.

This chapter aims at understanding and modeling groups and organizations, and the related processes, as a basis for an interaction design, which focuses on what we call *sociable HCI*. The goal is to develop guidelines for displaying different views of shared information in virtual communities, thereby supporting social awareness and social immersion of community participants. The next section of this chapter describes a sociological framework mirroring the dynamics of stabilization and destabilization of social systems in general. Such a framework constitutes a basis for disclosing indications of the ever-changing state of a *living social system* like a virtual community.

On the basis of such a framework, the third section of this chapter illustrates architectural principles for a *sociable HCI Design,* as well as basic principles and early attempts toward sociable HCI.

The fourth section discusses the application of novel techniques (data mining and collaborative filtering) "which automatically identify emergent communities of interest in user populations of networked computer systems, enabling enhanced group awareness and communication" (Balabanovic & Shoham, 1997, p. 66). In such a context, these techniques introduce new opportunities and challenges. Apart from the severe social concerns about privacy and data security in applying these new technologies in distributed computer systems, a more important concern is that, it is unlikely for the desired results to be produced if technology is not applied in the light of the social commitments that stabilize social relationships. This could lead to the argument that, by adopting a sociological perspective, we can improve the quality of recommender systems from a technical point of view. However, in making the realm of the social in virtual communities accessible and suitable for the broadest possible population, HCI designers cannot rely on purely technical approaches. As we point out, this would require a software system for textual context analysis capable of identifying and distinguishing four types of commitments with respect to social knowledge, which are relevant for tracking group dynamics. With regard to this, we share Terry Winograd's view, that "it is impossible to formulate a precise correspondence between combinations of words and the structure of commitments listened to in a conversation" (Winograd, 1987, p. 157).

The fifth section of the chapter, therefore, proposes a guideline-oriented *sociotechnical* approach, in which social moderators supported by software agent and novel data-mining technologies care about the "social design" of virtual communities. The moderator's task is to "make the user aware of this structure and to provide tools for working with it explicitly" (Winograd, 1987, p. 157), according to guidelines derived from a conceptual model of group stabilization.

UNDERSTANDING THE DYNAMICS OF SOCIAL SYSTEMS

The conceptual framework adopted in this chapter relies on the structure-functionalistic system theory (Parsons, 1937, 1960; Parsons, Shils, Naegele, & Pitts, 1965), which provides a dynamic phase model of the functional stabilization versus destabilization of social groups. Psychological requirements necessary for gaining social mutuality, whose enfolding has to be supported by sociable HCI, can be defined in relation to the distinct phases of this model.

Functionalist and Interpretive Paradigm

Structure-functionalistic systems theory arose out of the need felt by sociologists to develop theoretical and methodological tools adequate for dealing with the interrelatedness of various traits, institutions, groups, and so on, within a total social system, and to overcome previous atomistic and descriptive methods. Functionalism was brought into sociological thought by borrowing directly from, and developing analogies for concepts in the biological sciences. Biology investigates the structure of an organism, by identifying relatively stable arrangements of relationships between its cells, and refers to the contribution of the various organs for the survival of the whole organism as their function.

In Talcott Parsons' structural-functional theory (Parsons, 1937, 1960; Parsons et al., 1965), sociological functionalism has attained its most systematic formulation. Parsons argued that the significance of function is primarily attributed to its potential for linking structural categories. The most important features of structural-functionalism are: (a) the delineation of boundaries between the social and other relevant systems, notably the cultural, personality, and biological systems, (b) an abstract and transhistorical delineation of the major structural units of the social system, with a heavy emphasis upon the normative relationships between these units, and (c) a concern with the conditions of stability, integration, and maximum effectiveness of the system depicted in abstract form.

The functional orientation runs through all these characteristics, one of its most influential aspects being the idea of functional imperatives. The term refers to the four basic problems of a social system, which every social system faces and must cope with, if it is to be adequately maintained. The four functional imperatives are: (a) adaptation to other systems and to the physical environment, (b) attainment of system goals, (c) integration, and (d) maintenance of stability and consistency.

Thus, structure-functionalistic systems theory distinguishes four basic functional preconditions to be fulfilled in order for a social system to remain internally stabilized:

- Adaptation.
- Goal attainment.
- Integration.
- Latency or pattern maintenance.

More precisely, *adaptation* refers to the fact that, in attaining goals, a social system, like all living systems, is an open system continually interacting with, and adapting to an ever-changing external situation or environment, through interchanging input and output. One example of such an interaction may be the need for social care in a community of disabled and elderly people, and the community's enforced reliance to self-help groups in case social services have been cut by welfare organizations due to economic austerity measures.

Other structures and functions become specialized with reference to the internal states of the social system, in ways that are relatively independent from the immediate impact of external situations. These structures and functions are concerned, in the first place, with maintaining the states of the units (i.e., group members) serving as conditions of their effective interaction with others and with the external situation; and second, with the interaction of units in relation to each other, in terms of their mutual compatibility and reinforcement (i.e., with what is called "pattern-maintenance and integration").

Academic sociology long criticized the functionalist theory of reifying social processes by ignoring the creation of structure and by recasting individual actions into fixed properties and boundaries (Zey-Ferrell & Aiken, 1981). "In this type of reification," Linda Putnam (1983) wrote, "social structures exist prior to individual actions" (p. 35). In opposition to this traditional positivist belief about social science, arguing that knowledge about social behavior concerns observable phenomena, experience, and generalizable causal laws, the so-called "interpretive" approach grew up, which "treats structures as sets of complex, semiautonomous relationships, that originate from human interactions" (Putnam 1983, p. 35). The interpretive approach views structures as follows:

> Structures can be seen as an outgrowth of sets of relationships, that have real consequences on everyday interaction. Both the chart and the behaviors that create it are in states of change.... Therefore, the chart is symbolic in that it represents previous and potential relationships, but it is also structural in that the use of, or reference to, it impacts on daily actions among members. (Putnam, 1983, p. 35)

The interpretive paradigm of inquiry in social science argues that causal laws and generalizations are not necessary to an understanding of human actions and institutions. Rather, a process of interpretation is required, in which the meaning of an action is uncovered by analyzing the action in the light of the agent's particular situation, and of the conventions, practices, and rules of the agent's community.

In this view, social phenomena are fundamentally different from natural phenomena in that they are intentional; that is, they express the purposes and ideas of social actors. Therefore, actions, including also linguistic actions, conceived as "speech acts" (Austin, 1962), have an essentially meaningful or symbolic character. They must be understood in terms of the conventions of the community, which specify the meaning or significance of particular acts.

Albeit the interpretive approach was developed as a reaction to functionalism, these two approaches share some basic attributes of their conceptual frameworks for analyzing social systems. Much more so, we argue, they depend on each other for their existence. The interpretivist approach is particularly appealing to communication researchers. By treating social communities as the social construction of reality, the stabilization of *virtual* social communities can be seen as a process of communicating. Likewise, communication is not simply another social activity; but, it creates and recreates the social structures that form the essence of group dynamics. Thus, interpretivism is just a more precise way of operationalizing the functionalist concept of social systems, with the aim of merging its theoretical assumptions into sociotechnological solutions for maintaining virtual communities.

Consolidation of Social Groups

In explicitly dealing with interpersonal communication as it effects the petrifying reciprocity of intentions among social actors, group dynamical interaction theory generalizes findings of empirical small-group research into a layered four-phase model of increasingly consolidating orientation of social interaction (Tuckman, 1965):

1.   The first phase, called *forming,* is characterized by group members trying out what social activities are accepted or cause resistance. Common objectives and possible ways to attain common goals are reflected for the first time. Thereby, opinion leaders, who are believed to share an already sufficient amount of agreed values and beliefs of the whole group, are followed by those actors who experience, at least subjectively, some degree of conformity pressure.

2.   During the second phase, called *storming,* internal conflicts occur within a group. Certain group members counteract others, and cast doubts about the justification of concerted action and agreed ruling. Corporate identity is not yet entirely developed. Group objectives are rejected, because they restrict personal freedom.

3.   Group coherence enfolds during the third phase, called *norming.* Each group member accepts others, even self-willed individuals, and is helpful in the continuation of group processes. Goal attainment is prepared by information exchange and allocation of individual resources.

4.   The fourth and final phase, called *performing,* is dedicated toward completion of the groups mission; that is, reciprocity of social interaction petrifies; the solution of the group's problems is within reach.

Thus, over time, emergent norms become dissociated from the situation in which they first arose, and are generalized to cover broad areas of organized social life. As such, norms are seen, by persons subject to them, as existing outside themselves, and as binding on their actions. Norms may even become moral directives defining how one must act in a given situation. Adherence to norms is then not left up to individual choice or discretion, but is a social obligation.

However, if the laws of a community conflict sharply with more basic folkways and mores, norms will tend to be either ignored or changed. Thus, each phase of this layered approach of incrementally consolidating interaction patterns requires certain functional preconditions. If the functional preconditions for a certain phase are missing, tendencies toward destabilization will occur. Interaction patterns have to be consolidated by the group dynamics of the phase preceding the one whose preconditions are missing. Figure 11.1 depicts the four-phase model and the necessary preconditions for each phase:

- Enfolding of *empathy* is crucial for the forming phase of group stabilization. To find out which social activities are accepted or cause resistance within the group, group members have to reestablish intentions, intu-

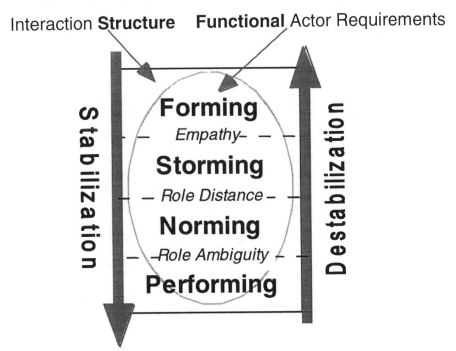

FIG. 11.1.   The four-phase model.

itions, problem-oriented insights, personal values, and beliefs of others as adequate for themselves (to the degree that this is possible).

- To modify concerted action and agreed ruling in the storming phase, *role distance* becomes important, that is, the subjective ability to evaluate and compare intentions, values, and beliefs of others with individual convictions on how to enact a certain social role. Role distance is thus exactly the already mentioned ability of persons to see the norms they are subject to as existing outside themselves, and binding on their actions.

- The incremental petrifaction of interaction patterns during group norming and performing has a considerable influence on the development or suppression of specific elements of individual personality. Important here are the conditions under which individual needs, interests, and beliefs can be coordinated with formally expected behavior. Sociology refers to this mechanism as *role ambiguity,* resulting from a balance between role conformity and role distance (Gullahorn & Gullahorn, 1963; Inkeles & Levinson, 1963; Kahn, Wolfe, Quinn, Snoek, & Rosenthal, 1964; Siegel, 1968; Stouffer & Toby, 1951; Sutcliffe & Habermann, 1956; Zurcher, Sonnenschein, & Metzner, 1966). This balance is only possible if the conformity pressure exerted on the individual to adjust to group requirements is restricted in such a way that sufficient freedom for the development of a personal identity is left. If parts of the community enforce norms rigidly, severe personal strains are generated.

Computer conferencing systems have already provided technological facilities for a completely new type of negotiations whose impact on group dynamics has proved to be enormous, especially with respect to the avoidance of social pressure to adjust to prevailing patterns of attitude and behavior, and also with respect to privacy issues. Computer conferencing provides facilities for expressing one's own opinion anonymously, and receiving statements under a pseudonym (the so-called pen-name). Thus, the argumentation of a negotiating party is freed from any expectations placed in advance on some person by other group members knowing the person's name, personality, and value positions within the group. This may have a favorable effect on the variety of ideas and opinions to be produced in the norming and performing phase of group stabilization. On the other hand, it may cause insufficient support to integrate the generated variety of suggestions and opinions into a goal attainment strategy during group norming and performing (Hiltz & Turoff, 1978).

BASIC PRINCIPLES AND EARLY ATTEMPTS TOWARD SOCIABLE HCI

As has been pointed out, subjective abilities of enfolding empathy, role distance, role conformity, and role ambiguity are important preconditions for the stabilization of social mutuality. These demanding subjective abilities must be supported by sociable HCI in virtual social environments. Is there any general principle, which, besides reflecting the circumscribed framework of dynamic stabilization and destabilization of social systems, has to be taken into account to forecast new interface technologies, or to generate at least rough ideas for new devices, or software processes supporting information interfaces? Most generally, the process of gaining social mutuality can be described in terms of a conceptual framework of social psychology referred to as *symbolic interactionism*.

Symbolic interactionism is an approach of inquiring social behavior that stresses linguistic and gestural communication, and especially the role of language in the formation of the mind, the self, and society. This school of thought has its roots in the concept of the *self* (or *personality*), as developed by George Herbert Mead (1934), who argued that reflexivity is crucial to the self as a social phenomenon. Social life depends on the ability of individuals to imagine themselves in other social roles, and on the capacity of making up one's *mind* by an internal conversation. Mead conceived society, as well as communities, as the outcome of an exchange of gestures that involves the use of symbols. Thus, symbolic interaction is the study of the self–society relationship as a process of symbolic communication between social actors, who construct their behavior in the course of its execution.

According to this conception, interaction partners serve mutually in social roles of communicators and recipients. Essentially, the process of gaining social mutuality results from the receptive ability to assign adequate meaning to communicated information. Intentions of others can adequately be understood by recipients only when the latter are able to reestablish the communicators' insights into communicated matters as completely as possible, by interpreting communicative behavior as symbolizing hidden values and beliefs. Social psychology refers to this ability to slip into the shoes of others as role taking. Role taking, in turn, is the indispensable precondition for role making, that is, performing the role of a communicator in social interaction, aiming at inducing desired reactions by others. Nevertheless, desired, or at least expected, reactions of others can be induced only by role making on the grounds of a previous reconfiguration, as complete as possible, of others' intentions by role taking. Social mutuality, or structural group stability, is thus gained by a progressive adjustment of perceived intentions in role-taking/role-making cycles. It is in the role-taking ability, which has to be enhanced by sociable HCI, that computer-mediated interaction takes place instead of immediate face-to-face meetings.

Early attempts to make this theoretical concept of social psychology fruitful for software technological solutions can be divided into two schools. Both of them tried to merge the outlined theoretical assumptions into formal, and therefore programmable, algorithms for the retrieval of contextual information in computer-conferencing systems.

The first approach to a formal algorithmic model of interpersonal relations was based on the basic structure of so called P-O-X equations, in which P is a *person*, who utters a statement (X) and O is an *other* person, who perceives this statement (see Fig. 11.2). Based on the mathematical principle of transitivity, "agreement" and "disagreement" between P and O can be identified under additional preconditions (Heider, 1946, 1958; Pieper, 1986). Other approaches tried to apply formal speech act theory (Austin, 1962; Searle, 1969, 1975) to unlock theoretical assumptions of social psychology, in order to trace back argumentation chains in groupware systems (Winograd, 1987; Winograd & Flores, 1987). However, all these approaches failed with regard to end-user acceptance. This was caused by the conceptual inflexibility of the formal logic underlying the corresponding HCI design.

Both these approaches retrieve structural information about interpersonal relations in networked computer systems by modeling such relations as a graph. Nodes of a graph represent individuals, and edges between nodes indicate that a direct relationship between individuals has been discovered. These interrelations can be illustrated by means of a model of interpersonal negotiations whereby P utters a statement (X), which is perceived by O. The way *Other* perceives this statement against the background of personal insights into the matter under discussion—which may also result from comparable suggestions from the same

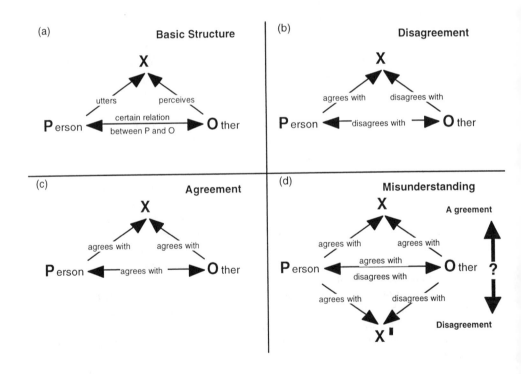

FIG. 11.2.   P-O-X structure of interpersonal relations.

person concerning the same matter—establishes a certain relationship between *person* and *other,* as regards the uttered statement (see Fig. 11.2a).

This relationship follows the mathematical principle of transitivity determining that, when two entities are equal to a third one, they are equal to each other; or, vice versa, when one of two entities is not equal to a third one, to which the other is equal, they are not equal to each other.

The former is the case in the situation of *agreement.* When both *person* and *other* agree with a certain statement, the relationship between person and other is obviously *agreement* (see Fig. 11.2c). The latter is the case in the situation of *disagreement.* When *person* agrees and *other* disagrees with a certain statement, the relationship between person and other is *disagreement* (see Fig. 11.2b). In both these cases, formal transitivity is underlying the situation.

However, a third situation can be identified, in which the formal structure of interpersonal relations becomes at least partially undetermined with respect to transitivity—that is when *person* and *other agree* with one statement, but they *disagree* with a second statement concerning the same matter. In this case, there is neither total agreement nor disagreement regarding the matter both statements refer to. The graph between *person* and *other* is qualified by undetermined ambivalence. One can say that *person* and *other* have to deal with interpersonal *misunderstanding* (see Fig. 11.2d).

There is evidence for the assumption that the greater the number of individuals involved, and the greater the number of statements uttered in interpersonal conversations, the more the group members are overburdened in identifying *misunderstandings.* Role taking for gaining mutual understanding could, therefore, be supported by transient software algorithms checking for ambivalent relations between *person* and *other* in common information spaces. From, this however, follows the stated conceptual inflexibility of corresponding HCI designs.

As a technological precondition, an argumentative reference structure allowing one to process corresponding (in)transitivity checks has to be established as an overlay to group information spaces. This has to be done by assigning additional indications to each statement being expressed. Though marking statements with the complex categories of speech acts theory is unfeasible, at least four indications have to be assigned to each statement: (a) the sender, and (b) receiver of a statement, and references to other statements, marked as (c) general agreement, or (d) disagreement (Pieper, 1986). Even though end-users can automatically be notified as senders based on login information, the problem is that, in any case, sociable HCI in such systems has to be burdened by additional system-driven dialogues asking for the remaining indications. Pure software systems are up to now inappropriate to reliably qualify speech acts as required in order to provide overlays to public information spaces. Already in the early 1980s, Terry Winograd admitted:

> We are not proposing that a computer can "understand" speech acts by analyzing natural language utterances. It is impossible to formulate a precise correspondence between combinations of words and the structure of commitments listened to in a

conversation. What we propose is to make the user aware of this structure and to provide tools for working with it explicitly. (Winograd, 1987, p. 107).

TECHNOLOGIES FOR BUILDING SOCIAL INTERACTION ENVIRONMENTS: DATA MINING AND COLLABORATIVE FILTERING

Technological advances offer new opportunities to reveal social relationships in distributed computer systems by relying on sources that do not require additional behavioral compliance with a poor user interface, designed from the point of view of an ambitious formal logic aiming at social networks modeling. Instead, the user interface may remain unobtrusively unchanged. Highly intentional behavior may be evaluated in the background with regard to its social relevance.

Social relationships in interaction environments can be monitored by data mining public information spaces and applying corresponding filtering technology. A distinction should be made between *collaborative* and *content-based* filtering (Balabanovic & Shoham, 1997). Whereas *content-based* filtering tries to identify items of *information* similar to those a given user has liked in the past, *collaborative* filtering identifies *users* with similar informational needs or tastes. It is collaborative filtering that appears to be best suited to support the outlined preconditions for consolidating social interaction, although it has to be further elaborated.

For instance, analyzing E-mail logs or measuring the degree of overlapping between different people's bookmark folders provides rich sources of such relationships. This kind of collaborative filtering supports the outlined group dynamics in different ways. Analyzing bookmark overlaps mainly supports the forming phase of group dynamics, in that linkages of users with similar informational interest may be promoted. Analyzing E-mail logs relates to performing, as it indicates the degree of reciprocity or—in other words—the stability of social interaction. However, it should be pointed out that accessing such information may raise severe concerns of privacy and data security.

A general software architecture of collaborative filtering consists of three main processes (Terveen, Hill, Amento, McDonald, & Creter, 1997):

- Search: data mine documents for a specified pattern, and extract contextual information surrounding each instance of the pattern.
- Categorization: apply rules that classify each instance of the pattern.
- Disposition: reflect the categorized information in the user interface.

To classify affinity between users with regard to particular informational topics is a much more difficult problem, which remains largely to be solved. Affinity relates to the storming and norming phases of the outlined four-phase model of social interaction. Dislike, as well as interest, for an item has to be revealed, because they both indicate a low social distance or—vice versa—high socioemotional involvement in justifying the degree of agreed ruling as the foundation of corporate group identity. For this purpose, some kind of reciprocal rating has unavoidably to be applied to common information sources. With this lies a problem, which has long been revealed by the behavioral-economics approach of social exchange theory (Ekeh, 1974; Homans, 1956). Users required to rate information before receiving predictions about the current degree of corporate identity in virtual social networks often "abandon the system before ever receiving benefits from it because they perceive effort without reward. Early adopters find there are not many other users available to rate and therefore they receive predictions for only a fraction [of the sources of the group information space]" (Konstan et al., 1997, p. 86). Because rating is such an onerous task for users, the fewer ratings required, the better. Implicit rating may reduce or eliminate perceived efforts, thereby making a continued use of sociable systems more likely.

The basic idea of implicit rating is to reuse indications for controversial impressions and opinions from online recommendations about all kinds of items, including Web pages. The PHOAKS system (People Helping One Another Know Stuff) is an experimental recommender system, whose general architecture for information filtering distinguishes four different indications of controversial judgements (Terveen et al., 1997). In recommending what the information on a Web page is useful for, and how useful it is, PHOAKS searches E-mail messages for occurrences of URLs, and counts an occurrence as an individually rated potentially controversial judgment if:

1. The message is not cross-posted to too many receivers, because then it can be assumed to be so general that no doubts arise about its relevance.
2. The URL is not simply part of a poster's signature.
3. The URL is not part of a quoted section of a previous message.
4. The textual context around the URL contains word markers that indicate it is recommended.

However, the general group dynamics concerned with storming and norming cannot be deduced from such a system, because it remains hidden to what degree recommendations filtered like this meet the receiver's tastes, interests, or opinion profiles, or whether they score highly against these profiles.

The revelation of such indications is especially important in problem-solving groups, where the generated variety of suggestions and opinions has to be incrementally integrated into a goal attainment strategy. These structuring problems still remain to be technically supported by sociable HCI. Following certain psychological assumptions of cognition theory (Huber, 1983; Minsky, 1968; Robey, 1983), architectural concepts for corresponding system design will have to refine the textual context analysis of implicit rating, until it is possible to identify and distinguish four types of knowledge relevant to the goal attainment of social groups: (a) *deontic knowledge,* (b) *factual knowledge,* (c) *explanatory knowledge,* and (d) *instrumental knowledge.*

Deontic knowledge refers to the description of a desired or nominal state of affairs relevant to the goals a group likes to attain. Factual knowledge describes the actual undesired state of affairs to be changed by the group. The discrepancy between deontic and factual knowledge, that is, the difference between the actual and the nominal state of affairs, obviously implies a problem. Explanatory knowledge includes the facts explaining the differences between actual state of affairs and the nominal one. Instrumental knowledge is concerned with the possible approaches to achieve the desired state of affairs.

If it becomes possible to reconstruct the pertinent global context of information according to such a knowledge classification at any point of argumentation within a virtual social environment, it may also be possible to detect those group dynamic issues that delay performing and goal attainment. Regarding public information spaces, factual statements, explanations and instrumental suggestions, and deontic visions could conceptually be incorporated into a comprehensive model of hierarchically layered reference structures (see Fig. 11.3). Transient P-O-X processing—as described earlier—could then be applied not only to each single layer but also to infringements upon layers (dotted lines). Thus, indications about the degree of factual, instrumental or—in the end—deontic petrification of group performance could be derived from such a conceptual model and accordingly be reflected in the user interface, for example, for further norming.

THE TECHNOLOGICAL CHALLENGE: MODERATORS OR AGENTS?

Software agents, which data mine public information spaces on grounds of collaborative filtering, should be able to evaluate the social traces of actors in virtual communities, in that they reveal the ambiguity, contradictions, and reciprocity of intentions and beliefs of community members. However, demands on software agents should not be overstrained. In critical application domains, agent technology can be replaced, or supplemented, by human beings acting as social moderators. At least as long as in virtual social environments reliable agent technology does not sufficiently support the outlined preconditions for consolidating interaction patterns, moderators will additionally have to care about the "social design" of the "ritual reality" (Anderson, 1997) which has already been described by the layered four-phase model of group stabilization.

Especially in the forming phase, the social moderator has to care about new group members, and to clarify how things—including supporting technology—work, what participants can and cannot do, and how they can or cannot benefit from group participation.

In the transition from the forming to the storming phase, a social moderator has to pay special attention to the nature and clarity of the group purpose. Social groups that do not meet a real need do not survive. Especially, under certain conditions of time and space independent interaction in virtual environments, it is necessary to rediscover the group's original purpose, and to examine whether that purpose has inappropriately changed or needs to. Under such circumstances, the handling of disputes and disruptions by the social moderator is often the key to the group survival. Accordingly, the moderator has to clarify policies or rules of conduct to prevent certain conflicts, or to guide their resolution. Any hidden or formal rules the group has already developed for conflict resolution have to be made adequately accessible to its members.

In the norming phase, it is important to allow members to increase their level of participation in the group, thereby promoting gratified behavior. Established members should be encouraged to play a mix of social roles. The willing and qualified should be involved in creating, maintaining, or extending group activities. Acknowledging membership and offering membership benefits motivates people to become members and provides an important sense of connection.

Performing of a group can best be measured by the regularity of group activities. Of great importance to the success of a community is the cyclic occurrence of some social rituals around which group members can structure a portion of their

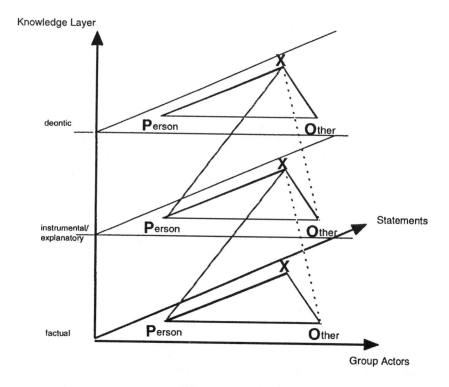

FIG. 11.3. Layered P-O-X model of knowledge-based group performance.

lives. This cyclic occurrence does not necessarily need to be organized by a social moderator, but can probably be executed by agent technology.

CONCLUSIONS AND FUTURE PROSPECTS

The way that the academic discussion about HCI design has been faltering is closely analogous to the misinterpretation of current folk sociologists when they attempt to come to an understanding of the realm of the social. The concept of society, as being drawn out by functionalist and interpretive approaches, has virtually disappeared from the sociological discourse. Accordingly, the primary focus of HCI design is the individual or, after Goffman (1969), the actor. As a consequence, HCI design is rapidly losing any ability to characterize these group-level actors in other than emotional, individual-level terms like *"user*-modeling," *"mental* models," and so on.

Emerging Technologies: New Concepts for Sociable HCI

A lot of new facilities are currently being conceived or developed with regard to networked multimedia, virtual reality environments, and interface agent technologies. These new technical developments make the inflexibility of conventional retrieval systems no longer an unavoidable constraint. Social interaction is always related to content and relations. Content can be structured by a complex and logical syntax. In terms of interactionism, however, relationships are structured by the semantics of symbolic clues, which have to be (re)interpreted dynamically in the aforementioned cycles of role taking/role making.

However, these new technologies have not come about as a result of a sociological perspective of analysis. This is mainly due to the fact that social research, up to now, has investigated the effects of technology on social phenomena, but not the effects of social phenomena on technology. In terms of variable analysis, usually adopted for sociological analysis, this amounts to saying that in analysis schemata, networked telematic systems are represented as independent variables, and social effects of such systems as dependent variables, and not the other way around. For example, according to the intervening concept of immediacy (Short, Williams, & Christie, 1976), the higher readiness to disagree with and advocate risky decisions under circumstances of group teleconferencing (c.f. Hiltz & Turoff, 1978) can be explained by less conformity pressure due to the less immediate and intimate mode of technologically mediated interaction.

The problem is due to the fact that concepts like social stratification, social status, social deviance, or even conformity are considered as phenomena resulting from the *combination* of known factors, and not to be derived from the known attributes of the *separate* constituent elements. This constitutes an obstacle to embedding social concepts into conceptual micromodels for sociable HCI design, because the latter would require further differentiation. Albeit facing criticism of being too reductionist, the so-called methodological individualism (Etzioni, 1968; Homans, 1964) may lead a way out of this dilemma. This school of sociology holds for the doctrine that all sociological explanations are reducible to the characteristics of individuals.

According to George Herbert Mead, social interactions may be viewed as a gesture-based conversation. This can occur at a preverbal level, among humans as well as animals—and one could also take into account avatars. A gesture-based conversation consists of a mutual adjustment of behavior, in which each participant uses the initial gesture of another participant as a cue for his or her own action, and his or her response becomes a stimulus to the other, encouraging either a shift toward his or her attitude, or the completion of the originally intended action.

Multimedia and virtual reality technologies support the relationship aspects of virtual communities in a brand new way. For instance, they make conceivable operations with symbolic objects in application scenarios like the diverse rooms of so called multiuser dungeons (MUDs), in which avatars connected to real end-users by means of gesture recognition techniques may symbolically interact as virtual group actors on behalf of their masters in real life (Damer, 1998).

Based on the outlined framework of dynamic group stabilization, design guidelines for sociable user interfaces, and corresponding prototype development, should take particularly into account the facilities of these innovative technologies. It is virtual reality technology, which for the first time in the history of information technology offers sufficient degrees of freedom to rapid prototype working models or detailed conceptual scenarios of augmented social environments, that are best suited to demonstrate or prove sociable HCI concepts through evaluation and testing.

Participatory Evaluation

In order to ensure the suitability of such concepts and interface prototypes, principles of participative systems design should be applied by an action-research related approach, not only investigating, but also actively influencing the virtual community under investigation by a social moderator. This refers to the process of detailing and refining the social dynamics of a community by observational methods related to the taxonomy outlined previously, as well as to technological processes of incremental rapid prototype development, for example, by object-oriented development tools to create MUD environments, and finally to issues of usability evaluation of the prototyped end product. The moderator, for instance, may ask users to comment on evaluations of prototyped user interfaces with regard to their revelations of the social dynamics of a virtual community.

By such a procedure, concept generation, prototype development, interface evaluation, and a possible redesign will all be made cultural issues to be discussed within the virtual community under investigation. These aspects, therefore, become submissive to the outlined phase model of increasing consolidation or decreasing destabilization of orientation for social interaction. With regard to participatory evaluation, consolidation is equivalent to social valuation, and destabilization to social devaluation of concepts and prototypes.

In fact, such a mutual understanding is probably most needed in designing and applying user interfaces to an information society for all. In contrast, attempts to reconstruct the motivational context of personal and, therefore, collective acts by identifying purely individual psychological variables and then summarizing or averaging them, simply reduces the apparently complex and more sophisticated phenomena to the less so. In any case, it first needs sociological reasoning, which as a leading concept may then be reduced to the characteristics of individuals for pragmatic reasons.

REFERENCES

Anderson, J. R. (1997). Local SIGs: the social design of a local special interest group. *ACM-SIGCHI Bulletin, 29*(2), 16.

Austin, J. (1962). *How to do things with words.* Cambridge, MA: Harvard University Press.

Balabanovic, M., & Shoham, Y. (1997). Content-based, collaborative recommendation. *Communications of the ACM, 40*(3), 66–72.

Damer, B. (1998). *Avatars! Exploring and building virtual worlds on the Internet.* Berkeley, CA: Peachpit Press.

Dodd, S. C., & Nehnevajsa, J. (1954). Physical dimensions of social distance. *Sociology and Social Research, 38,* 287–292.

Ekeh, P. (1974). *Social exchange theory. The two traditions.* Cambridge, MA: Harvard University Press.

Etzioni, A. (1968). Basic human needs, alienation and inauthenticity. *American Sociological Review, 33,* 870–885.

Goffman, E. (1969). *The presentation of self in every day life.* New York: Garden City.

Gullahorn, J. T., & Gullahorn, J. E. (1963). Role conflict and its resolution. *Sociological Quarterly, 4,* 32–48.

Heider, F. (1946). Attitudes and cognitive organization. *Journal of Psychology, 21,* 107–112.

Heider, F. (1958). *The psychology of interpersonal relations.* New York: Wiley.

Hiltz, S. R., & Turoff, M. (1978). *The network nation. Human communication via computer.* Reading, MA: Addison-Wesley.

Homans, G. C. (1956). *Social behavior: Its elementary forms.* New York: Harcourt Brace.

Homans, G. C. (1964). Bringing men back in. *American Sociological Review, 29,* 808–818.

Huber, G. (1983). Cognitive style as a basis for MIS and DSS designs: Much ado about nothing. *Management Science, 29,* 97–115.

Inkeles, A., & Levinson, D. J. (1963). The personal system and the sociocultural system in large-scale organizations. *Sociometry, 26,* 217–229.

ISO 9241. (1995). *Ergonomic requirements for office work with visual display terminals.* Geneva, Switzerland: International Standards Organisation.

Kahn, R., Wolfe, P., Quinn, R., Snoek, D., & Rosenthal, R. (1964). *Organizational stress: Studies in role conflict and ambiguity.* New York: Wiley.

Kautz, H., Selman, B., & Shah, M. (1997). Combining social networks and collaborative filtering. An interactive system for restructuring, visualizing and searching networks on the Web. *Communications of the ACM, 40*(3), 63–65.

Konstan, J. A., Miller, B. N., Maltz, D., Herlocker, J. L., Gordon, L. R., & Riedl, J. (1997). GroupLens: Applying collaborative filtering to Usenet news. *Communications of the ACM, 40*(3), 77–87.

Mead, G. H. (1934). *Mind, self and society from the perspective of a social behaviorist.* Chicago: University of Chicago Press.

Minsky, M. (Ed.). (1968). *Semantic information processing.* Cambridge, MA: MIT Press.

Parsons T. (1937). *The structure of social action.* New York: McGraw-Hill.

Parsons T. (1960). *Structure and process in modern societies.* New York: The Free Press.

Parsons, T., Shils, E., Naegele, K. D., & Pitts, J. R. (1965). *Theories of society.* New York: The Free Press.

Pieper, M. (1986). Computer conferencing: The gap between intention and reality. In L. Quortrup, C. Ancelin, T. Frawley, T. Hartley, F. Piehault & P. Pop (Eds.), *Social experiments with information technology and the challenges of innovation* (pp. 133–148). Dordrecht, Netherlands: D. Reidel.

Putnam, L. L. (1983). *The interpretive perspective: An alternative to functionalism.* In L. L. Putnam & M. E. Pacanowski (Eds.), *Communication and organizations. An interpretive approach* (pp. 31–54). Beverly Hills, CA: Sage.

Robey, D. (1983). Cognitive style and DSS design: A comment on Huber's paper. *Management Science, 29*(5), 580–582.

Searle, J. R. (1969). *Speech acts.* Cambridge, England: Cambridge University Press.

Searle, J. R. (1975). A taxonomy of illocutionary acts. In K. Gunderson (Ed.), *Language, mind and knowledge* (pp. 344–369). Minneapolis: University of Minneapolis Press.

Short, J., Williams, E., & Christie, B. (1976). *The social psychology of telecommunications.* New York: Wiley.

Siegel, J. P. (1968). Managerial personality traits and need satisfaction: The effects of role incongruity and conflict. In *Proceedings of the 76th Annual Convention of the American Psychological Association* (pp. 565–566). Washington, DC: American Psychological Association.

Stouffer, S. A., & Toby, J. (1951). Role conflict and personality. *American Journal of Sociology, 56,* 395–406.

Sutcliffe, J. P., & Habermann, M. (1956). Factors influencing choice in role conflict situations. *American Sociological Review, 21,* 695–703.

Terveen, L., Hill, W., Amento, B., McDonald, D., & Creter, J. (1997). PHOAKS: A system for sharing recommendations. A collaborative filtering system that recognizes and re-uses recommendations. *Communications of the ACM, 40*(3), 59–62.

Tuckman, B. W. (1965). Development sequence in small groups. *Psychological Bulletin, 63,* 384–389.

Winograd, T. (1987). A language/action perspective on the design of cooperative work. *Human–Computer Interaction, 3*(1), 3–30.

Winograd, T., & Flores, F. (1987). *Understanding computers and cognition. A new foundation for design.* Reading, MA: Addison-Wesley.

Zey-Ferrell, M., & Aiken, M. (1981). Introduction to critiques of dominant perspectives. In M. Zey-Ferrell & M. Aiken (Eds.), *Complex organizations: Critical perspectives* (pp. 1–21). Glenview, IL: Foresman.

Zurcher, L. A., Sonnenschein, D. W., & Metzner, E. I. (1966). The Hasher: A study of role conflict. *Social Forces, 44,* 505–514.

12

"Generating" Design Spaces: An NLP Approach to HCI Design

Margherita Antona
Demosthenes Akoumianakis
Constantine Stephanidis

This chapter proposes a grammar-based approach to the population of design spaces in the context of computer-based interactive systems. This new approach exhibits properties of multiple-metaphor environments, and allows the generation of alternative design specifications, combining elements from different toolkits, on the basis of task models. It builds, on the one hand, on the frequently occurring parallelism between user interfaces and sign systems, and on the other hand, on concepts and techniques recently elaborated in the field of Natural Language Processing, and in particular in the Head-Driven Phrase Structure Grammar theory. The grammar constitutes a simple and effective instrument for classifying and structuring the complex knowledge underlying the design of multiple-metaphor environments, as well as of capturing a level of abstraction suitable for a principled 'generation' of design spaces, and can be integrated in design tools for specification-based development environments.

The notion of metaphor in interface design is increasingly becoming a critical aspect in the attempt to provide more effective and higher quality interaction between humans and artifacts. In the past, the predominant metaphor for computer-based applications has been that of the office environment, which was visually encapsulated as the desktop in various graphical user interface (GUI) toolkits (e.g., Windows95, OSF Motif, Athena Widget Set). The Rooms (Henderson & Card, 1986) and the Book (Moll-Carrillo, Salomon, March, Fulton Suri, & Spreenber, 1995) metaphors are examples of alternative embodiments of real-world metaphors into a user interface. With the advent of the World Wide Web, the conventional visual desktop has been enriched with concepts from hypermedia, such as links, browsing, and so on. In the future, it is expected that neither the conventional desktop, nor the currently prevalent Web metaphors will provide adequate solutions to designing for the broadest possible end-user population and the variety of contexts of use (Stephanidis et al., 1998). In fact, the number and diversity of application domains in which the use of metaphors is critical continuously increase (examples include education software, home-based interaction environments, virtual and augmented realities, and electronic health care records).

223

Whereas, in the past, the use of metaphors was at the discretion of the designer, or in the best case, bound to what the existing development toolkit offered (e.g., trashcans, form filling), today, embedding metaphors to interface design is compelling for the wide adoption and user acceptance of (the) applications, in particular for nontraditional/nonbusiness applications such as the ones mentioned previously.

Metaphors may be applied at various levels, ranging from the overall interactive environment offered by an application, to the task level (i.e., how users engage and perform specific goal-oriented activities), to the physical level of interaction (i.e., representations used to convey intended meaning). Each of those levels may not involve the articulation of the same real-world metaphor, but variants of different ones (Stephanidis & Akoumianakis, in press). Thus, at the level of the overall interactive environment, users may be exposed to a booklike (Moll-Carrillo et al., 1995) or roomslike (Henderson & Card, 1986) metaphor, whereas in order to accomplish specific tasks (such as, e.g., the deletion of a file) alternative metaphors (e.g., deleting a file from a folder) may be used. Progressively, interactive computer-based applications move toward a state that can be characterized as *multiple-metaphor environments* (Akoumianakis, Savidis, & Stephanidis, 2000; Stephanidis & Akoumianakis, 1999; see also chap. 13, this volume).

In this perspective, human–computer interaction (HCI) design is anticipated to become a much more complex activity than in the case of single-metaphor design, and to require appropriate supporting tools (Stephanidis & Akoumianakis, in press). Design tools will be required, among other things, to facilitate designers in the population and exploration of the design space of multiple-metaphor environments, that is, the (sets of) design alternatives, or possible combinations of interface artifacts, integrating elements from different toolkits, which in turn may implement potentially different interaction metaphors. Furthermore, design tools will be required to facilitate designers in shaping such design spaces in a principled and coherent way on the basis of the user's characteristics, abilities, and preferences, the context of use, and the input/output modalities required or recommended as suitable to a given task. Population of design spaces and support for multiple metaphors are important aspects of designing *user interfaces for all* (see chap. 21, this volume).

Taking into account the preceding requirements, this chapter proposes a grammar-based approach to the population of design spaces in multiple-metaphor environments, which allows the generation of alternative design specifications on the basis of task models. Such an approach builds, on the one hand, on the frequently occurring parallelism between user interfaces and *sign systems* (Goguen, 1998; see also chap. 13, this volume), and on the other hand, on concepts and techniques recently elaborated in the field of *natural-language processing* (NLP), and in particular in *unification-based* approaches to natural language. The grammar is claimed to provide a simple and effective instrument for classifying and structuring the complex knowledge underlying the design of multiple-metaphor environments, as well as of capturing a level of abstraction suitable for a principled "generation" of design spaces. The approach described in this chapter relates to the notion of user interface in the context of user interfaces for all, because it provides support for informed choice and principled combination of metaphor(s) in multiple-metaphor environments (see also chap. 21, this volume).

This chapter is organized as follows. The next section introduces issues related to user interfaces as *sign systems,* whereas the third section briefly outlines the main characteristics of one of the more recent and widely applied language theories of the *unification-based* family, namely *head-driven phrase structure grammar*—HPSG (Pollard & Sag, 1987, 1994) and of its underlying formalism, namely *typed feature structures*—TFS (e.g., Carpenter, 1992; Carpenter & Penn, 1998). The same section also briefly describes how unification-based grammars can be used for the generation of text from abstract representations of meaning. Subsequently, the fourth section discusses the population of multiple-metaphor design spaces in HCI as a semantic-driven generation process relying on principles and mechanisms of the previously mentioned NLP approaches. The final section presents a discussion on the proposed approach.

THE USER INTERFACE AS A SIGN

User interfaces are often characterized in HCI literature as *signs* in a structuralist sense, that is, as associations of perceivable forms with some type of meaning (Goguen, 1998; see also chap. 13, this volume). Forms (or presentations) of user interface signs correspond to elements in a presentation domain, such as the (graphical representation of) interaction objects provided by conventional toolkits for the design and implementation of GUIs. Meaning, on the other hand, corresponds to elements in a function domain, that is, the machine-oriented version of the functional core of a system (functions of a computer environment, or *user tasks*). The form–meaning relationship in interface signs is typically *metaphoric;* that is, it relies on some nonarbitrary mapping between the elements in the presentation domain and the elements in the semantic domain, based on some characteristics of the presentation that correspond to some characteristics of the meaning. Commonly adopted presentation elements (e.g., buttons, text fields, icons) are usually representations of objects in everyday life domains (such as the office, the library, etc), and represent computer functions in terms of common activities already known to the user. For example, the menu interaction object class, as commonly encountered in popular user interface development toolkits, follows the "restaurant" metaphor, whereas push buttons, potentiometers, and gauges follow the "electric device" metaphor. Toolkits, which provide reusable collections of presentation elements and facilities for associating the desired functions to presentation elements, are often characterized as computational embodiments to *design languages* (Winograd, 1995).

Another important characteristic of interface signs, which can be characterized as the syntactic dimension, is the possibility to combine them to form complex signs whose meaning is determined by the meaning of the components. It can be plausibly argued that such a syntactic combination is determined by the underlying structure of the complex tasks to be represented in an interface, and that it obeys to (often implicit) general rules and principles that are, to a large extent, independent from specific metaphors or toolkits.

Furthermore, interface signs can be investigated, in a *pragmatic* dimension, with respect to their capability of conveying an intended meaning through some form in a given context. The notion of context is intended here in a very broad sense, including user abilities, requirements, and preferences, type of use, type of available modalities, and so forth (Stephanidis & Akoumianakis, 1999).

In the context just presented, the population of design spaces can be conceived as a process of mapping concepts in a function domain to symbols in a presentation domain, and vice versa (Akoumianakis et al., 2000; Stephanidis & Akoumianakis, 1999). Such a process should also be intended as compositionally combining single-interface artifacts into more complex ones on the basis of general principles also including context-dependent factors.

Research in the recent past has considered some of these issues. Some work has investigated the form–meaning relationship in user interfaces, and has proposed techniques for matching computer functions with representations on the basis of shared characteristics between the source and the target domain (Goguen, 1998; see also chap. 13, this volume). Other approaches, mainly in the field of model-based HCI, have proposed techniques and tools for task-driven interface design, for example, MASTERMIND (Browne, Davilla, Rugaber, & Stirewalt, 1998) and TLIM (Paternò & Meniconi, 1998). These efforts, however, though strong at capturing task semantics and interrelationships, do not account for alternative mappings between abstract task structures and concrete interaction elements. For example, let us assume that a menu is to be accessed by a sighted and blind user in order to delete a file from a list. Adopting a task-based approach, alternative options such as those depicted in Fig. 12.1 would typically be obtained.

All options in Fig. 12.1 share a common task organization with differentiation at the level of leaf nodes. Existing task-based techniques rarely provide the constructs to explicitly model the differentiated elements, thus resulting in a single design, and consequently a single implementation of the interaction. Therefore, the issue is to provide representations that do not limit the design space, but instead facilitate the selection of the maximally preferred alternative given the constraints of a particular context of use, thus facilitating the design of adaptable interfaces (see also chap. 21, this volume).

The approach presented in this chapter is inspired from recent computation-oriented approaches to language theory that establish a framework and provide an expressively adequate formalism for capturing the syntax, semantics, and pragmatics of natural languages as sign systems in a declarative and mutually constraining way. The aim is to provide a mechanism that allows the principled combination of elements from different toolkits in a task-driven fashion while imposing context-dependent constraints.

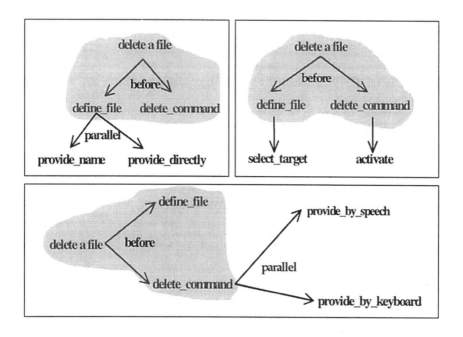

FIG. 12.1.   Alternative options for the task "delete a file."

LINGUISTIC THEORY AND NATURAL-LANGUAGE GENERATION

The inherent complexity and ambiguity of natural language, as well as the emergence of new applications such as *natural-language generation* (NLG), has driven research in both linguistic theory and NLP toward the development of approaches that allow capturing linguistic phenomena in concise, easily maintainable, computationally effective, and processing algorithm–independent ways. This trend emerged with the appearance of *unification-based* representation formalisms for the representation of linguistic knowledge (Shieber, 1986), and developed into a family of language theories (and underlying formalisms) based on such a notion of unification. Unification-based grammars are characterized by the declarative representation of linguistic objects, such as words, phrases, and sentences, by means of (recursive) *feature structures* (i.e., attribute-value pairs), which may be only partially instantiated (i.e., some values may be variables). A basic operation over feature structures, called *unification,* is used for checking their consistency and merging them. In essence, unification combines (partial) feature structure descriptions of a linguistic object into a unique (more informative) description that includes all the attributes and values of the unified descriptions, provided that they are consistent, that is, no attribute is explicitly assigned different values in those descriptions. Such an approach is also called *con-*

straint-based, because it emphasizes the notion that a grammar constitutes a set of constraints that the admissible signs of a language should satisfy.

One of the more recent and widely adopted unification-based knowledge representation languages is the TFS formalism (e.g., Carpenter, 1992; Carpenter & Penn, 1998). In TFS, elements in the modeled domain are declared as *types* organized in an inheritance hierarchy. This means that types may have subtypes (more specific instances of types), which inherit and further specify their properties. A unique most general type is assumed as the root of such a hierarchy. Thus, for example, the type hierarchy shown in Fig. 12.2 could be defined for modeling sentence structure in English.

Each typed feature structure is thus composed of a type and a collection of (possibly empty) attribute-value pairs for that type. Appropriateness of attributes for a type must be explicitly declared, and is inherited by its subtypes. Values of attributes are also typed (i.e., for each feature, the type of its possible values must be declared), and they may be either atomic or feature structures. Feature structures may also *share* values; that is, the values of two attributes in a feature structure may be declared as identical. For example, TFS 1 as follows is a TFS representing the sentence "John runs":

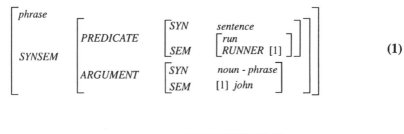

$$(1)$$

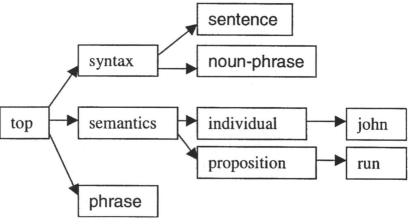

FIG. 12.2.   A sample type hierarchy for modeling sentence structure in English.

The structure in TFS 1 represents the fact that a linguistic entity of type *phrase* has a feature named *SYNSEM* that has a feature structure value composed of the features *PREDICATE* and *ARGUMENT*. The *ARGUMENT* feature represents the syntactic and semantic characteristics of the sentence subject, and namely that it is of syntactic category *noun-phrase* and denotes the individual *john*. The feature *PREDICATE* refers to the sentence as a whole and states that it is of syntactic category *sentence* and denotes a proposition whose predicate is *run*. Finally, such a proposition has a semantic argument, represented by the feature *RUNNER,* which shares the value of the sentence argument, "john." The sharing is represented graphically through the notation [1].

TFSs are inherently partial with respect to the information they provide, and may be ordered according to how specific is the information that they provide in the inheritance hierarchy. Such an ordering is called *subsumption.* Accordingly, a feature structure *subsumes* another one if: the type of the first is more general than the type of the second; both are assigned the same attributes and the values of these attributes in the first feature structure are more general than those in the second; and finally, if two attributes share their value in the first feature structure, they also share it in the second. Unification of typed feature structures is thus defined in terms of subsumption: The result of unification amounts to the most general feature structure subsumed by input feature structures. Unification is a very important characteristic of typed feature structure formalisms, which makes them suitable for representing not only linguistic knowledge, but also other types of knowledge for which the merging of partial representations is required, as, for example, in the case of the approach presented in the next section of this chapter, the knowledge associated to the syntax and semantics of interface signs.

TFSs, through their ability to represent, combine, and further specify partial information, can be used for modeling linguistic phenomena according to a variety of language theories.[1] HPSG (Pollard & Sag, 1987, 1994), a recent theory that has emerged as one of the best suited for NLP purposes, directly builds on the notion of TFS. Linguistic information in HPSG is modeled through types, type hierarchies, and inheritance. Such a theory views linguistic expressions (such as words, phrases, and sentences) as signs in the structuralist sense, relating a (phonological) form to a meaning.[2] Linguistic signs in HPSG also specify information related to the *utterance context.*[3] The notion of context includes the speaker of an utterance, the hearer, the utterance time, and so on.

HPSG is a *lexicalist* theory, as it takes the view that the largest part of linguistic information is encoded in the lexicon. In particular, the lexicon also contains the major part of the syntactic information of a language. Some words, the so-called

---

[1]For a brief review and further references, see Uszkoreit and Zaenen (1996).

[2]In HCI design, this can be assumed to be equivalent to mapping a task (e.g., database search) to an appropriate perceivable representation (e.g., icon, picture, figure, text, music, voice) or combination of representations.

[3]This chapter does not discuss the HPSG view of natural-language semantics and pragmatics. Pollard and Sag (1994) presented a detailed discussion of such a topic.

heads, are considered as determining the syntactic structures of the phrases in which they occur, because they impose constraints on other phrase components. These constraints are defined as the *valence* of a sign, that is, a list of specifications of possible complements the head can take. Thus, for example, the word *runs* is assigned the (partial) representation[4] depicted in TFS 2 as follows:

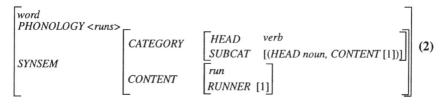

In TFS 2, the feature *PHONOLOGY* corresponds to the form of the lexical sign, whereas the feature *SYNSEM* contains the syntactic and semantic information of the sign, respectively specified in the *CATEGORY* and *CONTENT* features. In particular, the feature *HEAD* refers to the syntactic category of the sign, in this case *verb.* The feature *SUBCAT* represents the sign valence, that is, the fact that the word *runs* combines with a sign of category noun, whose content is identical with the semantic argument of "run," represented in the feature *RUNNER.*

The encoding of syntactic information in the lexicon does not introduce redundancy or unnecessary complexity, because properties of lexical entries can be described in a concise and principled fashion through inheritance hierarchies that allow the cross classification of words into word classes according to their properties. Thus, each word inherits the characteristics of all classes to which it belongs. For example, the word *runs* inherits features of finite verb forms, of all verbs, of words whose meaning is a unary relation, and so on.

Phrases (i.e., phrasal signs) are seen in HPSG as *projections* of their heads, constructed on the basis of the valence information encoded in head lexical entries. Phrasal signs are represented as *mother* feature structures that include the specification of their daughters, that is, of the (lexical or phrasal) signs that compose the phrase. For example, the sentence "John runs" would be (partially) represented as TFS 3:

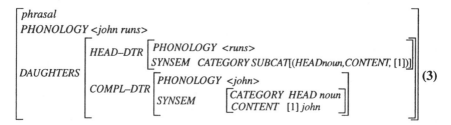

<hr />

[4]Examples of HPSG representations in this chapter omit information not related to the present discussion.

The feature structure in TFS 3 shows how the valence information encoded in lexical entries is used in the representation of phrases: A phrase is composed of a head sign (*HEAD-DTR*) and a (list of) complement sign(s) (*COMPL-DTR*), which share their feature values with the specifications in the head *SUBCAT* list. This mechanism allows for a very concise and elegant formulation of phrase structure rules, that is, of rules that recognize, or generate, phrasal signs. Rules in HPSG are called *schemata*, because they do not depend on the specification of syntactic categories. A single schema can combine a noun with its article, or a verb with its complements, and so on, by simply unifying with: (a) the head of a phrase, and (b) the elements required by the head's valence specification. A phrasal sign is complete (*saturated*) when all complements required by its head are included in its *DAUGHTERS* feature.

Schemata are constrained through *general principles* that apply to the information contained in the mother and in the head of a phrase and complement daughters. Two particularly important principles are the *head feature principle* and the *subcategorization principle*. The first states that a mother and its head daughter always share the same syntactic category (i.e., the value of the *SYNSEM/CATEGORY/HEAD* feature in the previous examples). This amounts to say that the phrasal projection of a verb is of category verb, and so forth. The second principle states that the valence of the head daughter of a phrase is equivalent to the valence of the mother with the addition of the identified complements: complement daughters come to (partially) fill the slots foreseen in the valence of the head. The slots that remain unfilled when a schema is applied constitute the valence of the mother. Other, less general, principles may concern: (a) construct specific phenomena, for example, phenomena concerning specific types of phrases such as coordinate phrases, and so on; (b) language-specific phenomena, that is, phenomena typical of specific natural languages, for example, the case of sentence complements.

Summing up, HPSG is characterized by: (a) a sign-based architecture, (b) the organization of linguistic information via types, type hierarchies, and constraint inheritance, (c) the projection of phrases via general principles from rich lexical information, and (d) the organization of such lexical information via a system of lexical types (Sag & Wasow, 1999). The HPSG approach to natural language has been applied in a variety of NLP applications. Its strength is due to the simplicity of the rules, to the easiness of specifying lexical information in multiple inheritance hierarchies and, in particular, to the possibility of working with partial, *underspecified* descriptions of linguistic entities. HPSG grammars are completely declarative, and *reversible;* that is, they can be used for both parsing and generation purposes[5] (Neumann & van Noord, 1993). The *semantic head-driven generation algorithm* (Shieber, Pereira, van Noord, & Moore, 1990), developed

---

[5]Reversible grammars can be complied in different ways according to whether they are to be used by a parsing or a generation algorithm. Generation takes as input a (semantic) representation of the text content and produces as an output the actual natural-language text corresponding to such a content.

for generating text with unification-based grammars, is the most commonly used algorithm for NLG with HPSG grammars, because it directly relies on the notions of syntactic head and valence, and exploits them in order to perform the generation process.[6]

A GRAMMAR FOR GENERATING DESIGN SPACES

In this section, concepts and mechanisms from HPSG are applied to facilitate the population of the design space in multiple-metaphor environments. Relying on the assumption that user interfaces can be studied as sign systems, the approach outlined here aims at investigating how the HPSG framework could be adopted for defining a *design grammar* that describes interface signs as the combination of a form (i.e., an interface object) with a content (i.e., a specification of an application function, or user tasks), as well as the principles that constrain the combination of such signs into phrasal signs (i.e., composite interface elements such as dialogues, or complete interfaces). A generation algorithm applied to such a grammar would produce specifications of design alternatives in multiple-metaphor environments, in the form of TFS. It is assumed that such specifications can be interpreted by the run-time libraries of a user interface development toolkit that undertakes the task of realizing the specifications into a user interface implementation. For this to be attainable, however, the toolkit should exhibit API functionality that allows interoperation. Example toolkits that serve this purpose have been developed, and are described in Savidis, Stergiou, and Stephanidis (1997) and Savidis, Vernardos, and Stephanidis (1997).

The grammar-based approach relies on the representation of interface objects, their syntactic properties, and their semantics into a TFS inheritance hierarchy. Figure 12.3 depicts a simplified example of such a hierarchy. Lexical interface signs, that is, representation of toolkit objects, as the lexicon of natural languages in HPSG, can be cross-classified on the basis of their ability to represent different tasks, and to combine with each other, independently from the specific metaphor in which they are embedded. Such an assumption is similar to the concept of *abstraction* over interaction objects in (Savidis, Stephanidis, & Emiliani, 1997). In a TFS hierarchy, abstraction can be elegantly captured through the extensive use of underspecification: Features are assigned as values that are not the most specific task types in the hierarchy, in such a way that a unique description subsumes a set of (more specific) descriptions.

The semantics of interface signs can be straightforwardly modeled in the grammar as a hierarchy of user and system tasks, based on the notion of *task context* (Akoumianakis & Stephanidis, 1997). A task context is considered as a characterization of "dialogue states," that is, indicators of what the user interface is doing at

---

[6]Semantic head-drive generation combines top-down and bottom-up techniques to connect phrases in sentences. The algorithm recursively looks for a semantic head on the basis of the semantic specification to be generated, generates the head daughters, and then connects the head to its mother.

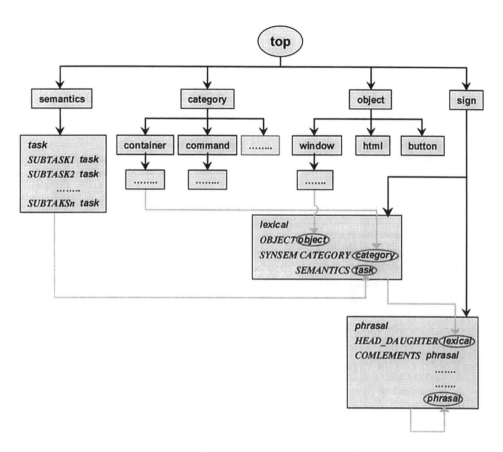

FIG. 12.3.   A simplified example of the type hierarchy for the design grammar.

any point in time. Task contexts encompass a degree of *abstraction* that entails that the same task context may be performed differently by different users, or the interface may exhibit differentiated behavior in the same task context depending on the current user. As a consequence, the same task context may be mapped onto several interface objects. In the grammar, tasks can have subtasks, which are either tasks (i.e., they can have their own subtasks), or lists of tasks of the same type. So, for example, a *search* task could be represented as TFS 4 or 5 as follows:

$$
\begin{bmatrix}
search \\
SUBTASK1 \ search\_condition \\
SUBTASK2 \ result\_presentation
\end{bmatrix}
\tag{4}
$$

$$
\begin{bmatrix}
search & & \\
SUBTASK1 & search\_condition & \\
SUBTASK2 & \begin{bmatrix} presentation\_list \\ HD \quad page\_presentation \\ TL \quad presentation\_list \end{bmatrix}
\end{bmatrix} \tag{5}
$$

In TFS 4 the task *search* is "decomposed" into two subtasks, namely the formulation of a query and the presentation of the result. In TFS 5 the second subtask of *search* is a list of tasks (all of type *page presentation*) whose number is not specified.

Additionally, the proposed approach assumes that interface objects combine in head-driven fashion, that is, some of them behave like heads that specify a valence partially determining the category and semantics of the other interface objects they can combine with. Abstractions over properties of interface objects as for their combination with other objects are adopted for modeling interface sign syntax. A typical example of such an abstraction is the concept of *container* (Savidis, Stergiou et al., 1997), which captures the characteristics of objects like windows, books, html pages, and so on.[7] Containers represent the main interaction areas for tasks, and may encompass other objects representing subtasks like menus, buttons, and so on. In the grammar, containers play the role of heads; that is, they are assumed to have a (partially specified) valence, which becomes further instantiated according to the task assigned to a container. Other examples of heads could be menus, which subcategorize for menu items according to the subtasks of the main task at hand, and toolbars, which subcategorize for buttons. The phrasal projection of a menu, that is, a menu containing a specified list of items corresponding to commands, can constitute one of the daughters of a container phrase. Categories such as container, menu, toolbar, and so forth, are therefore used in the grammar in a fashion similar to that of syntactic categories in natural-language grammars.

The inheritance hierarchy of the devised grammar also includes interface objects, simply classified according to their physical properties.[8] Thus, for example, the classification includes window objects, book objects, button objects, and so on. Furthermore, the hierarchy subdivides interface signs into lexical and phrasal signs, and assigns feature description to signs. Lexical signs, which provide the structure of entries in the grammar lexicon, have features representing constraints on the interface object of a sign and on its syntactic and semantic properties. Phrasal signs, on the

---

[7]Containers differ from other composite interaction elements insofar as they exhibit additional interactive behaviors such as navigation, access policy, and topology of contained elements.

[8]This chapter does not deal with issues related to spatial and temporal relationships of interface objects in interfaces, because the goal of the grammar is not to produce full interface specifications, but recommendations that can be interpreted and applied by the run-time libraries of toolkits, through appropriate application programming interface (API) functions. However, it is assumed that the inheritance hierarchy and the grammar can be extended to represent these relationships.

other hand, constitute the structures produced by phrase structure (generation) rules, that is, complex interface signs composed of lexical signs.

In the lexicon, semantic, and, to a lesser degree, syntactic properties of signs are underspecified. Lexical signs are assigned a general task type, so that they are compatible with all the specific tasks subsumed by such a type. This approach, though ensuring that the right type of task is associated with the right type of interface object (e.g., buttons are not associated with overall tasks), avoids the necessity of specifying for each interface object all the tasks it can be assigned, because this is defined implicitly through the inheritance hierarchy. TFS 6 is an example of lexical entry for a *book* container object:

$$
book \begin{bmatrix} lexical \\ INTERFACE\_OBJECT\ book \\ SYNSEM \begin{bmatrix} CAT \begin{bmatrix} HEAD & top\_level\_container \\ SUBCAT & [(CAT\ HEAD\ simple\_container,\ SEMTASK\ [1]), \\ & (CAT\ HEAD\ simple\_container,\ SEMTASK\ [2])] \end{bmatrix} \\ SEM\ TASK \begin{bmatrix} top\_level\_container\_task \\ SUBTASK\ 1\ non\_top\_level\_container\_task\ ,\ [1] \\ SUBTASK\ 2\ non\_top\_level\_container\_task\ ,\ [2] \end{bmatrix} \end{bmatrix} \end{bmatrix} \quad \textbf{(6)}
$$

The feature structure in TFS 6 represents a lexical sign, relating an interface object, namely a book, as specified by the value of the feature *INTERFACE_OB-JECT,* with a specification of syntactic and semantic properties, that is, the complex value of the feature *SYNSEM*. In particular, the sign is defined, as being a *top_level_container (SYNSEM/CAT/HEAD),* that is, a container representing an overall task. Its valence, represented by *SYNSEM/CAT/SUBCAT,* subcategorizes for two *non_top_level_* containers (corresponding to the two visible pages in an open book). Finally, the sign receives an underspecified semantic value *(SYNSEM/SEM/TASK)* corresponding to *top_level_container_task,* with two subtasks also underspecified. This means that the sign can represent any individual task of type *top_level_container_task* with two subtasks of type *non_top_level_container_task* (e.g., *search*). These two subtasks correspond to the semantics of the non-top-level container (e.g., *search_condition*) required by the sign valence. Their values are shared in the feature structure. Figure 12.4 provides a graphical representation of an abstract "book" object as captured in TFS 6.

A very small number of general phrase structure rules is sufficient for generating design alternatives. For example, a set of three schemata is sufficient to cover headed constructs, including those allowing an (initially) unspecified number of subtasks for some components. Schemata work as follows:

1.   The first schema combines a head with a specified set of daughters to generate a mother phrase by applying the head feature principle and the subcategorization principle.

top_level_container_task

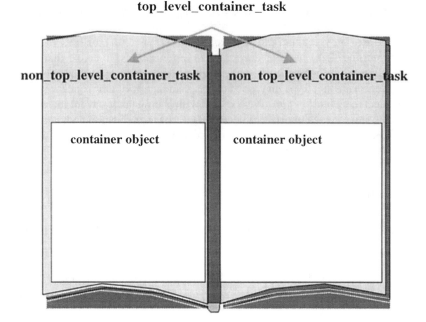

FIG. 12.4.   An underspecified "book."

2.   The second schema combines a head with an unspecified set of daughters, all characterized by a subtask of the same type (e.g., *page_presentation*) and an interface object of the same type (e.g., *non_top_level_container*), by applying the head feature principle and the subcategorization principle. A dummy interface object can constitute the head of such a phrase (e.g., an "invisible" composite container made up of several pages).
3.   The third schema simply generates phrases from heads with empty valences, and therefore no daughters.

The described grammar-based approach exploits the underspecification and multiple inheritance mechanism built in to the formalism and grammatical framework adopted. When a structured user task is given as input, the grammar produces all (multiple-metaphor) interface signs that can convey such a task. In the generation process, the input task and its subtasks are recursively unified with phrasal interface signs starting from the head of each construct in a head-driven fashion. The semantic underspecification of lexical entries allows the direct unification of the input with all interface signs whose semantics unify with the task at hand. So, for example, supposing that the grammar contains an entry for a *book* top-level container such as the one depicted earlier, as well as appropriate entries for two non-top-level containers, namely a *window* and an *html_container,* when requested to generate a phrasal sign for a *search* task, it will: (a) unify the semantics of book to *search,* with the two subtasks *search_condition* and *result_presentation;* (b) produce four alternatives, with all possible combinations of *window* and *html_container* as non-top-level containers instantiating the search subtasks. Figure 12.5 graphically represents the four generated alternatives.

A grammar of this type may also handle redundancy in user interface signs, that is, the degree to which alternative objects for the same subtask(s) are included in an interface sign. For example, a window may provide navigation commands in the form of menu items, grouped into a menu, or buttons grouped into a toolbar, or both. Various degrees of redundancy may be obtained, for example by encoding two (or more) lexical entries for the same object, which respectively enforce or disallow redundancy. In the window example, two entries for the object *window* would be encoded, both with an underspecified semantics constituted by a *top_level_container_ task,* which has as one of its subtasks an underspecified *command_list.* The *SUBCAT* of the two entries would differ: One entry would subcategorize for one phrase repre-

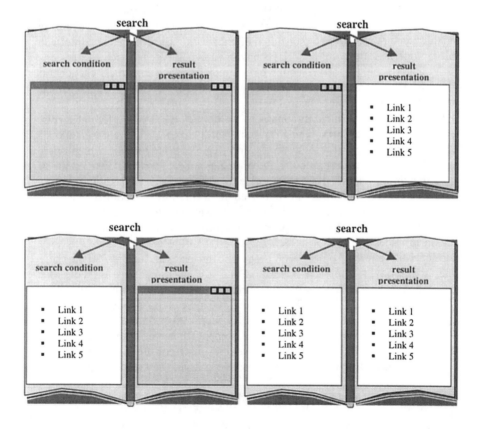

FIG. 12.5. Generating multiple-metaphor design alternatives.

senting the command list (i.e., sharing its semantics), whereas the second entry would subcategorize for two phrases representing the command list. Supposing that the grammar allows only buttons and menu items to represent commands, in generation the first entry would produce two output results, namely a window with a toolbar and a window with a menu. Both the toolbar and the menu would contain as many buttons or menu items as are the commands specified in the input semantics. The second entry would produce one output, namely a window with a toolbar and a menu. Globally, the generation of an input task involving navigation commands would produce the three alternatives mentioned previously. In summary, generated design alternatives may vary with respect to: (a) semantically equivalent objects from the same toolkit, (b) semantically equivalent objects from different toolkits, and (c) the degree of redundancy.

DISCUSSION

Reconsidering the key ideas underlying HPSG, as summarized at the end of the third section, from the viewpoint of HCI design, the outlined grammar-based approach appears to be characterized by: (a) a sign-based account of interface objects and their semantics, (b) the organization of knowledge concerning tasks, interface object properties, and combinatorial behavior via types, type hierarchies, and constraint inheritance, (c) the projection of complex interface constructs via general principles from rich lexical information, and (d) the organization of such lexical information via a system of lexical types.

The advantages of such an approach can be viewed at two different levels, namely a "representational" and an "architectural" level. At the representational level, the grammar constitutes a concise and elegant device for modeling properties of interface signs. In particular, the grammar appears well suited to establish an abstract representational level for modeling the syntax and semantics of interface objects in a metaphor/toolkit-independent fashion. In the adopted approach, abstraction is used to characterize not an object itself, but its combinatorial behavior in interface signs. The underspecification mechanism built in to the TFS formalism provides a mean of associating interface objects with tasks at a very abstract level, relieved from the toolkit-specific intrinsics and details.[9] At the same time, instantiated interface specifications are produced by unifying the underspecified lexical items with input semantic structures that correspond to specific instances of their semantics. Therefore, it can be claimed that the described approach is capable of generating interface specifications in multiple-metaphor environments without requiring an explicitly encoded pairing of interface objects with specific tasks, which would be time and resource consuming in an environment characterized by a wide variety of possible tasks and metaphors. This leads to the conclusion that design tools based on such an approach would minimize the effort required for encoding, maintaining, and updating the knowledge necessary for

---

[9]These can be accounted during interface implementation, when mapping design specifications to concrete implementations.

automatically populating the design space of multiple-metaphor environments. The grammar-based approach would also maximize the tool extendibility to account for new applications and metaphors/toolkits. In fact, the modeling of new applications affects only the subhierarchy of tasks without requiring changes to the lexicon or the rules, whereas new interface objects can be added to the lexicon without concern about the low-level content of the task hierarchy.

From an architectural point of view, the outlined grammar-based approach offers the possibility of investigating the interrelationships between syntactic, semantic, and also pragmatic characteristics of interface signs within a single framework. This brings into HCI design in general, and in user interfaces for all in particular, the insight, typical of contemporary computation-oriented language theories, that levels of analysis such as lexicon, syntax, semantics, and pragmatics, are not to be viewed as a sequence of separate models, but instead as a set of mutually constraining properties of signs. Such a perspective is particularly attractive for the purposes of investigating how different metaphors can be combined and integrated into user interfaces on the basis of user tasks in a principled and contextually grounded fashion. In fact, an interesting aspect of such an approach is the possibility of enforcing various types of constraints on the generation of design alternatives. Intrinsic grammar constraints are the already mentioned head feature principle and the subcategorization principle, which govern the combination of interface signs at an abstract level. Other principles may apply to specific types of constructs. For example, some interface signs can enforce *metaphor coherence* or *object identity* in some of their component parts: A toolbar can contain only buttons, a menu only menu items, and so forth. Another important category of constraints could be of a completely different nature, that is, extragrammatical constraints related to the *context of use* of an interface, such as user abilities, requirements, and preferences, and factors related to the type of usage and access device. For example, constraints could ensure that consistency is maintained in the specification of a design alternative with respect to the preferred user dialogue style, language, and so on. Following the distinction between "universal" and "language-specific" principles in HPSG, it could be argued that principles in an interface grammar can be classified into general principles whose validity holds across different toolkits or metaphors, for different users, and so on, and principles related to more specific aspects of user interfaces. Context-related constraints would be introduced in the grammar in the form of additional principles, which would then be applied in generation along with grammar internal principles. In such a way, context-related principles would "filter out" incoherent or contextually inappropriate design alternatives.

The design grammar proposed in this chapter is not meant to produce interface implementations, but, rather, interface specifications. It is intended to be applied in design tools integrated into specification-based development environments (Stephanidis & Akoumianakis, in press). In such environments, a design tool based on the proposed grammar would act as a source of design recommendations, in the form of specifications, which, once compiled, can be interpreted and applied by the run-time libraries of the user interface development system, thus allowing the instantiation of alternative interactive behaviors. In particular, in the framework of *unified user interface* development, described in Part VI of this volume, such a design tool could contribute orthogonally by providing semantic-based design specifications to be integrated with physical-level specifications produced by other tools (see chap. 23, this volume). In order to capture and apply context-related principles, the grammar could cooperate with other tools, such as user-modeling (Akoumianakis et al., 1997; see also chap. 23, this volume) or decision-making (Karagiannidis, Koumpis, & Stephanidis, 1997) tools, which would provide information concerning user preferences, dialogue style to be used, modality, device, and so forth. The produced specifications would then be implemented according to unified user interface implementation (see chap. 22, this volume), which allows the instantiation of different design alternatives in a single interface.

SUMMARY AND CONCLUSIONS

This chapter has presented a sign-based approach to the population of design space in multiple-metaphor environments. Borrowing from recent developments in language theory and in NLG, the chapter has proposed a grammar for the generation of design alternative specifications that is capable of mapping abstract user tasks to interface specifications combining elements from different toolkits. The main features of the design grammar, which is inspired from the HPSG theory, are: (a) hierarchical structuring of knowledge through typed feature structures, (b) highly underspecified (abstract) representation of lexical knowledge, (c) head-driven approach to the generation of specifications from tasks, and (d) introduction of principles constraining the combination of interface objects in complex constructs.

The advantages of the proposed approach have been identified in its representational conciseness and ease of modification and extension, as well as in its capability of imposing a variety of constraints on the specification of design alternatives, including user- and context-related constraints. The proposed design grammar is claimed to introduce a new approach to the population of design spaces in multiple-metaphor environments, capable of supporting the designer in a task up to now largely carried out without system support, or at best, with artifact-oriented rather than user- and context-oriented support. Furthermore, the proposed approach has been claimed to be both compatible and useful to the unified user interface development framework, elaborated in Part VI of this volume.

REFERENCES

Akoumianakis, D., Savidis, A., & Stephanidis, C. (2000). Encapsulating intelligent behaviour in unified user interface artefacts. *Interacting With Computers* (Special issue on *The Reality of Intelligent Interface Technology, 12*(4), 383–408.

Akoumianakis, D., & Stephanidis, C. (1997). Preference-based human factors knowledge repository for designing accessible user interfaces. *International Journal of Human–Computer Interaction, 9*(3), 283–318.

Browne, T., Davilla, D., Rugaber, S., & Stirewalt., K. (1998). Using declarative descriptions to model user interfaces with MASTERMIND. In P. Palanque & F. Paternò (Eds.), *Formal methods in human–computer interaction* (pp. 93–120). Berlin: Springer-Verlag.

Carpenter, B. (1992). *The logic of typed feature structure* (Cambridge Tracts in Theoretical Computer Science 32). Cambridge, England: Cambridge University Press.

Carpenter, B., & Penn, G. (1998). *The attribute logic engine user's guide* (Version 3.0 Beta) [Online]. Available: http://www.sfs.nphil.uni-tuebingen.de/~gpenn/ale.html

Goguen, J. (1998). An introduction to algebraic semiotics, with application to user interface design. In C. Nehaniv (Ed.), *Proceedings of Computation for Metaphors, Analogy and Agents* (pp. 54–79). Aizu-Wakamatsu, Japan: University of Aizu-Wakamatsu. Available: http://www-cse.ucsd.edu/users/goguen/projs/semio.html

Henderson, D. A., & Card, S. K. (1986). Rooms: The use of multiple virtual workspaces to reduce space contention in a window-based graphical user interface. *ACM Transactions on Computer Graphics, 5*(3), 211–241.

Karagiannidis, C., Koumpis, A., & Stephanidis, C. (1997). Modelling decisions in intelligent user interfaces. *International Journal of Intelligent Systems, 12*(10), 753–762.

Moll-Carrillo, H., Salomon, G., March, M., Fulton Suri, J., & Spreenber, P. (1995). Articulating a metaphor through user-centred design. In *Proceedings of the Human Factors in Computing Systems Conference* (pp. 566–572). New York: ACM Press.

Neumann, G., & van Noord, G. (1993). Reversibility and self-monitoring in natural language generation. In T. Strzalkovski (Ed.), *Reversible grammar in natural language processing* (pp. 59–95). Amsterdam. Kluwer.

Paternò, F., & Meniconi S. (1998). TLIM, a systematic method for the design of interactive systems. In P. Palanque & F. Paternò (Eds.), *Formal methods in human–computer interaction* (pp. 241–260). London: Springer-Verlag.

Pollard, C., & Sag, I. A. (1987). *Information-based syntax and semantics: Vol. 1. Fundamentals.* Chicago and London: University of Chicago Press.

Pollard, C., & Sag, I. A. (1994). *Head-driven phrase structure grammar.* Chicago and London: University of Chicago Press.

Sag, I. A., & Wasow, T. (1999). *Syntactic theory: A formal introduction* [Online]. Stanford: Center for the Study of Language and Information Publications. Available: http://hpsg.stanford.edu/hpsg/wasow.html

Savidis, A., Stephanidis, C., & Emiliani, P. L. (1997). Abstract task definition and incremental polymorphic physical instantiation: The unified interface design method. In *Proceedings of HCI International '97* (pp. 465–468). Amsterdam: Elsevier, Elsevier Science.

Savidis, A., Stergiou, A., & Stephanidis, C. (1997). Generic containers for metaphor fusion in non-visual interaction: The HAWK interface toolkit. In *Proceedings of INTERFACES '97 Conference* (pp. 194–196). EC2 & Development.

Savidis, A., Vernardos, G., & Stephanidis, C. (1997). Augmenting the Windows Object Library with embedded scanning techniques for motor-impaired user access. In *Proceedings of INTERFACES '97 Conference* (pp. 233–234). EC2 & Development.

Shieber, S. M. (1986). *An introduction to unification-based approaches to grammar.* Chicago and London: University of Chicago Press.

Shieber, S. M., Pereira C. N., van Noord, G., & Moore, R. (1990). Semantic-head-driven generation. *Computational Linguistics, 16*(1), 30–42.

Stephanidis, C., & Akoumianakis, D. (in press). Computational environments for human factors knowledge management in HCI design. In W. Karwowski (Ed.), *International encyclopaedia of ergonomics and human factors*. London: Taylor & Francis.

Stephanidis, C., & Akoumianakis, D. (1999). Knowledge-based interaction design. In A. Kent & J. G. Williams (Eds.), *The Encyclopaedia of Library and Information Science, 66*(29), 186–217.

Stephanidis, C. (Ed.), Salvendy, G., Akoumianakis, D., Bevan, N., Brewer, J., Emiliani, P. L., Galetsas, A., Haataja, S., Iakovidis, I., Jacko, J., Jenkins, P., Karshmer, A., Korn, P., Marcus, A., Murphy, H., Stary, C., Vanderheiden, G., Weber, G., & Ziegler, J. (1998). Toward an information society for all: An international R&D agenda. *International Journal of Human–Computer Interaction, 10*(2), 107–134.

Uszkoreit, H., & Zaenen, A. (1996). Grammar formalisms. In R. A. Cole (Ed.), *Survey of the state of the art in human language technology* (Sponsored by the National Science Foundation and the European Commission) [Online]. Available http://cslu.cse.ogi.edu/HLTsurvey/

Winograd, T. (1995). *Bringing design to software*. Reading, MA: Addison-Wesley.

IV Software Technologies and Architectural Models

13 The FRIEND21 Framework for Human Interface Architectures

Hirotada Ueda

This chapter describes the results of FRIEND21 project, funded by the Japanese government. FRIEND21 introduced the notion of adaptive change in user interfaces, and developed a set of frameworks facilitating the development of user interfaces adaptive to change through the dynamic combination of system functions and their presentation (metaphors). The FRIEND21 approach is based on: (i) the notion of Metaware, as the cognitive principle underlying the systematic description and control of adaptive change; and (ii) the Agency Model, as a mechanism for embodying the Metaware principle. The chapter also discusses the relevance of the FRIEND21 project results to the concept of User Interfaces for All.

FRIEND21 (future personalized information environment development) was a Japanese government–funded six-year-long project, which started in 1988 and concluded on March 31, 1994. The project aimed at investigating the nature of human interfaces for the 21st century, and at elaborating a methodology for their design and development. When it first began, the term *media fusion* was still novel, and the idea that information, telecommunications, publishing, broadcasting, and various other media sectors would be "fused" as the 21st century approached was new.

At that time, cognitive psychology was gaining acceptance in human interface research, and people began to advocate the idea that cognitive engineering should be applied to system design focused on the user (Object Management Group [OMG], 1993). Furthermore, research efforts in the field of user interface management systems (UIMS) had recently introduced the notion of separation between the functional and the presentational domains of computer system (Pfaff, 1985). The aim of the FRIEND21 project was to ensure the possibility for everyone to employ computers in daily life in the 21st century. The initial goal of the project was to provide "systems that anyone can use, anywhere, anytime." However, our research gradually developed a somewhat different objective, that is, the creation of a computerized society imbued with sympathy and care for the user, based on collaboration and coevolution between human beings and computers.

FRIEND21 research focused on the development of basic technologies for software that would be commercially available 5 years after the completion of the project. FRIEND21 moved from the starting point that computing in the emerging information society undergoes continuous change, and that an adaptive approach to the development of user interfaces should be adopted for coping with change. The project developed a set of frameworks (Nonogaki & Ueda, 1991) facilitating the development of user interfaces adaptive to change through the dynamic combination of system functions and their presentation (metaphors). The proposed approach is based on: (a) the notion of metaware, as the cognitive principles underlying the systematic description and control of adaptive change, and (b) the agency model, as a mechanism for embodying the metaware principle. The proposed framework allows handling user intentions, tasks, context, polysemy, synonymy, and multiplicity in an integrated manner.

This chapter is organized as follows. The next section introduces the notion of adaptive change in user interfaces. The third section describes metaware and its underlying principles and characteristics. The fourth section introduces the agency model as a framework for the implementation of metaware. The final section discusses the relevance of the FRIEND21 project results to the concept of *user interfaces for all*.

ADAPTIVELY CHANGING INTERFACES

A fundamental characteristic of the emerging information society is its continuous change and evolution (e.g., new application areas, evolving usage patterns, new user categories, etc.). As we move toward the information society, computers will increasingly serve as tools for access to all types and forms of information. For example, computers will provide applications for reading electronic publications, conversing with people at remote locations over networks, shopping by means of electronic catalogues, performing remote medical diagnosis, accessing multimedia databases, and so on. Computers will no longer appear in their present form, but as either invisible, or embedded components in popular information appliances, such as telephones and televisions. Moreover, their contemporary productivity-oriented usage patterns will be increasingly complemented with communication-oriented use, for example, everyday appliances for sending and receiving information (Ueda, 1992).

Thus, the provision of support for social and individual activities in the information society is progressively becoming more important than increasing the speed and efficiency of computers. The dynamic adaptation (Nonogaki et al., 1991) of user interfaces to facilitate this changing environment is a particularly significant requirement. Of course, computer hardware and software technologies will continue to advance. Still, because it is the human interface that mediates between computers and the human users, it is natural to think that the requirement for adaptation will be met with a suitable human interface that can adapt to a changing environment.

Attaining adaptability requires a thorough study of what changes in the environment, why, and how. Similarly, we must examine how the system can understand the changing situation, what should be changed inside the system, and how. Focusing on the causes of change, it is necessary to look beyond the system alone, and consider the broad context that includes users, system developers, society, and the environment that encompasses all these.

The FRIEND21 approach to modeling and explaining user behavior in human–computer interaction is based on Norman's model (Norman, 1988). According to such a model (Fig. 13.1), a user activity is divided into seven steps: (a) first, the user decides on a goal, that is, sets the objectives to be met with the use of the computer; (b) these objectives are then "translated" into intentions; (c) intentions are "translated" into a mental sequence of specific acts; (d) next, the foreseen acts are executed; when the computer processing is finished, the user (e) perceives the execution results; and (f) interprets what has been perceived; finally, (g) the user evaluates the interpretation to determine whether or not the results match the desired goals. If the results do not match the goals, the user reconfirms the objectives and reformulates intentions.

The aforementioned model regards interaction as the bridging of two gulfs: the gulf of execution and the gulf of evaluation (Norman, 1986). Designers of the human interface have striven to narrow these two gulfs. However, users, who do not share the thinking patterns of the designers, suffer from the heavy mental load imposed on them by unfamiliar and/or unusable design.

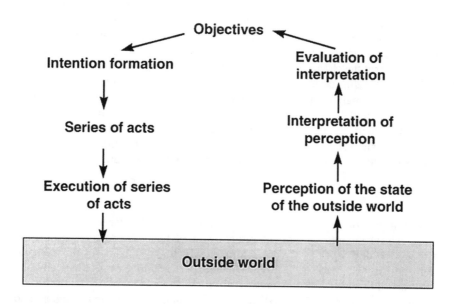

FIG. 13.1.   Norman's model of user behavior in human–computer interaction. From Norman (1988). Adapted by permission.

To reduce such a load, the system (i.e., the human interface architecture) must be equipped with a mechanism for adapting to the user's environment and purpose of use, by identifying user intentions related to task accomplishment. In order to be adaptable and have the ability to identify user intentions in relation to task accomplishment, a system must continuously incorporate information about the environment, the context, user characteristics, and task models. These requirements are shown in Fig. 13.2.

In our view, task models represent alternative possible schemes for carrying out the (sequences of) tasks foreseen by the system designers. The mechanisms for adaptation and intention identification can shoulder some of the user's mental burden (the shaded parts in Fig. 13.2) by making appropriate decisions according to context, user information, and task models. For example, by performing a task such as searching, the system evaluates what the user wants to do on the basis of the user's operations, and adapts to that situation.

The FRIEND21 project considered the following to be fundamental requirements concerning adaptation and change in user interfaces: (a) the possibility of customization through combining modules, (b) the adaptation to user's preferences, character, and purpose of use, (c) the interpretation of intentions underlying user operations, (d) the provision of support for the user-training process, and (e) the capability of conveying the state of the system to the user in an adequate manner. The common aspect of these requirements is that they focus on the user's perspective of the human–computer communication process. Metaware and the agency model have been proposed as integral constituents of an interface architecture capable of satisfying the preceding requirements. Metaware provides principles for what the interface architecture should describe and how, and for what kind of algorithm should control the change so as to allow flexible adaptation to individual users according to changing circumstances. It aims at providing a cognitive science-inspired and cognitive engineering-oriented approach to the aforementioned requirements. The agency model, on the other hand, has been proposed as a framework for the run-time implementation of the metaware principles. Both metaware and the agency model separate system functions from their presentation to the user. In the following sections, metaware and the agency model are presented in detail.

METAWARE

The term *metaware* results from the combination of the terms *metaphor* and *ware*. It has been introduced in literature by the FRIEND21 project to refer to the framework, developed by the project, for expressing the cognitive principles underlying the systematic description and control of adaptive change in the human interface. The objective of metaware is to provide a methodology for the design of interface metaphors that facilitate dynamic change according to context, and the creation of new meaning.

Metaware defines interface metaphors by analogy to cognitive (language) metaphors (Institute for Personalised Information Environment, 1995). Metaphors in

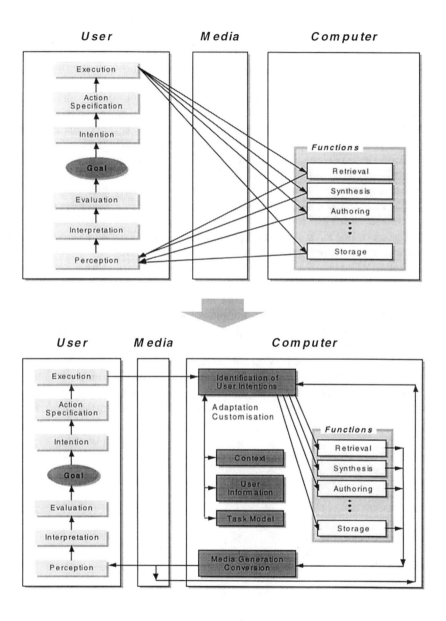

FIG. 13.2. Reduction of user's mental load by the user interface architecture.

human communication are essentially mechanisms for explaining concepts by example. In metaphors, concepts in a source domain (a familiar knowledge domain) are mapped to concepts in a target domain (the domain of the unknown task or problem), on the basis of some similarity between the two domains. Metaphor interpretation (hermeneutics) is the process of understanding metaphors in context through experience and culture-specific commonsense knowledge (Institute for Personalised Information Environment, 1995). In metaware, the source domain corresponds to the graphic images used for conveying tasks in the interface, whereas the target domain corresponds to the tasks to be presented to the user. In the desktop metaphor, for example, the "desk" concept is part of the knowledge domain of the "office" already known to the user, and constitutes the source domain. The target domain, in this case, comprises computer system functions, such as file management, editors, electronic mail, and so on. In metaware, metaphor hermeneutics is provided through a mechanism for dynamically establishing equivalence between images and tasks in context. Figure 13.3 represents the relationships between the three components of metaware.

The metaphor environment proposed in metaware is called a *multiple-metaphor environment,* because it allows multiple source domains to be appropriately combined. The source domains in metaware make up the *world model.* Metaware makes very active use of context sensitivity in selecting the appropriate interpretation of user actions and the appropriate presentation of system states, thus helping the user to communicate with the system in a way similar to human-to-human communication. Furthermore, in metaware, the role of interface metaphors is not limited to facilitating understanding of functions, but also extends to the creation of new meaning. The process that leads to the creation of meaning in metaware is similar to the extension of word meaning in the lexicon through the very frequent use of some metaphor (i.e., when a metaphor is used very often in language, the words used to express it lose their metaphoric meaning

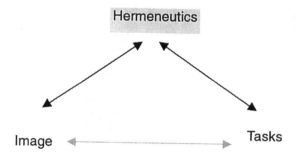

FIG. 13.3.   The three basic elements of metaware.

assume new semantics). Furthermore, metaphors extend and maintain polysemy of pictorial representations in the interface.

Three Properties of Metaware

Metaware is characterized by three fundamental properties: (a) synonymy/ polysemy of presentations, (b) temporality/spatiality, and (c) context sensitivity.

The properties of synonymy and polysemy concern the semantics of interface metaphors. In communication between people, a word can have more than one meaning (polysemy), and different words can have the same meaning (synonymy). This allows expressing "content" from different perspectives, and contributes to smooth communication. Polysemy and synonymy do not normally create problems in human understanding of natural language, because the latter takes place in a multimodal context: Intonation, gestures, facial expressions, and so on, help humans to identify the correct interpretation of natural-language utterances. By analogy to human-to-human communication, interfaces should employ metaphors that involve polysemy and synonymy (i.e., provide polysemous and synonym representations of system functions), and perform dynamic control of metaphor interpretation according to the context. In metaware, polysemy can be defined as one-to-many mapping of source domain elements to target domain elements. Synonymy, on the other hand, can be defined as a many-to-one mapping of source domain elements to target domain elements.

Metaware, therefore, introduces polysemy and synonymy in the process of conveying the system states to the user (the gulf of evaluation—see Hirose, 1992; Norman, 1986).

The interpretation of user operation intentions also introduces polysemy and synonymy in the process of conveying intentions to the system by the user (the gulf of execution—see Hirose, 1992; Norman, 1986; Ricoeur, 1973).

The properties of temporality and spatiality concern interface syntax (relations/arrangement of representation elements as they are perceived by users). Two syntactic interface styles are taken into account, the "temporal arrangement" and the "spatial arrangement." In temporal arrangements, representation symbols are placed sequentially following the flow of time, as in spoken language. In spatial arrangements, they are displayed all at once in space, as in a picture or photograph.

The third property of metaware is context sensitivity. The appropriateness of an interface metaphor to a given user task depends on the context. Context sensitivity in Metaware concerns both the semantic and the syntactic level of the interface. At the semantic level, it is called context-sensitive selection, whereas at the syntactic level it is called context-sensitive disclosure. Context-sensitive selection means that a given combination of presentation and function domains is selected so that it is the most appropriate for a specific context; the selection is made within the degrees of freedom allowed by polysemy and synonymy. Context-sensitive disclosure means that interface metaphors are presented to the user in an arrangement that is appropriate to the spatial and temporal context.

Four Variations of the Multiple-Metaphor Environment

Currently available multiple-metaphor environments do not necessarily give equal weight to the three properties described in the previous section. There are a number of variations, each of which emphasizes a different property.

Assuming that the properties of polysemy/synonymy and temporality/spatiality are orthogonal, the range of multiple-metaphor environments can be expressed in a two-dimensional space (Fig. 13.4). In this space, most of the currently used interface metaphors, which typically manipulate figures and tasks by one-to-one correspondence, correspond to a very small region characterized as spatial metaphor environment.

Metaware expands the currently limited metaphor methodology to different regions in the two-dimensional space, by allowing the application of different communication techniques. To give an example, one technique often used in the field of education consists in explaining a concept through more than one metaphor, in

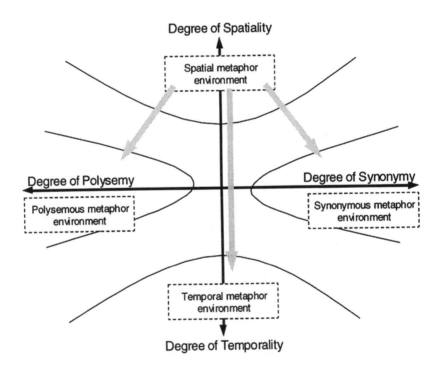

FIG. 13.4.   Variations of multiple-metaphor environments.

order to capture different aspects of the concept by means of different metaphors. This technique lies in the synonymy region of our space. Another technique, which lies in the polysemy region, can be compared to the actions of an assistant who can execute instructions differently according to the situation. Examples concerning temporality/spatiality are plays and movies. They involve both dimensions, but are more appropriately placed in the region that mainly features temporality.

Learning from these real-life examples, metaware allows expanding out from the limited region to which current interface metaphors are restricted into the other regions in this space. This expansion is described in detail in the following subsections. The metaware methodology has elaborated design guidelines for multiple-metaphor environments, exploiting the various dimensions in the two-dimensional space in such a way as to narrow the gulfs of evaluation and execution as much as possible. Guidelines concern semantic aspects of the design space (on the synonymy/polysemy axis) as well as syntactic ones (on the spatiality/temporality axis). Guidelines at the semantic level are summarized in the following sections, and include: (a) similarity, (b) synonymy, (c) source domain selection, (d) world model, (e) polysemy, and (f) target domain selection. Guidelines (a)–(c) aim at narrowing the gulf of evaluation, whereas guidelines (d)–(f) aim at narrowing the gulf of execution.

Similarity

The first guiding principle for context-sensitive selection in metaware is that "design should consider the similarity between the source domain and the target domain."

The basis of an interface metaphor is the existence of a *motivation* between the source domain and the target domain. Motivation is a term borrowed from semiotics to denote the relationships between a symbol representation and the related content. In metaphors, the similarity between the source domain and the target domain is the most relevant motivation (Ortony, 1979). Similarity can be defined as the sharing of attributes, or relationships between two objects. The treatment of similarity in FRIEND21 is based on common attributes. Regarding relationship-based similarity, structure-mapping theory (Institute for Personalised Information Environment, 1995) proposes various methods to solve the problem of the trade-off between the number of mapping degrees of freedom and the amount of computation.

In order to compute similarity between domains in the interface, a source domain module (SDM) and a target domain module (TDM) should be compared through a representation of relevant object attributes. An important aspect of attribute-based similarity is the selection of attributes as clearly definable as possible. Actually, similarity is a matter of how many of the most relevant attributes are shared. Attributes should be physically measurable, or clearly definable properties. At the same time, they should be variables that are relevant to the interface and that can be directly controlled by the system designer.

As a specific method for selecting attributes that satisfy these requirements, we propose a method that employs the results of a taxonomic analysis of the independent factors of the human interface. In such a process, as many attributes as possible should be considered, including, for example, the type of information (whether the information is visual or auditory), the type of information processing (information generation, information storage, or communication), and the number of information items (the amount of information involved). These attribute examples are a subset of the classification by Gawron, Drury, Czaja, and Wilkins (1989). Attribute values are determined by two methods. The attribute values in the SDM are determined on the basis of knowledge about the user's source domain, obtained through interviews, experiments, and so forth. Attributes are assigned subjective probabilities. In the TDM, attributes, selected from task classification, are assigned context-sensitive probabilities of alternative presentations.

Once the attributes and their values are determined, a measure of the degree of similarity can be introduced. The degree of metaphor similarity based on attributes is given by Tversky's calculation formula.[1]

Synonymy

The second guideline for context-sensitive selection in metaware is that "one target domain should correspond to multiple source domains." If it is possible to combine multiple source domains for one target domain, then it is possible to select among them the one that is most appropriate for a particular context.

For that purpose, it is first necessary to allow the maintenance of synonymy in context. Synonymy does not imply identity of meanings: In human language, different words have different meanings, although the difference may be only slight. What is important in synonymy is difference in similarity, that is, how similar are the different synonyms to the concept they are synonym to. A possible method for maintaining synonymy is through the establishment of *common terms*. Given an attribute representation for a set of source domains and a target domain, it is considered that synonymy can be maintained between those source domains that have relatively high similarity values with respect to the target domain module. Thus, that attribute representation is a common term for maintaining synonymy. Because similarity values change according to context, maintaining synonymy allows the selection of the most appropriate representation of system tasks in the interface according to the context.

Source Domain Selection

The third principle for context-sensitive selection in metaware is that "a source domain that is appropriate with respect to the context should be chosen" (Fig. 13.5).

---

[1]For a detailed analysis of how the measure of similarity is calculated, refer to Institute for Personalised Information Environment (1995, chap. 4, section 4.2, p. 61).

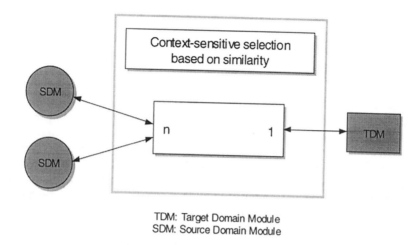

TDM: Target Domain Module
SDM: Source Domain Module

FIG. 13.5.   Context-sensitive selection based on "similarity." TDM = target domain module. SDM = source domain module.

This amounts to selecting among synonym representations of target elements in context, and breaks down to: (a) selecting among several source domains on the basis of their degree of similarity to the target domain, (b) managing the switching, or appending of source domains in context, and (c) managing the integration of the source domains that have been switched, or appended.

The dynamic rearrangement method has been developed to allow context-sensitive selection. It consists of three steps. First an SDM is selected for a given TDM among a set of synonyms by picking up the one with the highest measure of similarity value. As a second step, attribute values are continuously monitored, and similarity values recalculated accordingly. If, due to changes in context, a mismatch is detected, that is, an SDM different from the one currently in use comes out with a highest similarity to the TDM, the third step is executed. In this last step, metaphor switching is triggered and the SDM, which caused the mismatch, is replaced with an SDM that has a higher similarity value. This method is one of the solutions for the interface mismatch problem presented by Carroll, Mack, and Kellogg (1988).

A multiple-metaphor environment based on synonymy allows overcoming the limits of a single-metaphor environment, because it allows alleviating the problems associated with mismatches. Furthermore, such an environment makes it possible to offer to the user interesting new metaphors, by looking for source domains that are semantically distant from the target domain, while maintaining similarity of attributes.

World Model

The fourth guideline for context-sensitive selection in metaware is that "the world model should be similar to the user's work environment." Environments that are developed by a long-standing culture, such as the home, the office, or the city, are constructed so as to make it easy for the members of that culture to perform their roles. Accordingly, the use of a culturally shaped environment that the user is accustomed to as the source domain is a necessary condition for facilitating the realization of user goals, thus narrowing the gulf of execution.

Nevertheless, simply presenting an environment with which the user is familiar is not sufficient. When defining the relationships among multiple objects that make up the environment, the system designer must also describe the consequences of user operations on the environment. For example, it is necessary to specify what happens when the user places a document into a folder.

In other words, the designer must describe a part of the source domain as a model for the interface metaphor. Often, the source domain model comprises a microworld that copies part of the actual environment.

When designing interface metaphors, definitions such as "documents can be put into folders, but folders cannot be put into documents" must be provided somewhere in the program. This kind of definition is part of the world model. Designers need support for creating appropriate world models, in order to avoid possible problems such as lack of consistency in the model. The metaware methodology includes guidelines for world model description, based on three types of knowledge: object world definitions, meta-world definitions, and causal rules.

The object world definition is the definition of the objects that compose the microworld (imitating a part of the actual environment). The meta-world definition is the definition of meta-level objects for the manipulation of objects in the world, such as, for example, "hand." The causal rules define how objects behave in the world.

Individual objects defined by object world definitions have features, constituted by attributes and related values, as well as relationships, again with related values. Attributes are organized hierarchically. For example, a "part-of" hierarchy represents part–whole relationships among objects, whereas an "is-a" hierarchy represents membership relations among object classes.

The merit of a world model that resembles the user's work environment is that intentions in the user's mental model can be simply represented in the system. The merit of metaphors is that, although they mainly serve to facilitate understanding, they also help to narrow the gulf of execution, because functions can be executed through simple operations according to user intentions.

The advantage of organizing world model descriptions using the three separate types of knowledge (object world definition, meta-world definitions, and causal rules) is that it ensures the flexibility required for easily modeling new source domains, so that they can be freely added and replaced in the multimetaphor environment. This will ensure that the world model easily adapts to the changing environment of the information society. In the future information society, a wide variety of new computer applications and services will be available. Furthermore, these applications and services may be radically different to the ones we know and use today. The environment that the user is familiar with is also changing greatly. It would be futile to try to predict that course of change in advance and to thrust upon future users a fixed world model that embodies an ordinary work environment that existed at some point in time. The world model should have the flexibility to adapt to changes in the outside world. The construction of the world model with interchangeable modules as described here makes such flexibility possible for the first time.

Polysemy

The fifth guiding principle for context-sensitive selection in metaware is that "one source domain should have multiple corresponding target domains" (see Fig. 13.6). If it is possible to group multiple target domains for one source domain, then the target domain that is most appropriate for a given context can be selected from among them.

We propose a method for maintaining polysemy based on the use of the world model. When the user performs an operation, the desired target domain in the specific context is inferred from the operation and the knowledge described in the world model. This inference is similar to obtaining multiple solutions with the inference engine of a production system. Though it is possible to construct the system so as to obtain a single solution by making the conditions of the inference rules

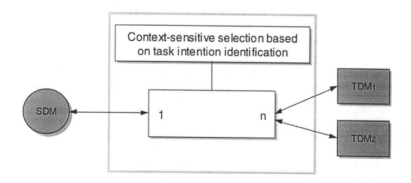

TDM: Target Domain Module
SDM: Source Domain Module

FIG. 13.6.   Context-sensitive selection based on "polysemy." TDM = target domain module. SDM = source domain module.

adequately strict, this is not desirable, as the result would be a singular operation as in conventional human interfaces.

The world model maintains polysemy through the use of causal rules. Consider, for example, a metaphor environment in which the newspaper serves as the source domain. In this environment, when the user performs the operation of opening a newspaper icon, his intentions may be various. He may want to read the front page, continue reading where he stopped the previous time, or perhaps read some other, yet unread, article. In the world models, all these possibilities are described as causal rules, designed in such a way that a number of them fire when the user opens the newspaper icon. Then the most appropriate target domain should be chosen dynamically, depending on the given task context. The selection method is described in the next subsection.

Target Domain Selection

The sixth guideline for context-sensitive selection in metaware is that "a target domain that is appropriate for the context should be selected." Appropriateness in this case means that the selected target domain matches the user's task intention.

For this purpose, a user task model is required. User task models can be described through syntax/semantics models, state transition diagrams, GOMS models, or production rules (Croft, 1984; Hirose, 1992; Nii, 1989; Rich, 1983; Uyama, 1993).

The model proposed in metaware includes a task intention tree based on a syntax/semantics model, a task description, a weighted directed graph, and state transition diagrams. The graph and the networks represent a task model that describes how the task is achieved. The second important requirement is a criterion for selecting target domains. Two criteria have been proposed for context-sensitive selection: (a) the degree of matching between the partial task intention tree learned by the system, and a series of user operations; and, (b) the degree of matching between the task descriptions provided by the designer, and the history of the messages issued by TDM and SDM agents. Figure 13.6 illustrates the first criterion.

Here, we give an example of context-sensitive selection according to the degree of matching between the partial task intention tree learned by the system and a series of user operations. This method can learn a user task model from a series of user operations, and in parallel infer the user's intended function in context.

In learning the task model, the functions executed as a result of the user's operations are chronologically arranged in a history, and frequently occurring patterns are extracted from the history. The hierarchical tree constructed bottom-up through this learning process is called a "partial task intention tree." On the other hand, inference of the user's intended function is triggered when a user operation fires multiple causal rules. In this case, a library of learned partial task intention trees and the user's operation history are compared. The function with the highest frequency of past execution is inferred from that comparison.

Conventional design guidelines for the human interface have not involved identification of the user's intentions, because the ambiguity-generating demerit of polysemy has been emphasized. Metaware, on the contrary, has given importance to making positive use of polysemy in order to reduce the semantic distance of the gulf of execution.

AGENCY MODEL

The agency model has been proposed by the FRIEND21 project as a run-time architecture for the implementation of multiple-metaphor environments according to the guidelines of the metaware methodology. The model consists of an integrated environment including a multiagent system, and related mechanisms for user–agent cooperation. The agency model (Fig. 13.7) has been developed taking into account generic requirements for interactive systems, such as extensibility, module reusability, and support for multimodal processing, as well as requirements specific to metaware, such as separation between SDMs and TDMs, context-sensitive selection, context-sensitive disclosure, and adaptation to individual users. In the following subsections, the concept of an "integrated environment" is introduced as an interactive system based on such requirements. As a second step, we examine related work on integrated environment models, and introduce the agency model, a run-time architecture for the multiple-metaphor environment. Finally, we describe an actual implementation of the agency model.

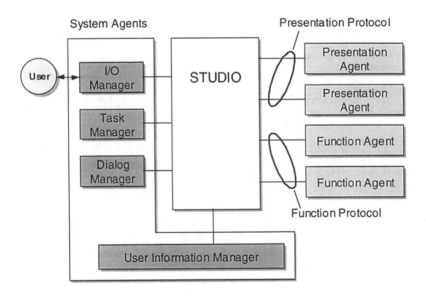

FIG. 13.7. Overview of the agency model.

The Integrated Environment

The Multiple-Metaphor Environment as an Integrated Environment.
Computer users often require a variety of interactive applications to accomplish
their tasks. In the office environment, for example, involved applications include
software for word processing, spreadsheets, electronic mail, and so on. Supporting
user tasks requires, therefore, the development of integrated systems comprising a
number of independent modules created by different designers. Some modules perform interaction with the user, whereas others execute functions in the task domain.
Furthermore, some of these modules may be familiar to the user, whereas others
may be less familiar. In this section, we define as "integrated environment" an interactive system in which different function modules cooperate to achieve the user's
tasks and provide a consistent system image.

A multiple-metaphor environment is an integrated environment. It consists of
SDMs and TDMs. These modules are designed independently and should be interchangeable, that is, it should be easy to add to, or remove from, the environment
both source and target modules. Furthermore, in order to facilitate design, modules
should be designed in such a way as to allow reusability of existing functions in the
design of new ones. Module reusability also preserves the consistency of the system image, which is advantageous to the user.

Related Work on Integrated Environments.   Several methods have been
developed for implementing integrated environments. A classification of integration methods was provided by Reiss (1990) as follows:

1. Integration at the file level (e.g., UNIX).
2. Integration based on a single programming language (e.g., Smalltalk).
3. Integration based on a common program database (tools share common
   data structures that represent aspects of the programs and their execution).
4. Integration based on the message facility (e.g., Field environment).

The Field environment employs message passing to achieve clarity of the
framework, as well as to facilitate reuse of existing tools, expansion, and distribution over the network. The message-passing approach is also used in the HP
SoftBench Environment (Cagan, 1990), the Envoy Framework (Palaniappan et al.,
1992), and the Common Object Request Broker Architecture, of the OMG (1993).

Reiss (1990) pointed out four requirements for the communication among different tools in the design of integrated environments. The following three are relevant for the present discussion:

1. Tools must be able to interact with each other directly.
2. Tools must share dynamic information.
3. The environment must make static specialized information available to all
   tools that need it.

These requirements can be satisfied by a software architecture in which management of static information, such as user and task models, is supported, each component (tool) is provided with an open command interface, and messages are broadcast by a message-passing facility. For example, selective message broadcasting allows any tool to send command messages to any other tool via a message server. In selective message broadcasting, each tool registers the patterns of messages that it is interested with the message server. Subsequently, all tools send their messages to the message server. The server broadcasts each message to those tools whose registered patterns match the message under discussion.

An integrated environment architecture of the type just described is represented in Fig 13.8.

Agency Model as a Run-Time Architecture

The previous section described integrated environments designed for achieving cooperative processing among tools. Building on such an architecture, the agency model has been defined as a run-time architecture for multiple-metaphor environment implementation.

The Protocols and Task Manager.   In conventional integrated environments (Fig. 13.8), user interfaces are statically linked to tools by means of graphical user interface (GUI) libraries. In a multiple-metaphor environment characterized by synonymy and polysemy, according to the metaware guidelines, the SDMs and

Message Server

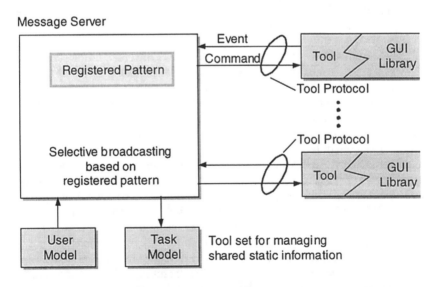

FIG. 13.8.   Integrated environment based on selective broadcasting.

the TDMs are separated. Therefore, a command interface for SDMs and TDMs must be defined and made open (i.e., easily extensible, modifiable, etc.).

In the remaining of this chapter, adopting the terminology of the UIMS field, TDMs are called *function agents,* and SDMs are called *presentation agents.* Function and presentation agents are essential components of the agency model. Agents are independent, autonomous software modules. They are provided with their own independent command interfaces *(protocols).* The message-passing interfaces of function and presentation agents are called *function* and *presentation protocol* respectively.

Because protocols are defined independently for each agent, the run-time architecture must have a mechanism for converting between them, on the basis of knowledge concerning user tasks (task models). Protocol conversion and task modeling are performed by a special agent called the *task manager.*

Cooperation Among Agents.   Context-sensitive processing requires a mechanism for obtaining a global understanding of the task execution context. Such a mechanism should also support the maintenance of a broad view of user task accomplishment by multiple agents. This implies that the local context information managed by each agent must be shared with the other agents, through an appropriate facility. The dynamic information sharing used in the HP SoftBench Environment (Cagan, 1990), for instance, which is based on event message passing, is not sufficient for this purpose.

Context-sensitive processing also requires conditional branching of code. If this is implemented in individual agents, extendibility and reusability will be lower, because modifications would have to be effected in each of the individual agents.

The "Process Specification" component of HP SoftBench, or the "Mission" component of Envoy enable the users themselves to register patterns of task accomplishment in the system. Users can also include in those descriptions logic for branching according to different contexts. When new modules are added, however, this conditional branching logic needs to be modified by the users themselves. This imposes a significant mental burden on the user, associated with the addition of new modules.

Context-sensitive processing involving multiple modules through the sharing of local context has been studied in the field of distributed artificial intelligence. One of the developed methods for maintaining context employs the blackboard framework (Nii, 1989). This framework manages context information as a global data structure. A module, called the Scheduler, accomplishes context-sensitive processing by controlling the execution of modules called "knowledge sources."

In the contract network protocol (Smith & Davis, 1981), a module broadcasts a task request, and modules that can execute the task present their own context to the requesting module in the form of a bid. The agent that performs the broadcasting selects the appropriate bidding modules, on the basis of context information, thus achieving context-sensitive processing.

In the agency model, a combination of the aforementioned approaches, appropriately modified, has been adopted for context-sensitive processing.

The Dialogue Manager.   The implementation of context-sensitive disclosure, that is, disclosure of new information to the user in an arrangement appropriate to the spatial and temporal context (Uyama, 1993), in a run-time architecture implies that the usage pattern selected by the cooperative processing of multiple modules can provide the user with the opportunity to try out unknown solution methods.

In the agency model, this amounts to the specific requirement of supporting the context-sensitive integration of new services (introduced by new modules) with respect to user tasks. A specific module, called the *dialogue manager,* supports cooperation with the user in context-sensitive disclosure, by identifying problems that agents cannot solve and by interacting with the user in order to solve them.

The dialogue manager facilitates user interaction with the system by responding to user inquiries concerning the context-dependent content of the studio, informing users about agent failures, and cooperating with them to find an appropriate solution.

The existence of the dialogue manager is an important difference between the agency model and previously mentioned software architecture models, whose main goal is to support design. In the case of the human interface, we consider that, in addition to ensuring extendibility, meaning that new modules can be designed and added, it is also necessary to support the integration of new services provided by the new modules into the user's own tasks. The implementation of the dialogue manager as an independent agent is a manifestation of this necessity.

The User Information Manager (Adaptation to Individuals).   According to metaware principles, user interfaces should adapt to individual differences. This requires the extraction and learning of individual user characteristics, such as preferences and task accomplishment patterns. The agency model provides a component for the global observation of the user's task accomplishment, namely the *user information manager.*

In the agency model, user tasks are accomplished by multiple agents that exchange data. Data exchange takes place in the studio, which also infers from interagent communications the short-term task accomplishment context, and the management of context information.

The user information manager identifies and learns task accomplishment patterns that are characteristic of individual users, by monitoring the short-term contexts generated in the studio. The results of such a learning process are accumulated in the form of user models and task models, which are made available to all agents in the architecture. User models and task models are established as long-term models that are exploited for the understanding of user operations. The user information manager maintains the user models and the tasks models. The user modules contain information such as the user's skills, knowledge, preferences, and interests. Part of this information is encoded at agent design time, whereas part is added directly by the user, or automatically learned through the monitoring of user interaction with the system. The task manager has the function of converting user intentions to low-level commands, defined by the function and presentation agents' communication protocols.

Providing an independent agent for user information management increases the portability of the integrated environment. The user can make use of a personal environment at any place and at any time.

The Input/Output Manager (Multimodal Processing).   The guideline for multimodality is expressed as follows: "Consistency should be maintained by an integrated interpretation of the dialogue between user and system, via multiple agents."

For the integrated understanding of inputs among multiple agents, cooperative processing is required: the user's intention is spread over multiple modalities, and various interpretations of it are possible, depending on the context. Accordingly, a mechanism for cooperation is required in order to achieve an integrated interpretation of the multiple inputs. On the output side, it is necessary to maintain the synchronization of time-dependent media, and the consistency of information among different media.

In the agency model, each input/output (I/O) device (including input devices such as keyboard and mouse, voice recognition equipment, pen input character recognition; and output devices such as displays, voice response systems, and music output devices) is handled by an I/O agent. The I/O manager is an agent that allows the user to participate in the dialogues between those I/O agents in the studio.

Details of Context-Sensitive Processing

Communication among agents in the agency model takes place in the studio. The studio provides the following four functions:

- Support communication (i.e., exchange of messages) among the agents connected to the studio.
- Determine the task accomplishment context form the history of agent communications.
- Support agent cooperation for context-sensitive selection and multimodal processing.
- Maintain consistency in multiple-view displays.

The agency model employs a two-layer structure for the studio. The two layers are called *task execution layer* and *cooperative processing layer*. The task execution layer manages the operation of the multiple-metaphor environment at the concept level. To accomplish context-sensitive and multimodal processing, the task execution layer uses the services provided by the lower cooperative processing layer. The cooperative processing layer provides facilities such as the selective broadcasting of messages, the blackboard scheduler, and the contract network protocol.

Message passing in the task execution layer is illustrated in Fig. 13.9. Cooperative processing is shown in Fig. 13.10. The studio consists of the studio manager and the scheduler. The task manager is realized as a set of partial task agents.

The Task Execution Layer.   Here we describe the type of messages used in the task execution layer of the studio and the tasks that can be executed. As shown in Fig. 13.9, the studio integrates user inputs (through various modalities) and passes control over to the presentation agents.

When receiving a DeviceEvent message, which indicates that a user operation has been performed through some input device, the presentation agent sends a MetaphorEvent message to the task manager according to the user's operations. The presentation agent changes the appearance of the presentation, when it receives a MetaphorCommand message from the task manager, and sends a MetaphorReply message to the task manager as a response. The format of MetaphorEvent, MetaphorCommand, and MetaphorReply messages are specified by the presentation protocol.

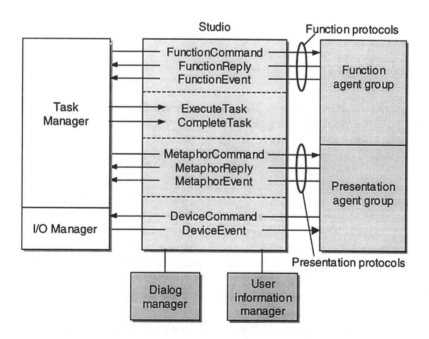

FIG. 13.9   Task execution layer of the studio.

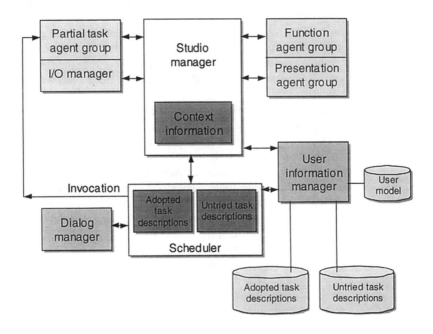

FIG. 13.10   Cooperative processing layer of the studio.

The function agent invokes primitive services in response to a FunctionCommand message from the task manager, and returns a FunctionReply message. The function agent also spontaneously issues FunctionEvent messages, according to its internal state. The format of FunctionEvent, FunctionCommand, and FunctionReply messages are specified by the function protocol.

The task manager performs operations according to the task model upon receiving a MetaphorEvent, MetaphorReply, FunctionEvent, or FunctionReply message. An operation can correspond to sending a FunctionCommand message, sending a MetaphorCommand message, changing the internal state of an agent, or initiating a subtask. Subtask initiation is performed by calling up a local procedure. The task manager sends task phase description messages to the studio at the beginning and at the end of task execution. Task phase description messages include ExecuteTask messages and CompleteTask messages.

The Cooperative Processing Layer.   As shown in Figure 13.10, cooperative processing is performed by the studio manager, the scheduler, and other agents. Here, an agent is defined as a process that possesses the following facilities:

- A message interpreter.
- An incoming message queue.
- A selective receive procedure for selecting messages from the queue.
- Single-casting procedure for communication between agents, performed through the following steps: (a) processing by the sender agent is blocked until the return of execution results, (b) the receiver agent interprets the context when the interpreter is idle, (c) the sender agent does not wait for execution results, and (d) the receiver agent queues the messages.
- A message-logging mechanism.

In the cooperative processing layer, the studio consists of two special agents, the studio manager and the scheduler. The task manager comprises multiple subtask agents, which are activated by the scheduler. When activated, they break up the task into operations according to the operation part of the task description.

All messages in the task execution layer are handled via the studio manager in the cooperative processing layer. The studio manager maintains a record of all kinds of messages (DeviceEvents, MetaphorEvents, MetaphorCommands, MetaphorReplies, ExecuteTasks, CompleteTasks, FunctionEvents, Function-Commands, and FunctionReplies) and uses it as context information that indicates the user's task accomplishment context. The studio manager adds sent messages to the context information, and then redistributes them to the interested recipients.

The task models comprise sets of task descriptions. The software designers predict how the user will use the designed software product and provide the predicted usage patterns as a task description.

The scheduler compares the context information with the task descriptions, and selects the task that is appropriate for the context. Context-sensitive selection is triggered when an event message is sent to the studio manager. The scheduler checks the initiation conditions for each task description in the set of adopted task descriptions and in the set of untried task descriptions. It then selects the task descriptions whose trigger fields match the event message in question.

Next, the scheduler checks the context condition fields of the selected task descriptions. Usually, there are several task descriptions whose context condition matches the current context information. The scheduler establishes a priority order for these task descriptions, based on context specificity. It also evaluates the context specificity of each task description as indicated by the degree of similarity between the context condition and the context information stored by the studio manager. The result is a list of task descriptions that satisfy the context conditions, sorted by context specificity.

If the task description on the top of that list has already been adopted, then a subtask agent is invoked. The agent executes actions in the way specified in the action description part of the task description. If the task description with the highest context specificity is as yet untried by the user, then the dialogue manager discloses the content of that task description to the user for his or her judgment.

SUMMARY AND DISCUSSION

The starting point of the FRIEND21 project was the idea that information technology is undergoing a rapid evolution involving a wide range of technical and social concerns. Society is more and more dependent on information technology, and it is placing many and diverse expectations on its development. Among these expectations, there is the development of software products that are oriented toward the needs of all members of the society. In this perspective, information technology is called to contribute to a more affluent society, a more fulfilling life, and a more enjoyable environment for everybody. As a consequence, information technology must cover all aspects of life, from the production of goods and services to their consumption and disposal. In the information society of the 21st century, there will be a much broader range of activities in which individuals make use of information, from daily living and entertainment, to cooperative work. This implies that information technology must be brought closer to people in their daily life. Technical issues involved in this process include:

- Enjoyable and useful information must be made available by the system on a daily basis, and anyone must be able to use it in a simple manner.
- Anyone must be able to easily access such information from any place at any time.
- People-to-people communication must be broadened through information, and communication means must become more versatile, user friendly, and appropriate to individual needs.
- Anyone should be able to use communication means without difficulty, and when new forms or information, or new functions are provided, it should be possible to learn them with very little effort.

This brings up a fundamental question concerning the evolution process of information technology from goal-oriented activities, toward social activities of everyday life. In the approach proposed by FRIEND21, the involvement of computers with people and society in general is taken up as a problem of expression. The fundamental question of concern is how the internal functions of a computer should be manifested to the user. According to FRIEND21, this is the first issue for investigation concerning human–computer interfaces.

Computer information and functions are usually visualized through some method that enables them to be named and retrieved. The FRIEND21 approach consists of substituting currently adopted techniques for making information and functions visible to the user with the metaphor technique.

The FRIEND21 project has proposed a two-tiered architecture for human–computer interfaces, which handles interaction between the user and the computer as a symbolic process, and helps making computer information and functions closer and more accessible to the user. In the FRIEND21 architecture, the design method that represents this cognitive process is called metaware, and the implementation architecture is called agency model.

The purpose of metaware is to adapt the computer to the environment, so as to support the wide range of situations appearing in everyday life. To enable the computer to expand its range of applications to a variety of activities such as word processing and electronic mail, household account keeping, video program recording, painting, and education, a coordinated relationship must be provided between functions and expressions that is consistent with everyday life where these activities occur. To achieve this, a framework is needed that allows representations (such as icons, windows, and other visual entities) to be polysemous, and, at the same time, provides a theory for the use of multiple representations (multiple metaphors) and for the combination of representations in "contexts." The framework should also be able to operate the computer in an adaptive manner, and select and display appropriate images to assist the user in executing tasks based on: (a) identification of task intentions and (b) context established through the user's personal operation history, preferences, and dislikes. In addition, the framework should be able to notify the user of the exact state of the system.

The metaware framework can be implemented through the agency model. The agency model defines representations and functions as objects and drives them, but also configures a problem-solving mechanism by having them cooperate in a mutually dependent manner. The agency model is a system that solves problems in the human interface by having three functionally distinct groups of agents cooperating with each other in a distribute fashion via a shared blackboard interface called "studio." These three agent groups are: (a) metaphor-environment agents that drive representations, (b) function agents that execute tasks, and (c) management agents that coordinate the interaction between the first two agent groups. The management agents provide relationships between representations and functions as an environment model, and dynamically interpret and control the relationships at the time of execution. The studio, where the agents exchange information, acts as a framework for achieving a multimedia and multimodal operation environment. It can adapt to system extensions through partial system modifications and plug-in upgrades.

The proposals of metaware and the agency model that came out of this project are based on bold hypotheses. However, the FRIEND21 Project's greatest accomplishment was the proposal of a new problem-solving methodology. Moreover, we have demonstrated at home and abroad that this methodology could be used as a vehicle to change our concepts, and thereby help us see things that until now had been invisible to us. These models will provide a foundation for a new problem-solving method in the future. We believe our proposals and results (Institute for Personalised Information Environment, 1995) will provide a foundation for a new problem-solving method to achieve user interfaces for all.

REFERENCES

Cagan, M. R. (1990). The HP SoftBench environment: An architecture for a new generation of software tools. *Hewlett-Packard Journal, 41*(3), 36–47.

Carroll, J. M., Mack, R. L., & Kellogg, W. A. (1988). Interface metaphors and user interface design. In M. G. Helander, T. K. Landauer, & P. V. Prabhu (Eds.), *Handbook of human–computer interaction* (pp. 67–85). Amsterdam: North-Holland, Elsevier Science.

Croft, W. B. (1984). The role of context and adaptation in user interface. *International Journal of Man–Machine Studies, 21,* 283–292.

Gawron, V. J., Drury, C. G., Czaja, S. J., & Wilkins, D. M. (1989). A taxonomy of independent variables affecting human performance. *International Journal of Man–Machine Studies, 31,* 643–672.

Hirose, M. (1992). Strategy for managing metaphor mismatching. Poster presented at the *Human Factors in Computing Systems Conference,* Monterey, CA (p. 6). New York: ACM Press.

Institute for Personalised Information Environment. (1995). *Human interface architecture guidelines.* Tokyo: Author.

Nii, H. P (1989). Blackboard systems. In A. Barr, P. Cohen, & E. Feigenbaum (Eds.), *Handbook of artificial intelligence* (Vol. 4, pp. 1–82). Reading, MA: Addison-Wesley.

Nonogaki, H., & Ueda, H. (1991). FRIEND21 Project: A construction of 21st century human interface. In *Proceedings of the ACM Conference on Human Factors in Computing Systems on "Reaching Through Technology"* (pp. 407–414). New York: ACM Press.

Norman, D. A. , & Draper, W. S. (Eds.). (1986). *User centered system design: New perspectives on human computer interaction.* Hillsdale, NJ: Lawrence Erlbaum Associates.

Norman, D. A. (1988). *The psychology of everyday things.* New York: Basic Books.

Object Management Group. (1993). *The common object request broker: Architecture and specification 1.2* (OMG, TC Doc 93-12-43). MA: Object Management Group, Inc.

Ortony, A. (1979). The role of similarity in similes and metaphors. In *Metaphor and thought* (pp. 186–201). Cambridge, England: Cambridge University Press.

Palaniappan, M., Yankelovich, N., Fittzmaurice, G., Loomis, A., Haan, B., Coombs, J., & Meyrowits, N. (1992). The Envoy framework, An open architecture for agents. *ACM Transactions on Information Systems, 10*(3), 233–264.

Pfaff, G. E. (Ed.). (1985). *User interface management systems* (Eurographics Seminars). New York: Springer-Verlag.

Reiss, S. P. (1990). Connecting tools using message passing in the field environment. *IEEE Software, 7*(4), 57–66.

Rich, E. (1983). Users are individuals: Individualizing user models. *International Journal in Man–Machine Studies, 18,* 199–214.

Ricoeur, P. (1973). Creativity in language: Word, polysemy, metaphor. *Philosophy Today, 17,* 97–111.

Smith, R. G., & Davis, R. (1981). Frameworks for cooperation in distributed problem solving. *IEEE Transactions on Systems, Man and Cybernetics, 11*(1), 61–70.

Ueda, H. (1992). FRIEND21 (Aiming for the personalized information environment). *ACM SIGGRAPH VIDEO REVIEW (Issue 79), in Human Factors in Computing Systems (CHI '92) Special Video Program (No.10), Monterey, CA.*

Uyama, M. (1993). A blackboard-based architecture for filtering new software features. In *Proceedings of ACM (COCS '93) Conference on Organizational Computing Systems* (pp. 210–215). New York: ACM Press.

14 User Modeling for Adaptation

Judy Kay

This chapter provides an overview of existing practices and foreseeable trends in user modeling, and of how user modeling may be utilized to account for designing interfaces for diverse user groups. User modeling provides the basis for a system to meet the particular needs and preferences of the individual user. In this chapter, we describe the form of user modeling needed for adaptation. We also describe the processes involved in building user models and maintaining them: elicitation, monitoring, stereotypic reasoning, community-based inferences, inference based on domain knowledge, and management of uncertainty. We describe the trends associated with increasing user modeling and adaptation of systems, and discuss some of the associated issues. These include the impact of keeping long-term models of users, the need for standards, the importance of reuse of user models, as well as the issues associated with ensuring users have proper control over, and access to this information. Finally, we discuss the role of user modeling and interface adaptation for the diverse needs of different user groups.

There is a broad range of systems where it is valuable to provide adaptation to the particular needs of the individual user. These include, for example:

- Advisors.
- Consultants.
- Help systems.
- Recommender systems that filter information on behalf of the user.
- Systems that tailor the output they produce to the particular needs of the individual.
- Systems that tailor the interaction and modality to match the user's preferences, goals, task, needs, and knowledge.
- Intelligent teaching systems that aim to teach, as a good teacher does, matching the teaching content, style, and method to the individual student.

The overarching goal of user-adapted systems is to improve the efficiency and effectiveness of interaction. They aim to make complex systems more usable, present the user with what they want to see, as well as speed up and simplify interactions (Malinowski, Kühme, Dieterich & Schneider-Hufschmidt, 1992). For example, the User Modeling Conference's *Reader's Guide* (Jameson, Paris, & Tasso, 1997) identifies the following purposes for user modeling: (a) helping the user find information, (b) tailoring information presentation to the user, (c) adapting an interface to the user, (d) choosing suitable instructional exercises or interventions, (e) giving the user feedback about their knowledge, (f) supporting collaboration, and (g) predicting the user's future behavior.

The Role of User Models for Adaptation

There are three main ways in which a user model can assist in adaptation. These are illustrated in Fig. 14.1, where the double vertical lines delimit the interaction between the user and a system.

Consider first the upper horizontal line, which indicates a user action at the interface. Note that this may be any action that can be accomplished through the devices available, including, for example: a mouse action, typing at the keyboard, data from an active badge worn by the user, the user's speech via audio input to the system, and data from other, more exotic input devices. A user model can assist the system in *interpreting* such information. So, for example, if the user input is ambiguous, the user model might enable the system to disambiguate that input. This is one important area in which user models have been used for natural-language understanding. Equally, the user model might help the system to interpret incorrect user actions. Existing systems already display nonadapted cases of this. For exam-

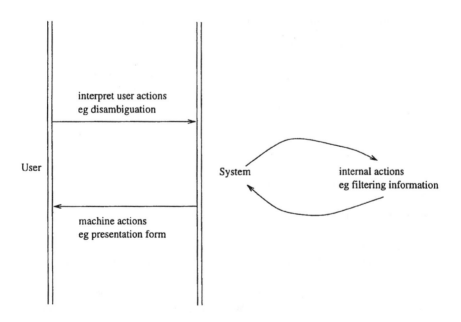

FIG. 14.1.   Role of the user model in adaptation.

ple, command interpreters can often deal with typing errors in command names, so that a request to change directory can work correctly even when the user mistypes the directory name. More sophisticated cases have also been explored, such as the case of spelling errors by dyslexic users (Spooner & Edwards, 1997). Another example where user modeling is intended to improve interpretation of the user's intended actions was described by Trewin and Pain (1997), whose system monitored the typing problems displayed by users with motor difficulties, so that it could identify difficulties such as long key depression errors.

Now consider the lower horizontal line representing the system's actions at the interface. These can be controlled by a user model to improve the quality of the interaction. A very simple example might involve the system tailoring its presentation form to the user. For the user who prefers or needs large print, information would be presented in a larger font than that preferred by other users. More sophisticated cases involve adaptation of the content, as well as of the form of the presentation. For example, teaching systems may match their actions to the user's modeled level of knowledge and competence, with simpler tasks and information for less able students, and more demanding material for the most able. Systems like PEBA-II (Milosavljevic, 1997) make use of comparisons between concepts to explain information, so that new concepts are presented in relation to concepts the user knows. A considerable body of work on adaptive hypertext aims to match the information and its presentation to the user (as described, e.g., in Brusilovsky, 1996).

Finally, the user model can drive the internal actions of the system. This is the goal of systems that filter information on behalf of the user. Similarly, an agent that operates on behalf of the user, perhaps interacting with other systems, may perform the required actions on the basis of the user model.

Many systems combine these elements. So, for example, a teaching system might monitor the user's attempts at set tasks and employ the user model to interpret these. Sometimes, an incorrect answer might be a simple slip. At other times, it may indicate lack of understanding. The user model may help the system distinguish two such cases. The same teaching system might then perform a complex series of internal actions to select a suitable teaching goal and the teaching method it will apply. This process would be heavily influenced by the system's model of aspects such as the user's knowledge and learning preferences. As the final stage of each interaction cycle, the machine produces an action at the interface. The form chosen for such an action can also be defined by the user model.

WHAT IS A USER MODEL FOR ADAPTATION?

At present, the term *user model* is used in many different ways. There are synonyms or near synonyms, like student model and learner model. In addition, there is an array of related terms, including cognitive models, conceptual models, mental models, system models, task models, user profiles, and others. We follow an emerging use of terms, with those for user-adapted systems coming from Wahlster and Kobsa (1986) and those for human–computer interaction (HCI) from D. A.

Norman (1983). We need to clarify the nature of user models suited to adaptation of interfaces for diverse populations of users.

Our definition of a user model is illustrated in Fig. 14.2. The "real world" depicted at the top includes actual entities in the world (like animals) as well as artificial and abstract notions like mathematics. Within that world, there is a part that constitutes the *context* of the modeling task at hand.

Both the real world and the context are depicted as irregular, untidy shapes. This reflects their complexity and "messiness." Below the real world and current context, we show two models of the context: On the left is the user's model, and on the right, the model of the programmer who is constructing both the user model and the associated user-adapted interactive system. These too are "real" in the sense that they exist in the minds of real people. These are also untidy shapes because they represent models hidden within human minds, and, therefore, may be quite complex and messy. Aspects of models that relate to the user's (and programmer's) knowledge and understanding are frequently described as *mental models*. Other

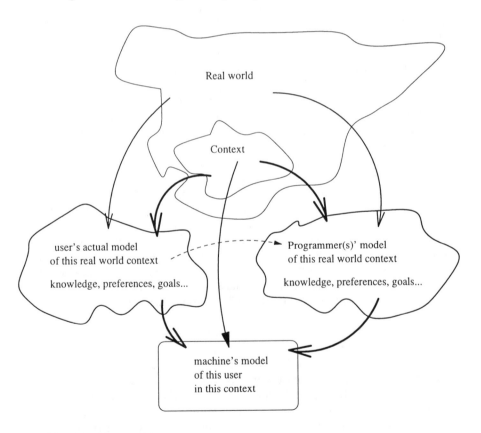

FIG. 14.2.   User model.

aspects that are important for user-adapted interaction include the user's preferences, interests, goals, and plans.

Our focus is on the user model shown at the bottom of the figure. This is a tidy, artificial model. We can distinguish two major forms of user models: *cognitive* and *pragmatic*. Cognitive user-modeling research builds user models that are intended to match the way that people actually think and know. Indeed, this aspect of user modeling is of importance for psychologists and educators. An exemplar is the ACT–R theory of skill acquisition (Anderson, 1993), which can be used to define user models for teaching systems (Corbett & Anderson, 1994). Of course, once the theory has done its job, the resultant user model is a well-defined artificial construction created by the psychologist-programmer. Our definition of user model includes those with claims to cognitive validity, as well as more pragmatic user models that serve to support adaptation of the interaction.

The arcs in the figure indicate the way the user model is formed. First, users must build their own model of the context, through experience and learning within that context in particular, and the rest of the real world in general.

Similarly, programmers build their model, but based on experience and learning that will almost inevitably differ from those of any particular user. This difference is very important for the design of effective user interfaces. A major concern for HCI designers is to avoid the pitfalls of assuming users are much like themselves. Rather, the programmer strives to develop an improved understanding of the user. This involves the programmer developing a model of the user. The dotted arc from the user's actual thoughts to the ones the programmer perceives as such indicates this process, with the programmer improving their understanding of the user.

The heavy lines feeding from the context indicate the primary influences on the models. The thinner lines from the rest of the real world remind us that people's understandings are often richly connected and any single context will be modeled and interpreted in the light of broader understandings. This is an important aspect for interface design, because, as we have already noted, programmers tend to assume, incorrectly, that users are much like themselves. Users can experience many problems when trying to use an interface, and it is very common for those problems to be due to differences in the user's and programmer's knowledge, goals, and preferences. HCI techniques, such as cognitive walk-throughs and heuristic evaluation, are intended to assist programmers in thinking of the problems that users will face. However, such techniques are based on a fundamental assumption of normative user models; for diverse user populations with extremely varied needs, such techniques should be adjusted. However, it is unclear how to adjust them to account for the considerable range of different user needs that are associated with various disabilities. For example, users with low vision have very different needs from users with motor difficulties, and even from users with more extreme vision impairments. Moreover, the exact needs of users belonging to one category can be quite difficult to assess. For example, users with a similar level of vision impairment may respond differently to various colors of text and background. The critical point made here is that disability is extremely varied. Yet, HCI techniques are largely based on design for a homogenous user population.

Once the programmer and user have models of the context, the heavy arcs from these to the user model show the two major influences on it. The arc from the user's model reflects the transfer of some of the user's beliefs, preferences, goals, interests, and so on. The heavy arc from the programmer's model reflects the fact that the model is controlled by programmers. Their influence will be indelibly cast into the model: It is the programmer who will define the limits of what can be modeled, how it is represented, how the user can contribute to it, and how it can evolve. This is why adaptation should be considered from the very early stages of designing a system.

Finally, there is a thin arc from the context to the user model. This captures the user attributes relevant to the current context. These are real-world aspects, rather than part of the user's beliefs and knowledge. For example, we may need to model the user's typing speed, an attribute of the user. This may differ from the user's beliefs about their typing speed. Such user attributes can be important in determining the most appropriate interface for the user, with one interface being superior for the user with excellent keyboard skills, and another being better for the user who types slowly or inaccurately. This difference is partially accounted for in GOMS keystroke analyses (Card, Moran, & Newell, 1980, 1983), which have a range of time estimates for keyboard actions. This range supports different usability assessments of an interface and takes into account the user's keyboard skill and the nature of the task being executed.

Because we are concerned with adaptation for the individual user, we focus on user models that are *individual,* in the sense that the system is able to construct different models for each user. So the model at the bottom of Fig. 14.2 will generally be different for different users. This is not the case in much of the user-modeling work in the HCI community, where the goal is to model a canonical or generic user. In that work, programmers may observe, or interview users as part of the process of defining a good canonical user model. Once this has been accomplished, the programmer can use that model to inform the creation of the interface to match the canonical user. For example, user-centered design (D. A. Norman, 1986) involves the system designers working with the users throughout the design of the system. The underlying assumption is that these users are truly representative of the whole population that will use the interface.

User-adapted interaction is fundamentally concerned with tasks for which a canonical user model is inadequate and even a small number of built-in models is not enough. For example, a customized newspaper is intended for a particular user. Individualized teaching should take account of the user's existing knowledge, goals, and learning preferences. Interfaces for users with various disabilities may need to be quite different according to the particular disability.

We further limit our scope to *dynamic* user models, which can change during interaction. This is necessary for user-adapted interaction, and it means that we cannot hard-code the model into the machine as in much of the HCI community's user modeling. For example, a user-adapted teaching system adapts its operation to take into account the user's learning over time.

Finally, we restrict ourselves to *explicit* user models, which are separable from the rest of the system. Once again, this differs from much work in the HCI community, where the programmer, who creates a user model, applies this to the interface design process, so that the user model is implicitly built into the system in the form of the multitude of decisions made during the system design.

If a system is to adapt significantly to different users' knowledge, needs, and preferences, it follows that the quality of that adaptation relies critically on the user model because that constitutes the basis for individualization. Before we explore the processes involved in creating and managing user models, we briefly consider some of the challenges in establishing what should be modeled, to what extent, and in what detail.

For some forms of adaptation, the user model will need to be detailed, modeling many aspects of the user. For example, customization of a newspaper could be improved if there existed a very detailed model of many aspects of the user. These would include the user's knowledge and interests across the full range of news areas. A very detailed model would enable a system to provide comprehensive coverage of the areas of primary interest to the user, and information could be presented at a level that would match the user's knowledge. Unfortunately, it is difficult to collect all the information needed to create and maintain a large, detailed, comprehensive, and accurate model.

We can expect that a small amount of information about the individual user can drive the adaptation of some of the high-impact aspects of the interface. For example, consider the case of users with poor keyboard skills, which makes typing very slow and error prone. A small amount of information about the exact nature of an individual user's problem can drive significant improvements in the interface. This may involve a quite modest user model. However, if there is a large variety of possible problems that cause typing errors, this will correspond to a large range of aspects to be assessed as potential elements of the user model. This means that there will be many possible elements in the user model, but any single user will have a very sparse model, with a very small number of those aspects being relevant for customizing their interfaces. This can be further complicated if such users are likely to exhibit significant changes in their keyboard skills. Changes may concern short-term aspects, such as levels of fatigue, or longer-term aspects, such as slow deterioration of motor skills.

One of the critical tasks for the programmer of an adaptive system is the choice of elements to include in the user model. As Fig. 14.2 indicates, the actual world and the user's mind are complex and messy. An effective adaptive system will be based on a judiciously chosen user model, including all the aspects that will enable the system to perform effective adaptations. At the same time, the programmer will need to be very pragmatic, modeling only those aspects that serve the two essential requirements for effective adaptation: First, there must be the possibility of collecting information of adequately high quality to provide an accurate user model; second, the aspects modeled must be those that the adaptive system is able to use to effect adaptation.

Having explored the nature of the user model, we need to consider its role in adaptation. At the simplest level, the user model has information about the user and a program can use this to alter its interactions with users. Fig. 14.1 and its associated discussion identified three main roles for the user model: (a) adapting the interpretation of user actions at the interface, (b) adapting the machine's actions on behalf of the user, and (c) adapting the presentation of information. Essentially, the user model depicted at the bottom of Fig. 14.2 exists as part of the "system," shown in the middle of Fig. 14.1. Whereas the rest of the system is more or less "constant," regardless of the particular user, the user model contains user-specific information, which drives the adaptation activities of the system.

THE PROCESSES OF USER MODELING

This section gives an overview of the processes that must occur for the creation and maintenance of user models. Each can range from the simple to the extremely sophisticated.

Elicitation of User-Modeling Information

This is the most straightforward method for acquiring information about the user: Simply ask. Many user-modeling systems involve some form of elicitation. At its simplest level, this might call on users to fill in a multiple-choice form, to give information about their preferences or knowledge. There is also scope for more sophisticated elicitation tools, such as a concept-mapping interface (Kay, 1991), which assists the user in externalizing their conceptual understanding of a domain.

Where adaptive interfaces are to be used by very diverse user populations, this becomes a particularly valuable source of information about the user. Consider, for example, users with low vision. They will often have a very good understanding of the exact nature of their difficulties, including subtle aspects, such as their ability to read particular foreground colors. If users volunteer such information directly, it can be used to customize elements of the interface. Given the diversity of disabilities, it seems likely that direct elicitation may be the best mechanism for capturing a high-grade user model, which can reflect such important but subtle needs and preferences.

Modeling Based on Observing the User

It is very easy to create application programs that monitor user activity. This has been described as clandestine modeling (Zissos & Witten, 1985) in a study involving users of EMACS. The quality of such monitor data is typically quite low. However, to counter this, it is easy to collect extremely large quantities of monitor data without making any demands on the user. With increasingly reliable and powerful systems, it becomes quite feasible to monitor large numbers of users over many years, as was done for users of the sam text editor (Kay & Thomas, 1995). The invisibility of such monitoring processes has the advantage of placing no load on the user. Of course, the corresponding disadvantage is that the user may be unaware of the monitoring and its implications.

In the context of diverse user populations, observation-based modeling has a critical role in those cases where the user may have difficulty in enunciating their knowledge, preferences, and needs. For example, users may not be able to say whether they know certain concepts well: It may be simpler to model their knowledge of those concepts from their behavior. Even a simple aspect like an assessment of the user's typing skill may be best assessed by observing the user. For the broad population (where users are not trained typists), users may not be able to state their typing speed. Yet, observations of their typing activity over some time can be used to determine this, in the way required by the user model and at varying levels of detail.

Stereotypes

Stereotypes are one of the common elements in much user-modeling work, and seek to capture default information about groups of people. For example, KNOME (Chin, 1989) modeled users' knowledge of the Unix operating system. Once a user was classified as an expert, it stereotypically inferred that this user knew many sophisticated aspects of Unix. This inference was intended to be statistically valid: For a large number of "guru" users, many would know most items assumed by the stereotype. Of course, any particular expert may not match the stereotype in its entirety.

Important to note, stereotype inferences are intended to serve as default values. Later, when more reliable information becomes available, the model will be refined, perhaps reflecting the fact that a particular user knows many, but not all, of the aspects included in the initial stereotype inference. This simple but powerful idea was introduced by Rich (1979, 1983, 1989), who used people's descriptions of themselves to deduce the characteristics of books that they would probably enjoy. Stereotypes are part of many user-modeling systems. An indication of their importance is that they were incorporated in several generalized user-modeling tools, including GUMS (Finin, 1989), BGP-MS (Kobsa, 1990a), UMT (Brajnik, Guida, & Tasso, 1990; Brajnik & Tasso, 1992), and the um toolkit (Kay, 1990). Although the user-modeling community has made considerable use of stereotypes, the exact nature of the reasoning described by that term differs considerably. We now explore the special character of stereotypes, and clarify what distinguishes them from other types of knowledge-based reasoning.

The essential elements of a stereotype are:

- *Triggers* that activate the stereotype; for example, the fact that the user is an expert Unix user would activate the expert-stereotype.

- The *inferences* that, in the case of KNOME, were the many aspects of Unix that an expert is assumed to know.
- An (optional) *retraction* facility to deactivate a stereotype when its essential triggers are violated, as, for example, in the case where a stereotype for a programmer has been activated and it is subsequently determined that the user cannot program.

Essentially, the stereotype mimics intuitive human reasoning, using a small amount of information about a person to arrive at a large number of default assumptions about them. As more information becomes available about individual assumptions, these are revised. Meanwhile, the overall initial classification of the user and most of the default assumptions continue to hold, unless information is acquired that indicates that the initial classification of the user was incorrect.

Constructing stereotypes means defining the trigger, the associated inferences, and any retraction conditions. The highly statistical nature of the enterprise immediately suggests the techniques that are most likely to be effective. Machine learning and statistical analysis are basic tools for building useful stereotypes.

Stereotypes are a special form of reasoning that seems to be particularly useful when reasoning about people. They are best used to establish default beliefs about the user, while the system waits to collect something better. They may also offer a shortcut in building a user model: Users can simply choose the stereotype trigger sets that they like best. This may provide a good enough model to let the adaptive system work effectively for them.

A notion, which is similar to that of stereotypes, is that of *community*. In Doppelganger (Orwant, 1993) communities constitute groupings of users. A user may be classified as belonging to several communities, and when the user's model has no explicit information about some aspect, its value is calculated across the communities the user belongs to. For example, suppose a particular user's music preferences correlate 30% with the community of users under the age of 30, 80% with the community of users aged 30 to 50, and 60% with users over the age of 50. Note that if the user's actual age is 90, their known music preferences will have contributed to the model for the over-50 community. However, when a system needs to predict a new music preference for the user, it will weight the under-30 model value by 30%, the 30 to 50 community model by 80% and the over-50 model by 60%. This is a very interesting notion whose predictive power is still to be evaluated.

This notion of community offers considerable promise for user modeling of diverse user populations. As data are collected for large numbers of users, it becomes feasible for new users to opt for a relevant collection of community models as their initial user model. There appears to be the potential for a community of users to develop its own models, based on costly and slow modeling techniques, such as elicitation and observation. These then become a community resource that can be shared with other users to assist them. For example, a community of users, who have low vision, might develop detailed models of preferences for particular fonts and presentation formats. They might need to develop such models empirically, with many users experimenting with various forms of presentation and comparing them. Just this type of information is currently shared in various forms of publications by special interest groups. Capturing it in a community model would make it more readily available for providing adaptations in interfaces.

Knowledge-Based Reasoning

The statistics-based approaches to reasoning, whether based on stereotypes or community membership, appear to be particularly important for user modeling. However, the more usual forms of knowledge-based reasoning are also important for reasoning about users. For example, if the user indicates they do not know anything about the notion "square root," we can infer that they do not know about complex numbers, because an understanding of square root is a prerequisite for understanding complex numbers. This follows from the structure of knowledge in mathematics.

Knowledge-based reasoning has much to offer for the types of communities defined by user disability. For example, suppose the user volunteers that they have low vision. Knowledge-based reasoning can generate many assumptions about interface adaptations that are likely to be helpful. These could be based on expert knowledge of the particular interaction requirements introduced by this specific type of disability. They might also include allowances for future prognoses, as, for example, in the case of degenerative illnesses. Note that this is subtly different from the case of community models and stereotypes. These latter types of reasoning about users are based on statistical information and, although they hold for large populations of users, they may be only approximate for an individual. By contrast, knowledge-based reasoning should have a predictable level of reliability for any individual and the expert who provides the knowledge should be able to assess this reliability.

Managing Uncertain and Changing Information About the User

There are many sources of uncertainty in user modeling. Consider first the example of user changes: At one point in time, the users do not know a concept, but a little later they begin to learn it and after a little more time, they know it well. The corresponding user model should reflect such changes. The problem is that it is difficult to collect accurate enough information about the user at each point in time. Typically, in such a situation, the user-modeling system will initially receive information suggesting that the user does not know the concept. It may later begin to receive mixed information, some suggesting that the user knows the concept and some that the user does not. Later again, the information available for user modeling may consistently indicate that the user knows the concept. This makes it necessary for the user-modeling system to be able to interpret the information and alter its conclusions about the model.

It can be difficult for the system to deal with such situations, especially at the point when the indications are contradictory. Moreover, this situation may be difficult to distinguish from other possible causes of a similar pattern of evidence about the user. For example, the user may simply have inconsistent beliefs and so may appear to know the concept in some situations but not in others. This problem has been carefully studied in the THEMIS system (Kono, Ikeda, & Mizoguchi, 1992, 1994).

There are many other sources of uncertainty about the user. A thesaurus of uncertainty and incompleteness lists over 30 forms (Motro & Smets, 1997) and most of them are encountered in user modeling. Moreover, user modeling may often involve the challenging problem of combining evidence from varied sources, each involving different forms of uncertainty.

TRENDS IN USER MODELING

In this section, we discuss the prospect of an increasingly important role for *long-term* user models, exploring the implications of maintaining such models.

Once we have a user model, it makes sense to reuse it in a range of applications. For example, consider a user model that can be used to customize a newspaper. This might well model the user's music preferences so that it can include news about music. Another system might teach music appreciation and this may also need some of the same information about the user's music preferences.

Reuse of the user models across applications is sensible from the user's point of view. Suppose, for example, the user spends time interacting with one application, all the while helping it develop a better user model and, hence, an improved ability to interact effectively. It would be very reasonable for the user to expect a second application to use relevant aspects of the same user model. This improves consistency across applications and saves the user the tedium of training new systems. As constructing and maintaining user models for adaptation can be costly, it is appealing to amortize some of the costs over different systems, thus reducing the cost of user modeling for a particular interactive system. For users with a disability, the effective cost of training a system may be very high, especially if interaction is slow and difficult. So, for example, a user with motor difficulties would probably be particularly frustrated if required to repeat the training of systems.

A level of user model reuse can be achieved by a database of user models that can be accessed by various applications. The importance of such a facility for diverse user populations has motivated the AVANTI project's development of such a resource, which is available across the network (Fink, Kobsa, & Nill, 1997). This means that access mechanisms need to be defined, for example, through the definition of user-modeling performatives (Kobsa, 1996; Paiva, 1996) based on the KQML protocol (Finin, McKay, Fritzson, & McEntire, 1994).

For a model to be reusable, it should be expressed in terms of an agreed upon ontology. This would make it possible for programmers of different systems to describe the user by means of standardized terms. The definition of such ontologies must also be associated with metadata standards.

Metadata is used to describe objects. For example, a piece of music may be described as "classical" and we may define terms like *composer* to associate with pieces of music. Suppose that a standard for metadata about music has been defined and used to classify collections of music. This will be valuable for a system that attempts to filter music on behalf of a user, provided that the system has access to a user model that corresponds to that metadata. This means that the creation of standards for metadata ontologies effectively imposes standards on user models. There is considerable activity to establish metadata standards. Consider, for example, the activity in this area for supporting teaching and learning. The ARIADNE project's[1] and the IMS metadata definitions (developed in collaboration with ARIADNE) provide a standard for describing learning objects. Similarly, a collection of standards efforts from the IEEE involve several working groups, and, in particular, one for metadata (LTSC).[2] The same IEEE standards effort has a Learner Model Working Group,[3] which aims to "specify the syntax and semantics of a 'Learner Model', which will characterize a learner (student or knowledge worker) and his or her knowledge/abilities," including aspects "such as knowledge (from coarse to fine-grained), skills, abilities, learning styles, records, and personal information." All these characteristics are to be represented "in multiple levels of granularity, from a coarse overview, down to the smallest conceivable sub-element." The goals of this work included enabling "learners (students or knowledge workers) of any age, background, location, means, or school/work situation to create and build a personal Learner Model, based on a national standard, which they can utilize throughout their education and work life." This work represents an important trend in user modeling and adaptation.

USER ACCESS AND CONTROL

We now discuss the need for enabling users to have final control over their own user model. As a prerequisite, users must have adequate access to the model because, otherwise, they would be unaware of what might be controlled.

Access to and Control Over Personal Information

Perhaps the most compelling reason for the user to have access to their user model is the right of individuals to know what information a system maintains about them. It can be argued, as, for example, Kobsa did (Kobsa, 1990b, 1993), that long-term user-modeling information should be viewed in the same way as databases of personal data.

One response to this is to avoid keeping any long-term modeling information at all. In many of the types of systems for which adaptation can be valuable, this approach is infeasible. For example, consider a teaching system: One would think lit-

---

[1]For more information on the *ARIADNE metadata,* see: http://ariadne.unil.ch/metadata.htm

[2]IEEE P1484.12 Learning Objects Metadata Working Group, http://grouper.ieee.org/groups/ltsc/wg-12.htm

[3]IEEE P1484.2 Learner Model Working Group, http://www.manta.ieee.org/p1484/wg-2.htm

tle of a personal teacher who never remembered their student from one learning session to the next. As most substantial learning goals can be achieved only over a substantial time frame, and in many separate sessions, user models within a teaching system need to include long-term representations of the learner's developing knowledge. Such reasoning is not limited to teaching systems. For example, a personalized newspaper will require a rich user model. This can be developed only over some time and the user would expect it to be useful over the long term.

Correctness and Validation of the Model

The user should be able to check and correct the model. This relates to the users' right to know what the computer stores about them. This is especially pressing when the user model has been inferred or built from observations of the user.

For example, a coaching system might well construct its user model on the basis of monitoring data from the use of a text editor. This process presents the risk for creating incorrect user models. Consider, for example, that the user allows another person to use their machine account: The inferences are no longer about the individual intended to be modeled. The model's accuracy can also be affected by people giving advice to the user: Someone might direct the user to type a series of complex commands. This could result in a user model with an inaccurate picture of the user's sophistication. If the user can easily check the user model, they can correct similar erroneous assumptions.

A similar argument is central to Csinger's thesis on scrutable systems (Csinger, 1995). In that case, the system applied user models to customize video presentations. The scrutability of the system was improved by presenting the relevant parts of underlying user model to the users, and allowing them to alter the values of those parts.

One objection to user-adapted systems is that they infer too much from too little information (Henderson & Kyng, 1995). This concern is greatest when the base information has low reliability, as in the case of data derived from monitoring the user. Making the model accessible ensures the user can see and control the bases of the system's inferences about them. In this way, access to the user model can help address this objection to user-adapted systems. A closely related issue was addressed by Browne, Totterdell, and M. Norman (1990), who observed that the longer the inference chain, the more difficult it is to be confident of the dependability of the final conclusions.

Of course, there is a potential problem in allowing users to see and alter the user model. They might just decide to lie to the system and create a model of themselves, as they would like to be (e.g., they might want the system to show them as experts). Another possibility is that playful and curious users might simply like to see what happens if they tinker with their user model. Giving users control implies that they also have the opportunity to corrupt their user model. This must be taken into account in designing user-adapted systems. It is also an important consideration in the design of experiments aiming to assess the effectiveness of systems with user modeling. However, the possibility of users purposely damaging their own user models is not a sufficient reason for denying them the right to such control.

Machine Determinism and Asymmetry of the Human–Machine Relationship

There are inherent differences between machines and people. These differences introduce the possibility for user-adapted systems to be especially effective in some interactions. In particular, the processes controlling a user-adapted computer system are normally deterministic. By contrast, in interactions between humans, the processes behind actions may not be accessible, even to the person who performed the actions (Nisbett & Wilson, 1977). Essentially, we argue that although the machine cannot really know the user's beliefs, it should be possible for the user to know the machine's beliefs, especially the beliefs contained in an explicit model of the user.

In systems intended to be cooperative, the accessibility of the user model has the potential to play an important part in helping the user understand the system's goals and view of the interaction. This may reduce user misunderstandings, which may be caused by expectations and goals held by the system and not appreciated by the user. So, for example, in a coach, the aspects modeled signify the elements considered important by the programmer of the teaching system.

This may be useful in dealing with another of the objections to user-adapted systems: It has been argued (Browne et al., 1990) that if the system and user both attempt to model each other, a situation described as "hunting" could arise. This means that users try to model the system but, as they do so, the system changes because it is trying to model them. User access to the model can have a stabilizing influence in this respect.

In general, we can expect the quality of collaboration between a user and a system to be enhanced, if both parties can be aware of the beliefs each holds about the other: An accessible user model can at least help the user become aware of the machine's beliefs.

DEALING WITH DIVERSE USER GROUPS

We now explore the current state of the art in user interface design. We discuss the predominance of systems that cater to a single, homogeneous user population. Then we explore the simpler forms of customization that are becoming widespread. Clearly, these offer some promise for satisfying the diverse needs of end-users. We review the relationship between such customization and user modeling and identify potential benefits of user modeling, even for the quite simple cases of customized interfaces.

First, however, we note that, currently, interface design is typically based on the assumption of a "canonical" user. Techniques such as user-centered design and participatory design derive from the view that the programmer should not assume that "the user" is just like them. Instead, the programmer uses various techniques to build the interface for the particular user community for which it is intended. Standard approaches to usability do acknowledge differences in users. For example, GOMS keystroke analyses of an interface can allow for differences in typing speed for experienced users. However, GOMS is intended for experienced users of the interface, and typically involves comparisons of the likely speed of different methods for doing a task. It is unsuitable for predicting usability for novice, casual, or intermittent users. In addition, the standard empirical data used for keystroke analysis do not account for users with disabilities.

Usability assessment and interface design are often based on the assumption that the programmer needs to understand the user community for which the software is intended. Larger and more diverse user populations effectively "contain" multiple user communities. The programmer needs to develop an understanding of each of these, and create interfaces usable for all of them. User models and user-adapted interaction can help because the programmer can separate the elements of the interface, linking the operation of each to associated user assumptions. So, for example, information that is presented to the novice user should be linked to the component of the user model that models the user's level of expertise. Similarly, other parts of the system can be conditionally connected to other parts of the user model. For example, font selection could be controlled by components of the user model that represent the user's visual acuity, so that users with excellent vision could be presented with information in smaller fonts, whereas users with various degrees of vision impairment could be presented the same information with larger, suitably selected fonts. Taking this example further, users with extreme vision impairment might be presented the information in a form that would not rely as heavily on vision, communicating some information using auditory or tactile channels. Such an approach to designing adaptive interfaces requires a shift in the way interfaces are constructed. Programmers need to move beyond developing an understanding of a single-user population. Instead, they need to carefully define their software to distinguish how each part relies on assumptions about the user. Then the software can readily take advantage of a user model to adapt to an individual user. The goal of user-adapted interaction is that a system be able to cater for a diverse user population, with different users having different needs, preferences, interests, background, and so on.

Customization Flags or User Models?

One straightforward and common approach for dealing with diverse user populations is to enable the user to specify preferences (or flags) that control some aspects of the system. One example of such preferences is the font used for the presentation of text in an interface. The user can engage in a dialogue with the application to set preferred fonts. So, for example, the user with a significant vision impairment might select a large font. By contrast, a very small font might be selected by a user with excellent visual acuity, along with an option to display large amounts of information concurrently. The application typically holds a record of such preferences, and uses it from session to session. Such a record might be regarded as a very simple form of user model. It might even be available to other applications, so that they too can match the user's font preferences.

Let us take this one stage further. Suppose that an application allows users to specify several simple aspects of the interface. This might go well beyond fonts. For example, users might specify typing errors they are aware of making frequently (e.g., "thier" instead of "their"), so that the system can automatically interpret these correctly and amend them. We can go on, defining many small customizations to aspects of the interface. An extensive set of such customizations might provide extremely useful improvements to the interface for an individual user. It could also be valuable as a resource for various applications. This would mean that the user could have the benefits of customizations applied to the new application, without taking the time and effort to go through the setup of each new application. Consider also the user who needs to move around and make use of many different computers. If each one has to be customized, even modest customizations could become tedious to set up. Such a multiuse set of customizations is effectively a simple user model: It should be associated with the user, not just one application on one machine.

However, we draw a distinction between flags for a program and elements of a user model. To better clarify this, consider the situation where one application keeps customization in one form and another application uses a different form. Even worse, there might be no simple mapping between the two. Consider how this works in a very simple case. Suppose a user has a vision impairment and sets a particularly large font as the preferred one for a particular application. Perhaps that font is not available in a second application. It is likely that the user selected the particular font arbitrarily; they simply wanted a large font. In this case, it would be more helpful if the user could specify in a user model that they prefer large, clear fonts. Then, the second application could use this information to select a suitable font. The distinction is that flags to a program are a less flexible form of user model.

There is an important effect of separating assumptions about a user into a user model. That model becomes an independent object associated to the user. As a consequence, it may be made available to a variety of applications that may operate on different computer systems, and access the user model from a suitable server (Fink et al., 1997).

At this point, we observe another shortcoming of adaptation based on program flags. Setting up customizations requires conscious effort from the user as well as time. Mackay (1991) studied users of Unix and observed that they did little customization. This may have been due to the effort required to effect such customizations. Indeed, in that study, the major reasons for not customizing were the lack of time (63% of users), and that users found it too difficult (33% of users). Although the second reason might be attributed to the design of Unix, we should note that the degree of sophistication and the range of possible customizations in such an environment makes it hard to design a simple interface for effecting arbitrary customizations.

This highlights an important reason for user models as a basis for customization. In particular, customization should be applicable given a relatively small number of high-level preferences expressed by the user. From these, the system needs to generate the large number of detailed preferences that must be established in order to customize (adapt) the system for the individual user. Mackay (1991) observed: "Users want information about their own use and that of other people with similar job responsibilities and attitudes from which they can base their customisation decisions" (p. 159).

Nielsen (1993) identified several problems related to customization: (a) novices are disinclined to customize interfaces, (b) the customization interface constitutes another part of the system, which must be designed well and the user must learn, and (c) users may not make the most appropriate customizations. He also noted a rather different type of shortcoming of adaptation: Different users may well see quite different interfaces. He observed that this makes it harder for users to help each other. Although this is an important concern, it becomes less significant as users become more expert in the use of a particular interface that they use regularly. Such interfaces can benefit most from adaptation to the particular user. In the context of users with special needs, the potential benefits of different adaptations are likely to outweigh the disadvantages of apparent inconsistency between different users' interfaces. For example, adaptations for the user with a vision impairment might be very different from those needed for a user with motor disabilities. In both cases, there are considerable benefits from different adaptations for users with these different disabilities.

Acquiring Individual User Models Within Diverse User Populations

To facilitate the deployment of user-adapted systems, one important concern is the pragmatics of constructing a user model to drive customization. We illustrate the mechanisms in Fig. 14.3, using the domain of filtering music tracks as an example.

The user is shown at the left. The heavy vertical line represents the interface between the user and the system. Two main types of interface tools are shown as the round-edged boxes at the interface. We describe these later.

The user model is shown as the upper-middle and lower-middle boxes. The upper one represents the user preferences for various attributes of music, for example, styles (like Baroque or Bebop jazz), performers (like Joshua Bell, the Sydney Symphony Orchestra), and so on. The lower box models the user's preference for objects in the application domain, in this case particular tracks of music in the filtering system. The figure shows an example of track tr1773. The figure shows the user model for tracks in bold because it represents the goal of the filtering system. The ultimate task of the filtering system is to select music the user will like, and this becomes part of the system's model of the user's preferences. The figure labels the three mechanisms to achieve this goal:

1. Elicitation of the user model information from the user via an interface is shown in the upper left and indicated by the arc labeled 1. This is simple but tedious for the user. However, elements of it are likely to be encountered in any filtering system that can collect feedback from the user about objects recommended by the system. This is the classic relevance feedback, which has been shown to improve performance in information retrieval (Salton & Buckley, 1990).

2. The knowledge-based approach is indicated by the six arcs with the label 2. Consider the endpoint, the user model for the user's preference for tracks of music. It combines the model of the user's preferences for music attributes with the attributes of the new track to be filtered, and decides whether the track should be selected for the user, or not. This route is closest to classic information retrieval approaches. The approach relies on knowledge of the domain. It must have knowledge of the attributes of each track, as identified in metadata about the track. It also needs a user model for these attributes. Then it reasons that the user will like tracks whose attributes match the user's preferences and will dislike tracks with few or no matches.

This method requires a model of the user's preferences for various attributes of music tracks. The simplest way to do this is via an interface tool that elicits the user's preferences, just as in Mechanism 1, but this time seeking the user's preferences for attributes. This will be less tedious than Path 1 if the number of attributes modeled is smaller than the number of tracks.

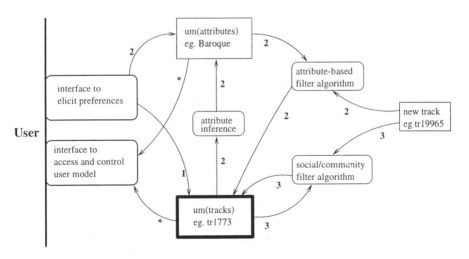

FIG. 14.3. Mechanisms for building and applying user models.

The other mechanism shown starts with the model of track preferences. An inference tool can use this approach by analyzing the model, identifying common attributes in tracks the user likes, and concluding the user likes music with these attributes. Similar reasoning can be used to infer attributes common in music the user dislikes.

3.   This final mechanism is a stereotypic, or statistics-based approach. Essentially, it is based on reasoning exemplified as follows: Many users who love tracks 1773 and 1775 love track 19965. In the information-filtering areas, it goes by names like social, collaborative, community, or clique filtering (Goldberg, Nichols, Oki, & Terry, 1992; Karunanithi & Alspector, 1996; Konstan et al., 1997; Orwant, 1995; Shardanand & Maes, 1995).

The last element of the figure is the interface shown at the lower left of the figure. It gives the user access to the user model and the possibility to control it. The figure marks this flow of information with an asterisk. This should enable the user to find answers to questions like:

- Why did the system select that recommendation?
- Why did the system *not* recommend a particular track?

Now consider Fig. 14.3 from the point of view of users with diverse needs. First consider the case where it is possible to identify a community of users who have well-defined but specialized needs, preferences, and the like. In this case, it may be feasible to construct helpful stereotypes that can prime the individual's user model. The user would have the benefits of a good starting-level adaptation with minimal individual effort, and could use the elicitation interface (Path 1 in the figure) to refine the details of the model.

We can expect that increasingly individualized adaptation of interaction will require detailed and sophisticated user modeling. If this is so, there may be a substantial cost in creating an entirely individual user model from scratch. Perhaps it would require the user to engage in a long elicitation interaction. Or it may require a long period of time during which the user is observed, and the model is built and slowly refined. A stereotype can short-circuit such lengthy processes.

We can see how this operates if we consider the communities served by various newspapers and magazines. Each such publication is created for a community of readers and the authors of articles have informal models of those readers. For example, a magazine for car enthusiasts is written on the basis of many assumptions about readers' interests and knowledge. If a new reader identifies with the community of car enthusiasts, they might initially accept a default stereotype for that community. Essentially, this type of reasoning can provide a good-enough, low-cost, initial user model. The usual mechanisms of elicitation and observation can operate from this basis toward a more refined model. Creators of various print media develop implicit models of their "users." Similarly, we can envisage specialized communities defining explicit stereotypic community models.

SUMMARY AND CONCLUSIONS

As machines become more powerful, it becomes practical to support more sophisticated degrees of adaptation to the individual. The user model can drive the adaptation of interaction by supporting:

- Interpretation of user actions.
- Internal machine actions on behalf of the user.
- Customization of presentation of the content and form of information at the interface.

These elements of interaction can be applied to a range of adaptive systems.

To support such adaptation for a diverse user population, we have argued that the user model should be: *individual, dynamic,* and *explicit.*

The processes of user modeling may involve generic inferences based on knowledge of the domain. They may be stereotypic, being statistically useful predictions for a population or community of users. Processes that provide individualized aspects are based on either elicited information or observations of the user. Pragmatic considerations mean that stereotypic reasoning will be important for helping diverse populations to develop an initial model within a particular domain. Then the individualization process can deal with details in which each individual differs from that initial model. They can also be valuable in updating the user model when the user changes.

For user models to be most effective, they need to be reused across applications. This creates the need for standards. The latter enable programmers of adaptable applications to exploit user-modeling information that is more likely to be available because it is shared with other applications.

Taking the view that the user and system should work in a partnership to achieve the user's goals, we have argued the need for user access to, and control of, their own user model. In addition to the usual concerns for private information, we have argued that the asymmetry in the human–machine relationship also lends support to the argument for access and control. This reduces some problems in user-adapted interaction.

The first example of large-scale deployment of individual user modeling has been within the Microsoft Excel user support product (Horvitz, 1997). The user models in this product track a small number of the user's recent actions and make inferences using a Bayesian net. Also widely deployed are systems that enable the user to customize various aspects of the presentation of information in the interface. We have discussed the relationship between simple customizations and adaptation driven by a user model. The appeal of the user model has three main foundations. First, it offers the potential for reuse. Second, it can enable sophisticated adaptations that require only modest time and effort from the user. Finally, the explicit representation of the user model facilitates provision of proper user access and control over their model.

As hardware costs drop and computers become increasingly widely available, the population of users grows both in number and diversity. The user model is the basis for adapting systems to cater for the varied needs, interests, preferences, background, and other characteristics of that diverse user population.

REFERENCES

Anderson, J. R. (1993). *Rules of the mind*. Hillsdale, NJ: Lawrence Erlbaum Associates.

Brajnik, G., Guida, G., & Tasso, C. (1990). User modeling in expert man–machine interfaces: A case study in intelligent information retrieval. *IEEE Transactions on Systems, Man and Cybernetics, 20*(1), 166–185.

Brajnik, G., & Tasso, C. (1992). A flexible tool for developing user modeling applications with non monotonic reasoning capabilities. In E. André, R. Cohen, W. Graf, B. Kass, C. Paris, & W. Wahlster (Eds.), *Proceedings of the 3rd International Workshop on User Modeling* (pp. 42–66). Kaiserslautern, Germany: Deursches Forschungszentrum fur Kunstliche Intelligenz.

Browne, D., Totterdell, P., & Norman, M. (1990). *Adaptive user interfaces*. London: Academic Press.

Brusilovsky, P. (1996). Methods and techniques of adaptive hypermedia. *User Modeling and User-Adapted Interaction, 6*(2–3), 87–129.

Card, S., Moran, T. P., & Newell, A. (1980). *The* keystroke-level model for user performance time with interactive systems. *Communications of the ACM, 23*(7), 396–410.

Card, S., Moran, T. P., & Newell, A. (1983). *The psychology of human–computer interaction*. Hillsdale, NJ: Lawrence Erlbaum Associates.

Chin, D. (1989). KNOME: Modeling what the user knows in UC. In A. Kobsa & W. Wahlster (Eds.), *User models in dialog systems* (pp. 74–107). Berlin: Springer-Verlag.

Corbett, A. T., & Anderson, J. (1994). Knowledge tracing: Modeling the acquisition of procedural knowledge. *User Modeling and User-Adapted Interaction, 4*(4), 253–278.

Csinger, A. (1995). *User models for intent-based authoring*. Unpublished doctoral thesis, University of British Columbia, Vancouver.

Finin, T. (1989). GUMS—A general user modeling shell. In A. Kobsa & W. Wahlster (Eds.), *User models in dialog systems* (pp. 411–431). Berlin: Springer-Verlag.

Finin, T., McKay, D., Fritzson, R., & McEntire, R. (1994). KQML: An information and knowledge exchange protocol. In K. Fuchi & T. Yokoi (Eds.), *Knowledge building and knowledge sharing* (Ohmsha and IOS Press) [On-line]. Available: http://umbc.edu/~finin/papers/kbks.pdf

Fink, J., Kobsa, A., & Nill, A. (1997). Adaptable and adaptive information access for all users, including the disabled and the elderly. In A. Jameson, C. Paris, & C. Tasso (Eds.), *Proceedings of the 6th International Conference on User Modeling* (pp. 171--73). New York: Springer-Verlag.

Goldberg, D., Nichols, D., Oki, B. M., & Terry, D. (1992). Using collaborative filtering to weave an information tapestry. *Communications of the ACM, 35*(12), 61–70.

Henderson, D. A., & Kyng, M. (1995). From customisable systems to intelligent agents. In R. M. Baeker, J. Grudin, W. Buxton, & S. Greenberg (Eds.), *Readings in human–computer interaction: Toward the year 2000* (2nd ed., pp. 783–792). San Francisco: Morgan Kaufmann.

Horvitz, E. (1997). Agents with beliefs: Reflections on Bayesian methods for user modeling. In A. Jameson, C. Paris, & C. Tasso (Eds.), *Proceedings of the 6th International Conference on User Modeling* (pp. 441–442). New York: Springer-Verlag.

Jameson, A., Paris, C., & Tasso, C. (1997). Reader's guide. In *Proceedings of the 6th International Conference on User Modeling* [Online]. New York: Springer-Verlag. Available: http://zaphod.cs.uni-sb.de/~UM97/guide.html/

Karunanithi, N., & Alspector, J. (1996). Feature-based and clique-based user models for movie selection. In S. Carberry & I. Zukerman (Eds.), *Proceedings of the 5th International Conference on User Modeling* (pp. 29–34). Kailua-Kona, Hawaii: User Modeling, Inc.

Kay, J. (1990). um: A user modelling toolkit. In *Proceedings of the 2nd International User Modelling Workshop* (pp. 251–261). Honolulu: University of Hawaii.

Kay, J. (1991). An explicit approach to acquiring models of student knowledge. In R. Lewis & S. Otsuki (Eds.), *Advanced research on computers and education* (pp. 263–268). Amsterdam: Elsevier Science.

Kay, J., & Thomas, R. C. (1995). Studying long term system use. *Communications of the ACM, 4*(2), 131–154.

Kobsa, A. (1990a). Modeling the user's conceptual knowledge in BGP-MS, a user modeling shell system. *Computational Intelligence, 6*(4), 193–208.

Kobsa, A. (1990b). User modeling in dialog systems: Potentials and hazards. *AI and Society, 4,* 214–240.

Kobsa, A. (1993). User modeling: Recent work, prospects and hazards. In M. Schneider-Hufschmidt, T. Kühme, & U. Malinowski (Eds.), *Adaptive user interfaces—Principles and practice* (pp. 111–128). Amsterdam: North-Holland, Elsevier Science.

Kobsa, A. (1996). *A standard for the performatives in the communication between applications and user modeling systems* [Online]. Available: ftp://ftp.informatik.uni-essen.de/pub/UMUAI/others/rfc.ps

Kono, Y., Ikeda, M., & Mizoguchi, R. (1992). To contradict is human: Student modelling of inconsistency. In C. Frasson, G. Gauthier, & G. McCalla (Eds.), *Intelligent tutoring systems* (pp. 451–458). Berlin: Springer-Verlag.

Kono, Y., Ikeda, M., & Mizoguchi, R. (1994). THEMIS: A non monotonic inductive student modeling system. *Journal of Artificial Intelligence in Education, 5*(3), 371–413.

Konstan, J. A., Miller, B. N., Maltz, D., Herlocker, J. L., Gordon, L. R., & Riedl, J. (1997). GroupLens: Applying collaborative filtering to Usenet news. *Communications of the ACM, 40*(3), 77–87.

Mackay, W. E. (1991). Triggers and barriers to customising software. In *Proceedings of the ACM SIGCHI Conference on Human Factors in Computing Systems* (pp. 153–160). New York: ACM Press.

Malinowski, U., Kühme, T., Dieterich, H., & Schneider-Hufschmidt, M. (1992). A taxonomy of adaptive user interfaces. In A. Monk, D. Diaper, & M. D. Harrison (Eds.), *People and computers VII* (pp. 391–414). Cambridge, England: Cambridge University Press.

Milosavljevic, M. (1997). Augmenting the user's knowledge via comparison. In A. Jameson, C. Paris, & C. Tasso (Eds.), *Proceedings of the 6th International Conference on User Modeling* (pp. 119–130). New York: Springer-Verlag.

Motro, A., & Smets, P. (1997). *Uncertainty management in information systems.* Boston/London/Dordrecht: Kluwer Academic Publishers.

Nielsen, J. (1993). *Usability engineering.* New York: Harcourt Brace, Academic Press.

Nisbett, R. E., & Wilson, T. D. (1977). Telling more than we can know: Verbal reports on mental processes. *Psychological Review, 84,* 231–259.

Norman, D. A. (1983). Some observations on mental models. Preface in D. Gentner & A. L. Stevens (Eds.), *Mental models.* Hillsdale, NJ: Lawrence Erlbaum Associates.

Norman, D. A., & Draper, W. S. (Eds.). (1986). *User centered system design: New perspective on human computer interaction.* Hillsdale, NJ: Lawrence Erlbaum Associates.

Orwant, J. (1993). *Doppelganger goes to school: machine learning for user modeling* (MIT MS 26 Thesis). Cambridge, MA: MIT Media Laboratory.

Orwant, J. (1995). Heterogenous learning in the Doppelganger user modeling system. *User Modeling and User-Adapted Interaction, 4*(2), 107–130.

Paiva, A. (1996). Towards a consensus on the communication between user modeling agents and application agents. *Workshop on Standardisation of User Modeling Shells in the International Conference on User Modeling* [Online]. Available at: http://cbl.leeds.ac.uk/amp/home-page.html

Rich, E. (1979). User modeling via stereotypes. *Cognitive Science, 3,* 355–66.

Rich, E. (1983). Users are individuals: individualizing user models. *International Journal of Man–Machine Studies, 18,* 199–214.

Rich, E. (1989). Stereotypes and user modeling. In A. Kobsa & W. Wahlster (Eds.), *User models in dialog systems* (pp. 35–51). Berlin: Springer-Verlag.

Salton, G., & Buckley, C. (1990). Improving retrieval performance by relevance feedback. *Journal of the American Society of Information Sciences, 41,* 288–297.

Shardanand, U., & Maes, P. (1995). Social information filtering: Algorithms for automating "Word of Mouth." In *Proceedings of the ACM SIGCHI Conference on Human Factors in Computing Systems* (pp. 210–217). New York: ACM Press.

Spooner, R. I. W., & Edwards, A. D. N. (1997). *User modelling for error recovery: A spelling checker for dyslexic users.* In A. Jameson, C. Paris, & C. Tasso (Eds.), *Proceedings of the 6th International Conference on User Modeling* (pp. 147–157). New York: Springer-Verlag.

Trewin, S., & Pain, H. (1997). Dynamic modelling of keyboard skills: Supporting users with motor disabilities. In A. Jameson, C. Paris, & C. Tasso, C. (Eds), *Proceedings of the 6th International Conference on User Modeling* (pp. 135–146). New York: Springer-Verlag.

Wahlster, W., & Kobsa, A. (1986). Dialogue-based user models. *Proceedings of the IEEE, 74*(7), 948–960.

Zissos, A., & Witten, I. (1985). User modelling for a computer coach. *International Journal of Man–Machine Studies, 23,* 729–750.

15 Interface Agents: A New Interaction Metaphor and Its Application to Universal Accessibility

Annika Waern
Kristina Höök

The principles of Universal Access *require that Information Technology is made accessible to all users, and not only to the "average" user. In this chapter we will try to show that interface agents are not just an extension of the user modeling vision of adapting to different users; instead, interface agents may bring about new interaction styles that provide a new design insight into the issue of* Universal Access. *In particular, the capabilities of interface agents for natural dialogue, voluntary user support, delegation of responsibility and representation of subjective perspectives may turn out to be not just important as novel interaction styles, but also crucial in attracting attention and in raising user awareness and motivation, thus improving the user's experience with the interactive system.*

WHY INTERFACE AGENTS?

Empathy is basic to human thinking. We usually assume that other human beings, animals, and sometimes even inanimate objects are intentional creatures with emotions and interesting relationships to ourselves and other beings. A large part of what we talk about during the course of day concerns discussing why people have performed certain actions, and what their intentions could have been. We try to figure out who are still our friends and who is friends with whom, and we reestablish our position in the ranking scale; in fact, some researchers (see, e.g., Dunbar, 1997) claim that this social "grooming" along with its evolutionary advantages is the reason why human beings have such a large brain and such a complex language.

We might ask ourselves what happens when we, these social, interaction-intensive, emotional human beings, are placed in front of a computer. Of course, despite the fact that we know that computers are dumb machines, we sometimes treat them as fellow beings (Reeves & Nass, 1996). We relate to them, we attribute intentions and emotions to them. Up to now, this effect has largely been ignored in the design of interfaces—if anything, human–computer interaction (HCI) has traditionally demanded that this is avoided. However, this axiom has recently been reconsidered. Is it really only a negative feature, or is it something that designers could actively take advantage of? That is, what would be gained from designing interfaces that behave like intentional, emotional actors? Could this potentially be a route to reaching a larger group of users—enable *user interfaces for all*? Our capacity for empathic reasoning is deeply built into us, and thus actor-based interfaces could draw on very basic human abilities; or will they just raise faulty expectations on systems that will lead users and designers astray?

In this chapter, we scratch the surface of these issues, presenting examples of systems that use the agent metaphor, and the goals they aim to accomplish. Finally, we discuss the potentials this approach brings to *universal access:* a new space of design solutions that may cater for user needs and requirements in ways that are difficult to support using mere direct manipulation.

Computers as Actors Is a Natural Metaphor

We can distinguish between two ways of dealing with computer systems or services: as tools, or as independent actors. The direct-manipulation interface was designed with the specific purpose of enhancing the tool view of the interface. Nevertheless, it is also possible to consciously design the interface to appear as an independent actor, rather than as a tool. This approach has occurred in many forms and has been given many names in research literature, such as interface agents, user facing agents, synthetic characters (see, e.g., Elliott, 1992; Maes, 1994), not to mention the robots and androids of science fiction literature (Asimov, 1950). This way, a number of aspects of human–human interaction can be exploited in human–machine interaction. In this chapter, we discuss the following usages of interface agents, all well represented in literature and existing systems:

- *Agents for help and learning.* Agents can "sit at the side" of an application, providing help and guidance on its usage. Agents can also act as tutors or colearners in a learning application.
- *Delegation.* Users can delegate tasks to agents, such as activities that should be done when the user is away from the computer, monitoring events from other sources, or performing activities at distant locations in the network.
- *The subjective focus.* Agents provide an anchor for subjective evaluations that follow a presentation (e.g., who it is that gives a helpful suggestion).
- *The dialogue partner.* Agents provide a counterpart in natural-language dialogue.
- *Emotional behavior.* Agents can both show emotions and rise emotions in the user (tools are not supposed to have such effects).
- *Agents as user representatives.* Agents can be instructed to behave in certain ways and then go out into a world and represent us—avatars are simple examples of this.

The easiest way to think about an interface agent is to envision it as a character, a face, a little dog, and so on, that is visible on the screen. However, this is not a necessary requirement. An interface agent need not be a synthetic character, visible on the screen, with natural-language capabilities. Many interfaces that adhere to the interface agent metaphor still implement this interaction using traditional windows with menus and clickable buttons, and in many cases this is perfectly appropriate. The interface agent metaphor is, in those cases, indicated only by the fact that the system exhibits autonomous functionalities, which run in the background while the user is doing other things, and now and then call for the user's attention.

In such simple examples of interface agents, the distinction between the agent metaphor and the tool metaphor becomes blurred, because the user may choose to view such functionalities as tools. Designers should, therefore, be careful, because the tool view may become rather unnatural, leaving the users at a loss with how to deal with the provided functionality. Most users will relate to aspects like active help, or natural-language capabilities, as if an independent actor provided them— even if the system designer aimed for a pure tool metaphor for the system.

It should be emphasized that the view of an interface as being a tool or an actor resides entirely with *the user*. The designer can intend the system to be treated as a tool, or as an agent, but the user may take the opposite view. A good design is, from this perspective, one that adequately reflects the designer's intentions, and influences the user to take the intended view.

USING INTERFACE AGENTS IN INTERACTION DESIGN

As discussed in the introduction, there are many interface functionalities that are most easily interpreted when provided by an independent actor, rather than when being manipulated and controlled directly by the user. In this section, we discuss such functionalities in more depth, providing examples from literature on systems that have exploited this aspect of human–agent communication.

Agents as Dialogue Counterparts

A problem with the tool view of interfaces is that direct manipulation does not lend itself easily to some interface tasks. The manipulation of objects that are not presented at a particular time of the interaction is one such problem. Because there is no way for all objects to always be immediately accessible, direct manipulation requires a structure, an information space (menus, links, folders, etc.) in which the user must navigate to find the object he or she wants to manipulate. Another well-known problem is quantification: If a user wants to select not one, but *all* objects with certain characteristics, direct manipulation becomes cumbersome. In addition, the objects must exist at the time of selection: Referring to an object that used to exist, creating a new object, or scheduling an activity for an object that will be created 2 days from now requires more advanced interaction than pure direct manipulation allows. The usual way of solving these problems is to introduce dialogue boxes, buttons, and menus, in which some rudimentary language interaction takes place. It should be noted, however, that, even if these stay within a pure point-and-click interaction style, they gradually move away from pure direct manipulation, and, depending on how advanced the functionality is, they may influence users to view aspects of the interface as an actor. In addition, all direct-manipulation solutions to these problems require that still more information is presented visually onscreen. This emphasis on visual appearance creates an acute problem for universal access, drastically limiting the attainable solutions for visually impaired users.

Perhaps the most crucial problem is performing selections among a large number of objects in a direct-manipulation interface, that is, when the visual appearance gets too cluttered. In general, the whole visualization field aims to provide users with tools that enable them to visualize and manipulate large information spaces, and is thereby ignoring the visually impaired users.

These problems have motivated researchers to look for ways of reintroducing interaction in natural, or seminatural language. Systems that are capable of natural-language dialogue are typically interpreted as actors. In general, maintaining a dialogue requires a dialogue counterpart: There is something or someone that you talk to, and which, or who responds to your utterances (Bretan, 1995).

The dialogue counterpart may be more or less humanoid. Many studies show that the more anthropomorphic a character is, the more naturally the user will respond to it, and the more "human" the dialogue becomes (King & Ohya, 1995; Koda & Maes, 1996; Reeves & Nass, 1996; Sproull, Subramani, Kiesler, Walker, & Waters, 1996). This can be both advantageous and disadvantageous. If the agent is made very anthropomorphic in character, users may believe it to possess capabilities that it does not actually have, such as generic language skills, or expertise in areas of competence that lie at the side of the agent's true capabilities (Shneiderman, 1997). On the other hand, if the dialogue is made more human, users will also adapt to the dialogue counterpart in terms of words used in the dialogue, length of sentences, and even the speed of speech and the pitch of voice. This is a form of language game that goes on in all human–human interaction (Bretan, 1995) and is very useful in dialogue systems because it is a partly automatic, well-understood way for us to adapt to our dialogue counterpart. So, if the interface agent uses words, sentence structures, and ways of speaking that the system itself has little problem in interpreting, users have a means to adapt to system limitations, without even having to think about them.

There are few examples of systems that use interface agents solely as dialogue counterparts, but one interesting example is the DIVERSE system (Bretan, 1995). DIVERSE was an early prototype for a virtual reality system that used an agent as a dialogue counterpart. In this system, the agent was seen as a mediator between the system and the user. It continuously followed the user around in the environment, and was able to parse commands from the user that were effected in the environment. A screen shot from the system is shown in Fig. 15.1. In DIVERSE, the role of the agent as a natural-language interpreter was emphasized by explicitly showing a list of "objects in focus" above the head of the character, which was used to interpret vague utterances.

Agents for Help and Learning

The most well known usage of interface agents is perhaps as a source for help on computer applications. Typically, such agents do not actually perform actions, but provide helpful hints to a user on how to use a particular system. In general, whenever a system contains a separate help facility that can be accessed in parallel with the main system functionality, the former can be seen as a separate interface agent. In practice, users will maintain a tool view of the help system, unless it is either active, context- or user-adaptive, or provided by a synthetic character.

Help is an important contributor to universal access, as it can provide support to both novice and expert users, and allows users to gradually learn a system while using it. Help can be provided to the individual user in many ways. First, we can differentiate between active and passive help: help that is provided when the system has inferred a need for it, or help that is provided only on demand.[1] Second, help can be adapted both to the context in which it is provided (e.g., the operations that the user most recently performed), and to a model of the user (e.g., a model of

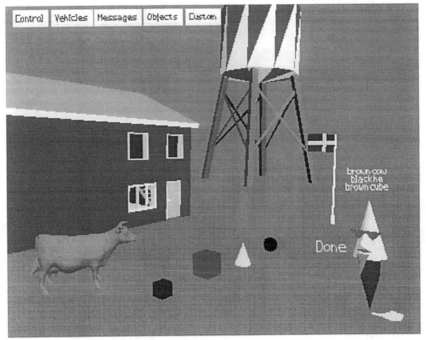

FIG. 15.1.   A screen shot from the DIVERSE system.

[1]A third alternative is to continuously provide help; see, for example, RUA (Meyer, 1994).

the user's current knowledge, or misconceptions). If no adaptation takes place, the user must have a way of posing help queries to the system. This can be done in some free-format way (e.g., entering free-text search queries), or by navigating through help menus.

Active help is a quite problematic issue. On the one hand, it can be extremely beneficial if posed at the right moment to the right user. On the other hand, users seldom like it—if you are involved in a difficult task, you will dislike being disturbed by a help message, even if it is appropriate, because it interrupts your current train of thought. One system that recognized this effect was the Flexcel demonstrator (Oppermann, 1994), which provided help on a spreadsheet application. In Flexcel, proactive help was provided in a very unobtrusive manner—the system only displayed a small lightbulb in a corner of the window; the user could then activate help, when he or she wished, to review the system's suggestion (see Fig. 15.2).

One system that combines virtually all types of help in an interface agent is the Microsoft Office Assistant (Horvitz, Breese, Heckerman, Hovel, & Rommelse, 1998). This help system was first provided with the release of Microsoft Office 97, and combines active and passive help, user- and context-sensitive help, as well as free-text search queries. Help is provided through text and menu windows that are anchored to a cartoonlike, animated agent, which resides in its own window. The reactions to the Office Assistant have been varied. On the one hand, it was one of the main selling features of Office 97. On the other hand, users often complain about its intrusiveness and the inadequacy of provided help.

In general, individualized help is a quite difficult problem. To be effective, the system must infer what the user wants to do *next*. As the system does not control

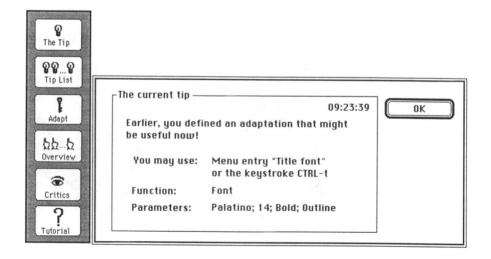

FIG. 15.2.   The help menu bar and an example help window from Flexcel.

what the user is doing, the space of alternatives is usually very large. It is seldom possible to base any good suggestions solely on tracing user actions in the interface, as these provide very low level information compared to the kind of goals that the user needs help on (Waern, 1997). Querying the user is one way of deriving additional, higher level input toward this goal, but one should be aware of the fact that when users need help, they also have great difficulties in formulating a good help query, and they may even have difficulties navigating in a help menu. There exists a lot of research in the area of providing helpful answers to "incorrect" help queries (see, e.g., Pollack, 1986), that essentially boils down to the fact that a collaborative user–system dialogue is needed. But the example systems that achieve such dialogues typically deal with small domains, and it is unclear if the proposed solutions scale to larger domains.

Tutors and Learning Companions

A special type of help agents is those that occur in educational software. In these applications, the primary purpose of the agent is not to help the user with his or her current task, but to help the user in learning a domain of expertise. Agents here occur in two forms: as teachers or helpers, and as learning companions. The tutoring agent is frequently represented in educational software. A perhaps more interesting variant of the educational agent is the *learning companion* agent. A companion who is not a teacher but a comate might provide a more compelling, motivating style of interaction. Collaborative learning of this form was proposed by Pierre Dillenbourgh and John Self (1992) as one way of moving away from the instructivist style of teaching to a more constructivist style.

One example of an interface agent that combines the learning companion and tutor roles is "Herman the bug" (Lester, Converse, Stone, Kahler, & Barlow, 1997) developed in the context of the Intellimedia project at North Carolina State University. In this system, children learn ecology by constructing a plant that is to survive in a particular environment. The agent will help the child out by explaining why a particular solution is not optimal, or why a plant construction was successful. Herman is an animated and strongly anthropomorphic character, but appears as a playful robot rather than as a teacher (see Fig. 15.3). This adds a "game" character to the system, and puts children in a more sympathetic mode toward the character, making them less reluctant to listen to its advice. This kind of "infotainment" usage of the agent metaphor may seem to be of little importance, but in fact, the main focus for the Intellimedia project is to study the role of interface agents in educational software. The wish is to create tools that support learning in novel ways, leveraging between students with different backgrounds, motivations and learning styles.

Delegation

Delegation is perhaps the most obvious example of a functionality that is difficult to accomplish through pure direct manipulation. Delegation has always been used

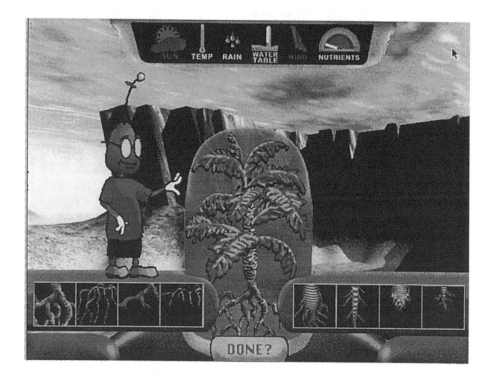

FIG. 15.3.   Screen shot of "Herman the bug." Copyright by James Lester.
Reprinted by permission.

in computer systems, often with vaguely anthropomorphic representations. An extremely simple example is the Macintosh "fetch" program, which searches for and retrieves documents from remote servers, and uses a picture of a running dog to indicate that it is active. More current examples include Web retrieval agents, which search the Web for pages that fit the user's interest profile (Lieberman, 1995), filtering agents that filter incoming mail and sort it into different folders, and shopping agents that roam the network for the best price for a record (several examples of such systems are discussed in Maes, 1994). These types of agents are typically not given anthropomorphic representations at all, even if, in some cases, they are given quite advanced means for inferring the user's interests and preferences. Typically, the agent will instead display an explicit representation of its controls (such as a set of filter rules) that the user can set up and change at wish. This is often true even for agents that infer user preferences from their behavior (see, e.g., Lieberman, 1995; Waern, Tierney, Rudström, & Laaksolahti, 1998).

Delegation is perhaps the most important usage of interface agents with respect to universal access. The reason is that attaining access through direct manipulation is sometimes very hard, or even impossible for the visually impaired, the spastic, or, quite simply, for the person whose hands and eyes are occupied for a given period of time (e.g., when driving the car, changing diapers, etc.). One generic approach to delegation is the creation of interface agents that take on the role of personal assistants, "butlers." The personal assistant is an agent that serves an individual user, keeping track of the user's current task, interaction device, and general preferences and capabilities. The personal assistant agent could be an excellent way of dealing with temporary and permanent handicaps, as it could be used to adapt the interaction with various services to the user's capabilities on the fly.

There is also a completely different usage of agents for delegation, and that is when the agent serves as a *user representative* in an environment where other users and/or agents are present. In this case, the most important interface is not directed toward the represented user, but toward *other* users (or their agents). Shopping agents take on this role, as they may interact with selling agents that allow them to close deals with them. However, the most striking examples of agent representatives are avatars, user representations in multiuser virtual reality environments. Avatars exist in both full three-dimensional environments, such as the "Meridian 59" game,[2] and in graphical MUD and CHAT environments, such as "Comichat."[3]

A problem with avatars is that the user needs to control all aspects of their behavior in real time. In real life, control over nonverbal communication (such as body language) is automatic. Approaching the same level of control in an avatar is almost impossible, so it becomes necessary to control avatars by delegation: The user can control the avatar's behavior only at a high level and the detailed body language must be planned autonomously by the avatar (Vilhjálmsson & Cassell, 1998). This provides a leveraging factor that can be used for increased access. Users may define "macros" for certain useful interaction schemata, prior to interaction. While participating in a discussion, the user can then invoke these behaviors through simple commands, acquiring a richer dialogue than would otherwise be possible in real time. One example of this kind of interaction would be to allow language-, or speech-impaired users to participate in discussions by generating full natural-language utterances from sign language.

Emotional Behavior

Emotional behavior has been little exploited in human–machine interaction. In effect, emotional behavior has rather been seen as an aspect of human behavior that is undesirable in human–machine interaction. The regained interest in emotional and affective sides of computing maintains that the intimate interaction between cognition and emotion is a fundamental aspect of everyday life, as well as of HCI

---

[2]For more information, see: http://meridian.3do.com/meridian/

[3]For more information, see: http://www.microsoft.com/windows/ie/chat/

(e.g., Morkes, Kernal, & Nass, 1998; Picard, 1997). A major factor in this change has been the rapid developments in the game and infotainment areas, where emotional behavior is exploited to the extreme.

The field of affective computing focuses on emotional components of human–machine interaction. Rosalind Picard (1997) went through a number of potential applications for such techniques. There are several possible components of such systems: First, systems can aim to raise emotions in the user; second, the system can be made to exhibit emotions (and presumably, that raises emotions in the user as well); and, finally, the system can aim to recognize the user's emotions and adapt to them. It is very difficult to see how a system could exhibit emotions without resorting to the actor metaphor (although music has been suggested as one such means). Similarly, a system that adapts to user emotions can hardly be viewed as a pure tool.

So far, systems (apart from games) that utilize affective computing remain very much on the drawing board. An early example are the agents developed within the affective reasoner framework (Elliott, 1992). The affective reasoner agents are capable of reasoning about and referring to the emotions of users and other agents. For example, if one agent was happy because another one was sad, the sad agent would recognize this feeling as "gloating." The affective reasoner allows agents to take on different personality traits in how they react to other agents' expressed feelings. Agents can express emotions in language and facial expressions (they are depicted using cartoonlike "smileys" on a screen). The inference schemata used in the affective reasoner are based on representing emotional expressions like "furious," "angry," "gloating," and the like, in relation to other agents' actions or emotions (see Table 15.1).

In a study by Elliott (1997), it was investigated to which extent this computer-generated face with spoken output (and music) could express recognizable emotions. The generated face was compared to a human actor. Overall, subjects did significantly better at correctly matching videotapes of computer-generated presentations with the intended emotion scenarios than they did with videotapes of a human actor attempting to convey the same scenarios.

Another example of "affective" agents are ones that express emotions rather than reason about them, such as the poker-playing agents developed by Koda (Koda, 1997; Koda & Maes, 1996).

The Subjective View of Information

An interface agent can also take on the role of presenting a subjective view (Schank, 1991). It can have an attitude and communicate contextualized, subjective issues, such as the value of a certain piece of information. For example, is it better to go via Manchester or London when traveling to York? Well, one attitude could be that if you want to have some fun, you should spend an evening in London rather than Manchester on your way to York. Another attitude could be that Manchester is closer to York, so the

TABLE 15.1

The Table of Emotion Categories

Private/Group	Specification	Category Label and Emotion Type
Well-Being	appraisal of a situation as an event	**joy**: pleased about an event distress: displeased about an event
Fortunes-of-Others	presumed value of a situation as an event affecting another	**happy-for**: pleased about and event desirable for another **gloating**: pleased about an event undesirable for another **resentment**: displeased about an event desirable for another **sorry-for**: displeased about an event undesirable for another
Prospect-Based	appraisal of a situation as a prospective event	**hope**: pleased about a prospective desirable event **fear**: displeased about a prospective undesirable event
Confirmation	appraisal of a situation as confirming or disconfirming an expectation	**satisfaction**: pleased about a confirmed desirable event **relief**: pleased about a disconfirmed undesirable event **fears-confirmed**: displeased about a confirmed undesirable event **disappointment**: displeased about a disconfirmed desirable event
Attribution	appraisal of a situation as an accountable act of some agent	**pride**: approving of one's own act **admiration**: approving of another's act **shame**: disapproving of one's own act **reproach**: disapproving of another's act
Attraction	appraisal of a situation as containing an attractive or unattractive object	**liking**: finding an object appealing **disliking**: finding an object unappealing
Well-Being/Attribution	compound emotions	**gratitude**: admiration + joy **anger**: reproach + distress **gratification**: pride + joy **remorse**: shame + distress
Attraction/Attribution	compound emotion extensions	**love**: admiration + liking **hate**: reproach + disliking

total trip will be more efficient if you go that way. These attitudes are easily communicated if we see them as originating from different characters.

One demonstrator system that uses the subjective view is the car saleswoman agent Jennifer[4] (Elliott & Brzeinski, 1998) (Fig. 15.4). Jennifer is just one of several agents developed by the Extempo company, all providing highly subjective views. These agents are heavily anthropomorphic. Jennifer uses gestures and a language that gives the user a distinct impression of "sales talk." She will even sometimes put in personal remarks (like "I like to drive fast cars") as means of establishing a strong subjective impression. This particular agent finds its role in advertising, and provides a route to combining traditional means of advertisements with the more personal touch of the salesman, or, as in the case of Jennifer, saleswoman.

Narratives

According to Bruner (1986), Schank (1991), and others, stories are fundamental to human cognition. This aspect is little used in direct-manipulation interfaces, which rather aim to provide parallel access to massive amounts of information with no narrative structure to connect them. A user-facing agent makes it easier to imagine a narrative style of interaction. For example, instead of browsing the Web, a character can tell a story that will guide the user through the information. The help facility of a word processor can be structured as a narrative rather than as an abstracted information database.

A narrative does not necessarily have to involve characters, but, as pointed out by Persson (1998), in order to understand a narrative, the listener has to make the causal, temporal, intentional links between the events in the narrative (as studied in reception studies). Persson said:

> Of course the strive for coherence and meaning is present in comprehension of all types of discourses, not only narratives (e.g., scientific papers, poems, recipes, manuals, news, etc.). What is characteristic about narratives, however, is that the events prototypically involve characters. Narratives deal with anthropomorphized individuals with some form of inner psychology (they think, feel, believe, and have intentions), who acts and reacts on the surrounding environment (which often includes other characters).

This means that characters, possibly implemented as interface agents, may help users to add a narrative structure, the "glue" between events that constitute a story that they then remember. Such stories may help both in memorizing actual information, and the route to finding it. In our research group, we are in particular exploring the usage of narrative structure as an aid for users that find it difficult to navigate in pure hypermedia structures. Studies show that the ability to navigate hypermedia varies greatly between different users (Höök, 1996).

---

[4]For more information, see: http://www.extempo.com/characters/jennifer.html

VIEWS OF USER–AGENT–SYSTEM COLLABORATION

What then is the role of the interface agent in HCI? One way of creating an interface agent is to design the entire interface as a dialogue between the user and an interface agent. Phone-based services, especially if they are voice controlled, can be experienced as interface agents, because they allow only one mode of interaction, which is distinctly command oriented. Banking services provide a perfect example of this, because the user is often given a distinct impression of delegating work to an agent ("transfer money between my accounts," "sell my shares at a minimum price of N dollars"). Sometimes, it may even be unclear whether this delegation is done to an automatic service, or if a human will eventually be involved in performing the task.

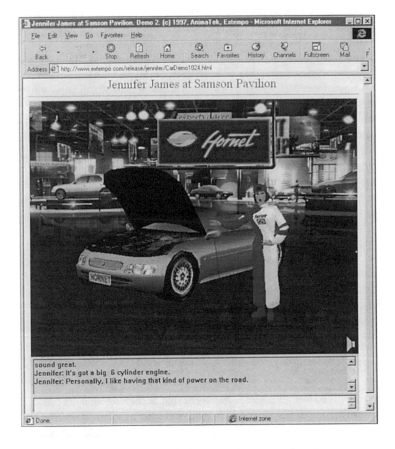

FIG. 15.4.   Screen shot from Jennifer. Supplied by extempo. Copyright by extempo. Reprinted by permission.

The "agent as the interface" solution is however the less commonly advocated case in literature. The reason is that such interfaces may become obscure. Users may find it difficult to grasp what functionalities are available in the system, and cumbersome to access them. Instead, agents are typically seen as a complement to direct manipulation. Seen from this perspective, we can distinguish two uses of the agent metaphor: as a "sidekick" to the user, or as a user representative.

The Sidekick

The most commonly advocated design for interface agents is to include them in a dual interface, where direct manipulation is also possible. The main purpose for this design is to provide a clear distinction between fixed, nonadaptive functionalities and those that are personalized. The direct-manipulation part is kept nonadaptive and stable, whereas the agent part provides all "intelligent" functionalities: It can be user-adaptive, provide natural-language capabilities, and so on. The agent can be able to interact both with the user and with the basic direct-manipulation system (Fig. 15.5). This design of interfaces has been advocated numerous times (see, e.g., Maes, 1994; Rich & Sidner, 1997), as it provides the advantage that the user can always circumvent the agent, and perform things "manually" in the basic system.

In the most generic case, both the user and the agent see and use the same direct-manipulation interface, and maintain a dialogue around what is happening and what the user wants to do. There also exist two more limited cases of the generic sidekick agent design that are frequently represented in agent literature. The first is when the agent cannot perform actions in the basic system, but only observe the user's interactions. In this case, the agent is restricted to giving *help* on the system. The other special case is when the agent cannot observe the user's interactions with the basic system, but is restricted to receiving commands from the user and performing them in the basic system. In this case, the agent realizes delegation as an alternative route for the user to perform things in the basic system (in this context, the term *indirect management* is often used, to denote an interaction model complementary to direct manipulation). The DIVERSE agent discussed in the Using Interface Agents in Interaction Design section of this chapter is one example of a sidekick agent that realizes delegation.

Embedded Agents

A variant of the sidekick agent is the agent that is embedded in the basic interface, rather than running in parallel with it. In this case, the agent resides in the interface to a particular subsystem or subservice, rather than being accessible all the time that the service or system is used. The user can access it only if he or she navigates to this service, and its functionality is typically closely related to the service. Embedded agents are most commonly found in virtual reality applications, but they can also be used in hypermedia. The Jennifer saleswoman agent is an example of an embedded agent. Embedded agents can act as pure sidekick agents, if they attach themselves to a user and follow him or her around as the system is being traversed.

One interesting difference between the embedded and the pure sidekick agent is that embedded agents can more easily take on roles other than pure help provision. Embedded agents can be used to embody a particular functionality, or to represent other users or organizations, and take on attitudes that reflect their role. For example, assume that a sidekick agent is asked to retrieve information about CD records. Because the agent is not provided by any particular retailer, the user will assume that it presents the results based solely on the user's preferences. If the agent always prefers records from a particular store, the user will first become confused and possibly fooled, and eventually start to reject the agent's suggestions. But an embedded agent acting as a clerk in an online record store may very well bias its suggestions based on what the store has in stock. The user will anticipate this subjective selection, and is thus able to place the correct amount of trust in the agent's suggestions.

The Representative

In multiple-agent systems, the interface agent can take on a different role. Because there are several actors and users available, agents can have two user interfaces: one toward its "own" user, and one toward other users (or their agents). This way,

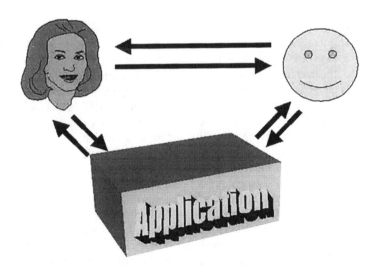

FIG. 15.5.   The 'agent as a sidekick' model of interaction.

agents can represent users or organizations toward other users or agents (see Fig. 15.6). Typically, the user can delegate tasks to his or her personal agent. But as discussed earlier, the agent also represents the user's attitudes and intentions toward other users. The agent also becomes a way for the user to present his or her subjective views in rich ways, with a clear reference to who maintains these views.

The representative agent is often designed as an embedded agent, the obvious example being the avatar. But the representative agent need not necessarily be embedded. The personal assistant agent discussed previously may very well take on the role as a user representative, toward services in an agent architecture. In this case, the personal assistant could handle things like the initial adaptation of new services to user characteristics, decide what information about the user a service can share, close payment deals with services, and contact services proactively, as it anticipates that the user will want to access them.

CRITICISMS AND RISKS

The main techniques for interface agent design (anthropomorphic interfaces, natural-language functionalities, and indirect-management models of interaction) have all raised much controversy in the HCI field. The reason is that they may violate many of the usability principles developed for direct-manipulation systems. These principles include giving the user *control* over the system, making the system *predictable,* so that it always gives the same response given the same input, and making the system *transparent,* so that the user can understand something of its inner workings (Höök, 2000).

This critique is true—interface agents do have these effects—but it may also be seen as completely irrelevant. The whole purpose of introducing interface agents is to move away from a tool-based view on interface design, so it can no longer be crucial that the user has control over the entire interaction. Instead, the system may sometimes take complete control—initiate the dialogue, force the user to respond in certain ways, and so on (see, e.g., "You Don't Know Jack,"[5] a quiz show game). An agent should take its own initiatives and act on its own goals. For the same reasons, the requirements on predictability and transparency do not apply. After all, we interact with multiple agents in our everyday environment, which are neither predictable nor transparent (friends, colleagues, children, and dogs), and we not only learn to live with it, we enjoy it.

The more serious critique of interface agents deals with two different issues. First, is it really possible to implement systems that support the interface agent metaphor with adequate functionality? And second, how will the use of interface agents affect our perception of computers? Will it cause us to place too much trust in our systems, shifting the responsibility from the individual user to the system?

---

[5]For more information, see: http://www.bezerk.com

FIG. 15.6.   Agents as representatives for users in a space of users and agents.

Interaction Reserved for Human–Human Communication?

The first critique was well voiced by Lucy Suchman (1987, 1997). She proposed that the term *interaction* might best be reserved to describe what goes on between persons, rather than being extended to encompass relations between people and machines. Her argument is that any conceptual model, which a designer will build into an interactive system, will have its limitations, and, in many cases, will not at all reflect what users actually use the system for. Any agent or adaptive system that makes assumptions about the user's task will fail to interact properly with the user, because the assumptions will be limited to those preconceived usages of the system. Suchman proposed instead that machines should be designed to be "readable," that is, that users can always inspect the system's state and interpret it themselves and then act.

Even if we can find examples of intelligent user interfaces that fulfill the "readable" criteria discussed by Suchman, a main point with agent interfaces is that they reintroduce the dialogue between user and system (the indirect-management interface). The Suchman critique is thus difficult to counter. However, it is not clear that the critique singularly applies to the indirect-management style of interaction. Most computer applications provide users with a very obscure picture of the system's current state—at least for those who are not computer scientists, or particularly knowledgeable in how the system works. As pointed out by Maes (1997), the same argument could be made against driving cars, when you do not have a complete model of how the engine or the brakes work. The point is that people do drive around in cars and manage really well without these models. When the car breaks down, we willingly hand over the car to a car mechanic agent, whom we trust to have the necessary skills for fixing it. The main task for drivers is to use the car to get somewhere—they cannot be bothered with a complete model of the car's inner workings. The same goes for computer systems—if they can provide me with, for example, relevant high-quality information, I would not necessarily be bothered by exactly how it is done (just as I do not usually bother myself with how the journalists of my daily newspaper find their information).

Anthropomorphism Considered Harmful?

The second critique has been well voiced by, among others, Shneiderman (1997) and Lanier (1996). Shneiderman and Lanier both argued against anthropomorphic agents, because they give users the impression that the system will be able to take responsibility for its actions and that they will act rationally, similar to human beings. This may change the users' understanding of computers as stupid machines, and shift users' perception of responsibility from users to systems. It is of crucial importance that the design of interface agents is such that it creates the right expectations in the user: neither too high, nor too low. Here we must be careful of imitating human–human communication and assuming that that would be the best model of performance.

When anthropomorphic interfaces have been studied it has been foremost with the aim to see which expectations they raise in their users. The general conclusion is that the more "natural" the interface, the higher expectations on intelligence in the system. Many studies show that the more anthropomorphic a character, the more naturally the user will respond to it, and the more humanlike becomes the dialogue. In a study by Sproull et al. (1996), a career-counseling program was studied. Two interfaces were used: one textual and one facial interface. It turned our that users were more aroused (less confident, less relaxed) with the face, presenting themselves in more positive light. There were also gender differences: Women preferred text, and men preferred the face. Brennan and Ohaeri (1994) showed that users talked more to the anthropomorphic interface. King and Ohya (1995) showed that users attributed more intelligence to anthropomorphic interfaces. Koda and Maes (1996) showed that realistic faces are liked and rated as more intelligent than abstract faces.

This can be both advantageous and disadvantageous. If the agent is made very anthropomorphic in character, users may believe that it possesses capabilities it does not actually have, such as generic language skills, or expertise in areas of competence that lie at the side of the agent's true capabilities (Shneiderman, 1997). On the other hand, if the capabilities of the agent are well communicated, so that the expectations are met, a potentially powerful interface can be created.

On a deeper level, we need to understand how the humanlike communication capabilities affect users. Reeves and Nass (1996) conducted a series of experiments to further our understanding of the difference between human–human communication and HCI. They showed that, even people who state that they do not consider a PC (personal computer) a social actor, do, in fact, respond to PCs as if they were social actors. They show that politeness to a computer, or perception of a television's expertise are not marks of inexperience or ignorance, but simple reflexive responses.

People respond to media the way they respond to people because it is simpler to respond naturally. Moreover, only minimal cues are enough to elicit these social responses, which mediate perceptions as diverse as personality of the medium and production quality.

Another question is what is *improved* by an anthropomorphic interface? Is there such a thing as a "persona effect"? The effects of adding an anthropomorphic character to an otherwise stable interface have been studied by, among others, van Mulken, André, and Müller (1998) and Wright, Belt, and Lickorish (1998). The first of these studies showed no objective results (such as learning effects, or improved task performance) from introducing the character, but it revealed a positive effect on the subjective estimate of whether the explanation was difficult or not. The second study showed a negative learning effect of introducing an animated character to an explanation.

These conflicting results point to the need for a better understanding of how animated, or synthetic characters should be designed to not divert attention from the content, or the main tasks of users. As pointed out by Andrew Stern (Hayes-Roth, Ball, Lisetti, Picard, & Stern, 1998) (designer of the "Catz and Dogz"[6] system), the artistic design and practical understanding of the creation of synthetic characters is crucial in determining the success of a system.

So far, we have yet to see more studies on whether these interfaces can be designed to aid or appeal to particular groups of users. Possibly, they will be directed to selected groups of users, much in the same way that television shows and movies are aimed at particular age and cultural groups.

Requirements on a New Methodology for Design

One reason for the vehement critique of interface agents is that it is rooted in preconceived notions of what the human activities are, which computer interfaces are intended to support. The design of direct-manipulation interfaces was motivated by the usage of computers as fast and efficient tools supporting users in task-oriented behavior. But today, computer interfaces have a much wider range of applications; they are used to support users learning about novel domains, controlling real-time processes, communicating with other people, and simply for pure entertainment. This has already led to novel design cultures, which, in many cases, violate the basic principles of interface design. One example is the stability of Web pages. A very basic principle of traditional interface design is that the interface should be kept stable, so users can learn it. However, in the Web, we expect Web sites to change both their design and their internal structure quite often. Indeed, we will often suspect that a site that has not changed its design for a considerable period of time will contain old and outdated information.

Many of our example agents have been developed with objectives that are vastly different from just providing a user with the correct functionality, as effi-

---

[6] For more information, see: http://www.petz.com

ciently as possible. Instead, it is the experience of the tool, whether it is a delightful, motivating, or arousing experience that matters. The move to interface agent interaction is thus largely motivated by a change in the *usage* of interfaces.

Finally, we would like to point out that the eventual success of interface agents is largely a matter of culture, if they "catch on." If they do, a culture will develop that will spell out when they are appropriate and how they should be designed, so that users can anticipate what functionalities agents support and how they are controlled. It is very important to develop principles for the design and evaluation of interface agents, but, as with direct manipulation, these methods and principles will be confined to the culture in which they were formulated.

THE IMPACT OF INTERFACE AGENTS ON UNIVERSAL ACCESSIBILITY

It is of paramount importance that information technology is made accessible to all users, and not only to the average, so-called "normal" user. This is the principle of universal access. The population as a whole is an extremely heterogeneous user group, in terms of individual abilities (knowledge, preferences, cognitive and physical abilities, etc.), social and cultural background, and their reasons for turning to a system. In this chapter, we have tried to show that interface agents are not just an extension of the user-modeling vision, that is, to adapt to different users. Instead, interface agents may bring about new interaction styles that provide a new design insight into the issue of accessibility. In particular, the capabilities of interface agents for natural dialogue, voluntary support, delegation of responsibility, and representation of subjective perspective may turn out to be important not just as novel interaction styles, but crucial in attracting attention, and raising user awareness and motivation, thus improving the user's experience with the interactive system.

Interface agents may also introduce a new design philosophy. Traditionally, information-processing psychology has had a minor impact on the study of accessibility of interactive computer-based software. Interface agents, on the other hand, constitute a theme, which is more compliant with developmental perspectives in HCI, based on frameworks such as distributed cognition, activity theory, and language/action perspectives. This is due to the focus on cooperative dialogue engagement, delegation of responsibility, and so on. Thus, it can be expected that the aforementioned emerging HCI design perspectives, when articulated through suitably constructed interface agents, offer a richer set of tools to explain, describe, and facilitate a wider range of interactions (e.g., social interaction). Along these lines, we believe that interface agents have the potential to contribute significantly toward the goal of user interfaces for all.

REFERENCES

Asimov, I. (1950). *I, robot* (1st ed). New York: Gnome Press.

Brennan, S. E., & Ohaeri, J. O. (1994). Effects of message style on users' attributions toward agents. In *Conference Companion of Human Factors in Computing Systems* (pp. 281–282). New York: ACM Press.

Bretan, I. (1995). *Natural language in model world interfaces* (Research Rep. No. 95-017). Stockholm: Stockholm University.

Bruner, J. (1986). *Actual minds, possible worlds.* Cambridge, MA: Harvard University Press.

Dillenbourgh, P., & Self, J. A. (1992). A framework for learner modelling. *Interactive Learning Environments, 2*(2), 111–137.

Dunbar, R. (1997). *Grooming, gossip, and the evolution of language.* Cambridge, MA: Harvard University Press.

Elliott, C. (1992). *The affective reasoner: A process model of emotions in a multi-agent system* (Report No. 32). Institute for the Learning Sciences Tech, Northwestern University, Illinois.

Elliott, C. (1997). Hunting for the Holy Grail with "emotionally intelligent" virtual actors. *ACM Intelligence Magazine* [Online]. Available: http://www.depaul.edu/~elliott/papers/intelligence/

Elliott, C., & Brzeinski, J. (1998). Autonomous agents as synthetic characters. *AI Magazine, 19*(2), 13–30.

Hayes-Roth, B., Ball, G., Lisetti, C., Picard, R., & Stern, A. (1998). *Panel session: Affect and emotion in the user interface. International Conference on Intelligent User Interfaces* (pp. 91–94). New York: ACM Press.

Höök, K. (1996). *A glass box approach to adaptive hypermedia.* Doctoral thesis, SICS Dissertation Series 23, Stockholm.

Höök, K. (2000). Steps to take before IUI becomes real. *Interacting With Computers, 12(4), 409–426.*

Horvitz, E., Breese, J., Heckerman, D., Hovel, D., & Rommelse, K. (1998). The Lumiere Project: Bayesian user modeling for inferring the goals and needs of software users. In *Proceedings of the 14th Conference on Uncertainty in Artificial Intelligence* (pp. 256–265). San Francisco: Morgan Kaufmann.

King, W. J., & Ohya, J. (1995). The representation of agents: A study of phenomena in virtual environments. In *Proceedings of the 4th IEEE International Workshop on Robot and Human Communication* (pp. 199–205). Piscataway, NJ: IEEE Press.

Koda, T. (1997). *Agents with faces: A study on the effects of personification of software agents.* Unpublished master's thesis, MIT Media Laboratory, Cambridge, MA.

Koda, T., & Maes, P. (1996). Agents with faces: The effects of personification of agents. In H. Thimbleby & A. Blandford (Eds.), *People and computers XI: HCI '96 Adjunct Proceedings* (pp. 98–103). London: The British HCI Group.

Lanier, J. (1996). My problems with agents. *Wired, 4*(11). Available: http://www.wired.com/wired/archive/4.11/myprob.html.

Lester, J., Converse, S., Stone, B., Kahler, S., & Barlow, T. (1997). Animated pedagogical agents and problem-solving effectiveness: A large-scale empirical evaluation. In *Proceedings of the 8th World Conference on Artificial Intelligence in Education* (pp. 23–30). Amsterdam: IOS Press.

Lieberman, H. (1995). Letizia: An agent that assists web browsing. In *Proceedings of the International Joint Conference on Artificial Intelligence* (pp. 924–929). San Francisco: Morgan-Kaufmann.

Maes, P. (1994). Agents that reduce work and information overload. *Communications of the ACM, 37*(7), 31–40.

Maes, P. (1997). Direct manipulation vs. interface agents (Position statement). In J. Moore, E. Edmonds, & A. Puerta (Eds.), *Proceedings of the 1997 International Conference on Intelligent User Interfaces* (pp. 41–43). New York: ACM Press.

Meyer, B. (1994). Adaptive performance support: User acceptance of a self-adapting system. In *Proceedings of the 4th International Conference on User Modeling* (pp. 65–70). Bedford, MA: The MITRE Corporation.

Morkes, J., Kernal, H., & Nass, C. (1998). Humor in task-oriented computer-mediated communication in human–computer interaction. In *Proceedings of Human Factors in Computing Systems* (pp. 215–216). New York: ACM Press.

Oppermann, R. (1994). Adaptively supported adaptability. *International Journal of Human–Computer Studies, 40,* 455–472.

Persson, P. (1998). Navigation, narrative and emotion. *Presentation at 13 annual review conference* [Online]. Available: http://www.i3net.org/ser_pub/annualconf/abstracts/avatars/navigation.html

Picard, R. (1997). *Affective computing.* Cambridge, MA: MIT Press.

Pollack, M. E. (1986). *Inferring domain plans in question answering.* Unpublished doctoral dissertation, University of Pennsylvania, Philadelphia.

Reeves, B., & Nass, C. (1996). *The media equation: How people treat computers, television, and new media like real people and places.* Cambridge, England: Cambridge University Press.

Rich, C., & Sidner, C. L. (1997). Segmented interaction history in a collaborative interface agent. In *Proceedings of the International Conference on Intelligent User Interfaces* (pp. 23–20). New York: ACM Press.

Schank, R. (1991). Where's the AI? *AI Magazine, 12*(4), 38–48.

Shneiderman, B. (1997). Direct manipulation for comprehensible, predictable and controllable user interfaces. In *Proceedings of the International Conference on Intelligent User Interfaces* (pp. 33–39). New York: ACM Press.

Sproull, L., Subramani, M., Kiesler, S., Walker, J. H., & Waters, K. (1996). When the interface is a face. *Human–Computer Interaction, 11*(2), 97–124.

Suchman, L. A. (1987). *Plans and situated actions: The problem of human–machine communication.* Cambridge, England: Cambridge University Press.

Suchman, L. A. (1997). From interactions to integrations: A reflection on the future of HCI. In S. Howard, J. Hammond, & G. Lindegaard (Eds.), *Proceedings of Human–Computer Interaction Conference.* Chapman & Hall. On-line abstract. Available: http://www.acs.org.au/president/1997/intrct97/suchman.htm.

van Mulken, S., Andre, E., & Muller, J. (1998). The persona effect: How substantial is it? In *Proceedings of Human–Computer Interaction Conference* (pp. 53–66). Berlin: Springer.

Vilhjálmsson, H., & Cassell, J. (1998). BodyChat: Autonomous communicative behaviors in avatars. In *Proceedings of ACM Autonomous Agents '98* (pp. 269–276). New York: ACM Press.

Waern, A. (1997). Local plan recognition in direct-manipulation interfaces. In *Proceedings of the International Conference on Intelligent User Interfaces* (pp. 7–14). New York: ACM Press.

Waern, A., Tierney, M., Rudström, Å., & Laaksolahti, J. (1998). *ConCall: An information service for researchers based on EdInfo* (SICS Tech. Rep. No. T98-04). Available: http://www.sics.se/~mark/papers/t9804/t98-04.htm.

Wright, P., Belt, S., & Lickorish, A. (1998). Animation, the fun factor and memory. In *Workshop on computers and fun* (organized by the British HCI Group). York, England: The British HCI Group.

16 Accessibility in the Java™ Platform

Peter Korn
Willie Walker

In the past, each time a new computing paradigm became established in industry, it raised barriers for people with disabilities at the very time that it was providing new utility to mainstream users. The transition from character-based systems to the Graphical User Interface environments of Macintosh, X, OS/2, and Microsoft Windows provided tremendous new utility to the majority of existing users, while effectively shutting out access for a portion of the user population, for many years in each case, until the necessary specialized assistive technologies were developed. For the first time in computing environments, the Java platform is establishing a new computing paradigm, at the same time that accessibility support is developed directly into the core of that platform's user interface mechanisms.

BACKGROUND

Shortly after the first full release of the Java™ platform, Sun Microsystems Inc. began investigating the accessibility issues surrounding the Java platform (Trace R&D Center, University of Wisconsin-Madison, 1997). The culmination of this work was a detailed set of requirements for accessibility, which can be summarized into the following three overriding requirements:

1. Build accessibility support into the user interface primitives from the outset.
2. Support existing screen access techniques.
3. Support direct accessibility.

Based on these requirements, Sun undertook two concurrent efforts: (a) development of the Java Accessibility Application Programming Interface (API), and (b) development of the Pluggable Look and Feel architecture. Both of these are part of the Java Foundation Classes (JFC), a set of user interface primitives around which mainstream developers are creating and shipping rich and complex products today.

THE JAVA ACCESSIBILITY API

One of the key challenges in gaining access to graphical user interface (GUI) applications has been deciphering the meaning behind the drawn images that to a human eye are clear representations of buttons, check boxes, menus, and so on, but to an assistive technology product are nothing more than a constellation of pixels. The Java Accessibility API addresses this problem by being a standard, supported contract between assistive technologies and GUIs of mainstream software. This contract defines how mainstream software shall describe and make accessible all of the GUI elements utilized by that mainstream software, so that assistive technologies can present the GUI to users who cannot directly use it in its original form.

Overview of the Java Accessibility API

The Java Accessibility API defines a contract between the individual user interface elements that make up an application or applet written in the Java programming language, and an assistive technology that provides access to that Java technology-enabled application/applet. If a Java technology-based application/applet fully supports the Java Accessibility API, then it should be compatible with, and friendly toward assistive technology products (such as screen readers, screen magnifiers, etc.). The accessibility contract defines a taxonomy of the various user interface elements used in Java technology-based programs, and further defines a set of queryable attributes—as well as the mechanism for making these queries—that exist on these user interface elements.

The Java Accessibility API is supported by individual user interface elements in three steps:

1.   The element declares that it is accessible by implementing the `Accessible` Java programming language interface (essentially it simply tags itself as accessible), which is a promise to return an `AccessibleContext` object when asked for one.
2.   The element provides an `AccessibleContext` object, that describes the core accessibility attributes common to all user interface elements (e.g., name, description, locale, role, states, parent, children)
3.   The element provides additional objects, that implement additional aspects of the Java Accessibility API, as appropriate for that element (e.g., `AccessibleAction` for those elements that have actions, such as buttons, check boxes, etc.; `AccessibleValue` for those elements that take on a value, such as sliders and scroll bars)

The Java Accessibility API in Detail

The Java Accessibility API consists of over a dozen classes and interfaces, which are used to fulfill various aspects of the accessibility contract. Many of these classes encapsulate specific state information into a Java programming language object, which can easily be translated into multiple languages, and extended to encompass new values in the future (e.g., objects to represent an element's *role* and its *states*). The key portions of the API are described in Table 16.1.[1]

TABLE 16.1.
The Java Accessibility API

Object	Use
Accessible	The top-level interface of the accessibility contract, for all user interface elements that are compatible with assistive technologies. This interface is implemented directly on the element, so it is instantly recognizable as adhering to the accessibility contract. This interface returns an instance of the AccessibleContext object (see below), which contains all of the accessibility contract information for this element.
Accessible Context	The core object of the accessibility contract, for all elements that are compatible with assistive technologies. This class provides the minimum information required from all elements—name, description, locale, role, states, parent, children, etc. In addition, it provides a mechanism for getting the additional accessibility information that some elements provide (e.g., AccessibleComponent, AccessibleSelection). This additional information is encapsulated in the objects listed below, which are returned by the AccessibleContext object.
Accessible Component	For elements that are drawn onto the screen, such as virtually all standard user interface elements. This interface provides a standard way to query the visual attributes of the element, such as foreground and background color, bounding rectangle.
Accessible Selection	For elements that contain other elements that can be selected within them, such as list boxes, trees, menus, etc. This interface provides a way to get the set of currently selected child elements, select and deselect individual child elements, and select all, or deselect all child elements.
Accessible Text	For elements that contain rich, editable text, such as simple text entry fields, word-processing documents, etc. This interface provides a way to map pixel locations to the letters, words, and sentences; get the bounding rectangles of any portion of text within the element; and get detailed attribute information about individual characters within the text.
Accessible Hypertext	For elements that contain hypertext, such as World Wide Web documents, online help systems, etc. This interface extends the AccessibleText interface above, and adds the ability to enumerate a set of AccessibleLink objects with the text and the ability to map those AccessibleLink objects to character runs within the text.
Accessible Value	For elements that take on one of several values, such as check boxes, sliders, scroll bars, etc. This interface provides a standard way to get the current, minimum, and maximum values of the element, as well as to set a new value for the element.

[1]For more details, please visit the Sun™ Accessibility Web site at: http://www.sun.com/access

The Java Accessibility API—Example

Figure 16.1 illustrates how the Java Accessibility API works in practice with a particular user interface element from the JFC, namely the `JSlider` user interface element. `JSlider` implements the `Accessible` interface, and so provides the Java programming language method `getAccessibleContext`. Calling this method returns an `AccessibleContext` object that details the core common information for all accessible user interface elements—namely things like name, description, locale, role, states, and so on. The `AccessibleContext` object returned can then be further queried to determine whether or not this object takes on values, is a visually drawn object, and so forth. In the case of `JSlider`, the object does take values and is drawn visually on screen, so `AccessibleValue` and `AccessibleComponent` objects are returned in response to those queries, which detail value and component information respectively.

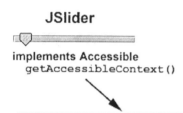

JSlider

implements Accessible
  `getAccessibleContext()`

```
                    Accessible Context
                    implements AccessibleComponent
                    implements AccessibleValue

                    String getAccessibleName()
                    String getAccessibleDescription()
                       . . .

 AccessibleValue                          AccessibleComponent
   int getCurrentAccessibleValue()           Color getBackground()
   int getMaximumAccessibleValue()           Color getForeground()
   int getMinimumAccessibleValue()           . . .
```

FIG. 16.1.   An example of a user interface element implementing the Java Accessibility API.

Helping Build Assistive Technologies: The Java Accessibility Utilities

The Java Accessibility Utilities are a Java programming language package freely available from Sun, for use by assistive technology vendors in their products. These utilities provide the support needed by assistive technologies, in order to use the Java Accessibility API. The subsections that follow detail this support.

Key Information About the Java Technology-Enabled Application(s).
The Java Accessibility Utilities contain methods for retrieving key information about the Java technology-based application(s) running in the Java virtual machine.[2] This support provides a list of the top-level windows of all of the Java technology-based applications, an event-listener[3] architecture to be informed when top-level windows appear (and disappear), and means for locating the window that has the input focus, as well as for locating the mouse position.

Automatic Loading of Assistive Technologies.   In order for an assistive technology to work with a Java technology-based application, it needs to be loaded into the same Java virtual machine as the Java technology-based application it is providing access to. This support is built into the Java 2 platform (i.e., the Java Development Kit—JDK™—version 1.2). Because many implementations of the Java platform are based on the JDK 1.1 Tool, which does not inherently support this feature, the Java Accessibility Utilities provide this support for the Sun implementations of the JDK 1.1 Tool.

Event Support.   The Java Accessibility Utilities provide two Java programming language classes for monitoring events in the Java virtual machine. The first class supports monitoring of all interaction events in all of the Abstract Windowing Toolkit (AWT)[4] components running in the Java virtual machine. This class provides system-wide monitoring of AWT events, and is based on registering an individual listener for each AWT event type on each AWT component that supports that type of listener. These events include windows being activated, check boxes getting checked, and keyboard focus moving from one user interface element to another. Thus, an assistive technology can register a "focus listener" with each and every AWT component in each and every Java technology-based program in the Java virtual machine. Those individual listeners will funnel the respective events to the assistive technology that registered the listener in the first place. Thus, whenever a component gains or loses focus (e.g., when the user hits the TAB key), the assistive technology will be notified.

The second Java programming language class for monitoring events in the Java virtual machine is an extension of the approach discussed previously and is intended to provide additional support for monitoring the Java Foundation Classes software, including events supported by the Project Swing components (see subsections Support for the Java Accessibility Architecture and the section The Pluggable Look and Feel, later in this chapter). These events include move-

---

[2]The Java virtual machine is part of the Java Runtime Environment and is responsible for interpreting (and executing) intermediate code (i.e., bytecodes).

[3]An event listener is an object that has registered its interest in receiving notifications when specific (types of) events take place. The listener usually registers with, and receives notifications from the event source, that is, the object that "generated" the event.

[4]The Abstract Windowing Toolkit (AWT) is a set of user interface components, embedded into the Java programming language.

ment of the text caret, menus appearing/disappearing, and changes to the models underlying the various elements that make up the pluggable look and feel of Project Swing (see The Pluggable Look and Feel section, later in this chapter).

AWT Translators.   With the release of the JFC, many developers, who were using the AWT to build the user interfaces of their Java technology-based applications, are diverting to the Project Swing classes in the JFC. Many are also expected to update their existing AWT programs to Project Swing. Still, a significant number of Java technology-based applications that make use of (some) AWT components for displaying their user interfaces will remain. The Java Accessibility Utilities contain a set of classes that implement the Java Accessibility API on behalf of AWT components—in effect translating for them! These translators work in concert with the support for finding `Accessible` elements in the first place. If an object is not an actual instance of `Accessible`, the respective method looks for a `Translator` that will implement the `Accessible` interface, on behalf of that component.

Like much of the rest of the Java Accessibility support, the translator architecture is completely extensible. Any programmer can create a translator. As long as the user's environment is configured properly, the Java Accessibility utility classes will automatically find the new translator and engage it. This means that both mainstream developers and assistive technology vendors can create and distribute new translators, making formerly inaccessible user interface elements accessible in the process.

The Java Access Bridge to Native Code.   In order for existing assistive technologies that are available on (or *native* to) specific host systems (e.g., Microsoft Windows, Macintosh, OS/2) to provide access to Java technology-based applications, they need some way of communicating with the Java Accessibility support in those Java technology-based applications. The Java Access Bridge supports that communication, through a Java programming language class that contains "native methods" (i.e., methods that interface to platform-specific code). Part of the code for the class is actually supplied by a Dynamically Linked Library (DLL) on the host system—the Solaris™ Operating System, IBM's OS/2™, Microsoft Window™, Mac OS, and so forth. The assistive technology running on the host (e.g., a Macintosh screen reader) communicates with the (Macintosh) native DLL portion of the bridge class, which, in turn, communicates with the Java virtual machine, and from there to the Java Accessibility Utility support and the Java Accessibility API on the individual user interface objects of the Java technology-based application to which it is providing access. Figure 16.2 illustrates how all these components work together.

For example, in order for a screen reader for Microsoft Windows to provide access to Java technology-based applications running on that system, that screen reader would make calls to the Java Access Bridge for Microsoft Windows. If the user launched a Java technology-based application, the bridge would inform the screen reader of this fact. Then the screen reader would query the bridge about the Java technology-based application. The bridge would, in turn, forward those queries on to the Java Accessibility Utilities (that were loaded into the Java virtual machine), and, in many cases, on to the individual user interface object that implemented the Java Accessibility API. When answers to those queries came back to the bridge, the bridge would forward them to the screen reader for Microsoft Windows, which would then use the answers to tell the user what was going on in the Java technology-based application.

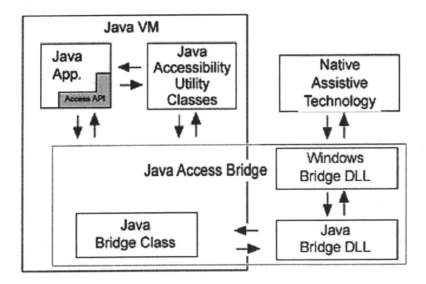

FIG. 16.2.   Architecture of the Java Access Bridge to native assistive technologies.

Development and Use of the Java Accessibility Architecture

Contributions to the Design of the Java Access API.   The Java Accessibility API is part of Sun's JFC and of the JDK version 1.2. Though formally a Sun product, the Java Accessibility API was designed in partnership with experts in the assistive technology community, including members of the IBM Special Needs Services group,[5] the Trace R&D Center,[6] and the Assistive Technology Resource Center of the University of Toronto.[7] These three groups, in particular, developed code alongside the development of the Java Accessibility API to test it and ensure it

---

[5]For more information, see http://www.austin.ibm.com/sns/

[6]For more information, see http://www.trace.wisc.edu/

[7]For more information, see http://www.utoronto.ca/atrc/

meets the needs of the disability community. IBM also contributed source code to implement portions of the Java Accessibility API and developed a set of programmer guidelines to assist mainstream developers in supporting the Java Accessibility API (IBM Corporation, 1999).

In addition to the core group of assistive technology experts, numerous assistive technology vendors provided feedback to the design and direction of the Java Accessibility API. Additional input came from members of the American Council of the Blind,[8] the American Foundation for the Blind,[9] the National Council on Disability,[10] the National Federation of the Blind,[11] the attendees of the Sun Microsystems presentations at the *Closing the Gap*[12] and *Technology and Persons with Disabilities*[13] conferences, and the members of the *java-access* USENET mailing list maintained at *java.sun.com*.[14] Some early input also came from a representative of Microsoft's Accessibility team,[15] in private meetings and also as part of an accessibility review of the Java platform, which Sun commissioned in 1996 just after the platform was released (Trace R&D Center, University of Wisconsin-Madison, 1997).

Support for the Java Accessibility Architecture.   The Java Accessibility API ships as part of the JFC, and is implemented by the Project Swing user interface component set, which is also part of the JFC. The Project Swing component set is an extremely rich and flexible collection of user interface components, which are completely platform-independent. These components include not only common GUI elements such as buttons, check boxes, radio buttons, menus, static and editable text fields, lists, and dialogue boxes, but also more complex elements, such as tabbed-panes, two-dimensional tables, tree views, styled text fields, and even an extensible HTML viewer. Furthermore, because all of these elements are Java programming language objects, they can be embedded within each other (e.g., a menu that contains a table, a tree whose nodes are check boxes, etc.).[16] The reception of Project Swing by mainstream developers has been very positive and a considerable number of applications that use it have already been deployed. Exemplary application areas include e-mail clients, climate modeling packages, text editors, application installation packages, and software development tools.

In addition to mainstream developers, those in the assistive technology community are also building products that use the Java Accessibility API. First among

---

[8]For more information, see http://www.acb.org/

[9]For more information, see http://www.afb.org/

[10]For more information, see http://www.ncd.gov/

[11]For more information, see http://www.nfb.org/

[12]For more information, see http://www.closingthegap.com/conf/

[13]For more information, see http://www.csun.edu/cod/

[14]The *java-access* USENET mailing list: java-access@java.sun.com

[15]For more information, see http://www.microsoft.com/enable/

[16]For more detailed and up-to-date information about the Project Swing component set, please refer to the "Swing connection" Web site at: http://java.sun.com/products/jfc/tsc/

these is IBM. The IBM Special Needs Services group is building a family of assistive technologies for the Java platform, including a screen reader for the blind. Their screen reader takes advantage of all of the features of the Java Accessibility API, and because of this, is able to provide far more fidelity than screen readers for OS/2, Macintosh, or Windows.

THE PLUGGABLE LOOK AND FEEL

One of the primary goals of the software accessibility community has been to allow mainstream applications to be directly accessible by people with disabilities, that is, a user should not need to employ a separate assistive technology to use an application. Furthermore, a mainstream application should permit direct accessibility without requiring a separate accessible version.

Introduction

The Project Swing component set, part of the JFC, is a new GUI toolkit written for the Java platform, which simplifies and streamlines the development of windowing components. With Project Swing, one can develop lean and efficient GUI components that have precisely the "look and feel" that you specify. For example, a program that uses Project Swing components can be designed in such a way that it will execute without modification on any system that supports the Java platform, and will always look and "feel" (i.e., behave) just like a program written specifically for the particular computer on which it is running.

When a Project Swing program is run under Windows, it has the appearance and behavior of a program written specifically for Windows. When that same program is run on a UNIX® workstation, it runs just like any program written for a UNIX platform. When run on an Apple Macintosh, it looks and behaves just like any program written specifically for the Mac, and so on.

All of this can be achieved thanks to Project Swing's Pluggable Look and Feel. The remainder of this chapter discusses the Pluggable Look and Feel and how it can be used to make applications more accessible to people with disabilities.

The Model-View-Controller Design Pattern

Many GUI toolkits intermix a component's data model, its presentation, and its reaction to input events. With these types of toolkits, component model and state information is often hidden from external view (e.g., from assistive technologies and testing harnesses), and it can be difficult to extend the toolkit later. Instead of following this design, the Project Swing architecture is rooted in the model-view-controller (MVC) software design pattern (Barkakati, 1991). The MVC pattern calls for an application to be broken up into three separate parts (see Fig. 16.3):

- A *model* that represents the data for the application.

- The *view* that is the representation of the model.
- A *controller* that takes user input on the view and translates it to changes in the model; changes in the *model* are also automatically propagated to the *view* by the *controller.*

For example, in the case of a simplified scrollbar, the model is the portion that holds the minimum, maximum, and current values, the view is the portion that draws the scrollbar on the display, and the controller is the portion that listens for input events. Although they are separate, the three parts work together. When a component is to be painted, the view refers to the values contained in the model and draws the scrollbar on the display. When an input event occurs (such as a click on the decrement button), the controller updates values in the model. Finally, when values in the model are changed, the model notifies the view and the view repaints itself.

Project Swing does not follow this software design pattern completely, however, and instead couples the view and controller together (see Fig. 16.4). For example, in the case of a scrollbar, the view draws the increment and decrement buttons as well as the scrollbar thumb. Because it handles input events, the controller needs to respond to mouse events occurring over particular locations in the view. For example, the controller may do one thing when the mouse is clicked over the decrement arrow, and another thing when the mouse button is clicked over the increment arrow. With the view and controller as separate objects, the controller would require significant knowledge of the view, which is something that the MVC architecture tries to avoid. By combining the controller and the view, the Project Swing architecture is simplified, and it does so without hiding state and model information.

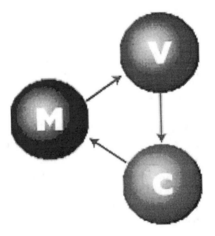

FIG. 16.3. The model-view-controller software design pattern.

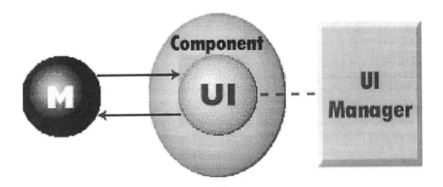

FIG. 16.4. Architecture of user interface components in Project Swing.

The UI (user interface) component, in the Project Swing variation of the MVC pattern, handles the duties of both the view and controller in the original pattern. To handle the view duties, the UI queries the model to determine the values needed to present the component. For the controller duties, the UI listens for input events and updates the model accordingly. This portion of the UI is exactly the same as it is in the MVC design pattern. Because the view and controller parts are combined in the UI, however, the controller can easily determine where input events occur in the view. Using the scrollbar example, it is much easier to determine where the user clicked the mouse in the view, because the view and controller are in the same UI object.

The use of this architecture renders Project Swing a highly flexible and powerful development tool. For example, because the model is a separate object, different components can share the same model. For instance, a slider and a scrollbar could share the same data model: When a change occurred in the slider, the scrollbar would update itself automatically.

In addition, because the UI is the only object responsible for the presentation of the component, it can easily be swapped out and replaced with a different UI. This is one of the key features of Project Swing that permits the concept of the Pluggable Look and Feel, which is discussed in more detail in the next section.

The Pluggable Look and Feel

As mentioned earlier, the Project Swing toolkit allows an application developer to write an application once, yet permit that application to take on a different appearance at run time. For example, when running on an X Windows System display, the application can appear as if it were written using Motif. When running on a Windows system, however, the application can appear as though it were written for

Windows. This can be done without modifying the application, and it is possible because the Project Swing toolkit supports the concept of a look and feel.

A look and feel defines a collection of UI classes (i.e., components that, when combined with the respective models, make up the individual user interface elements in Project Swing) to be used by different interface objects. For example, a look and feel would provide a UI class for Project Swing buttons, a different one for Project Swing scrollbars, and so on. Furthermore, the UI classes provided by a specific look and feel have similar appearance and behavior attributes. For example, the Motif look and feel that comes with Project Swing defines a set of UI classes that look and behave like Motif widgets, whereas the Windows look and feel defines a set of UI classes that look and behave like Win32 components. Other look and feels might be designed specifically to address issues of corporate identity, or for specific cultural backgrounds.

Figure 16.5 presents instances of an application running with different look and feels. In each of the images, the same application is running. The only thing that has changed is that the application is using a different look and feel class to obtain the UI classes for the components.

Just as the UI is a separate and replaceable part of a component, the look and feel is a separate and replaceable part of a Project Swing application. That is, an application is not bound to a particular look and feel; it can change, or *plug in* differ-

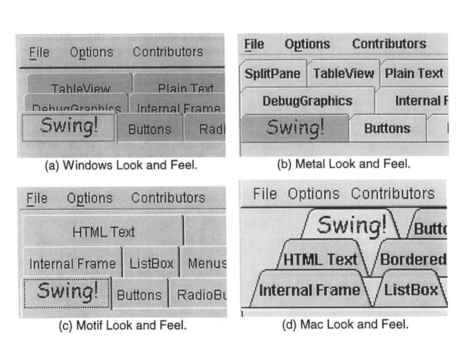

FIG. 16.5.   Examples of an application running with different look and feels.

ent look and feels at run time. The entire concept of replaceable UIs and changing look and feels in Project Swing is known as the "Pluggable Look and Feel of Project Swing."

In Project Swing, most of this work happens "behind the scenes." The application developer works with Project Swing toolkit components such as buttons and scrollbars on a higher level of abstraction, using actions such as "let me know when the user activates the button," instead of lower level ones such as "did the user right click on the button?". It is up to the component instances to obtain and work with the UI. As a result, the application developer is presented with a consistent API and is not concerned with the UI instance being used by the component. In general, an application should not be concerned with the particular look and feel in use at any particular time: the application should merely work directly with the component and let the component work with the UI.

Accessible Look and Feels

Prior to the Pluggable Look and Feel of Project Swing, users with disabilities had two options when it came to accessing applications (at least in terms of commercially available products): (a) specialized software, closely tailored to their "nonmainstream" modality (e.g., a talking Web browser), or (b) a specialized access application (e.g., a screen reader), which "interpreted" the interface of mainstream applications and provided a representation in alternative modalities (e.g., speech). With the Pluggable Look and Feel of Project Swing, there now exists a third option: presentation of the interface and information of mainstream applications directly in alternative modalities.

In order to get direct access to a Project Swing program via a nonmainstream modality (e.g., audio, Braille, etc.), the user would need a look and feel containing a set of accessible UI classes installed on their system. In setting this up, the user can either replace the default look and feel with a more accessible one, or use an accessible look and feel simultaneously with the default one.

In cases where the accessible look and feel actually modifies what is being drawn on the screen, the user would most likely want to replace the default (visual) look and feel. For example, a user with low vision may want to use a look and feel where the components draw themselves very large. Figure 16.6 presents two instances of a user interface that demonstrate this. The image on the left presents an application using a "regular" look and feel, and the image on the right presents the same application using a look and feel designed for users with low vision.

Users can force the run-time environment to always use a particular look and feel, which better serves their requirements and preferences, by setting appropriately respective properties in an environment initialization file. This, however, may not always work. The reason for this is that an application developer may force Project Swing to use a particular look and feel, regardless of what the default look and feel is set to. For example, an application may have been tested only against one particular look and feel, and the application developer does not

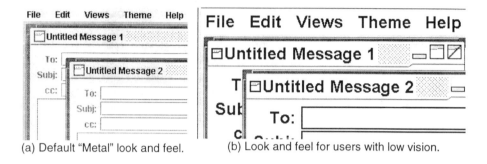

(a) Default "Metal" look and feel.   (b) Look and feel for users with low vision.

FIG. 16.6.   Example of accessible look and feel.

wish to permit users to run the application under conditions in which it may not have been tested.

In the cases where the user wants to use a look and feel that is primarily nonvisual (e.g., audio or Braille), the user would most likely want to use it in conjunction with the default look and feel. To do so, one can take advantage of the multiplexing look and feel that ships with Project Swing. The multiplexing look and feel is discussed in detail in the following section.

The Multiplexing Look and Feel

To make it easier to extend the capabilities of a user interface created in Project Swing, without needing to create a new look and feel, Project Swing supports the concept of a *multiplexing look and feel*. When a component asks for its UI instance, the multiplexing look and feel will transparently create and simultaneously support UI instances from several different look and feels (see Fig. 16.7). This enables the concept of more specialized look and feels, known as *auxiliary look and feels,* that can be combined to create an environment tailored to the end-user. For example, a user can take advantage of Project Swing's multiplexing look and feel to automatically load and maintain the default visual look and feel and several auxiliary look and feels such as one for audio and another for Braille. This will all be done at the same time and without requiring modifications to the visual look and feel.

Without the multiplexing look and feel, developers wishing to enhance a look and feel would need to create a subclass of that look and feel. For example, to add audio support to the Metal look and feel, without using the multiplexing look and feel, a developer would need to create a subclass of the Metal look and feel and add audio support to that subclass. Furthermore, to add this audio support to other look and feels, such as the Motif and Windows look and feels, developers would need to create subclasses of them as well. There are at least two problems with this approach. The first is that each subclass will essentially use a copy of the same code, creating a difficult support situation for the developer. The second is more severe for the end-user: some application developers may force a particular look and feel to be used. In these cases, the end-user will not even be able to use the enhanced look and feel subclass. The multiplexing look and feel solves these problems by allowing multiple look and feels to be combined. The first problem is solved because the developer can create a specialized look and feel that can be combined with other look and feels. The second problem is solved because the specialized look and feel can be used in conjunction with whatever default look and feel the application may have locked in place.

As is the case with alternative default look and feels, users can instruct the Project Swing run-time environment to automatically obtain a component's UI from the multiplexing look and feel, instead of obtaining it directly from the default look and feel. The resulting multiplexing UI is a small delegate that obtains and maintains UIs from the default and auxiliary look and feels. As a result, when a method is called in a multiplexing UI instance, the multiplexing UI will call the same method in each of the UIs obtained from the default and auxiliary look and feels.

Developing an Auxiliary Look and Feel.   An auxiliary look and feel is no different from any other look and feel subclass except that it does not have to provide all the support of a look and feel that would be used as the default one. For example, an auxiliary look and feel that supports just audio feedback does not need to provide any code for painting. As a result, developing an auxiliary look and feel can be easier than developing a visual look and feel and also permits the developer to concentrate solely on providing just specialized functionality. Note that the primary purpose of an auxiliary look and feel is to enhance the default look and feel. Therefore, auxiliary look and feels tend to be nonvisual. Because an auxiliary look and feel is a look and feel subclass, however, there is nothing to prevent the auxiliary look and feel from rendering information on the display. The following sections briefly provide recommendations for developing auxiliary look and feels.

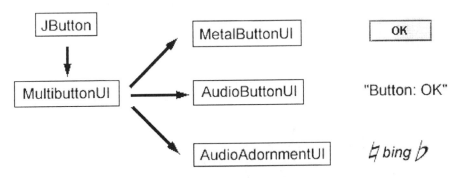

FIG. 16.7.   Example of the employment of the multiplexing look and feel for a button control.

1.   Use the `installUI` and `uninstallUI` Methods: In most cases, auxiliary look and feels will primarily be interested in the `installUI` and `uninstallUI` methods of the UI. These are the methods that are called when a component's look and feel is set, and give the UI a chance to register and remove listeners on the component and its data model.

2.   Do not extend visual look and feels: UI classes of an auxiliary look and feel should not be implemented as subclasses of the UI classes of a visual look and feel. The reason is that they may accidentally inherit code that installs listeners on a component instance, or renders the component on the display. This would result in the auxiliary look and feel competing with the default one, rather than cooperating with it. Instead, it is recommended that the UI classes of an auxiliary look and feel directly extend the abstract UI classes in the `swing.plaf` package. By doing so, the developer of an auxiliary look and feel can avoid the situation of competing with the default look and feel. In addition, it is also recommended each UI class of an auxiliary look and feel overrides the methods of the `swing.plaf` UI class it extends. The reasons for this are similar to the reasons for not extending a visual look and feel. For example, the `ComponentUI` class provides a default implementation for the update method. This default implementation draws on the display if the component is opaque. If a UI class from a nonvisual auxiliary look and feel did not override this method, it would cause all opaque components to appear as blank areas on the screen!

3.   Examining UI instances from other look and feels: In some rare instances, a UI instance from an auxiliary look and feel may be interested in the default UI instance used by the component. In these cases, the UI instance from the auxiliary look and feel can obtain the UI from a component by calling its `getUI` method. If the resulting UI is an instance of one of the multiplexing look and feel UI classes from the default multiplexing look and feel (e.g., `MultiButtonUI`), the UI instance from the auxiliary look and feel can call the `getUIs` method of this object to obtain an array containing a complete list of all UIs handled by the multiplexing factory. The first element is guaranteed to be the UI created from the default look and feel.

The Default Multiplexing Look and Feel.   The multiplexing look and feel itself is meant to be transparent to all developers and users. It should "just work" and is used only when the user tells Project Swing to use an auxiliary look and feel. There may be cases, however, where the implementation of the default multiplexing look and feel does not meet a developer's needs. This section describes the implementation of the default multiplexing look and feel.

The default multiplexing look and feel is shipped with Project Swing. When it is in use, each component requesting a UI instance will actually get an instance of a multiplexing UI, instead of an instance of a UI from the default look and feel. The default multiplexing UI obtains and maintains UIs from the default and auxiliary look and feels, and refers to these UIs in the following manner:

- The UI instance from the default look and feel is always the first to be created. After that, a UI instance will be created from each auxiliary look and feel in the order they were specified by the user.
- When a method that requests information from a UI instance is invoked, the multiplexing UI instance returns only the results from the UI that are obtained from the default factory.[17] For example, when the getPreferredSize method is invoked on a multiplexing UI, the UI returns only the results of invoking the getPreferredSize method on the UI obtained from the default factory. The rest of the UIs obtained from the auxiliary factories are ignored.
- When a method that does not request information from the UI instance is invoked, the multiplexing UI instance will invoke that method on the UI obtained from the default factory and all the UIs obtained from the auxiliary factories as well. For example, invoking the installUI method on a multiplexing UI causes the multiplexing UI to invoke the installUI method on the UI obtained from the default factory and the UIs obtained from the auxiliary factories.

In all cases, the UI instance obtained from the default look and feel is always acted upon first, and then the auxiliary look and feels are acted upon in the order that they have been specified by the user.

Although the behavior of the default multiplexing look and feel should be flexible enough to cater for most cases of combining look and feels (thus largely alleviating the need to develop an alternative multiplexing look and feel), Project Swing allows the user to specify the multiplexing look and feel to use (by editing the respective entry in a properties file). Nevertheless, users doing this should be careful, because the suppliers of auxiliary look and feels will most likely have developed and tested against Project Swing's default multiplexing look and feel.

CONCLUSIONS

The Java Accessibility API and the Pluggable Look and Feel of Project Swing address key requirements for accessibility support identified by experts in the accessibility field and organizations representing various disability constituencies: accessibility support designed in from the outset, support for existing screen access techniques, and support for direct accessibility. Companies and organizations are already building access solutions based on these technologies, including a screen reader from IBM based on the Java Accessibility API, and an audio look and feel from the Assistive Technology Resource Center of the University of Toronto.

---

[17]A "factory" is a software design pattern employed in Project Swing; factories are intended to centralize the production of objects and ensure transparency in the specific type of object created and the creation process itself. Factories in Project Swing are used to provide abstraction in the support for multiple/alternative look and feels.

The Java Accessibility API provides a path for users of traditional screen access solutions to move to the Java platform. With the Java Access Bridge, these users can get access to Java platform objects using existing screen access techniques, without even having to change computers or operating systems. Moving to the Java platform need not entail a new and arduous learning curve, like the one many users went through moving from character-based systems to today's GUIs.

The Pluggable Look and Feel of Project Swing is the first mainstream architecture to reach for the dream of directly accessible mainstream applications. Numerous mainstream applications are currently being written using Project Swing, and are specifically taking advantage of the Pluggable Look and Feel architecture to support the Motif, Windows, Mac, and the Java platform's native look and feels.[18]

This awareness of look and feel independence on the part of mainstream applications, coupled with current research into nonmainstream look and feels, is the most promising attempt to date toward the dream of directly accessible mainstream applications. We are on the verge of true user choice in how mainstream, off-the-shelf computer technology will behave.

By providing the Java Accessibility API and the pluggable look and feel of Project Swing, Sun Microsystems contributed toward the introduction of basic technologies for *user interfaces for all* (Stephanidis, 1995) in the mainstream software industry.

ACKNOWLEDGMENTS

REFERENCES

Barkakati, N. (1991). Model-view-controller (MVC) architecture of Smalltalk-80. In *Object-Oriented Programming in C++* (pp. 74–85). Carmel, IN: SAMS Publishing.

IBM Corporation. (1999). *IBM guidelines for writing accessible applications using 100% pure Java*™ (Document Version 1.1) [Online]. Available: http://www.austin.ibm.com/sns/snsjavag.html

Stephanidis, C. (1995). Towards user interfaces for all: Some critical issues. In Y. Anzai, K. Ogawa & H. Mori (Eds.), *Symbiosis of human and artifact - future computing and design for human-computer interaction*, Proceedings of the 6th International Conference on Human-Computer Interaction (vol. 1, pp. 137–142). Amsterdam: Elsevier, Elsevier Science.

Trace R&D Center, University of Wisconsin-Madison. (1997). *Java accessibility preliminary examination* (Document Version 2.0) [Online]. Available: http://www.trace.wisc.edu/docs/java_access_rpt/report.htm

---

[18]For more details on this, as well as other finer details of look and feels in Project Swing, please refer to the Project Swing documentation at: http://java.sun.com/products/jfc

V

Evaluation

17 User Interface Adaptation: Evaluation Perspectives

Demosthenes Akoumianakis
Dimitrios Grammenos
Constantine Stephanidis

This chapter seeks to provide an insight into the evaluation of adaptable and adaptive systems, and by this account, contribute towards an understanding of how evaluation can be planned and conducted so as to inform the design of User Interfaces for All. *To this end, the chapter outlines the premises of user interface adaptations and determines some of the conditions and criteria for selecting suitable evaluation instruments. It is shown that out of the broad range of available evaluation techniques, only a few can be applied in the context of adaptable and adaptive user interfaces. The basic qualification criterion, which classifies a technique as being appropriate, is the degree to which it offers the constructs to assess situational and context-specific aspects of user interface adaptations, thus determining its usefulness in a social context. Our claim is that evaluation of adaptable and adaptive behavior can only be informed by an analytical frame of reference, which offers a richer unit of analysis than that of the unaided and isolated users, with no access to other people or to artifacts and tools for accomplishing the task at hand.*

The advent of graphical user interfaces (GUIs), in the 1980s, contributed to the proliferation of a wide range of advanced interaction platforms enabling multimodal and multimedia user interactions across different contexts of use. More recent achievements, such as wearable equipment, three-dimensional visualization, and virtual and augmented realities have facilitated new virtualities (e.g., digital libraries, digital money, virtual market place), which are progressively forming the core of residential demand for advanced interactive products and services. As a result, the tasks humans have to perform with interactive computer systems have substantially changed in structure and content; they have become more complex, knowledge demanding, and interaction- and communication-intensive.

The distinctive characteristic of such a progress is the growth in the target customer base, which now includes virtually everyone as opposed to just the business worker or the researcher. Indeed, people with radically different requirements, abilities, skills, and preferences will increasingly demand customized, high-quality access to the emerging information space. In this context, user interface adaptation becomes a critical quality attribute of interactive software intended to meet the requirements of potentially different target user groups. It is precisely this aspect of adaptations that make them critical for the study of *user interfaces for all.*

Despite the broad challenges that adaptation poses to human–computer interaction (HCI) design and other related research communities (e.g., user modeling, domain modeling, user interface tool developers), it also challenges the user interface evaluation community. This chapter sets out to (a) investigate briefly how evaluation of adaptable and adaptive systems has been approached to date, (b) identify some of the shortcomings of prevailing practices, and (c) examine how such evaluations could be informed by recent developments in HCI so as to be planned and carried out more effectively. The chapter is structured as follows. The next section reviews the current state of the art in evaluation approaches and tools, in order to identify inherent assumptions, which may determine the conduct and result of evaluation of adaptable and adaptive systems. This helps establish several conditions and criteria for selecting methods and instruments when evaluating adaptable and adaptive features of an interactive system. In subsequent sections, we assess the suitability of existing evaluation approaches and tools based on the identified criteria and provide recommendations for work that is still needed. The chapter ends with a summary and conclusions.

EVALUATION PERSPECTIVES: AN OVERVIEW

In the short history of HCI, evaluation has always been a part of the development life cycle of interactive systems, though the stage at which it is performed, as well as its scope, goals, and methods vary significantly across different development approaches, reflecting the particular characteristics and requirements of each approach. R. E. Eberts and C. G. Eberts (1989) reviewed over 100 papers in the area of HCI, and clustered them into four main approaches, taking into account five stages of the user interface design process, namely task analysis, design, preevaluation, prototyping, and postevaluation. The approaches they identified are the *empirical approach,* the *predictive modeling approach,* the *anthropomorphic approach,* and the *cognitive approach.*

The empirical approach bases the choice of the best design on the ease of use. This choice takes place in early design phases and replaces intuition by empirical findings. Empirical designs, nevertheless, are not easy to plan and conduct and also unintentionally support the tendency of overgeneralization of results. Methods within the predictive modeling approach try to predict the performance of humans interacting with the computers, for instance, in terms of time or errors. For example, GOMS (Bonnie & E. D. Kieras, 1996a, 1996b; Card, Moran, & Newell, 1983) can be used to decide which design will be better than others, whereas state transition networks help to ensure that a design is complete and does not lead to nonrecoverable error states. Within the anthropometric approach the model of human–human communication is used for the design of human-computer interaction. Methods in this approach typically strive for abstract system attributes such as, for example, "user-friendly" or "natural" interaction. Finally, the cognitive approach applies theories in cognitive science and cognitive psychology to the human–computer interface to make the information processing easier and more efficient. This approach sees the user as a flexible, adaptive information processor who is actively trying to solve problems when interacting with a computer. It has also been used to suggest which designs may be appropriate and easy to use for users instead of merely testing the design after it has been finalized.

Evaluation mirrors these design efforts by assessing the quality of the design considerations, prototypes, or resulting systems along the aforementioned dimensions. The term *user interface evaluation* implies the set of activities, methods, and tools that are dedicated to assessing these various dimensions. In the short history of HCI, user interface evaluation has progressively scaled up, in terms of scientific goals and instruments, to develop and/or accommodate techniques and tools that best suit the changing paradigms of HCI.

Several scientific disciplines participated in the efforts to evaluate user interfaces, for example, human factors and cognitive psychology. The primary concerns of early evaluations were to fit the machine to the skills of the users or to provide predictive instruments for evaluating the cognitive coupling between the human and the user interface (see Table 17.1). Therefore, the user was at the center of any assessment intended to analyze available skills or to investigate prevailing cognitive patterns or strategies. Unfortunately, these results could not be used immediately by system designers, whose focus and background was on computer systems, but not on users. Provided that the system designers were able to transfer the results to the field of design without loss or change of the inherent wisdom, they were faced with a design space instead of specific, unambiguous design recommendations. Then, the final decision for or against a specific design within this resulting design space most often lacked a systematic approach and was highly dependent on the system designer's background.

Furthermore, theories were developed to understand and predict the cognitive behavior and the performance of users, again with the humans at the center of investigations. Early efforts in this direction focused on a wide spectrum of user tasks (e.g., text processing, graphics manipulation, telephone operation) and resulted in models (or theories) of users. Examples include the keystroke-level model (Card, Moran, & Newell, 1980), GOMS (Bonnie et al., 1996a, 1996b; Card et al., 1983), cognitive complexity theory (D. E. Kieras & Polson, 1985), and so on. However, these models exhibit several shortcomings that constituted the main subject of criticism in subsequent years. For instance, as discussed in chapter 1 of this volume, the aforementioned models do not account for context-oriented parameters, such as cognitive development, learning, and human error; instead, they describe how systems are to be used rather than how they are actually used, and demand highly skilled articulation and specialized knowledge on the part of the members of the design team.

The fact that the evaluation methods associated to each design cluster were very expensive or took place late in the system's development life cycle, raised the need for pragmatic approaches earlier in the design process. *User-centered design*

TABLE 17.1.

Shift of Evaluation Foci

Evaluation Approach	Underlying Knowledge Base	Questions Being Addressed by the Evaluation
Human factors evaluation methods	Human abilities and skills	Does the system fit to the skills of the user?
Cognitive science evaluation methods	Human cognitive processing	Does the system provide an effective cognitive coupling between user and system?
Evaluation methods following the user-centered design process	*System* properties and functionality	Is the system *usable* by users?

emerged from the realization of these shortcomings as an effort toward techniques and tools that focus on the requirements of end-users and provide early evaluation feedback. The prime objective of this approach is to reduce the cost of design defects and to meet specific usability objectives. User-centered design itself is not a new design cluster or an evaluation method, but it innovates the design and evaluation process by putting the system in the focus of investigations and by allowing user participation in early design phases. Evaluation methods within a user-centered design are taken from the variety of methods available within the identified four clusters with an emphasis on empirical studies, and are not necessarily restricted to any single cluster, thus avoiding any potential associated shortcomings. The term coined for this approach, namely *user-centered design,* appeared in 1986, as the main title of an edited collection of papers on the topic (Norman & Draper, 1986). The normative perspective of user-centered design is to fulfill the need for "usability now," by providing techniques that foster tight design-evaluation feedback loops, iterative prototyping, early design input, end-user feedback, and so forth. Figure 17.1 illustrates the iterative nature of user-centered design and some of the general categories of techniques that may be used to facilitate the various phases

In subsequent years, due to the compelling need to cost-justify usability throughout a product's life cycle, evaluation moved toward a variety of techniques, generally referred to as inspection-based evaluation (Nielsen, 1993), which, though inexpensive, are less formal in their conduct and deliverables. Following several success stories in the use and cost justification of these techniques and in particular inspections, the consolidated experience gave rise to a generally applicable process model for constructing human-centered systems. Human-centered design is documented in a draft document of the International Standards Or-

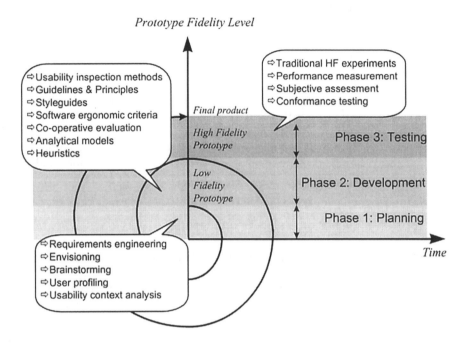

FIG. 17.1.   User-centered design—techniques, methods, and tools.

ganisation (ISO 13407, 1997), which provides a principled approach and guidelines for attaining usability.

CONVENTIONAL APPROACHES TO EVALUATION

Despite the growth of research efforts related to user interface adaptation, the evaluation of systems exhibiting adaptable and adaptive behavior remains a challenge. This is due to several reasons. First of all, adaptable and adaptive software constitutes a relatively new development in HCI and, thus, the available knowledge regarding the design and development of such systems has not yet been consolidated to present the evaluation community with a uniform collection of challenges. A second, and perhaps more important reason, is that, traditionally, adaptive behavior in HCI was largely attributed to the human partner, as computer systems did not embody the required sophistication to identify the need for, and subsequently implement, adaptive behavior during interactive "sessions." As a result, the human operator was required to compensate for the shortcomings of computer system behavior and to recruit the capabilities necessary to accomplish a set of designated tasks.

However, recent technological advances, in both hardware and software, have resulted in radical changes regarding the type, range, and scope of tasks penetrated by computers, and have enabled more intelligible interactive behavior on the part of the machine. This has improved the capability of interactive systems to exhibit adaptable and adaptive behavior across a wider range of computer-mediated human activities. The implication on HCI has been the growth of the research agenda to include an explicit effort to develop new insights toward the development of such systems. In this context, there have been efforts, though few and disjointed, targeted explicitly to the development of methodological guides and instruments, which would help designers of adaptable and adaptive interfaces evaluate their designs. Today, however, and despite the increasing focus on evaluation of adaptable and adaptive software, the results are hardly generalizable across design cases and application domains.

Comparative Evaluation and Its Shortcomings

Traditionally, the emphasis of evaluation in interactive systems has been on performance and, thus, many of the available instruments are performance oriented. This common thread was quickly adopted by early research efforts on adaptive systems, which aimed to illustrate the value of adaptive behavior by demonstrating performance improvements (Totterdell & E. Boyle, 1990). The traditional approach that characterized evaluation in these, as well as more recent efforts (see e.g., C. Boyle & Encarnacion, 1994; Brusilovsky & Pesin, 1995; Debevc, Meyer, Donlagic, & Svecko, 1996; Höök, 1998; Kaplan, Fenwick, & Chen, 1993), is the comparative assessment of a system with and without adaptations, or between the adaptive system and one (or more) nonadaptive counterparts. In most cases, such comparisons have been employed as a means of "isolating" the effects of introducing adaptation in an otherwise static user interface. This type of comparative studies is typically expensive, carried out late in the development cycle, and covers various performance-oriented targets, such as effectiveness, efficiency, error recovery, task completion times, and so on.

However, this approach poses some challenges that complicate both the conduct of, and the derivation of reliable results from, the evaluation. Some of these challenges are briefly discussed next with the intention to reveal the complexity that underlies such efforts, rather than to question the applicability of the approach.

First, for novel systems with adaptable and adaptive capabilities, it may not be possible to find an equivalent nonadaptive counterpart, thus rendering comparative evaluation difficult. One approach that may be used to remedy this shortcoming is to provide the adaptable and adaptive behavior as optional functionality. This means that the same system can be used both with and without adaptation capabilities. Such an approach raises several architectural requirements on the part of the adaptive system, which are not met by the vast majority of current implementations (Höök, 1998). Additionally, it raises a general concern, namely that the nonadaptive system may not have been designed "optimally" for the task. At least, this should be the case, if adaptivity is an inherent and natural part of the system—when taken out, the system is simply not complete (Höök, 1998).

Second, comparative evaluation requires as many nonadaptive systems (or nonadaptive instances of a system) as the number of different dimensions in which adaptations take place in the adaptive system (Totterdell & E. Boyle, 1990). To illustrate the point, let us consider a relatively simple example, whereby an adaptive system performs adaptations along two dimensions, namely: (a) *physical interaction* (e.g., it orders the options in a menu according to the frequency of their use) and (b) *syntactic adaptations* (e.g., it adapts the command order from function-object syntax to object-function-syntax and vice versa, depending on user-related information). Clearly, the adaptive system may undertake these adaptations in any order and combination, which increases the number of tests that are needed for a meaningful comparison of the adapted and nonadapted interface instances.

Third, comparative evaluation is primarily a performance-oriented method and, thus, it reveals operational differences by studying task completion times, errors made, error recovery, user actions, and other performance-related measures. It should be noted, however, that performance is not the only (and in some cases not even an important) quality target to be addressed, especially when one considers nonbusiness activities. In the majority of cases reporting evaluation of adaptive systems, other criteria seem to be more relevant. For example, in adaptive hypermedia systems, quality of search and the relevance of the search results to the user's intended objective are considered to be more important than task completion time[1] or other performance-oriented measures. Therefore, a question arises as to how comparative assessment can be used to assess the degree to which such non-performance-related quality targets are met by an adaptive system.

The preceding argument is also supported by the experience of the authors in developing adaptable and adaptive user interfaces targeted to various user categories and different contexts of use (such as the TIDE ACCESS and the ACTS AVANTI projects—see Acknowledgments), where several non-performance- oriented dimensions were taken into account during the design and evaluation phases of the user interface. In the majority of cases, these dimensions are context related and cannot be assessed objectively. Examples of such critical factors in the development of an adaptable and adaptive user interface include, but are not limited to:

- The speed by which the system becomes aware of the user's intended goal and the appropriateness of adaptations it initiates.
- The responsibility for initiating and effecting adaptations, which, in turn, relates to aspects of the application domain such as criticality, complexity, and so forth.
- The level and type of support and guidance that the system offers to the user.

---

[1]In any case, as pointed out by Gilmore (1995), one may be misled if task completion time is always used as the main criterion of success.

It is evident that such factors depend on the context of use and have a strong situational character. One approach to deriving such information is through making assumptions about the user and monitoring interactive sessions, so the system can reason about alternatives and accordingly revise its interactive behavior.

In the case of the AVANTI browser (see chap. 25, this volume), an alternative approach (i.e., not based on performance measures) has been used for the evaluation of the adaptation features of the user interface. The focus of the particular approach has been on the derivation of results that would serve as feedback into the iterative design phase. Specifically, the problems faced in the development of the AVANTI user interface related to the lack of empirical evidence on which to base the design of adaptation, and the inadequacy of existing methods to assess the way and extent to which the adaptation facilities of the interface affect interaction qualities such as accessibility, usability, and acceptability. These problems were partly compensated for through iterative, expert-based assessment cycles in the design of alternative dialogue components for each supported interaction task, and in the definition of the adaptation logic for materializing the required adaptable and adaptive behavior. These assessment activities were based on collective experience and consensus building among the participating experts, and offered adequate evidence and support regarding the suitability of the alternative interaction designs for the different user and usage context characteristics, as well as the appropriateness of the adaptation decisions.

End-user-based evaluation activities were also introduced in AVANTI. The intention was to assess the accessibility and overall usability of the user interface. The evaluation involved both field studies and laboratory experiments, whereas the data-collection methods used included usability questionnaires, observations, and interviews. The evaluation provided encouraging results regarding the overall acceptance of the user interface characteristics, but the focus of the evaluation (on the interface as a whole, rather than on specific adaptive characteristics of the interface) did not enable the acquisition of adequate evidence regarding the appropriateness, usefulness, usability, and acceptability of individual aspects of adaptation.

The experience gained in AVANTI provides additional evidence regarding the difficulties in evaluating interactive systems that exhibit dynamically changing and nondeterministic (from a user's perspective) behavior, using broadly available objective metrics and performance-oriented evaluation techniques. Moreover, the design of alternative interaction dialogue components to cater for the different user and usage context characteristics, and the definition of appropriate adaptation logic, indicated the existence of critical factors, other than performance, that influence the overall quality of the system.

Such critical quality targets become equally (if not more) important to performance, as computers continue to penetrate an increasingly broad range of residential and social human activities. For instance, in educational software products, computer-mediated communication and other multimedia applications, the critical attributes determining quality are usefulness, enjoyment, engagement, and so on, rather than mere performance. These, however, are not measurable with currently available techniques and tools.

Context as a Supplementary Evaluation Prerequisite

Summarizing the previous section, one may conclude that putting the emphasis of adaptive systems evaluation on objective assessment of performance-oriented qualities precludes any attempt to take context-oriented parameters and situational aspects of interaction into account. This, however, runs against the widely held view that adaptable and adaptive behavior has a strong situational character and, therefore, it should be studied in context (Grüninger & van Treeck, 1993); however, this is something that is only loosely addressed by existing instruments and previous research efforts.

Taking context into account during evaluation implies a departure from the tradition of laboratory-based experiment where users are assumed to be unaided individuals with no access to other people or to the tools available for accomplishing a task. Rather, it requires an account of the "joint" social system in which adaptation is an inherent property. Therefore, evaluation of adaptable and adaptive behavior is meaningful only when the designer is able to assess the context in which such adaptations take place and the usefulness resulting from such adaptations. The study of context in the evaluation of adaptable and adaptive systems, introduces two main requirements: The first relates to the need for early evaluation (as part of the design cycle) through real-world inquiries, whereas the second concerns the capability to capture the end-users' attitudes and opinion regarding the adaptations the system is performing.

The need for real-world inquiries is pertinent for designers in order to obtain a clear understanding of how users interact with other people in a social group as well as how specific artifacts are used to accomplish the task at hand. To this end, techniques that are based on how a system should be used rather than how it is actually used—for example, engineering models, following either the tradition of GOMS (Card et al., 1983) or subsequent developments like programmable user models (Young, Green, & Simon, 1989) or runnable user models (Blandford & Young, 1993)—offer a limited account of context, the qualitative changes that drive practice and the social dynamics that underlie human behavior (Nardi, 1996).

The second prerequisite for meaningful evaluation of adaptable and adaptive systems is that of capturing the end-users' attitudes and subjective opinion. The compelling need to formulate any evaluation results on the basis of the real experience and the respective opinion of end-users arises from the fact that they are the ones to whom the system adapts and, therefore, they know best how well the system performs these adaptations. However, capturing end-user opinion for an adaptable and adaptive system has proven to be a nontrivial task. In particular, it requires a deep understanding of the utility of adaptable and adaptive behavior and how it can be evaluated through measurement. Nevertheless, as discussed in the next section, there have been substantial efforts in this direction.

EVALUATING ADAPTABLE AND ADAPTIVE USER INTERFACES

Having identified the additional points of view required to evaluate adaptable and adaptive behavior of interactive computer-based systems, this section provides a methodological perspective on the issue, and indications for selecting suitable evaluation instruments. To this end, we are primarily concerned with methods and tools that can facilitate the prerequisites outlined in the previous section. Our intention is not to undervalue any particular approach to evaluation, but rather to reveal the complexity involved in the evaluation of adaptable and adaptive user interfaces and how it can be methodologically attained.

Real-World Inquiries

This condition raises implications on the overall methodological frame of reference on which design and evaluation of adaptable and adaptive systems should be based. In particular, it requires that a formative as opposed to a summative evaluation approach is adopted (see Table 17.2).

There are several emerging approaches, which utilize formative evaluation as the basis for inquiring into HCI. Examples include participatory design ("Participatory Design," 1993), activity theory (Bødker, 1989, 1991), situated action models (Suchman, 1987), distributed cognition (Hutchins, 1995), as well as theories such as the language/action perspective (Winograd, 1987).

As already pointed out in chapter 1 of this volume, the normative perspective adopted in these efforts is that interactions between humans and information artifacts should be studied in specific social contexts, taking into account the distinctive properties that characterize them. Despite this common commitment to the study of context, the aforementioned alternatives differ with regard to at least three dimensions, namely the unit of analysis in studying context, the categories offered to support a description of context, and the extent to which each treats actions as structured prior or during human activities (Nardi, 1996).

From the preceding, it follows that from a methodological point of view, the evaluation of adaptable and adaptive systems should be considered as an activity within a prescriptive rather than predictive framework, such as those outlined in the diagram of Table 17.2. Thus, it should be planned, designed, and conducted in such a way so as to facilitate real users' experience with artifacts, as they are to be encountered in real contexts of use. Moreover, the instruments considered suitable are those that allow such experiences to be documented, reflected upon, and analyzed. A critical element in this direction is to provide for explicit account of subjective end-user experiences.

User Attitudes

The second precondition for meaningful evaluation of adaptable and adaptive systems is that of capturing and analyzing end-user attitudes (e.g., the user's degree of

TABLE 17.2.

Characteristics of a Suitable Framework for Evaluating
Adaptable and Adaptive Interaction

Scientific goal	Prescriptive & user-involved	Predictive
Methodological approach	Formative experiment	Laboratory experiment
Frameworks	Activity theory Distribute cognition Language / action theory Situated action models Contextual design	Cognitive science
Techniques	Psychometric questionnaires Ethnographic methods Scenarios of use Participatory design Analytical HCI design	Keystroke-level model GOMS family Cognitive complexity
Underlying science base	Development psychology & social sciences	Information processing psychology

satisfaction). In the recent past, there have been several approaches and instruments developed to facilitate capturing and measuring end-user attitudes. Some of these techniques are user-centered design oriented and aimed at providing early design input through analysis of user responses, whereas others are usability evaluation specific. The former class includes techniques such as user-based surveys, diary methods, representations of use, and other instruments falling within the scenario-based perspective on system development (Carroll, 1995), as well as other analytical methods and tools such as argumentation and design rationale (Carroll & Moran, 1996), participatory techniques (Kensing, Simonsen, & Bødker, 1998), and so forth. All these methods, despite their design orientation, encompass strong evaluation elements as they unfold critical elements of context that influence the planning, design, and conduct of evaluation.

Another category of tools aims to provide reliable measurement of the user's degree of satisfaction. These instruments take the form of psychometric questionnaires. In the recent past, there have been several developments in this direction aiming to explicitly facilitate usability engineering. Examples include the QUIS questionnaire (Shneiderman, 1992), the SUS usability scale (Brooke, 1996), the IBM Computer Usability Satisfaction Questionnaires (Lewis, 1995), the SUMI questionnaire (Kirakowski & Corbett, 1993), and more recently, the PUTQ questionnaire (Lin, Choong, & Salvendy, 1997). Additionally, recent developments have led to similar instruments for subjective usability evaluation of multimedia software[2] and World Wide Web sites[3].

What is important to note is that these questionnaires are not intended to measure adaptability and/or adaptivity. Moreover, they do not include any information on the context of use. Consequently, their current suitability as measurement scales of adaptable and adaptive behavior is questionable, which perhaps explains why they have not been used in any effort reported in the relevant literature. However, this does not diminish the value and potential usefulness of the overall approach (e.g., development of psychometric instruments) for the evaluation of adaptable and adaptive behavior in the context of usability engineering frameworks.

DISCUSSION AND CONCLUSIONS

The chapter has briefly reviewed recent work on evaluating adaptable and adaptive computer-based interactive systems, with the intention to highlight the complexity involved and identify some of the reasons why traditional evaluation instruments may not be particularly effective for this purpose. Additionally, the chapter presented indications for selecting the scientific base and specific evaluation methods on the basis of recent practice and experience with adaptable and adaptive systems.

From this account it follows that, despite recent progress in the techniques and tools available for building systems exhibiting adaptable and adaptive behavior, their evaluation remains a challenge. This is due to the highly context oriented nature of adaptations and the situational parameters that influence the objective measurement of such systems. Clearly, performance-oriented measures do not suffice to convey a complete assessment of adaptable and adaptive system qualities and, frequently, they may mislead the overall evaluation. It is, therefore, of paramount importance for designers of such systems to appreciate the situational and context-specific aspects that determine the usefulness of adaptations, through appropriate methodology and evaluation instruments.

The primary shortcoming of traditional approaches within human factors and cognitive sciences perspectives on the evaluation of interactive systems exhibiting adaptation capabilities is their insufficiency to capture context in any meaningful manner. Thus, they cannot account for the situational aspects that are of highest importance for the usability and utility of adaptations (e.g., learning, competence, system awareness of user's situation). Emerging approaches rooted in either developmental psychology (e.g., activity theory) or the social and behavioral sciences (e.g., ethnography, situation action models), seem to offer richer insights and, thus, a more appropriate basis for studying context and developing prescriptive methodological frames of reference. However, these methods, with only a few exceptions, are still far from being popular within the HCI design community.

---

[2]Measuring the Usability of Multi-Media Systems (MUMMS). For more information see: http://www.ucc.ie/hfrg/questionnaires/mumms/

[3]Website analysis and Measurement Inventory (WAMMI). For more information see: http://www.ucc.ie/hfrg/questionnaires/wammi/, or http://www.nomos.se/wammi/index.html

Finally, regarding the choice of suitable instruments, it can be concluded that techniques that foster end-user involvement and that can capture and document user attitudes are very important, though the currently available collection may not be sufficient. There are, however, promising approaches that need to be further refined to meet the peculiarities and the distinctive characteristics of adaptable and adaptive interaction. Future work in the field should seek to address the challenge of appropriately enhancing existing methods and techniques and to assess practically their suitability for the evaluation of adaptation-capable user interfaces, and their applicability across application domains, user categories, and contexts of use.

REFERENCES

Blandford, A., & Young, R. (1993). Developing runnable user models: Separating the problem solving techniques from the domain knowledge. In J. Alty, D. Diaper, & S. Guest (Eds.), *People and computers VIII* (pp. 111–122). Cambridge, England: Cambridge University Press.

Bødker, S. (1989). A human-activity approach to user interfaces. *Human–Computer Interaction, 4*(3), 151–196.

Bødker, S. (1991). *Through the interface: A human activity approach to user interface design*. Hillsdale, NJ: Lawrence Erlbaum Associates.

Bonnie, E. J., & Kieras, D.E. (1996a). The GOMS family of user interface analysis techniques: Comparison and contrast. *ACM Transactions on Computer–Human Interaction, 3*(4), 320–351.

Bonnie, E. J., & Kieras, D.E. (1996b). Using GOMS for user interface design and evaluation: Which technique? *ACM Transactions on Computer–Human Interaction, 3*(4), 287–319.

Boyle, C., & Encarnacion, A. O. (1994). MetaDoc: An adaptive hypertext-reading system. *User Models and User Adapted Interaction, 4*(1), 1–19.

Brooke, J. (1996). SUS: A "quick and dirty" usability scale. In P. W. Jordan, B. Thomas, B. A. Weerdmeester, & I. L. McClelland (Eds.), *Usability evaluation in industry* (pp. 189–194). London: Taylor & Francis.

Brusilovsky, P., & Pesin, L. (1995). Visual annotation of links in adaptive hypermedia. In I. Katz, R. Mack, & L. Marks (Eds.), *Companion proceedings of the ACM Conference on Human Factors in Computing Systems* (pp. 222–223). New York: ACM Press.

Card, S. K., Moran, T. P., & Newell, A. (1980). The keystroke level model for user performance time with interactive systems. *Communications of the ACM, 23*(7), 396–410.

Card, S. K., Moran, T. P., & Newell, A. (1983). *The psychology of human–computer interaction*. Hillsdale, NJ: Lawrence Erlbaum Associates.

Carroll, J. M. (Ed.). (1995). *Scenario-based design: Envisioning work and technology in system development*. New York: Wiley.

Carroll, J. M., & Moran, T. (1996). *Design rationale: Concepts, methods, and techniques*. Hillsdale, NJ: Laurence Erlbaum Associates.

Debevc, M., Meyer, B. V., Donlagic, D., & Svecko, R. (1996). Design and evaluation of an adaptive toolbar. *User Modelling and User-Adapted Interaction, 6*(1), 1–21.

Eberts, R. E., & Eberts, C. G. (1989). Four approaches to human computer interaction. In P. A. Hancock & M. H. Chignell (Eds.), *Intelligent interfaces: Theory, research and design* (pp. 69–127). Amsterdam: North-Holland, Elsevier Science.

Gilmore, D. J. (1995). Interface design: have we got it wrong? In K. Nordby, P. H. Helmersen, D. J. Gilmore & S. A. Andersen (Eds.), *IFIP International Conference on Human–Computer Interaction.* Oxford, England: Chapman & Hall.

Gruninger, C., & van Treeck, W. (1993). Contributions of a social science based evaluation for adaptive design projects. In M. Schneider-Hufschmidt, T. Kühme, & U. Malinowski (Eds.), *Adaptive user interfaces—Principles and practice* (pp. 319–330). Amsterdam: North-Holland, Elsevier Science.

Höök, K. (1998). Evaluating the utility and usability of an adaptive hypermedia system. *Knowledge-Based Systems, 10*(5), 311–319.

Hutchins, E. (1995). *Cognition in the wild.* Cambridge, MA: MIT Press.

ISO 13407. (1997). *Human-centered design processes for interactive systems.* Geneva, Switzerland: International Standards Organisation.

Kaplan, C., Fenwick, J., & Chen, J. (1993). Adaptive hypertext navigation based on user goals and context. *User Models and User Adapted Interaction, 3*(3), 193–220.

Kensing, F., Simonsen, J., & Bødker, K. (1998). MUST: A method for participatory design. *Human–Computer Interaction, 13,* 167–198.

Kieras, D. E., & Polson, P. G. (1985). An approach to the formal analysis of user complexity. *International Journal of Man–Machine Studies, 22,* 365–394.

Kirakowski, J., & Corbett, M. (1993). SUMI: The software measurement inventory. *British Journal of Educational Technology, 24,* 210–212.

Lewis, R. J. (1995). IBM Computer Usability Satisfaction Questionnaires: Psychometric evaluation and instructions for use. *International Journal of Human–Computer Interaction, 7*(1) 57–78.

Lin, H., Choong, Y, & Salvendy, G. (1997). A proposed index of usability: Method for comparing the relative usability of different software systems. *Behaviour and Information Technology, 16*(4/5), 267–278.

Nardi, B. (1996). *Context and consciousness: Activity theory and human–computer interaction.* Cambridge, MA: MIT Press.

Nielsen, J. (1993). *Usability engineering.* Boston: Academic Press.

Norman, D., & Draper, W. (1986). *User-centered design: New perspectives on human–computer interaction.* Hillsdale, NJ: Lawrence Erlbaum Associates.

Participatory Design. (1993). [Special issue]. *Communications of the ACM, 36*(4).

Shneiderman, B. (1992). *Designing the user interface, Strategies for effective human computer interaction* (2nd ed.). Reading, MA: Addison-Wesley.

Suchman, L. A. (1987). *Plans and situated actions: The problem of human machine communication.* Cambridge, England: Cambridge University Press.

Totterdell, P., & Boyle, E. (1990). The evaluation of adaptive systems. In D. Browne, P. Totterdell, & M. Norman (Eds.), *Adaptive user interfaces* (pp. 161–194). London: Academic Press.

Winograd, T. (1987). A language/action perspective on the design of co-operative work. *Human–Computer Interaction, 3*(1), 3–30.

Young, R., Green, T., & Simon, T. (1989). Programmable user models for predictive evaluation of interface designs. In K. Bice & C. Lewis (Eds.), *Proceedings of SIGCHI Conference on Wings for the Mind* (pp. 15–19). New York: ACM Press.

18 Quality in Use for All

Nigel Bevan

Designing an interactive product or service for all possible users requires consideration of more than just physical accessibility. Even if there is sufficient physical accessibility, many systems will still present major barriers to their use, in terms of cognitive accessibility: the functionality, terminology, information structure and interface style frequently confuse the intended user. Usually, the main emphasis in systems design is on building systems that meet specific functional requirements, without a sufficiently detailed understanding of the cognitive and physical capabilities and expectations of the intended users, or a clear view of the context in which the system will be used. The problem is compounded by the difficulty that designers usually have in recognizing shortcomings or limitations in their own design. There are cost-effective procedures for dealing with this problem, which have recently been formalized in ISO 13407: Human-centered design process for interactive systems (International Standards Organisation [ISO] 13407, 1999).

User-centered design provides a framework which can potentially make Design for All a reality. The goal is to achieve quality in use for all. User-centered design encompasses processes, tools and techniques which can be used to identify and document the complete range of user requirements, including special needs arising from (dis-)abilities, skills, preferences, or any other characteristic of the end-user population. The tight design / evaluation loop advocated by user-centered design provides feedback to correct design deficiencies at an early stage while changes are relatively simple to make. Procedures for evaluating quality in use have been developed as part of approaches to usability evaluation, and now need to be extended to encompass procedures for evaluating accessibility.

In summary, Design for All entails both physical and cognitive accessibility. New hardware and software technologies are required to make it easier to provide physical accessibility. New integrated approaches to system development are required to make it easier to provide cognitive accessibility. Only by combining these activities can Design for All be achieved.

The objective of *design for all* is to provide accessible and easy to use technology for both professionals and the user population at large. It is no longer sufficient to just deliver technically excellent systems. There is increasing demand for computer systems that are widely accessible, easy to learn and use, and easy to integrate into work or leisure activities. Despite the rapid increase of computer power and the progress in the sophistication of systems development, these objectives are not being achieved: It is widely believed that the majority of computer users still cannot get their systems to do exactly what they want.

Currently only few interactive systems support the real needs of their users. For example: (a) most video recorders are unnecessarily complicated to use (Thimbleby, 1991); (b) few Web pages are suitable for use by blind users with a text browser (K. Bartlett, E-mail communication, March 29, 1999); and (c) most business users of computer systems are only productive for 30% to 40% of the time, with 60% to 70% of the time spent trying to understand how to use the system, or recovering from errors (Macleod, Bowden, Bevan, & Curson, 1997).

These examples of poor usability are a consequence of not following a design process that is sufficiently user centered. Many existing development processes focus exclusively on adherence to technical and process specifications. In order to produce systems that better match user needs, it is essential to enhance current design processes to incorporate techniques for usability and accessibility through a *user-centered* approach.

Norman and Draper (1986) introduced the term *user-centered design* to refer to the design of computer systems from the user's point of view. The ISO 13407 standard (1999) describes the activities required to apply a user-centered approach to the design of interactive systems. Although this standard focuses on computer systems, the principles it contains are equally applicable to any interactive system used by humans. This chapter explains the benefits of using user-centered design principles and practices to achieve *quality in use for all* users of interactive products and services.

QUALITY IN USE: BENEFITS AND BARRIERS

One of the major objectives of design should be to achieve quality in use for all, that is, to provide tools and facilities that support people's work and leisure activities and ensure that products can be used by the widest possible range of people, to achieve their real-world tasks. This requires not only easy-to-use interfaces, but also the provision of appropriate functionality and support for computer-mediated activities.

Quality in Use Benefits

Increased *quality in use* would bring significant benefits to industry, end-users, and society:

- *Improved accessibility.* User-centered design provides methods and tools for identifying, and catering to a wide range of user needs, thus helping define a broad range of accessibility requirements, including user interface and content accessibility, which can be accounted for in the course of subsequent design activities.
- *Increased efficiency.* A well-designed system incorporates good ergonomic design, is tailorable to the physical capabilities and preferred way of working of end-users, and will allow them to operate effectively and efficiently, rather than lose vital time struggling with a poorly designed user interface and badly thought-out functionality.
- *Improved productivity.* A good interface to a well-designed product will allow the user to concentrate on the task rather than the tool. If the interface is designed inappropriately, it may increase rather than reduce the time needed to perform a task, and have a deleterious effect on other aspects of user performance and the quality of task results.
- *Reduced errors.* A significant proportion of so-called "human error" can be attributed to a product with a poorly designed interface that is not closely matched to the users' task needs, or to their mental model of the task. Avoiding inconsistencies, ambiguities, or other interface design

faults, while adhering to user expectations in terms of task structure and sequencing has the potential to significantly reduce user error.

- *Reduced training.* A poorly designed user interface and dialogue can be a barrier to an otherwise technically sound system. A system designed with a focus on the end-user can reinforce learning, thus reducing training time and effort.

- *Improved acceptance.* This is particularly important where usage is discretionary. Users are more likely to use and trust a well-designed, accessible system, which has been designed so that information is both easy to find and provided in a form that is easy to assimilate and use.

With all these potential benefits, why are many systems still not designed for greater usability and accessibility? There are currently a series of practical barriers to achieving quality in use for all.

Barriers to Achieving Quality in Use for All

Incomplete Requirements. The first step in user-centered design is understanding the user requirements. Software development practices have placed an increasing emphasis on achieving quality, and ensuring that the delivered products meet the stated requirements. But a recent survey (Vintner & Poulsen, 1996) showed that the source of 80% of software defects is poor or missing requirements. Of these, only 15% are related to functionality, and of the remaining defects, 60% arise from usability errors. In addition, the requirements of users with special needs are frequently overlooked, partly because they usually represent only a small portion of the target user population, and partly because their needs are far from the typical experience of most design teams. Acknowledging and planning for the full range of user needs is an essential prerequisite for achieving *design for all.*

For a design process to take a user-centered approach to the identification of requirements, the process must include activities that can capture both usability requirements derived from the capabilities of the end-user groups, as well as accessibility requirements for users with special needs. This is an extension of the principles of user profiling and task analysis, which are a typical part of usability engineering (Hakos & Redish, 1998).

Commercial Barriers. A major obstacle to the wider adoption of practices that would facilitate the goal of quality in use for all is the perceived additional cost resulting from the user-centered activities that are required to achieve quality in use, as well as with the additional hardware and software features that are required to provide wide accessibility. The economic benefits of using user-centered design to improve quality in use for major user groups is now well established (Bias & Mayhew, 1994; Karat, 1992; Keil & Carmel, 1995), but there are several potential obstacles to extending these benefits to minority groups with special needs. These obstacles include: additional design and production costs, lack of a market that demands access for all (see also the subsection Demand Barriers of this chapter), and lack of legislation requiring access for all.

There are several potential solutions to this problem, for example:

- Legislation that identifies accessibility as an obligatory feature of systems developed in publicly funded projects, or acquired by public procurement.
- Development of hardware and software architectures and features that reduce the costs of providing access for all.
- Development and fostering of an expectation for access for all in the market, which would, in turn, progressively induce the incorporation of design for all into established professional practices.

Practical Barriers: What Is "All"?   A major practical barrier for the designer is that access for "all" is impossible, if taken literally. For each product and market the potential range of "all" has to be defined, and incorporated into the design. The final definition of "all" is usually a trade-off between the requirements of potential users, and commercial and legislative constraints. However, this trade-off is rarely made explicit.

Quality in use for all needs to be supported by a clear definition of:

- Who should be able to use the product and in what circumstances?
- What types of users will be excluded and why?

Technical Barriers.   Achieving wider accessibility is much easier if developers can adopt well-established general-purpose solutions. This requires:

- The availability of appropriate methods, techniques, and tools that can be used to improve quality in use.
- Software architectures and user interface tools that can facilitate the design and development of interactive applications adaptable to the widest possible range of cognitive and physical requirements.

Development Barriers.   Recently, there has been increasing awareness of the value of user-centered design, and increasing use of RAD-based methodologies, which are compatible with user-centered design. However, many current development processes do not take a user-centered approach, and thus fail to identify the full range of user needs and incorporate feedback from users.

Demand Barriers.   Surmounting all these barriers needs a cultural change in the expectations and demands of purchasers and users, and a change in priorities in development from meeting technical specifications to meeting the widest possible range of user needs. Such a process, which is usually referred to as *demand articulation,* is not only important, but also necessary to raise public awareness, create incentives for the industry and, ultimately, substantiate the demand for universal accessibility and design for all.

SPECIFYING QUALITY IN USE REQUIREMENTS

User Interface Requirements

Traditional approaches to achieving quality put emphasis on meeting specified requirements that are primarily functional. Attempts have been made to broaden the perception of quality, for example in ISO/IEC FDIS 9126 (2000), which defines quality from a user perspective as functionality, reliability, usability, efficiency, maintainability and portability (Fig. 18.1).

This approach was derived from the ISO 8402 (1994) definition of quality: *"Quality: the totality of characteristics of an entity that bear on its ability to satisfy stated and implied needs"*.

The ISO/IEC FDIS 9126 definitions acknowledge that the objective is to meet user needs. But ISO 8402 makes it clear that quality is determined by the presence or absence of particular attributes, with the implication that these are specific attributes that can be designed into the product. Thus, when referring to software, these would be attributes of the source code. When combined with an ISO 9001 (1994) compliant quality process, the most natural interpretation is that quality should be specified and evaluated at the level of source code attributes.

For the user interface, requirements at this level of detail can be specified using style guides, or design guidance such as is contained in ISO 9241-11 (1998). This standard also provides a potential means to specify and evaluate usability. Schemes to do this have been developed in Germany (e.g., Gediga, Hamborg, & Düntsch, 1999, Prümper, 1999).

In terms of accessibility, requirements at this level would specify the interface features required to ensure access by users with special needs. This approach is very valuable in identifying and minimizing low-level user interface shortcom-

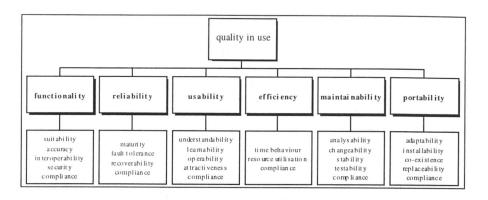

FIG. 18.1.   Software quality characteristics (ISO/IEC 9126 Quality Model).

ings and defects. However, a good interface does not necessarily guarantee a system that enables users to efficiently achieve their goals and effectively complete their tasks. This requires a broad, top-down approach to design and evaluation.

Quality in Use Requirements

ISO/IEC 14598-1 (1998) takes this broad view, explaining that quality in use is the users' view of the quality of a system, and it is measured in terms of the results of using the system, rather than properties of the system itself. Thus, quality in use is the combined effect of the system's quality characteristics, as these are experienced by the end-user. A product meets the requirements of a particular user if it enables the user to be effective, productive in terms of time and resources, and satisfied, regardless of the specific attributes the product possesses.

Quality in use in any particular system depends not only on its usability, but also on the appropriateness of the provided functionality for a given task, the performance that users can achieve using the system and the reliability of the system (Bevan, 1997a). Additionally, to provide quality in use for all, a system will need to provide both *physical* and *cognitive* accessibility for the intended users. Physical accessibility implies the provision of the physical means through which all potential users will be able to operate the system. Cognitive accessibility means meeting the cognitive needs of the users. The characteristics that are required to make software cognitively accessible are defined in general terms in ISO 9241-10, where it is stated that software should be suitable for the task, self-descriptive, controllable, in conformance with user expectations, error tolerant, suitable for individualization, and suitable for learning.

In this context, quality in use can be viewed as a black-box approach to specification and evaluation, assessing the extent to which a product or service meets the needs of users, irrespective of its specific attributes. In particular, quality in use depends on the circumstances in which a system is used, including such factors as the tasks, equipment (hardware, software, and materials), and the physical and social environment.

Consequently, quality in use is broader in scope than the usability of the interface. However, the term *usability* has several interpretations, and in some ergonomic standards (e.g., ISO 9241-11 and ISO 13407) the terms *usability* and *quality in use* are used synonymously. In this chapter, the term *usability* is used in the ISO/IEC FDIS 9126-1 (2000) sense, of understandability, learnability, operability and attractiveness.

EVALUATION OF QUALITY IN USE

Evaluation of quality in use should take place as part of a user-centered design process that starts with the specification of the requirements for quality in use, in specific contexts of use. The MUSiC Performance Measurement Method (Macleod et al., 1997), based on ISO 9241-11 (1998) documents all the necessary procedures one has to follow to evaluate quality in use, from deciding what and how to evaluate, to producing the final usability report. The steps involved are:

1. *Define the product to be tested.* The first step is to define which version and which components of a product or system are to be evaluated.

2. *Define the context of use.* The next step is to clarify the intended context of use of the product: What are the intended user groups, what skills and what cognitive and physical capabilities will the intended users have, what task goals can be achieved with the product, and what physical and social conditions will the product be used in?

3. *Specify the quality in use requirements.* Quality in use is measured in specific contexts of use. Measures of quality in use have three essential components:

- Effectiveness: Can users complete their tasks correctly and completely?
- Productivity: Are tasks completed in an acceptable length of time?
- Satisfaction: Are users satisfied with the interaction?

Quality in use requirements are often set by comparison with alternative means of achieving a task. For instance, a new communication aid for the disabled should enable users to communicate more effectively, in a shorter time, and with more satisfaction than when using existing solutions.

4. *Specify the context of evaluation,* so that the evaluation can be carried out in conditions as close as possible to those in which the product will be used. It is important that:

- Users are representative of the population of users who will use the product.
- Tasks are representative of the ones that the system is intended to support.
- Conditions are representative of the normal conditions in which the product is to be used.

5. *Design an evaluation* to meet the specified context of evaluation. In designing the evaluation, one should keep in mind that it should measure the performance and satisfaction of users as they perform set tasks within this context. Satisfaction can be measured with a validated questionnaire such as SUMI (Kirakowski, 1996).

6. *Perform the user tests and collect data.* When assessing quality in use it is important that the users work unaided, having access to only forms of assistance that would be available under normal conditions of use. In addition to measuring effectiveness, efficiency, and satisfaction, it is usual to document the problems users encounter, and to obtain clarification by discussing those problems with the users at the end of the session. It is often useful to record the evaluation on video, which permits more detailed analysis. It is also easier for users to work undisturbed, if they are monitored remotely by video.

7. *Analyze and interpret the data.* The data are used to calculate metrics for effectiveness, efficiency, and satisfaction. There is only limited experience in using these metrics to assess quality in use for all, and additional metrics are a topic for further research (Stephanidis et al., 1999).

8. *Produce a usability report.* This should give a description of the measures obtained for the system under test, and could be used to compare the system with initial requirements, with similar systems, or with the same system as it evolves over time.

From the preceding, it follows that the evaluation of quality in use should take place as part of a user-centered design process that starts with the specification of the requirements for quality in use in specific contexts of use. Thus, if accessibility is to be integrated within the evaluation of quality in use, it should first be introduced within user-centered design. The next section provides a brief summary of the phases of user-centered design, and the extensions needed.

USER-CENTERED DESIGN AND DESIGN FOR ALL

ISO 13407

ISO 13407 (1999) provides guidance on achieving quality in use by incorporating user-centered design activities throughout the life cycle of interactive computer-based systems. It describes user-centered design as a multidisciplinary activity, which incorporates human factors and ergonomics knowledge and techniques, with the objective of enhancing effectiveness and productivity, improving human working conditions, and counteracting the possible adverse effects of the use of computer-based systems on human health, safety, and performance.

There are four user-centered design activities that need to take place at all stages during a project. These are to:

- Understand and specify the context of use.
- Specify the user and organizational requirements.
- Produce design solutions.
- Evaluate designs against requirements.

The iterative nature of these activities is illustrated in Fig. 18.2. The process involves iterating until the objectives are satisfied.

The sequence in which the activities are performed and the level of effort and detail that is appropriate vary depending on the design environment and the stage of the design process.

Understand and Specify the Context of Use. The characteristics of the users, tasks, and organizational and physical environment define the context in which the product is used. It is important to understand and identify the details of this context, in order to guide early design decisions, and to provide a basis for evaluation.

The context in which the product is to be used should be identified in terms of:

1. *The characteristics of the intended users.* Relevant characteristics of the users can include knowledge, skills, experience, education, training, age, and

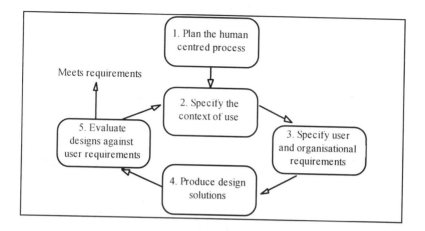

FIG. 18.2.   The interdependence of user-centered design activities.

physical abilities. If necessary, define the characteristics of different categories of users, for example, having different levels of experience, different roles, or different levels of cognitive capabilities.

2.   *The tasks the users are to perform.* The description should include the overall goals of use of the system. The characteristics of tasks that can influence quality in use in typical usage scenarios should be described, for example, the frequency and the duration of task performance for a specific task. Tasks should not be described solely in terms of the functions, or features provided by a product or system. Instead, the description should include the allocation of activities and operational steps between human and technological resources.

3.   *The environment in which the users are to use the product.* The associated hardware, software, and materials can either be identified as specific products, or be described in terms of their attributes or performance characteristics. Relevant characteristics of the physical and social environment also need to be described. For example, aspects that may need to be described include attributes of the wider technical environment (e.g., a local area network), the physical environment (e.g., workplace, furniture), the ambient environment (e.g., temperature, humidity), the legislative environment (e.g., laws, ordinances, directives, and standards), and the social and cultural environment (e.g., work practices, organizational structure and attitudes).

An effective way to collect and document the necessary information is to use an annotated checklist of user, task, and environmental characteristics, such as the one found in the Usability Context Analysis Guide (Thomas & Bevan, 1995). Such a "general-case" checklist would need to be used in combination with one intended to capture the special needs of end-users (e.g., Maguire, Heim, Endestad, Skjetne, & Vereker, 1998).

The output of this activity should be a description of the relevant characteristics of the users, tasks, and environment, which identifies what aspects of the overall context of use have significant impact on the system design. This description is unlikely to be a single output that is issued once. It is more often a "working document" that is first produced in outline terms and is then reviewed, maintained, extended, and updated during the design and development process.

Specify the User and Organizational Requirements.   In most design processes, there is a major activity dealing with the specification of the functional and other requirements for the product or system. For user-centered design, it is essential to extend this activity to include an explicit statement of user and organizational requirements, in relation to the context of use description and in terms of the usability and accessibility of the human–computer interface and the overall quality in use.

Objectives can be set, with appropriate trade-offs identified between different requirements. The requirements should be stated in a form that permits subsequent testing.

Including accessibility as a nonfunctional requirement implies a commitment on behalf of the stakeholders to invest the required resources to meet the requirements of users with special needs.

Produce Design Solutions.   The next stage is to create potential design solutions by drawing on the established state of the art and the experience and knowledge of the participants. The process, therefore, involves:

- Using existing knowledge to develop proposed multidisciplinary design solutions.
- Making those design solutions more concrete (using simulations, models, mock-ups, etc.).
- Presenting the solutions to users and allowing them to perform tasks (or simulated tasks).
- Using feedback from the users to improve the designs.
- Iterating this process until the user-centered design goals are met.

Though rapid prototyping techniques and tools have substantially improved the outcome of this stage, there is still much to be expected to facilitate accessibility. This is due to the fact that tools for accessible design are largely missing. In the recent past, with initiatives such as Microsoft's Active Accessibility[1] and Sun's Java Accessibility[2] the situation has been slightly improved. However, unless tools that

---

[1]For more information, please refer to: http://microsoft.com/enable/msaa.default.htm

[2]For more information, please refer to chapter 16 of this volume, or to: http://java.sun.com/products/jfc/accessibility/doc/

facilitate edit-evaluate-modify prototyping cycles become available, accessibility will remain a loosely addressed target. To justify this account, one can draw parallels with graphical user interfaces (GUIs) over the past two decades, or, more recently, Java. Specifically, it was not until tools for building GUIs became available, that graphical interaction took off. Similarly, the primary constraint of Java's diffusion is the lack of suitable tools to facilitate the development of Java applications (Regan, 1998).

Evaluate Designs Against Requirements.   Evaluation is an essential step in user-centered design and should take place at all stages in the system life cycle. Evaluation can be used to:

- Provide feedback that can be used to improve design.
- Assess whether user and organizational objectives have been achieved.
- Monitor long-term use of the product or system.

Early on in the design process the emphasis is on obtaining feedback that can be used to guide design, whereas later, when a realistic prototype is available, it is possible to measure whether user and organizational objectives have been achieved.

In the early stages of the development and design process, changes are relatively inexpensive. The longer the process has progressed and the more fully the system is defined, the more expensive the introduction of changes becomes. It is, therefore, important to start evaluation as early as possible.

In the recent past, there have been substantial developments in the area of techniques and tools for evaluation. Many of those can be used either directly, or with slight modifications, to assess the degree to which designs meet specified accessibility requirements. It should be mentioned, however, that the primary use of these techniques is in formative evaluation (i.e., provide feedback that can be used to improve design). Techniques for summative evaluation of the accessibility of an interactive product are still pending, as they would require an operational definition of accessibility to derive appropriate measures (see also chap. 17, this volume).

Relationship Between User-Centered Design and Design for All

The objective of design for all in the context of the emerging information society is as follows:

> [It is] the conscious and systematic effort to proactively apply principles, methods and tools, in order to develop Information Technologies and Telecommunications products and services which are accessible and usable by *all* citizens, thus avoiding the need for a posteriori adaptations or specialized design (Stephanidis et al., 1998, pp. 112–113).

This means that products and services should have quality in use for all citizens in all the relevant contexts of use.

As discussed in the previous sections, user-centered design can be extended to provide the framework for a design process within which this can be achieved; however, one shortcoming of current approaches to user-centered design is that (as in ISO 13407) there is no explicit mention of accessibility. Moreover, due to its commercial orientation, user-centered design typically emphasizes the need to identify the intended users, rather than exploring how access can be provided for *all* users. However, with the advent of the information society and the introduction of novel contexts of use, the concept of the "intended user" needs to be expanded, to include occasional, or "unforeseen" users. One of the important steps that need to be made in this direction is that requirements for accessibility should properly be considered as part of a user-centered design process. Formal integration of accessibility with user-centered design to achieve user-centered design for all is a matter for further research and development (Stephanidis et al., 1999). It should also be noted that design for all is a more ambitious objective than quality in use for all, which may currently be achieved to some degree using a posteriori adaptations or specialized design.

EXAMPLE OF USER-CENTERED DESIGN

To illustrate the approach, consider a hypothetical example of how user-centered activities could contribute to the design of a Web site that provides a database of the results of recent publicly funded research. The example is used to demonstrate some of the concepts put forward in the previous sections, both in terms of the process followed in the development of an interactive system, and in the focus of various activities within that process.

The first step is to identify the expected context of use. Analysis of existing users of technology research results identifies the main users of the Web site as professionals seeking information on new technology; they would be likely to access the site in two circumstances:

- While visiting technology information centers.
- While at their normal place of work.

Technology information centers are approached to provide information on the typical goals of their clients, and a profile of their range of clients. The profile includes people with poor eyesight and other physical disabilities.

Discussions with information center clients and other potential users leads to agreement on the goals for quality in use. This is a black-box approach. Rather than specifying the presence or absence of interface features, criteria are established for users' effectiveness, productivity, and satisfaction when using the product:

- At least 90% success when looking for information contained in the system.
- The average time to find information should not exceed 10 minutes.
- The user satisfaction with the Web site should be at least as great as with other sites providing similar services.

The design team maps out the potential structure of the site using self-adhesive notes for each intended page, and discusses the structure with potential users. A rough sketch of each page is then produced on a series of cards, which are presented to users in sequence, and users are asked to step through a series of tasks by selecting an option on each card. The feedback from these tests is used to refine the site structure until it meets the expectations of users.

Special consideration is given to the information and navigation requirements of users with poor eyesight and other physical disabilities, who will access the site with specialized browsers and adapted physical interfaces. Two design options are considered: (a) to produce an alternative version (or alternative versions) of the site for users with special needs (e.g., optimized for aural rather than visual presentation) and (b) to follow Web accessibility guidelines to produce pages that cater to a large range of special requirements of different end-user categories.

The decision whether to invest in alternative versions is based on the extent to which a generic site is expected to have negative effects on the quality in use for different user groups, and the degree of priority given to providing optimal access for all.

Parts of the site are then mocked up as Web pages. Color prints of these pages are again presented to physically and cognitively able users to find out whether they can successfully navigate the site. Users with special needs are also involved in the process using appropriate interface mock-up techniques. A trial implementation is then made of part of the site with sample contents in the database.

The success of the user-centered design process can be evaluated by comparing the quality of the product in specific contexts of use, with the original quality in use goals. This type of validation can be quite expensive to perform, so it is generally necessary to concentrate on key user groups and tasks. Based on the expected context of use, the site should be evaluated by users with and without access to assistance from an information center, as well as with and without disabilities.

Representative categories of users are selected: information center clients without disabilities, information center clients with specific types of disability, and researchers from companies producing innovative technology. Typical task scenarios are defined, and a minimum of eight users from each category are given a series of tasks to perform using the types of PCs and Internet access that are typically found in their working environment. To establish whether the initial quality in use goals have been met, measures are taken of the task time, the degree of success in achieving the tasks, and the rated satisfaction in using the Web site. For more information on a user-centered approach to design see, for example, Bevan (1997b) for Web site design guidelines; Bevan and Curson (1998) and Daly-Jones, Thomas, and Bevan, (1997) for information on methods for user-centered design; and Bevan (1995) and Macleod et al. (1997) for quality in use measurement.

CONCLUSIONS

Many systems are designed to be accessible only by users with specific physical abilities and skills. Similarly, many systems are designed to be usable only by users with specific cognitive abilities and skills. Currently, there are two separate professional communities concerned with improving product accessibility and product usability. Both of them share the objective of meeting user needs in order to achieve quality in use. However, whereas usability has been primarily concerned with the range of "typical," or "average" users (by implication able-bodied), accessibility is concerned with extending design to incorporate users with physical and cognitive disabilities. User-centered design can provide a common framework for enhancing current design practice, in order to meet the real needs of both these majority and minority user groups.

One of the most frequent failures of current design processes is a lack of understanding of real user needs. User-centered design requires a detailed understanding and specification of user requirements, and the active participation of users in an iterative process of evaluating whether proposed design solutions meet user needs.

Current requirements engineering and specification processes need to be enhanced to explicitly incorporate quality in use requirements related to the capabilities of the users who are expected to use the system, and the requirements for effectiveness, productivity, and satisfaction resulting from the system's use.

Many design teams lack the necessary skills to carry out user-centered design activities, particularly in smaller organizations. Large information technology suppliers are making substantial investments in usability and accessibility, particularly in the competitive office systems market. Much remains to be done to create the same demand for usable and accessible systems in other markets, which currently tolerate unacceptably poor quality in use for the end-user.

It is a common failing to provide new technology without understanding how it will support user needs. Procurement specifications need to demand quality in use for all, and evaluation of quality in use should be an essential part of the acceptance testing process.

A first step in this direction is a recent initiative by the U.S. information technology industry to make usability more visible in the procurement process (Bevan, 1999; Blanchard, 1998). Producers have agreed to make Industry Standard Usability Test Reports available to potential corporate purchasers. It is hoped that this will be an influential step toward establishing quality in use in its rightful place as the prime goal of systems design.

REFERENCES

Bevan, N. (1995). Measuring usability as quality of use. *Software Quality Journal, 4,* 115–130.

Bevan, N. (1997a). Quality and usability: A new framework. In E. van Veenendaal & J. McMullan (Eds.), *Achieving software product quality* (pp. 25–34). Den Bosch, The Netherlands: Tutein Nolthenius.

Bevan, N. (1997b). Usability issues in Web site design. In *Proceedings of HCI International '97* (pp. 803–806). Amsterdam: Elsevier, Elsevier Science.

Bevan, N. (1999). Industry standard usability tests. In S. Brewster, A. Cawsey, & G. Cockton (Eds.), *Human–Computer Interaction INTERACT '99,* (vol. 2; pp. 107–108). Swindon: British Computer Society.

Bevan, N., & Curson, I. (1998). Planning and implementing user-centered design. In *Adjunct Proceedings of the Human Factors in Computing Systems* (pp. 111–112). New York: ACM Press.

Bias, R., & Mayhew, D. (Eds.). (1994). *Cost-justifying usability.* New York: Academic Press.

Blanchard, H. (1998). The application of usability testing results as procurement criteria for software. *SIGCHI Bulletin, 30*(3), 16–17.

Daly-Jones, O., Thomas, C., & Bevan, N. (1997). *Handbook of user-centered design.* Teddington, England: National Physical Laboratory.

Gediga, G., Hamborg, K., & Düntsch, I. (1999). The IsoMetrics usability inventory: An operationalisation of ISO 9241-10. *Behaviour and Information Technology, 18,* 151–164.

Hakos, J. T., & Redish, J. C. (1998). *User and task analysis for interface design.* New York: Wiley.

ISO 8402. (1994). *Quality vocabulary.* Geneva, Switzerland: International Standards Organisation.

ISO 9001. (1994). *Quality systems—Model for quality assurance in design development, production, installation and servicing.* Geneva, Switzerland: International Standards Organisation.

ISO 9241-11. (1998). *Ergonomic requirements for office work with visual display terminals (VDTs)—Part 11. Guidance on usability.* Geneva, Switzerland: International Standards Organisation.

ISO 13407. (1999). *User-centered design process for interactive systems.* Geneva, Switzerland: International Standards Organisation.

ISO/IEC 14598-1. (1998). *Information technology—Evaluation of software products—Part 1. General guide.* Geneva, Switzerland: International Standards Organisation.

ISO/IEC FDIS 9126-1. (2000). *Software product quality—Part 1: Quality model.* Geneva, Switzerland: International Standards Organisation.

Karat, C. M. (1992). Cost-justifying human factors support on development projects. *Human Factors Society Bulletin, 35*(11), 1–8.

Keil, M., & Carmel, E. (1995). Customer developer links in software-development. *Communications of the ACM, 38*(5), 33–44.

Kirakowski, J. (1996). The software usability measurement inventory: background and usage. In P. W. Jordan, B. Thomas, B.A. Weerdmeester, & I. L. McClelland (Eds.), *Usability evaluation in industry* (pp. 169–177). London: Taylor & Francis.

Macleod, M., Bowden, R., Bevan, N., & Curson, I. (1997). The MUSiC Performance Measurement Method. *Behaviour and Information Technology, 16,* 279–293.

Maguire, M. C., Heim, J., Endestad, T., Skjetne, J. H., & Vereker, N. (1998). *Requirements specification and evaluation for user groups with special needs* (RESPECT Deliverable D6.2, HUSAT Research Institute) [Online]. Available: http://www.iboro.ac.uk/eusc

Norman, D. A., & Draper, W. S. (Eds.). (1986). *User-centered system design: New perspectives on human–computer interaction.* Hillsdale, NJ: Lawrence Erlbaum Associates.

Prümper, J. (1999). Test it: ISONORM 9241/10. In H.-J. Bullinger & J. Ziegler (Eds.), *Human-computer interaction: Ergonomics and user interfaces,* Proceedings of the 7th International Conference on Human-Computer Interaction (vol. 1, pp. 1028–1032). Hillsdale, NJ: Lawrence Erlbaum Associates.

Regan, B. (1998). *Java: One year out* [Online]. Available: http://www.webreference.com/content/java/

Stephanidis, C. (Ed.), Salvendy, G., Akoumianakis, D., Bevan, N., Brewer, J., Emiliani, P. L., Galetsas, A., Haataja, S., Iakovidis, I., Jacko, J., Jenkins, P., Karshmer, A., Korn, P., Marcus, A., Murphy, H., Stary, C., Vanderheiden, G., Weber, G., & Ziegler, J. (1998). Toward an information society for all: An international R&D agenda. *International Journal of Human–Computer Interaction, 10*(2), 107–134.

Stephanidis, C. (Ed.), Salvendy, G., Akoumianakis, D., Arnold, A., Bevan, N., Dardailler, D., Emiliani, P. L., Iakovidis, I., Jenkins, P., Karshmer, A., Korn, P., Marcus, A., Murphy, H., Oppermann, C., Stary, C., Tamura, H., Tscheligi, M., Ueda, H., Weber, G., & Ziegler, J. (1999). Toward an information society for all: HCI challenges and R&D recommendations. *International Journal of Human–Computer Interaction, 11* (1), 1–28.

Thimbleby, H. (1991). Can anyone work the video? *New Scientist, 129*(1757), 48–51.

Thomas, C., & Bevan, N. (Eds.). (1995). *Usability context analysis: A practical guide* (Version 4). Teddington, England: Serco Usability Services.

Vintner, O., & Poulsen, P. M. (1996). *Experience driven software process improvement.* Paper presented at Software Process Improvement '96, Brighton, England [Online]. Available: http://www.iscn.ie/news/sp96/o.vinter.html

VI
Unified User Interfaces

19 The Concept of Unified User Interfaces

Constantine Stephanidis

This chapter introduces the concept of Unified User Interfaces and highlights some of its distinctive properties that qualify it as a plausible and effective approach towards the goal of User Interfaces for All. In particular, Unified User Interfaces seek to convey a new perspective into the development of user interfaces aiming to provide a principled and systematic approach towards coping with diversity in the target user requirements, tasks and environments of use.

Today, software products support interactive behaviors that are biased toward the "typical," or "average" able-bodied user, familiar with the notion of the "desktop" and the typical input and output peripherals of the personal computer. This has been the result of assumptions (by product developers) regarding the target user groups, the technological means at their disposal, and the type of tasks supported by computers. Thus, the focus has been on "knowledgeable" workers, capable and willing to use technology in the work environment, to experience productivity gains and performance improvements.

Though the information society is still in its infancy, its progressive evolution has already invalidated (at least some of) the assumptions in the preceding scenario. The fusion between information technologies, telecommunications, and consumer electronics has introduced radical changes to traditional markets and complemented the business demand with a strong residential component. At the same time, the type and context of use of interactive applications is radically changing, due to the increasing availability of novel interaction technologies (e.g., personal digital assistants, kiosks, cellular phones, and other network-attachable equipment) that progressively enable nomadic access to information.

The aforementioned paradigm shift poses several challenges: Users are no longer only the traditional able-bodied, skilled, and computer-literate professionals; product developers can no longer *know* who their target users will be; information is no longer relevant only to the business environment; and artifacts are no longer bound to the technological specifications of a predefined platform. Instead, users are potentially *all* citizens of an emerging information society who demand customized solutions to obtain timely access to virtually any application, irrespective of where and how it runs.

This chapter introduces the concept of *unified user interfaces* and points out some of its distinctive properties that render it an effective approach toward the goal of *user interfaces for all*. In particular, unified user interfaces seek to convey a new perspective on the development of user interfaces that provides a principled and systematic approach toward coping with diversity in the target users' groups, tasks, and environments of use. In other words, unified user interfaces provide a pathway toward accommodating the interaction requirements of the broadest possible end-user population. In this respect, they also facilitate a concrete insight into how the principles of *design for all* can shape prevailing human–computer interaction (HCI) design and development practices so that the range and scope of interactive experiences offered to the end-user are broadened.

The notion of a unified user interface originated from research efforts aiming to address the issues of accessibility and interaction quality for people with disabilities. The intention was to articulate some of the principles of design for all in a manner that would be applicable and useful to the conduct of HCI. Subsequently, these principles were extended and adapted to depict a general proposition for HCI design and development, which was complemented by specific methodologies, techniques, and tools.

The objective of this part of the book is to unfold the rationale and sketch the outlines of unified user interfaces as these evolved through a series of collaborative research and development (R&D) projects. Thus, this introductory chapter serves the purpose of defining and elaborating some key concepts, so as to provide an account of the underlying rationale and premises of unified user interfaces. Subsequent chapters describe: (a) technical details, recommendations, and guidelines for the construction of unified user interfaces (see chaps. 20, 21, and 22), (b) specific development efforts targeted toward the provision of tools for design and implementation (see chaps. 23 and 24), as well as (c) a comprehensive case study that brings practical insight to the application of unified user interface principles in the domain of Web-browsing technologies (chap. 25).

In the remainder of this chapter, we first describe the main challenges that unified user interface development aims to address. Then, we define the concept and compare it with traditional approaches to implementing interactive software. Finally, we elaborate on the concept of adaptation in order to obtain an understanding of what is the need for, as well as what constitutes desirable adapted interactive behaviors in the context of unified user interfaces.

THE CHALLENGES: ACCESSIBILITY AND INTERACTION QUALITY

Accessibility

In the past, the term *computer accessibility* was usually associated with access to interactive computer-based systems by people with disabilities. In traditional efforts to improve accessibility, the driving goal has been to devise hardware and software configurations (or alternative access systems) that enable disabled users to access interactive applications originally developed for able users. There have been two possible technical routes to alleviate the lack of accessibility of interactive software products. The first is to treat each application separately and take all the necessary implementation steps to arrive at an alternative accessible version *(product-level adaptation)*. The second alternative is to "intervene" at the level of the particular interactive application environment (e.g., MS-Windows, X Windowing System) in order to provide appropriate software and hardware technology, so as to make that environment alternatively accessible *(environment-level adaptation)*. In effect, with the latter option, the scope of accessibility is extended to cover potentially all applications running under the same interactive environment, rather than a single application.

Product-level adaptation has been practically tackled as redevelopment from scratch. Due to the high costs associated with this strategy, it is considered the least favorable option for providing alternative access. As a result, environment-level adaptations, addressing a range of applications, have been acknowledged as a more promising strategy (Mynatt & Weber, 1994).

The architectural view of environment-level adaptations is outlined in Fig. 19.1. The server is the environment-supported interactive functionality, whereas the client denotes running interactive applications.

In the past, the vast majority of approaches to environment-level adaptations have focused on the issue of accessibility of graphical environments by blind people. In this context, the GUIB Project (Textual and Graphical User Interfaces for Blind People) (GUIB Project, 1995), partially funded by the TIDE Program of the European Commission (DG XIII), and the MERCATOR Project (Mynnatt & Edwards, 1992) in the United States, are considered the most sophisticated efforts toward this direction (Mynatt & Weber, 1994). Through such projects, it became apparent that any effort toward environment-level adaptations should be based on well-documented and operationally reliable software infrastructures, supporting effective and efficient extraction of dialogue primitives during user–computer interaction. Such dynamically extracted dialogue primitives are to be reproduced, at run time, in alternative input/output forms, directly supporting user access. Recent examples of software infrastructures that satisfy the aforementioned requirements are the Active Accessibility™ Technology, by Microsoft[1], and the Java Accessibility™ Technology, by JavaSoft[2] (see also chap. 16, this volume).

---

[1]http://www.microsoft.com/enable/msaa/

[2]http://java.sun.com/products/jfc/accessibility/doc/

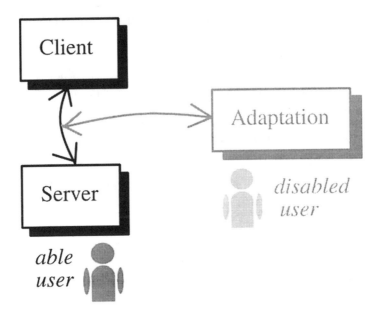

FIG. 19.1.   The general architectural model of accessibility-oriented environment-level adaptations.

Despite recent progress, the prevailing practices aiming to provide alternative access systems, either at the product or environment level, have been criticized for their essentially reactive nature. The critique is grounded on two lines of argumentation. The first is that reactive solutions typically provide limited and low-quality access, something that has already been identified in the context of nonvisual interaction, where non-visual user interfaces should be more than automatically generated adaptations of visual dialogues (Savidis & Stephanidis, 1995). The second line of critiquing tackles the economic feasibility of the reactive paradigm to accessibility. In particular, it is argued that continuing to adopt prevailing practices within the software industry leads to the need to reactively develop solutions for interactive computer-based products, accessible to users with situational, temporary, or permanent disability. Clearly, this is suboptimal for all parties concerned and renders the reactive approach to accessibility inadequate and inappropriate in the long run. Instead, what is required is a proactive approach to cater to the requirements of the broadest possible end-user population.

Interaction Quality

Having briefly described the accessibility challenge, we now shift to the second aspect that motivated unified user interface development, namely the compelling need for interaction quality. In the present context, interaction quality entails both end-user perceived quality, as well as certain process-oriented, nonfunctional attributes that are required to guarantee quality in use for interactive products. End-user perceived quality is a concept that has been developed and documented in the literature on user-centered design and software quality (Bevan, 1995). However, the available material addressing the latter perspective on quality does not offer any insight to the case of disabled and elderly people.

Process-oriented, nonfunctional quality attributes, on the other hand, concern engineering properties and requirements of the tools through which a product is developed. For example, the requirement for "linking to," rather than directly "calling" an interaction platform, is a nonfunctional quality attribute of a user interface, which guarantees platform independence. Platform independence, in turn, ensures that an interface can be used across different interaction platforms (e.g., visual and nonvisual), thus leading to increased quality in use, for a broader range of users. Whether or not such a nonfunctional quality is supported by a particular tool is an issue that can be determined by assessing the extend to which the tool possesses certain required and recommended properties (see chap. 22, this volume).

From the previous discussion, it follows that a posteriori adaptations are bound to deliver low-quality solutions for disabled and elderly people. This is attributed to several factors:

- First, the current generation of tools for building interactive products seldom support the nonfunctional quality attributes assumed by unified user interface development.
- Second, due to intensive competition, the focus of mainstream industries is on minimizing the time-to-market, which, in turn, leads to narrow usability budgets (if any) and low-quality products and services. Consequently, any a posteriori adaptations to such products are bound to be suboptimal.
- Finally, a posteriori adaptations suffer from the lack of an accumulated body of knowledge that can be used to inform assistive technology designers. With only a few exceptions, the vast majority of assistive technology projects are based on intuition, rather than good and proven human-centered practices. As a result, the adaptations developed are not those expected by the assistive technology customer base. This can be explained when one considers the structural characteristics of the industry that lead to a low rate of innovative activity, but also the lack of methodological ground to guide and inform adaptable and adaptive design. Consequently, the results are suboptimal for all parties concerned.

Accessibility and Interaction Quality in the Information Society

The shift of paradigm in the context of the emerging information society (see also chap. 1, this volume) emphasizes even further the compelling need for proactive approaches, as accessibility and interaction quality become global targets. This is also evidenced by recent changes in the patterns of demand and supply for information technology (IT) products and services.

In particular, the changing pattern in the demand for IT products is due to (a) the increasing number of computer users characterized by diverse abilities, skills, requirements, experiences, and preferences, (b) the requirement for product specialization to cope with the increasing knowledge-based nature of tasks, and (c) the requirement for increasingly nomadic access to information.

To respond to such a demand, the supply side engages in efforts to provide (a) a wide range of interaction technologies (e.g., new input/output devices and interaction techniques) incorporating facilities for multimodal and multimedia interactions, (b) novel interaction paradigms (e.g., virtual and augmented realities, ubiquitous computing, Java and World Wide Web, wearable equipment), and (c) support tools for communication-intensive tasks, collaboration, and cooperation.

In all these efforts, accessibility and interaction quality become prerequisites for delivering products and services that can be customized to individual contexts of use and facilitate access to community-wide information resources by the target user groups. Therefore, they need to be carefully planned and embedded into the product life cycle, from the early phases of design, through to implementation and testing. In order for this to materialize, a number of currently prevailing HCI assumptions need to be revisited. One such critical assumption relates to the popular notion of designing for the *average* user. With the shift of the paradigm from business computing to communication-intensive, group-centered, collaborative, and cooperative activities in a global information space, it is evident that designing for the average user is becoming more of an illusion, rather than a realistic design goal. This is already evident in the case of Web-based applications and services. Designers no longer "know" who their target users might be, as anyone with an Internet connection may potentially access their Web pages.

In this context, the study of people with disabilities is especially relevant, not only from the point of view of demographics and the shared social responsibility, but also due to the fact that the accessibility requirements posed by disabled and elderly people challenge the state of the art in usability engineering (see chaps. 29 and 30, this volume). It should be noted that in the past, the problem faced by disabled and elderly people was not only lack of access to interactive computer-based products and services, but also low quality of interaction, in the cases where access was granted.

THE CONCEPT OF UNIFIED USER INTERFACES AND UNIFIED USER INTERFACE DEVELOPMENT

Unified User Interfaces

A unified user interface is defined as an interactive system that comprises a single (i.e., unified) interface specification, targeted to potentially all user categories and contexts of use (see Fig. 19.2). Such a specification can be built using either a traditional programming language, or a dedicated language that allows the construction of the unified user interface as a composition of abstractions at different levels.

The distinctive property of a unified user interface is that it can realize alternative patterns of interactive behavior, at the physical, syntactic, or even semantic levels of interaction, by automatically adapting to accommodate specific user- and context-oriented requirements. Typically, such alternative interactive behaviors encompass interaction elements available in different toolkits or interaction platforms (e.g., Windows95, toolkit for nonvisual interaction), suitable for the different target user groups (e.g., sighted and blind users, respectively).

The unified implementation, which, as mentioned earlier, can be either explicitly programmed, or produced by compiling an interface specification, undertakes the mapping of abstract interaction elements to concrete/physical resources available in the target toolkits. This is achieved by utilizing specific functionality or tools (e.g., toolkit servers) to connect (or link) with the underlying platform(s) in order to utilize the available interaction resources in a manner that is independent from the platform.

The need for unified user interface development emerged in the process of constructing an appropriate technical framework, in order to address the *user interfaces for all* objective. Figure 19.3 depicts the trajectory that led to the unified user interface concept and the respective design and development paradigms. The starting point of this trajectory has been that, when we aim to design interactive software applications and telematic services accommodating the requirements of "all" users in different contexts of use, we need to take into consideration the diverse attributes that characterize the users and the envisaged contexts of use (Step

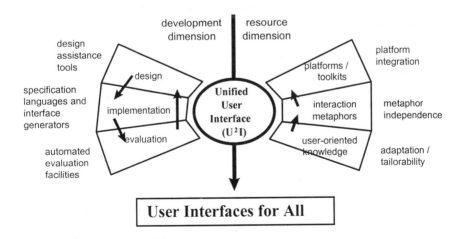

FIG. 19.2   The concept of unified user interfaces.

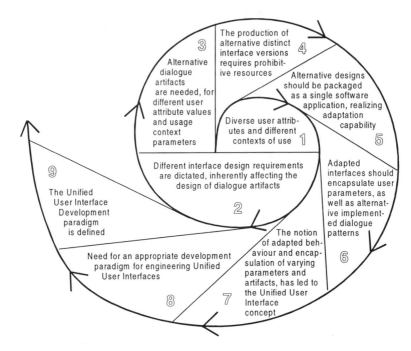

FIG. 19.3.   From diverse user requirements and usage contexts, to the unified user interface engineering paradigm.

1). These varying user- and usage-context-attribute values give rise to different design requirements, which, in turn, affect the design of dialogue artifacts (Step 2). As a result, alternative dialogue artifacts will have to be constructed, at various points of the interface design process, as dictated by the differing user- and usage-context- attribute values (Step 3).

When trying to map the outcomes of such design processes into an implemented interactive application, a key issue is how the various alternative dialogue artifacts will be "packaged." As discussed earlier, the production of alternative interface versions requires prohibitive resources for development, maintenance, upgrading, and distribution (because all distinct versions should potentially be made available for concurrent "execution"), which turns out to be practically unrealistic (Step 4). This is particularly evident in the case of nondesktop computer systems, such as public access terminals, that anyone should be able to use. Consequently, the packaging of the various alternative dialogue patterns into a single software application has been considered the most promising approach. In this context, packaging may not necessarily imply the construction of a monolithic software system incorporating all the various dialogue artifacts; rather, it can also be instantiated as a logical collection within a single resource. For example, a repository can be made directly accessible by a single software application that encompasses adaptation capabilities, thus being able to select the most appropriate dialogue patterns for a particular end-user and target usage context (Step 5).

In order to facilitate such a capability, interactive applications should encompass information about individual users, as well as alternative dialogue patterns in an implemented form (Step 6). This notion of encapsulation has led to the definition of the unified user interface concept, which proposes the realization of user- and usage-context adapted behavior, by encapsulating all the varying design parameters and alternative dialogue artifacts (Step 7). The need for an appropriate development strategy for unified user interfaces (Step 8) has led to the introduction of the unified user interface development paradigm, targeted to the development of unified user interfaces (Step 9). The unified user interface development paradigm entails interdisciplinary processes driving the production of automatically adapted software applications and services. It is general enough so as not to exclude particular design and implementation practices, while, at the same time, it offers sufficient details to drive the engineering process of unified user interface software. As any new development paradigm, it naturally requires some initial investment to be effectively adopted, assimilated, and applied. However, if the constructed software products are intended to be used by user populations with diverse requirements, operated in different usage contexts, it is argued that the gains will heavily outweigh the overhead of additional resources that need to be invested.

The concept of a unified user interface was developed in the context of TIDE-ACCESS, a European Commission–funded collaborative R&D project (see Acknowledgments) and has been described in a series of papers (Akoumianakis, Savidis, & Stephanidis, 2000; Akoumianakis & Stephanidis, 1997; Savidis, Paramythis, Akoumianakis, & Stephanidis, 1997; Savidis, Stephanidis, & Akoumianakis, 1997; Stephanidis, 1995). The respective development methodology has been presented in the course of mainstream HCI conference tutorials (Stephanidis, Akoumianakis, & Paramythis, 1999; Stephanidis, Savidis, & Akoumianakis, 1997). Furthermore, it has been applied in several design cases, such as the development of a hypermedia application accessible by blind people (Morley, Petrie, O'Neill, & McNally, 1999; Petrie, Morley, McNally, O'Neill, & Majoe, 1997), the development of two communication aid applications for speech-motor- and language-cognitive-impaired users (Kouroupetroglou, Viglas, Anagnostopoulos, Stamatis, & Pentaris, 1996) and the development of an accessible browser for Web-based interaction (see chap. 25, this volume) with a metropolitan information system, in the context of the European Commission–funded ACTS-AVANTI project (see Acknowledgments).

Unified User Interface Development

Schematically, the phases of unified user interface development are depicted in the diagram of Fig. 19.4. Unified design entails an early account of the broadest possible range of end-user requirements and contexts of use, so as to develop effective representations depicting the global task execution context. Unified implementation, on the other hand, requires the capability to encapsulate design alternatives into suitable dialogue patterns and to map abstract design components to corresponding implemented (interaction platform-specific).

Two distinctive requirements characterize unified user interfaces. The first is the requirement for an analytical design activity leading to the *representation* of the design knowledge required to reveal and differentiate among plausible design alternatives. The second requirement is that of *encapsulation* of the corresponding dialogue patterns into a (conceptually) single interactive entity. In this context, representation implies the use of suitable notations to capture and encode both design artifacts and accompanying design rationale. On the other hand, encapsulation entails the use of suitable dialogue specification techniques (programmatic, declarative, etc.) to manipulate interactive artifacts in a manner that is not dependent on a particular target interaction platform (e.g., by avoiding direct "calls" to the platform's interactive facilities).

The design of a unified user interface entails three distinctive iterative tasks, namely *enumeration* of design alternatives, *abstraction* toward reusable unified design components, and *rationalization* of the design space. Enumeration of design alternatives can be attained through techniques that foster an analytical design perspective (such as design scenarios, envisioning, ethnographic methods) and facilitate the identification of plausible design options for different user

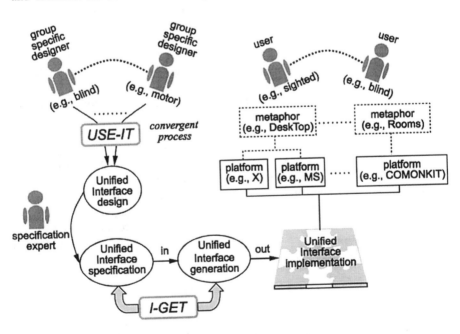

FIG. 19.4.   The design and implementation phases for unified user interfaces.

groups (i.e., design space). Thus, for example, given the task of designing a user interface accessible by both sighted and blind users, enumeration would reveal visual (e.g., menu) and nonvisual (e.g., three-dimensional acoustic sphere) candidate artifacts for making a choice from a list of elements.

Abstraction in design entails the identification of abstract interaction components that can be used to encapsulate alternative concrete artifacts. Such abstract components are decoupled from platform-, modality-, or metaphor-specific attributes to provide a kind of reusable design "library." In the context of our previous example, abstraction would lead the designer to identify an abstract "selector" possessing only the necessary attributes (e.g., size of selection set and choice indicator). Such an abstract element can be subsequently mapped to any particular concrete instance, given a specific user and context of use. Moreover, abstract components may be used to compile composite interface elements suitable for different users and contexts of use.

Finally, rationalization of the design space implies the explicit encoding of the rationale for mapping an abstract design element (e.g., the abstract selector) to a concrete artifact (e.g., a menu, or a three-dimensional acoustic sphere). This is typically achieved by assigning criteria to design alternatives and providing a method for selecting the maximally preferred option.

To facilitate encapsulation, unified user interface development requires techniques that enable (a) the grouping of alternative dialogue patterns (e.g., implemented design alternatives, catering to different user requirements) on the basis of an abstraction model and (b) the context-sensitive mapping of abstract components to suitable concrete artifacts. To this effect, the process of unified user interface implementation involves (a) the construction of a unified user interface as a composition of abstractions at different levels of interaction, (b) the manipulation and management of the physical resources (e.g., various toolkits), and (c) the establishment of the relationships between the involved abstractions and the available physical resources.

Unified User Interface Development Versus Traditional Practices

The premise of unified user interface development is that of studying the global execution context of tasks and human activities, to identify suitable alternatives to accommodate individual requirements. This calls for analytical insight and pluralism in the respective outcomes, as no single solution is likely to be acceptable to all users. Such a focus contrasts the prevailing HCI design philosophy and supporting methodologies, which are primarily single-artifact oriented. It follows that unified user interfaces require a broader scope of design to explicitly account for context-oriented phenomena, as well as a powerful development framework to enable the generation of user interface implementations through specifying, rather than programming, interactive dialogues.

Such an approach necessitates a shift of perspective, in relation to design and development practices. In particular, the design of unified user interfaces requires explicit means to account and model context-oriented parameters. However, such contextual insights can be facilitated only by adopting more suitable units for analyzing and modeling interactions (e.g., activity), than contextually isolated user actions (or keystrokes), which have been the primary focus of cognitive models. Additionally, the focus of design is on populating design spaces, rather than identifying a single best fit. Table 19.1 summarizes some of the major differences between unified and traditional design practices.

Moreover, unified development requires corresponding means to provide the basis for user interface implementation. This challenges traditional practices with regard to both the architectural model according to which unified artifacts become embedded into user interface implementations, and the mechanisms offered for context-sensitive processing of alternatives toward the selection of a maximally preferred option. In Table 19.2, we contrast traditional and unified user interface development.

Table 19.1.
Contrasting Traditional and Unified User Interface Design

Design Criteria	Traditional Development Paradigm	Unified User Interface Development
Focus	Single artifact that fits all	Analytical insights to populate design spaces
Outcome	Single-object hierarchy	Polymorphic task hierarchy
Process	Top down or bottom up	Middle out
Scope of design representation	Implicitly bound to the object hierarchy	Bound to rationalized design spaces; explicit in the run-time behavior

TABLE 19.2.
Contrasting Traditional and Unified User Interface Development

Design Criteria	Traditional Development Paradigm	Unified User Interface Development
Implementation model	Programming as the basis for generating the user interface implementation	Generation from specifications
Premise of run-time code	Making direct calls to a platform	Linking to the platform
Platform utilization	Multiplatform environments	Multiple toolkit environment
Platform independence	Generalization across platform properties	Platform abstraction mechanism

Unified User Interface Development Tools

The unified user interface development paradigm is supported by a set of development tools, which have been built to provide an integrated framework that efficiently supports the design and implementation of unified user interfaces. The main characteristics of this framework are:

- *Platform independence,* intended to address the pluralism of interaction platforms and graphical environments (e.g., MS-Windows™, the X Windowing System), offering the versatility required for the management of different environments.
- *Metaphor independence,* so as to cater to the interaction needs and characteristics of diverse target user groups, which may necessitate the coupling of different interaction metaphors to different categories of users and usage situations.
- *Automatic adaptation capabilities,* so that the resulting user interfaces are adaptable and adaptive to the individual user abilities, requirements, skills, and preferences.
- *Unified interface specification,* which aims to reduce the overall development costs for unified user interfaces through the introduction of specification-oriented (rather than implementation-oriented) interface construction techniques.

In order to efficiently support the implementation of unified user interfaces, the framework comprises:

1. The I-GET User Interface Management System, based on a high-level fourth-generation language for user interface specification (presented in chap. 24, this volume). The resulting interface implementations can run on different target platforms, which may correspond to particular interaction metaphors.
2. The USE-IT design support tool, through which adaptability of the user interface to the specific requirements of the target user group is achieved (presented in chap. 23, this volume). This tool takes the appropriate decisions regarding the lexical characteristics of the dialogue, based on (a) knowledge about the user characteristics, abilities, and preferences and (b) knowledge about the structure of the lexical level characteristics with respect to the various target user groups (i.e., interaction objects, interaction techniques, devices, etc).

The unified user interfaces that are developed with I-GET automatically inquire the adaptability decisions generated by the USE-IT tool, and apply these decisions during user–computer interaction. Subsequent chapters of this part of the book elaborate on selected key properties of these user interface development tools.

ADAPTATION DIMENSIONS IN UNIFIED USER INTERFACE DEVELOPMENT

In developing unified user interfaces, the primary challenge, though not the only one, is to envision, model, and deploy the required type and range of adapted interactive behavior. In the emerging interactive paradigm (characterized by nomadicity, ubiquitousness, seamless interactivity, etc.), context-oriented assessments are likely to determine both the type and range of adapted behavior that best suits a given human task or activity. For example, a unified user interface may adapt depending on the type of interaction technology available at the end-user's site. Thus, for instance, in the case of low screen resolution and the presence of audio output hardware, the interface may "decide" to provide fewer visual effects, while emphasizing audio feedback. Context of use is also important to the extent that it provides an account of primary and secondary tasks. For example, when driving, visual attention is primarily focused on the road, thus any type of secondary interaction should not disturb the user's focus of attention.

From the preceding, it follows that our concept of automatically adapted behavior denotes context-sensitive adaptation, which in turn, may be distinguished into:

- *User-adapted behavior,* where the interface is capable of automatically selecting an interaction approach more appropriate to the particular end-user (i.e., *user awareness*).
- *Usage-context adapted behavior,* where the interface is capable of automatically selecting an interaction approach more appropriate to the particular situation of use (i.e., *usage context awareness*).

Triggering interface adaptation is a run-time process that requires some kind of inference capability to reason about, and draw decisions on, several items (see also Fig. 19.5): (a) *which* are the parameters that necessitate the introduction of adaptations (i.e., user context and usage context), (b) *when* adaptation is to be applied (i.e., before or in the course of interactive episodes), and (c) *how* adaptation is to be realized by means of specific interaction artifacts, which may concern the semantic, syntactic, or lexical level of interaction.

In general, the overall adapted interface behavior can be viewed as the result of two complementary classes of system initiated actions: (a) adaptations driven from initial user and context information, usually acquired prior to initiating interaction[3], and (b) adaptations decided on the basis of information inferred/extracted by monitoring interaction.[4]

---

[3]Such information may indicate user-specific attributes (e.g., physical and/or cognitive abilities, domain expertise), platform specifications (e.g., terminal capabilities), and so on.

[4]This type of information would be useful in inferring that the user is tired, identifying dynamic user preferences in a particular interaction style, detecting changes in environmental parameters (e.g., noise), and so on.

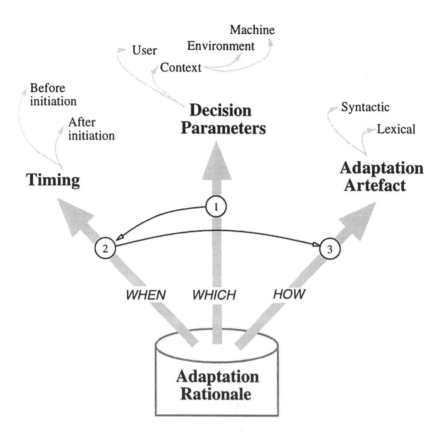

FIG. 19.5.   Resolving "which" (i.e., driving parameters), "when" (i.e., timing), and "how" (i.e., choice of adaptation artifacts) in the context of the adaptation-oriented run-time decision-making process.

The former behavior is referred to as *adaptability,* reflecting the interface's capability to automatically tailor itself initially to each individual end-user. The latter behavior is referred to as *adaptivity,* and characterizes the interface's capability to cope with the dynamically changing/evolving situation of use. It should be noted that adaptability is crucial to assure accessibility, because the essence is to initially (i.e., before initiation of interaction) provide a fully accessible interface instance to each particular end-user. Furthermore, adaptivity can be applied only on accessible running interface instances (i.e., ones with which the user is capable of performing interaction), because interaction monitoring is required for the identification of changing/emerging decision parameters that may drive dynamic interface enhancements. The complementary roles of adaptability and adaptivity are depicted in the diagram of Fig. 19.6, whereas the key differences among these two adaptation methods are illustrated in Table 19.3.

CONCLUSIONS

There are a number of conclusions that may be drawn from the preceding account. First of all, accessibility and interaction quality have become global requirements that do not relate only to disabled and elderly people, but to the population at large. As such, they need to be carefully planned and embedded into the development life cycle of an interactive product or service from the early phases of design to implementation and testing. To this end, the interaction requirements of disabled and el-

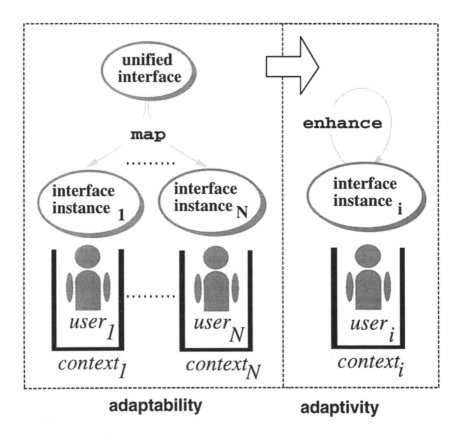

FIG. 19.6.   The complementary roles of adaptability (left) and adaptivity (right) in unified user interfaces encapsulating automatically adapted behaviors.

TABLE 19.3.
Key Differences Between Adaptability and Adaptivity in the Context of Unified
User Interfaces

Adaptability	Adaptivity
1. User and usage-context attributes are considered known prior to interaction	1. User/usage-context attributes are dynamically inferred/detected.
2. "Assembles" an appropriate initial interface instance for a particular end-user and usage-context.	2. Enhances the initial interface instance already "assembled" for a particular end-user and usage-context.
3. Works before interaction is initiated.	3. Works after interaction is initiated.
4. Provides a user-accessible interface.	4. Requires a user-accessible interface.

derly people constitute the "cutting edge" of scientific challenge and pose demands on additional technological developments in user interface software and technology. Furthermore, it is now widely acknowledged that improvements to facilitate the socioeconomic integration of people with disabilities in the information age are likely to bring about global benefits for all users (see chap. 29, this volume).

Second, unified user interfaces represent a concept that justifies the argument that designing user interfaces for the broadest possible end-user population is more of a challenge than a utopia (Stephanidis & Emiliani, 1999). This is grounded on the firm experiences within the context of various R&D projects, funded by national and transnational nonmarket institutions (e.g., the European Commission, the National Science Foundation in the United States, and the Ministry of International Trade and Industry in Japan), as well as from recent industrial developments, such as the various accessibility initiatives launched by major software vendors (e.g., Sun, Microsoft, IBM).

Finally, it should be pointed out that unified user interfaces are not considered to be the only possible approach seeking to address user interfaces for all. However, it is the first systematic effort that has been thoroughly developed and applied in practice. Subsequent chapters elaborate further on several technical details, thus refining the general principles into a practical interface development framework, and distilling recent experience and practice into a collection of useful recommendations. Though unified user interface development, as presented in the chapters that follow, assumes powerful tools and advanced user interface software technology, our intention is not to argue that such a realization is the only one possible, or indeed, superior to any alternative that may be developed in the future. Therefore, in our account of unified user interfaces, we have intentionally left out many implementation-specific details concerning the realization of unified user interface development tools. Nevertheless, it is argued that, in the long run, the availability of such tools is likely to be a critical factor of the concept's potential adoption and diffusion.

REFERENCES

Akoumianakis, D., Savidis, A., & Stephanidis, C. (2000). Encapsulating intelligent interactive behavior in unified user interface artifacts. *Interacting With Computers, Special Issue on The Realities of Intelligent Interface Technology, 12*(4), 383–408.

Akoumianakis, D., & Stephanidis, C. (1997). Supporting user-adapted interface design: The USE-IT System. *Interacting with Computers, 9*(1), 73–104.

Bevan, N. (1995). Measuring usability as quality of use. *Software Quality Journal, 4,* 115–130.

Edwards, A. (1995). Computers and people with disabilities. In A. Edwards (Ed.), *Extra-ordinary human–computer interaction—Interfaces for users with disabilities* (pp. 19–43). New York: Cambridge University Press.

GUIB Consortium. (1995). *Textual and graphical user interfaces for blind people* (The GUIB PROJECT—Public Final Report). London: RNIB Press.

Kouroupetroglou, G., Viglas, C., Anagnostopoulos, A., Stamatis, C., & Pentaris, F. (1996). A novel software architecture for computer-based interpersonal communication aids. In *Proceedings of 5th International Conference on Computers and Helping People with Special Needs* (pp. 715–720). Wien, Austria: Oldenberg.

Morley, S., Petrie, H., O'Neill, A.-M., & McNally, P. (1999). Auditory navigation in hyperspace: Design and evaluation of a non-visual hypermedia system for blind users. *Behavior and Information Technology, 18*(1), 18–26.

Mynatt, E. D., & Edwards, W. K. (1992). The Mercator environment: A nonvisual interface to the X Window system (Graphics Visualization & Usability Center, Tech. Rep. No. GIT-GVU-92-05). Atlanta: Georgia Institute of Technology.

Mynatt, E., & Weber, G. (1994). Nonvisual presentation of graphical user interfaces: Contrasting two approaches. In *Proceedings of ACM Conference on Human Factors in Computing Systems* (pp. 166–172). New York: ACM Press.

Petrie, H., Morley, S., McNally, P., O'Neill, A-M., & Majoe, D. (1997). Initial design and evaluation of an interface to hypermedia systems for blind users. In *Proceedings of Hypertext '97* (pp. 48–56). New York: ACM Press.

Savidis, A., Paramythis, A., Akoumianakis, D., & Stephanidis, C. (1997). Designing user-adapted interfaces: The unified design method for transformable interactions. In *Proceedings of the ACM Conference on Designing Interactive Systems* (pp. 323–334). New York: ACM Press.

Savidis, A., Stephanidis, C., & Akoumianakis, D. (1997). Unifying toolkit programming layers: A multi-purpose toolkit integration module. In M. D. Harrison & J. C. Torres (Eds.), *Conference Proceedings of the 4th Eurographics Workshop on Design, Specification and Verification of Interactive Systems* (pp. 177–192). Wien, Germany: Springer-Verlag.

Stephanidis, C. (1995). Towards user interfaces for all: Some critical issues panel session on user interfaces for all: Everybody, everywhere, and anytime. In *Proceedings of HCI International '95* (pp. 137–142). Amsterdam: North-Holland, Elsevier Science.

Stephanidis, C., Akoumianakis, D., & Paramythis, A. (1999). Coping with diversity In HCI: Techniques for adaptable and adaptive interaction. *Tutorial No. 11 in HCI International '99* [Online]. Available: http://www.ics.forth.gr/proj/at-hci/html/tutorials.html

Stephanidis C., & Emiliani, P-L. (1999). Connecting to the information society: A European perspective. *Technology & Disability Journal, 10*(1), 21–44.

Stephanidis, C., Savidis, A., & Akoumianakis, D. (1997). Unified interface development: Tools for constructing accessible and usable user interfaces. *Tutorial No. 13 in HCI International '97* [Online]. Available: http://www.ics.forth.gr/proj/ at-hci/html/tutorials.html

20 The Unified User Interface Software Architecture

Anthony Savidis
Constantine Stephanidis

This chapter describes the Unified User Interface architecture; the aim is to provide guidance to user interface developers on how to employ the Unified User Interface development methodology in order to implement the interactive component of applications and services. The rationale of the proposed architectural abstraction is explained by first reviewing the state of the art, and secondly identifying some of the prominent shortcomings of the currently available support for adaptable and adaptive interactive behavior. Subsequently, the basic software components of the Unified User Interface architecture are described in terms of their functional role and communication patterns establishing the details of information flow through these components, in order to perform user- and usage context- oriented adaptation.

In the emerging information society, an increasing range of human activities will ultimately become mediated by computers in various types, forms, and sizes. Such an evolution is driven by continuous efforts to provide less specialized, and interoperable, hardware in which software is increasingly being used to replace previously hardwired functions. Such a trend is already evidenced by the recent proliferation of interaction techniques and platforms, such as interactive TVs, TV interfaces to computers, Web phones, and so on. This strategy creates a wide range of opportunities while opening up the road toward automatic adaptations, because software may be adapted "on the fly" as required, whereas hardware has to be used "as is."

In view of the emerging technological infrastructure, the unified user interface development paradigm provides a new and promising technical framework to enable the construction of self-adapting software products, which are open, expandable, and compliant with other current and popular development paradigms, such as distributed object computing and component-based development. To this end, a central issue in unified user interface development (as indeed in any other system development effort) is the notion of software architecture. A software architecture can be used to study an existing system, as well as to guide the development of new ones.

389

This chapter has two main objectives. The first one is to review recent work on user interface software architectures, emphasizing their primary focus and technical objectives. The second is to elaborate on the components of the proposed unified user interface architecture, in order to guide user interface developers toward the construction of unified user interfaces. Our use of the term *architecture* follows the definition offered by Jacobson, Griss, and P. Johnson (1997): "The software architecture, first of all, defines a structure. Software components have to fit into some kind of design.... Second, the architecture defines the interfaces between components. It defines the patterns by which information is passed back and forth through these interfaces" (p. 38). Consequently, our elaboration of the unified user interface architecture seeks to provide an insight into the structure, interfacing, and information exchange among components of the proposed architectural abstraction.

SOFTWARE ARCHITECTURES AND ADAPTATION

Architectural Models of Interactive Systems

Early work following the introduction of graphical user interfaces had focused on window managers, event mechanisms, notification-based architectures, and toolkits of interaction objects. Such architectural models were quickly supported by mainstream tools, thus becoming directly encapsulated in the prevailing user interface software and technology. Today, all available user interface development tools support object hierarchies, event mechanisms, and callbacks as the basic implementation model.

In addition to these early attempts at architectural components of user interface software, there have been other architectural models, with a different focus, which, however, did not gain as much acceptance in the commercial arena as was originally expected. The *Seeheim* model (Green, 1985) and its successor, the *Arch* Model (UIMS Developers Workshop, 1992), have been mainly defined with the aim to preserve the so-called "principle of separation" between interactive and noninteractive code of computer-based applications. These models became popular as a result of the early research work on user interface management systems (UIMS) (Myers, 1995).

Apart from these two architectural models, mainly referring to the *interlayer* organization aspects of interactive applications, there have been two other more implementation-oriented models, with an object-oriented flavor: the model view controller (MVC) model (Goldberg, 1984) and the presentation-abstraction-control (PAC) model (Coutaz, 1990). Those models focus on *intralayer* software organization policies, by providing logical schemes for structuring the implementation code. All four models, though typically referred to as architectural frameworks, are today considered as metamodels, because they do not meet the fundamental requirements of a software architecture, as defined by Jacobson et al. (1997).

The Seeheim Model.   The Seeheim model has been the first related to the run-time structure of interfaces produced by UIMSs, and has reflected the orientation toward the physical and conceptual separation of the interaction-specific software from the noninteractive functional core.

It introduces four logical system components (see Fig. 20.1, left part): (a) *application interface model,* which holds information to internally interface the functional core, (b) *dialogue manager,* which holds information regarding the structure of dialogue and its syntactic rules, (c) *I/O manager,* which holds information about the structure and syntax of I/O (input/output) items, and (d) *representation manager,* which manages the mapping of I/O objects to internal objects, and vice versa. A discussion regarding the Seeheim model can be found in ten Hagen (1991).

The Arch Model.   This model was introduced in the UIMS Developers Workshop (1992), and is considered to be the successor of the Seeheim model. It enhances the basic Seeheim model by introducing explicit interfacing layers among the various components. It has been called the Arch model because it was first drawn in a way geometrically resembling the arch shape. It engages five logical components (see Fig. 20.1, right part), three of which come from the Seeheim model (i.e., *dialogue component, domain-specific component,* and *interaction toolkit component*), whereas the other two play the role of an interfacing layer (i.e., *domain adaptor component* and *presentation component*), adapting, as required, the constructs, or the functionality provided by the components drawn below it. The Arch model made more explicit the need for *intermediate software layers* connecting the dialogue control implementation with both the functional core and the underlying toolkit.

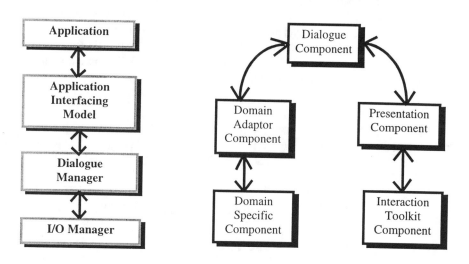

FIG. 20.1.   The Seeheim model (left part), and its successor, the Arch model (right part).

The MVC Model.   MVC is an architectural model, introduced in the context of Smalltalk-80 (Goldberg, 1984), for building interactive systems. The essence of this model is to invariably provide a clear separation between the internals of a system (functional core) and its corresponding interactive modules. According to MVC, an interface consists of three layers (see Fig. 20.2): (a) the *model layer,* which implements the application functionality and provides the internal information structures, (b) the *view layer,* which implements the mechanisms for presenting various aspects of the application layer to the user, and (c) the *controller layer,* which handles the user interaction with the application. The controller processes user input, and invokes the appropriate model function; when the work is done, the model sends back a message to the controller. The view updates the display in response to the model messages, accessing also directly the model for further information. Thus, the model comprises a view and a controller, but it never directly accesses any of them. The view and controller, on the other hand, access the model's functions and the data when required.

The PAC Model.   The PAC model recursively defines the organization of interactive software as a hierarchy of PAC agents. According to Coutaz (1990), a PAC agent defines competence at some level of abstraction. Referring to Fig. 20.3, it is a three-facet logical cluster that includes: (a) presentation, which is a perceivable behavior, (b) abstraction, which is the functional core that implements some internal services and defines an interface to other agents, and (c) control, which links abstraction to a presentation and maintains relationships with other agents.

The C-part of an agent may communicate with corresponding C-parts of hierarchically higher or lower agents. A presentation and its related abstraction never communicate directly, but exchange data in their own formalism via a common control. The main difference with the MVC model is that I/O specific requests are integrated in the presentation component (in contrast to MVC, where this functionality is embedded in the controller).

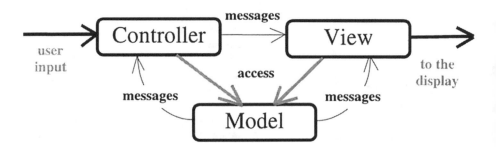

FIG. 20.2.   The MVC model. From Barkakati (1991). Reprinted by permission.

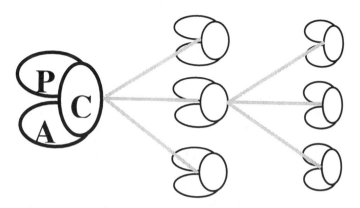

FIG. 20.3.   The PAC model, supporting hierarchical organization agents by connecting to their respective control parts. From Coutaz (1990). Copyright 1990 by IFIP. Reprinted by permission.

Adaptive Interaction

Most of the existing work regarding system-driven user-oriented adaptation concerns the capability of an interactive system to dynamically (i.e., during interaction) detect certain user properties, and accordingly decide various interface changes. This notion of adaptation falls in the *adaptivity* category, that is, adaptation performed after initiation of interaction, based on interaction-monitoring information (see also chap. 19, this volume); this has been commonly referred to as *adaptation during use*. Although adaptivity is considered a good approach to meeting diverse user needs, there is always the risk of confusing the user with dynamic interface updates. An example of potential interaction problems that may result is the *hunting* effect, described in Browne, Totterdell, and Norman (1990): The system tries to adapt to the user, the user tries to adapt to the system; this way, they may never reach a stable configuration. Work on adaptive interaction has addressed various key technical issues, concerning the derivation of appropriate constructs for the embodiment of adaptation capabilities and facilities in the user interface; some of these issues that are of relevance to unified user interfaces are discussed next.

- *Which architecture?* To start with, although concrete software architectures for adaptive user interfaces are not clearly defined, there exist various proposals as to what should be incorporated (in a computable form) into an adaptive interactive system. In Dieterich, Malinowski, Kühme, and Schneider-Hufschmidt (1993), these categories of computable artifacts are summarized as being the types of models that are required within a structural model of adaptive interactive software. In this summary, taken from Dieterich et al., an appropriate reference to Hartson and Hix (1989) is made, regarding structural interface models: "Structural models of the human–computer interface are descriptive of the general process of human–computer communication; that is, they theoretically and generically describe the structure of end-user exchanges with computers…. Such models guide a dialogue developer and help organize the dialogue development process (p. 18). However, developers require concrete software architectures in structuring and engineering interactive systems, and software systems in general. In this sense, the information provided in Dieterich et al. does not fulfill the requirements of an interface structural model, as defined in Hartson and Hix, nor of a software architecture, as defined in Jacobson et al. (1997) and Mowbray and Zahavi (1995). This fact leads to the initial argument that a concrete, generic architectural framework for adaptive interfaces, and automatically adapted interactions as a broader category, is completely lacking. This argument is further supported and elaborated on, as various aspects of existing work on adaptive interfaces are incrementally analyzed, in the subsequent sections.

- *User models versus user-modeling frameworks.* In all types of systems aiming to support user adaptivity in performing certain tasks, both embedded user models and user-task models have played a significant technical role. In Kobsa and Wahlster (1989), an important distinction is made between the user-modeling component, encompassing methods to represent user-oriented information, and the particular user models as such, representing an instance of the knowledge framework for a particular user (i.e., individual user model) or user group (i.e., a stereotype model). But even this distinction still explicitly associates the user model with the modeling framework, thus necessarily establishing a dependency between the adaptation-targeted decision-making software (which would need to process user models) and the overall user-modeling component. This remark reveals the potential architectural hazard of rendering an adaptive system "monolithic": Because the user model is linked directly with the modeling component, and decision making is associated with user models, it may be deemed necessary or appropriate that all such knowledge categories be physically located together.

- *Alternative dialogue patterns and the need of abstraction.* The need for explicit design, as well as run-time availability of design alternatives has been already identified in the context of interface adaptation (Browne, Norman, & Adhami, 1990). In view of the need for managing alternative interaction patterns, the importance of abstractions has been identified, starting from the observation that design alternatives constructed with an adaptation perspective are likely to exhibit some common dialogue structures. Cockton (1987) pointed out that "flexible abstractions for executable dialogue specifications" are a "necessary condition for the success of adaptable human–computer interfaces." This argument implies that an important element in the success of adaptive systems is the provision of implemented mechanisms of abstraction in interactive software, which allow the flexible run-time manipulation of dialogue patterns.

- *Dynamic user attribute detection.* The most common utilization of internal dialogue representation has involved the collection and processing of interaction monitoring information. Such information, gathered at run time, is analyzed internally (through different types of knowledge processing) to derive certain user attribute values (not known prior to the initiation of interaction), which may drive appropriate interface adaptivity actions. A well-known adaptive system employing such techniques is MONITOR (Benyon, 1984). Similarly, for the purpose of dynamic detection of user attributes, a monitoring component in conjunction with a UIMS are employed in the AIDA system (Cote Muñoz, 1993). An important technical implication in this context is that dialogue modeling must be combined with user models. Thus, as discussed earlier, it becomes inherently associated with the user-modeling component, as well as with adaptation-targeted decision-making software. Effectively, this "biases" the overall adaptive system architecture toward a monolithic structure, turning the development of adaptive interface systems to a more complicated task. It is argued that such an engineering complication, which potentially emerges when dynamic user attributes are to be detected, can be practically avoided.

    This argument is also supported by the fact that, in most available interactive applications, internal executable dialogue models exist only in the form of programmed software modules. Higher order executable dialogue models (which would remove the need for low-level programming), as those previously mentioned, have been supported only by research-oriented UIMS tools. Conversely, the outcome of interface development environments, like VisualBasic™, is, at present, in a form more closely related to the implementation world, rendering the extraction of any design-oriented context difficult or impossible. Hence, on the one hand, dynamic user attribute detection will necessarily have to engage dialogue-related information, whereas, on the other hand, it is unlikely that such required design information is practically extractable from the interaction control implementation.

- *Interface actions to perform adaptivity.* The final step in a run-time adaptation process is the execution of the necessary interface updates at the software level. In this context, four categories of actions to be performed at the dialogue control level have been distinguished (Cockton, 1993), for the execution of adaptation decisions: (a) *enabling* (i.e., activation or deactivation of dialogue components), (b) *switching* (i.e., selecting one from various alternative preconfigured components), (c) *reconfiguring* (i.e., modifying dialogue by using predefined components), and (d) *editing* (i.e., no restrictions on the type of interface updates). The preceding categorization represents a rather theoretical perspective, rather than an interface-engineering one. Furthermore, the term *component* denotes mainly visual interface structures, rather than referring to implemented subdialogues, including physical structure and/or interaction control.

In this sense, it is argued that it suffices to define only two action classes, applicable on interface components: (a) *activate* components and (b) *cancel* activated components (i.e., deactivate). These two actions directly map to the implementation domain (i.e., activation means "instantiation" of software objects, whereas cancellation means "destruction"), thus considerably downsizing the problem of modeling adaptation actions, and are fundamental in the unified interface software architecture.

- *Structuring dialogue implementation for adaptivity.* The notion of *interface component* refers to implemented subdialogues provided by means of prepackaged, directly deployable, software entities (Short, 1997). Such entities increasingly become the basic building blocks in a component-based software assembly process, highly resembling the hardware design and manufacturing process. The need for configurable dialogue components was identified in Cockton (1993), as a general capability of interactive software to visualize some important implementation parameters, through which flexible fine-tuning of interactive behaviors may be performed at run time, even from within the software layer itself (i.e., self-adaptation).

  However, the analysis in Cockton (1993) is based on a theoretical ground, and mainly identifies requirements, without proposing specific approaches to achieving this type of desirable functional behavior. For instance, the distinction among "scalar," "structured," and "higher order" objects proposed does not map to any interface-engineering practice. Moreover, the definition of adaptation policies as "changes" on different levels does not provide any concrete architectural model, nor reveals any useful implementation patterns. The results of such theoretical studies are good for understanding the various dynamics involved in adaptive interaction; however, they do not provide any added-value information for engineering adaptive interaction.

It can be concluded from the preceding that the incorporation of adaptation capabilities into interactive software is far from trivial and cannot be attained through the existing, traditional approaches to software architecture. Therefore, there is a genuine requirement for the definition of a new software architecture that can accommodate the adaptation-oriented requirements of unified user interfaces. Such an architectural framework has been defined and is described in the next section.

THE UNIFIED USER INTERFACE SOFTWARE ARCHITECTURE: BASIC COMPONENTS AND PROTOCOLS

The introduction of the unified user interface development paradigm has been supported with an appropriate architectural framework. This framework promotes an insight into user interface software, which is based on structuring the implementation of interactive applications by means of independent intercommunicating components with well-defined roles and behaviors. In this architecture, the notion of *encapsulation* plays a key role: In order to realize system-driven adaptations, all the various parameters, decision-making logic, and alternative interface artifacts are explicitly represented in a computable form, constituting integral parts of the run-time environment of an interactive system. It should be noted that the architectural properties of the unified user interface software architecture comply with the definition of architecture provided by the Object Management Group (OMG) (Jacobson et al., 1997; Mowbray & Zahavi, 1995). In accordance with these definitions, an architecture should supply components, description of functional role(s) per component, communication protocols among components, or application programming interfaces (APIs), as well as implementation and interoperability issues.

A unified user interface performs run-time adaptation to meet context-specific requirements as designated by a particular instance of the design space or the global task execution context. In this manner, it can practically attain the target of providing different interface instances for different end-users and usage-contexts. Figure 20.4 depicts the basic *orthogonal* components of the unified architecture, on the vertical dimension. These are: (a) the *dialogue patterns component* (DPC), (b) the *decision-making component* (DMC), (c) the *user information server* (UIS), and (d) the *context parameters server* (CPS). Additionally (see Fig. 20.4), the components of the *Arch* model are represented on the horizontal dimension. The dialogue patterns component is named dialogue control in the Arch metamodel, both playing a common functional role in their respective architectural framework; hence, even though two distinct blocks are drawn in Fig. 20.4 for clarity, they are merged into a single "black box," indicating that their architectural role is virtually identical.

The unified architecture enables the deployment of existing interactive software, by requiring only some additional implemented modules, mainly serving coordination, control, and communication purposes. In this context, the dialogue control module of a typical Arch-based interactive system may be expanded into a dialogue patterns component, without affecting its original functional role. This capability facilitates the vertical growth of existing interactive applications, following the unified architectural paradigm, so that automatically adapted interaction can be accomplished.

The rest of this section provides an account of the characteristic properties of each of the main components of the unified user interface architecture.

User Information Server

This module maintains the individual profiles of end-users, which, together with usage-context information (provided by the context parameters server), constitute the primary input source for the adaptation decision-making process (designated to the decision-making component). An initial end-user identification mechanism

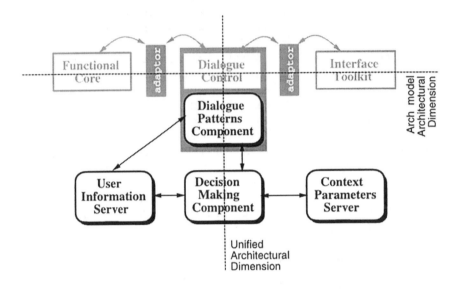

FIG. 20.4.   The unified user interface architecture (vertical dimension), being orthogonal with a typical run-time architecture of interactive systems (Arch metamodel, horizontal dimension).

(e.g., some kind of password upon start-up) is required by the user information server, for retrieving the corresponding user profile, or making any deductions based on such knowledge. In the case of desktop systems with dedicated users, all interactive applications may gain a hard-wired "user ID"; otherwise explicit user identification is needed. Alternative means of identification, such as smart cards, can be employed for systems deployed in public information points.

From a knowledge representation point of view, static or preexisting user knowledge may be encoded in any appropriate form, depending on the type of information the user information server should feed to the decision-making process. Moreover, additional knowledge-based components may be employed for processing retrieved user profiles, drawing assumptions about the user, or even updating the original user profiles. Systems such as BGP-MS (Kobsa, 1990), PROTUM (Vergara, 1994), or USE-IT (chap. 23, this volume) may be employed for such intelligent processing purposes.

Apart from such initial (i.e., prior to initiation of interaction) manipulation of user profiles, the user information server may also collect and process run-time interaction events, in order to draw (additional) inferences about the end-user. Such inferences may result in the identification of dynamic user preferences, loss of orientation in performing certain tasks, fatigue, inability to complete a task, and so on. In the communication with the rest of the architectural components, user-oriented information does not pose any special requirements, as it can be conveyed through message passing of sequences of attribute values.

Context Parameters Server

This component encompasses information regarding the usage environment and interaction-relevant machine parameters. During the interface design process, the identification of those important parameters, relevant to the context(s) of use, will need to be carried out (see chap. 21, this volume). This module is not intended to support device independence, but to provide *device awareness,* thus enabling the decision-making component to select those interaction patterns that, apart from fitting the particular end-user attributes, are also appropriate for the type of equipment available on the end-user machine. The usage-context attribute values are communicated through simple message passing to the decision-making component, before the initiation of interaction; additionally, during interaction, some dynamically changing usage-context parameters may also be fed to the decision-making component for decisions regarding *adaptive* behavior. For instance, let us assume that the initial decision for selecting feedback leads to the use of audio effects. Then, the dynamic detection of an increase in environmental noise may result in a run-time decision to switch to visual feedback (the underlying assumption being that such a decision does not conflict with other constraints).

Decision-Making Component

This module encompasses the logic for deciding the necessary adaptation actions, on the basis of the user and context attribute values, received from the user information server and the context parameters server, respectively. Such attribute values will be supplied to the decision-making component, prior to the initiation of interaction (i.e., initial values, resulting in initial interface adaptation, referred to as *adaptability*), as well as during interaction (i.e., changes in particular values, resulting in dynamic interface adaptations, referred to as *adaptivity*). The decision-making component is only responsible for deciding the necessary adaptation actions, which are then directly communicated to, and subsequently performed by, the dialogue patterns component.

As part of the unified user interface design process (see chap. 21, this volume), alternative dialogue design artifacts may need to be constructed to fit different user and usage-context parameters. In the implementation process, all these distinct dialogue artifacts become implemented interactive components. The run-time adaptation process, for a particular situation of use, is practically a context-sensitive selection of those components that have been designed to address that particular situation.[1]

---

[1]In case a situation has not been anticipated, then incremental updates should be introduced. The unified user interface architecture, being composed of orthogonal components, makes this task easier than traditional models.

In order to perform such a context-sensitive selection, the decision-making component encompasses information regarding all the various dialogue patterns present within the dialogue patterns component, and their specific design role. There are two categories of adaptation actions that are decided and communicated to the dialogue patterns component: (a) *activation* of specific dialogue components and (b) *cancellation* of previously activated dialogue components.

These two categories of adaptation actions suffice to express the various interface component manipulation requirements for realizing either adaptability or adaptivity (see Table 20.1). Substitution is modeled by a message containing a *cancellation* action (i.e., the dialogue component to be substituted), followed by the necessary number of *activation* actions (i.e., which dialogue components to activate in place of a canceled component). Therefore, the transmission of those commands in a single message (i.e., a *cancellation* action followed by *activation* actions) is to be used for implementing a substitution action. The need to send at once information regarding the canceled component, together with the components that take its place, emerges when the implemented interface requires knowledge of all (or some) of the newly created components during interaction. For instance, if the new components include a container and the various contained components, and if upon the creation of the container we need information on the number and type of the particular contained components, we have to ensure that all the relevant information (i.e., all engaged components) is received as a single message.

Dialogue Patterns Component

This module implements all the alternative dialogue patterns identified during the design process, on the basis of various user and context attribute values. The dialogue patterns component may employ predeveloped interactive software, in combination with additional interactive components. The latter case is normally

TABLE 20.1.
The User Interface (UI) Component Manipulation Requirements (left),
for *Adaptability* and *Adaptivity*, and Their Expression Via *Cancellation/Activation*
Adaptation Actions

	Adaptability	Adaptivity
Initial selection of UI components	• Via activation actions.	• Not for adaptivity.
Dynamic selection of UI components	• Not for adaptability.	• Via activation actions.
Dynamic cancellation of UI components	• Not for adaptability.	• Via cancellation actions.
Dynamic substitution of UI components	• Not for adaptability.	• Via messages containing a cancellation action, followed by the necessary number of activation actions.

expected, if the directly deployed interactive software does not provide all the required implemented patterns, addressing the target user and usage-context attribute values.

The dialogue patterns component should be capable of applying pattern activation/cancellation decisions originated from the decision-making component. Additionally, interaction-monitoring components may be attached to various implemented dialogue patterns, providing monitoring information to the user information server for further processing (e.g., keystrokes, notifications for use of interaction objects, task-level monitoring). The particular level of detail and frequency of monitoring are to be requested at run time by the user information server.

ADAPTABILITY AND ADAPTIVITY CYCLES

The completion of an adaptation cycle, being either adaptability or adaptivity, is realized in a specific number of distributed processing stages, performed by the various components of the unified architecture. During those stages, the components communicate with each other, requesting or delivering specific pieces of information. The overall communication requirements among the components are illustrated in Fig. 20.5. The final outcome of an adaptation cycle, being the activation/cancellation decisions, are emphasized with a large dashed arrow.

Figure 20.6 outlines the processing steps for performing both the initial adaptability cycle (to be executed only once), as well as the two types of adaptivity cycles (i.e., one starting from "dynamic context attribute values," and another starting from "interaction-monitoring control"). Local actions indicated within components (in each of the four columns) are either outgoing messages, shown in bold typeface, or necessary internal processing, illustrated via shaded rectangles.

For each component, the actions "flow" logically on the vertical direction, within its corresponding column. However, Fig. 20.6 is not to be used to compare the logical ordering of actions across components (i.e., the fact that an action in one component is vertically drawn lower than a particular action in another component does not imply that the former action precedes the latter in time).

DETAILED COMMUNICATION SEMANTICS

We present the communication protocol among the various components in a form mainly emphasizing the rules governing the exchange of information among various communicating parties, as opposed to a strict message syntax description. Hence, our primary focus is on the semantics of communication, regarding (a) the type of information communicated, (b) the content it conveys, and (c) the usefulness of the communicated information at the recipient component side.

In the unified software architecture (outlined in Fig. 20.4) there are four distinct bidirectional communication channels, each engaging a pair of communicating components. For instance, one such channel concerns the communication between the user information server (UIS) and the decision-making component (DMC); each channel defines two protocol classes, for example, UIS → DMC (i.e., type of mes-

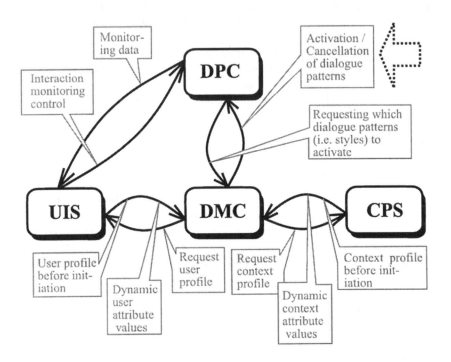

FIG. 20.5.   Communication requirements among the components of the unified architecture in order to perform interface adaptation cycles.

sages sent from UIS to DMC) and DMC → UIS (i.e., type of messages sent from DMC to UIS). The description of the four communication channels for the unified user interface architecture follows, defining four pairs of protocol categories.

Communication Between the User Information Server and the Decision-Making Component

In this communication channel, there are two communication rounds: (a) prior to initiation of interaction, the decision-making component requests the user profile from the user information server, which replies directly with the corresponding profile (as a sequence of attribute values); and (b) after initiation of interaction, each time the user information server detects some dynamic user attribute values (on the basis of interaction monitoring), it will communicate those values immediately to the decision-making component. In Table 20.2, the classes of messages communicated between the user information server and the decision-making component are defined, with simple examples.

Communication Between the User Information Server and the Dialogue Patterns Component

The communication among these two components aims to enable the user information server to collect interaction-monitoring information, as well as to control the type of monitoring to be performed. The user information server may request

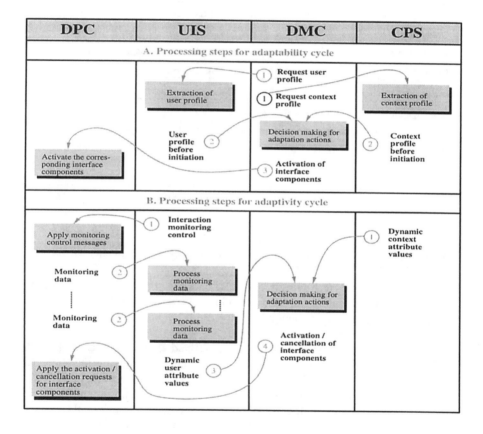

FIG. 20.6.   Processing steps, engaging communication among architectural components, to perform the initial adaptability cycle (A), as well as the two types of adaptivity cycles (B); the messages communicated among components are labeled with their logical ordering, whereas internal processing points are indicated with shaded rectangles.

TABLE 20.2.

Communicated Messages Between the User Information Server
and the Decision-Making Component

UIS → DMC	
Message class	Exporting an end-user profile.
Content structure	Sequence of <attribute, value> pairs.
When communicated	When requested by the decision-making component.
Example	`{ <domain knowledge, limited>, <visual ability,` `sighted>, <age, 35>, <motor ability, fine>, … }`
Message class	Dynamic detection of user parameters.
Content structure	Parameter, context.
When communicated	Each time an inference is made by the user information server.
Example	`{ user confused, Link Selection Task}`
DMC → UIS	
Message class	Requesting user profile.
Content structure	Message contains no data (only the message class header).
When communicated	Before decision making is initiated in the decision-making component.
Example	`{ }`
Message class	Requesting explicitly the value of a user attribute.
Content structure	Attribute.
When communicated	As needed by the decision-making component.
Examples	`{ age }, { domain knowledge }, {Web knowledge }`

monitoring at three different levels: (a) *task* (i.e., when *initiated* or *completed*), (b) *method* for interaction objects (i.e., which logical object action has been accomplished—like "pressing" a "button" object), and (c) *input event* (i.e., specific device event—like moving the mouse or pressing a key).

In response to monitoring control messages, the dialogue patterns component will have to (a) activate/disable the appropriate interaction-monitoring software modules and (b) continuously export monitoring data, according to the monitoring levels requested, back to the user information server (initially, no monitoring modules are activated by the dialogue patterns component). In Table 20.3, the classes of messages communicated between the user information server and the dialogue patterns component are defined, with simple examples.

Communication Between the Decision-Making Component and the Dialogue Patterns Component

The dialogue patterns are organized within the dialogue patterns component in the following manner. Suppose that alternative dialogue patterns are designed for a par-

TABLE 20.3.
Communicated Messages Between the User Information Server
and the Dialogue Patterns

UIS → DPC

Message class	Controlling the level of monitoring.
Content structure	Status, Level.
	Status = on or off.
	Level = TaskLevel or EventLevel or MethodLevel.
	EventLevel = event, ObjectName, EventCategory.
	TaskLevel = task, TaskName.
	MethodLevel = method, ObjectName, MethodCategory.
When communicated	As needed by the user information server inference mechanisms.
Examples	`{ on, task, Link Selection }`
	`{ off, event, HTMLPageWindow, KeyPress }`
	`{ on, method, BWDPageButton, Pressed }`
	`{on, method, HTMLPageScrollbar, Scrolled }`

DPC → UIS

Message class	Monitoring data.
Content structure	TaskBased or EventBased or MethodBased.
	TaskBased = Task, TaskName, TaskAction.
	TaskAction = initiated or completed.
	EventBased = event, ObjectName, EventCategory, EventData.
	EventData = Sequence of <Parameter, Value> pairs.
	MethodBased = method, ObjectName, MethodCategory.
When communicated	When the corresponding user actions are performed.
Examples	`{ task, Link Selection, initiated }`
	`{ task, Link Selection, completed }`
	`{ event, PageLoadingControlToolbar, KeyPress, <key, "a"> }`
	`{ event, BWDPageButton, MouseButtonPress, <Button, 2> }`
	`{ method, StopLoadingButton, Pressed }`
	`{ method, ReloadButton, Pressed }`

ticular user subtask, addressing distinct values of the decision parameters (i.e., user and context parameters). Then, as part of the software implementation, each such alternative dialogue pattern is associated with its respective subtask and is given an arbitrary indicative name, unique among the dialogue patterns of its corresponding subtask. Each implemented dialogue pattern for a particular subtask is referred to as *style*. Styles can thus be identified by reference to their designated name.

At start-up, before initiation of interaction, the dialogue patterns component will request (for each user task) the names of the *styles* (i.e., implemented dialogue patterns) that have to be activated, in order to realize the adaptability behavior. The decision-making component will trigger the adaptability cycle and will respond accordingly (at the end of the adaptation cycle, see Fig. 20.6, Part A).

After initiation of interaction, the decision-making component may "take the initiative" to communicate dynamic style activation/cancellation messages to the dialogue patterns component. Such a communication always occurs at the end of each adaptivity cycle (see Fig. 20.6, Part B). Table 20.4 defines the classes of messages communicated between the decision-making component and the dialogue patterns component, with simple examples.

Communication Between the Decision-Making Component and the Context Parameters Server

The communication between these two components is very simple: The decision-making component will request the various context parameter values (i.e., usage-context profile), and the context parameters server will respond accordingly. During interaction, dynamic updates on certain context property values are to be communicated to the decision-making component for further processing (i.e., possibly new inferences will be made).

As previously mentioned, the context parameters server aims to support two levels of functionality: It encompasses information regarding the available I/O facili-

TABLE 20.4.
Communicated Messages Between the Decision-Making Component
and the Dialogue Patterns Component

DPC → DMC	
Message class	Requesting active styles for a particular subtask.
Content structure	TaskName.
When communicated	Upon start-up, to initiate the necessary dialogue patterns.
Examples	{ Link Selection }, { Page Loading Control } { Page Display Control }
DMC → DPC	
Message Class	Posting decisions regarding style activation/cancellation.
Content structure	StyleDecision = DecisionType, StyleSignature. DecisionType = activation or cancellation. StyleSignature = TaskName, StyleName.
When communicated	• At the end of the adaptability cycle, prior to initiation of interaction. • At the end of each adaptivity cycle, after initiation of interaction.
Examples	{Activation, Link Selection, Direct} {Activation, Stop Loading, With-Confirmation} {Cancellation, Link Selection, Direct} {Activation, Link Selection, With-Confirmation}

ties (i.e., a type of system "registry") on the end-user machine; and it monitors particular environment parameters, such as environment noise or user presence in front of the terminal (e.g., via infrared sensors). The first category of information is necessary for identifying, on the decision-making component side, those interaction techniques that on the one hand conform to user attribute values, and on the other hand can be fully supported via the peripheral equipment at the end-user terminal. The second category is necessary for many purposes. For instance, it may be utilized in order to provide interaction not conflicting with the particular environment state (such as avoiding audio feedback, if high environment noise is detected).

Another scenario concerns supporting the inference process of making dynamic assumptions about the user. For example, if notification is received that the user moves away from the terminal, then that particular interaction session is terminated; otherwise, idleness for a long period of time, while the user is still in front of the terminal, could be interpreted as a potential "confusion," "loss of orientation," or "inability to perform the task." Table 20.5 defines the classes of messages communicated between the decision-making component and the context parameters server, with simple examples.

TABLE 20.5.
Communicated Messages Between the Decision-Making Component
and the Context Parameters Server

DMC → CPS	
Message class	Requesting context parameter values.
Content structure	This message contains no data (only the message header).
When communicated	Prior to initiation of interaction, beginning of adaptability cycle.
Examples	{ }
CPS → DMC	
Message class	Responding with a list of usage-context parameter values.
Content structure	Sequence of <attribute, value> pairs.
When communicated	When requested by the decision-making component.
Examples	{ <Screen Resolution, <640, 480>, <Mouse Available, YES>, <Terminal Position, <120, "cm">, <Software volume ctrl, YES>, <Speech Synthesis, YES> }
Message Class	Dynamic usage-context attribute values.
Content Structure	Attribute, value.
When communicated	Each time a usage-context attribute value is dynamically modified.
Examples	{ Environment noise, 65 dB } { User in front of terminal, NO } { User in front of terminal, YES }

COMPONENT IMPLEMENTATION ISSUES

In this section, we provide the type of development strategies/tools that we consider more appropriate for each architectural component. Such guidelines have been consolidated from real-life experience with medium- and large-scale unified user interface development projects, such as a hypermedia information system (Petrie, Morley, McNally, O'Neill, & Majoe, 1997) supporting dual interaction (Savidis & Stephanidis, 1995), an augmentative communication system for language-cognitive- and speech-motor-impaired people (Kouroupetroglou, Viglas, Anagnostopoulos, Stamatis, & Pentaris, 1996), and an adaptable and adaptive Web browser (see chap. 25, this volume). We briefly analyze the component implementation issues by means of the tool categories that are appropriate for each component, rather than by addressing specific algorithmic aspects, or software-engineering methodologies.

Additionally, for some of the components, more elaborated architectural patterns are proposed. These can be treated as generic design patterns (Gamma, Helm, R. Johnson, & Vlissides, 1995), and provide an implementation structure for building those components, so that the desirable functional behavior can be effectively accomplished.

User Information Server—Implementation Issues

A typical knowledge representation approach may be employed for representing user models and profiles, as well as for drawing assumptions about particular user attributes at run time (i.e., during interaction). An appropriate, local (to the user information server component) user representation method may be defined and adopted, both for storage of user information, as well as for manipulation via the various inference engines. However, the user representation should be always converted to the general form of attribute value pairs when communicated to the "external world" (i.e., the rest of the architectural components). In some cases, the user information server may need to merely play the role of a user profile repository. In such situations, a minimalistic implementation approach can be taken, where (a) a database of user profiles is maintained and (b) a small implementation shell is added to access the database, by processing communication requests. If more than one repository is needed, the user information server should be at least capable of receiving and processing (i.e., reasoning) interaction-monitoring information. The best candidate for this purpose is the employment of a logic programming language (e.g., Prolog), with support for interprocess communication.

From the analysis of various aspects of existing adaptive systems (Dieterich et al., 1993), with respect to the employment and utilization of user-modeling methods, an appropriate architectural design pattern for the user information server has been derived. This architectural pattern is illustrated in Fig. 20.7 and consists of the following four logical components:

1. The *monitoring logger,* which maintains interaction-monitoring events (i.e., stores events from the interaction history); the modules performing such monitoring reside in the interactive software.

2. The *user model,* which is needed to drive an adaptation-oriented decision-making process; the individual user models (or profiles) may be locally stored within this component.

3. The *design information,* providing design-related knowledge, which is necessary both for associating interaction-monitoring data with design context, as well as for supporting dynamic user attribute detection.

4. The *inference mechanism*, which encompasses all the necessary knowledge to derive, during interaction, user attribute values (e.g., preferences, disorientation, expertise).

Dialogue Patterns Component—Implementation Issues

This component should encompass all the dialogue patterns in an implementation form. In this sense, typical development methods for building interactive software may be freely employed; the unified architecture poses no restrictions on the type of interface tool being utilized. Additionally, some special-purpose functionality needs to be introduced "on top" of the interactive software implementing the designed dialogue patterns, realizing the "packaging" (of interactive software) as the dialogue patterns component of a unified system architecture.

Such extra required functionality is split into three categories (see Fig. 20.8): (a) *communication* with the rest of the architectural components (i.e., receiving/sending messages, following the intercomponent communication semantics), (b) *monitoring,* encompassing interaction-monitoring code, which is to be at-

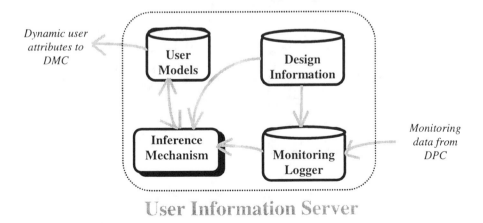

FIG. 20.7.   The architectural pattern for the user information server.

tached to various interface components, and (c) *coordination*, capable of applying activation/cancellation decisions on interface components (as received from the decision-making component), and enabling/disabling interaction-monitoring components (as received from the user information server).

This is the only type of functionality required to make existing interactive software available in the context of a unified user interface architecture. It should be noted that the employed interactive software may internally encompass architectural links in the context of a particular software framework, such as an Arch-based interactive structure, as shown in Fig. 20.4. All such architectural connections are not affected by the software expansion approach of Fig. 20.8, because the original interactive software is not modified in this respect (i.e., *orthogonality* principle). In practice, there may be some lower level issues that need to be addressed:

- Attaching interaction-monitoring software as independent software modules may present implementation barriers, requiring monitoring code to be "injected" within the original software implementation. This implementation barrier has been recently overcome, because all widely available user interface development tools are now built on top of software toolkits (e.g., Java AWT/Swing, Microsoft Foundation Classes™ or

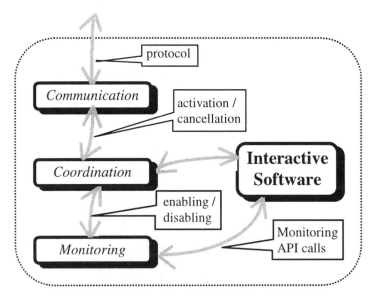

FIG. 20.8. Expanding existing interactive software to constitute the dialogue patterns component in unified system architecture.

Microsoft ActiveX™) that explicitly provide monitoring application programming interfaces (e.g., Active Accessibility™, by Microsoft, and Java Accessibility™, by JavaSoft).

- When expanding interactive software to encapsulate adapted behaviors, there are some designed dialogue patterns not made available in an implementation form. This is naturally expected, if new user or usage-context parameter values (not originally foreseen) are taken into consideration, thus resulting in the construction of new design artifacts. In such cases, a dedicated implementation process should be initiated to make all extra dialogue patterns available in an appropriate implementation form.
- It may be the case that some user interface components are built with non-programming-oriented tools, thus "lacking" an application programming interface, through which run-time activation/cancellation could be effected. This issue would be resolved if widely available tools in this category (e.g., Visual Basic, Java Beans interactive construction tools, ActiveX component builders) generated code through which programmatic access to interface components would be possible.

Decision-Making Component—Implementation Issues

This component encompasses the representation of design logic in a form that, for each subtask, appropriately associates user and context attribute values with the various implemented alternative styles (i.e., dialogue patterns), if any. Typical logic/knowledge programming languages are most appropriate for implementing this kind of functionality. For applications with simple decision-making logic, common programming languages may also suffice for hard-coding the decision process. However, this is only suggested in the case of simple "if–else" logic. In cases of more complex design relationships and design-based reasoning, a knowledge representation framework should be employed.

Context Parameters Server—Implementation Issues

This component is expected to be at the level of system software, accessing external peripheral equipment and providing information on the available I/O devices (i.e., some kind of a simple "registry" for installed peripheral equipment). This component is also responsible for usage-context monitoring. The idea of monitoring the environment within which the user interacts has been practically applied in the context of smart-home technology, in order to mainly identify alarm situations, by processing the input from various types of special-purpose sensors. Due to the lower level functional requirements of these components, a typical third-generation programming language is considered suitable for their implementation.

SUMMARY AND CONCLUSIONS

This chapter has pursued two complementary goals: first, to point out the inadequacy of existing architectural models of interactive software for the construction of self-adapting user interfaces, and, second, to propose and describe a new architectural framework that is capable of supporting the development of unified user interfaces. The main conclusions that can be drawn from the presented material can be summarized as follows.

There is a genuine requirement for the introduction of a new architecture to encapsulate the properties of self-adaptation in interactive software, as these are put forward by the unified user interface concept. This requirement cannot be met through the employment of existing UIMS metamodels (such as the Seeheim and Arch models) or general interactive software models (such as MVC and PAC), as both categories offer constructs that are too general with regard to self-adaptation, failing to address key issues related to its theoretical and practical underpinnings. Furthermore, work on adaptive interaction has not offered thus far concrete results regarding the structural and functional characteristics of adaptation-capable interactive software, especially as far as the architecture of such software is concerned.

As a consequence, the required support for self-adaptation in unified user interfaces has been addressed through the introduction of a respective architectural framework. The proposed framework promotes an insight into user interface software, which is based on structuring the implementation of interactive applications by means of independent intercommunicating components with well-defined roles and behaviors. In the unified user interface architecture, the notion of *encapsulation* plays a key role: In order to realize system-driven adaptations, all the parameters, decision-making logic, and alternative interface artifacts are explicitly represented in a computable form, constituting integral parts of the run-time environment of an interactive system. Along these lines, the framework comprises four main components:

- The user information server, which maintains the individual profiles of end-users. The knowledge available in the UIS may be derived either from preexisting user-oriented information, or through inferences concerning particular user attributes, or dynamic interaction states, drawn upon run-time interaction-monitoring information.
- The context parameters server, which maintains information regarding the usage environment and the machine parameters that are of relevance to interaction. This module is not intended to support device independence, but to provide device awareness into the decision-making process. The CPS is capable of retrieving the aforementioned parameters at start-up, as well as of monitoring modifications and dynamic updates at run time.
- The decision-making component, which encompasses the logic for deciding on necessary adaptations, on the basis of user and context attribute values, made available by the user information server and the context

parameters server, respectively. The DMC is responsible only for deriving decisions regarding necessary adaptation actions; these decisions are subsequently communicated to, and performed by, the dialogue patterns component.

- The dialogue patterns component, which implements the alternative dialogue patterns identified during the unified user interface design process. It may employ predeveloped interactive software, along with additional, dynamically assembled, interactive components. The DPC is capable of applying the adaptation decisions it receives from the decision-making component, and incorporates interaction-monitoring facilities that enable it to collect (and communicate to the user information server) information about the ongoing interaction.

The presentation of the aforementioned components has included (a) a description of their characteristic properties and functional role within the architecture, (b) the communication protocols between each pair of communicating components, and (c) issues related to the implementation of each of the components. The specific presentation approach was selected in order to provide the required support and guidance for the development of unified user interfaces, without, however, introducing unnecessary restrictions in the design and the implementation of the individual components, or the architectural framework as a whole.

In conclusion, it should be noted that one of the primary properties of the unified user interface architecture relates to the fact that it can be applied orthogonally to the overall architecture of interactive systems. This is evidenced by (a) its orthogonality to the components of the Arch metamodel, as well as (b) its suitability for the expansion of existing interactive software, so that properties of unified user interfaces are introduced therein. As a result, the proposed architecture offers itself naturally, not only to the development of new unified user interfaces, but also to the introduction of the related principles and techniques in existing interactive software.

The application of the unified user interface architecture in the development of a Web browser is discussed in chap. 25 of this volume.

REFERENCES

Barkakati, N. (1991). Model-view-controller (MVC) architecture of Smarttalk-80. In *Object-oriented programming in C++* (pp. 74–85). Carmel, IN: SAMS Publishing.

Benyon, D. (1984). MONITOR: A self-adaptive user-interface. In *Proceedings of IFIP Conference on Human–Computer Interaction: INTERACT '84* (Vol. 1, pp. 335–341). Amsterdam: North-Holland, Elsevier Science.

Browne, D., Norman, M., & Adhami, E. (1990). Methods for building adaptive systems. In D. Browne, M. Totterdell, & M. Norman (Eds.), *Adaptive user interfaces* (pp. 85–130). London: Academic Press.

Browne, D., Totterdell, P. , & Norman, M. (Eds). (1990). Conclusions. In *adaptive user interfaces* (pp. 195–212). London: Academic Press.

Cockton, G. (1987). Some critical remarks on abstractions for adaptable dialogue managers. In *Proceedings of the 3rd Conference of the British Computer Society, People & Computers III, HCI Specialist Group* (pp. 325–343). University of Exeter, UK.

Cockton, G. (1993). Spaces and distances—Software architecture and abstraction and their relation to adaptation. In M. Schneider-Hufschmidt, T. Kühme, & U. Malinowski (Eds.), *Adaptive user interfaces—Principles and practice* (pp. 79–108). Amsterdam: North-Holland, Elsevier Science.

Cote Muñoz, J. (1993). AIDA—An adaptive system for interactive drafting and CAD applications. In M. Schneider-Hufschmidt, T. Kühme, & U. Malinowski (Eds.), *Adaptive user interfaces—Principles and practice* (pp. 225–240). Amsterdam: North-Holland, Elsevier Science.

Coutaz, J. (1990). Architecture models for interactive software: Failures and trends. In G. Cockton (Ed.), *Engineering for human–computer interaction* (pp. 137–151). Amsterdam: North-Holland, Elsevier Science.

Dieterich, H., Malinowski, U., Kühme, T., & Schneider-Hufschmidt, M. (1993). State of the art in adaptive user interfaces. In M. Schneider-Hufschmidt, T. Kühme, & U. Malinowski (Eds.), *Adaptive user interfaces—Principles and practice* (pp. 13–48). Amsterdam: North-Holland, Elsevier Science.

Gamma, E., Helm, R., Johnson, R., & Vlissides, J. (1995). *Design patterns, elements of reusable object-oriented software.* Reading, MA: Addison-Wesley Professional Computing Series.

Goldberg, A. (1984). *Smalltalk-80: The Interactive Programming Environment.* Reading, MA: Addison-Wesley.

Green, M. (1985). Report on dialogue specification tools. In G. Pfaff (Ed.), *User interface management systems—Proceedings of the Workshop on UIMS.* (pp. 9–20). New York: Springer-Verlag.

Hartson, H. R., & Hix, D. (1989). Human–computer interface development: Concepts and systems for its management. *ACM Computing Surveys, 21*(1), 241–247.

Jacobson, I., Griss, M., & Johnson, P. (1997). Making the reuse business work. *IEEE Computer, 10,* 36–42.

Kobsa, A. (1990). Modeling the user's conceptual knowledge in BGP-MS, a user modeling shell system. *Computational Intelligence, 6,* 193–208.

Kobsa, A., & Wahlster, W. (Eds.). (1989). *User models in dialog systems,* Berlin: Springer.

Kouroupetroglou, G., Viglas C., Anagnostopoulos, A., Stamatis, C., & Pentaris, F. (1996). A novel software architecture for computer-based interpersonal communication aids. In J. Klauss, E. Auff, W. Kremser, & W. Zagler (Eds.), *Interdiscplinary aspects on computers helping people with special needs, Proceedings of 5th International Conference on Computers Helping People with Special Needs* (pp. 715–720). Wien, Austria: Oldenberg.

Mowbray, T. J., & Zahavi, R. (1995). *The essential CORBA: Systems integration using distributed objects.* New York: Wiley.

Myers, B. (1995). User interface software tools. *ACM Transactions on Computer–Human Interaction, 12*(1), 64–103.

Petrie, H., Morley, S., McNally, P., O'Neill, A-M., & Majoe, D. (1997). Initial design and evaluation of an interface to hypermedia systems for blind users. In *Proceedings of Hypertext '97* (pp. 48–56). New York: ACM Press.

Savidis, A., & Stephanidis, C. (1995). Developing dual interfaces for integrating blind and sighted users: the HOMER UIMS. In *Proceedings of the ACM Conference on Human Factors in Computing Systems* (pp. 106–113). New York: ACM Press.

Short, K. (1997, February). *Component based development and object modeling* [Computer software, version 1.0]. Texas: Texas Instruments Software.

ten Hagen, P. J. W. (1991). Critique of the Seeheim model. In D. A. Duce, M. R. Gomes, F. R. A. Hopgood, & J. R. Lee (Eds.), *User interface management and design, Proceedings of the workshop on UIMS* (pp. 3–6). Berlin: Springer-Verlag.

UIMS Developers Workshop. (1992). A meta-model for the run-time architecture of an interactive system. *SIGCHI Bulletin, 24*(1), 32–37.

Vergara, H. (1994). *PROTUM—A Prolog based tool for user modeling* (Bericht Nr. 55/94, WIS-Memo 10). Kostanz, Germany: University of Kostanz.

21 The Unified User Interface Design Method

Anthony Savidis
Demosthenes Akoumianakis
Constantine Stephanidis

This chapter describes the Unified User Interface design method. This new user interface design method has been developed to enable the "fusion" of different design alternatives, resulting from the consideration of differing end-user attributes and contexts of use, into a single unified form, as well as to provide a design structure which can be easily translated into a target implementation by user interface developers. The Unified User Interface design method is elaborated here in terms of its primary objective, underlying process, techniques used, representation, and overall contributions to Human-Computer Interaction design.

In its short history, human–computer interaction (HCI) has accumulated a substantial body of knowledge that provides insight into the design of user interfaces. There is an abundance of design techniques that differ with regard to at least two dimensions, namely (a) the underlying science base and (b) the type, range, and scope of design outcomes, as well as the feedback they offer into the (re)design process.

Regarding the science base, there are techniques, such the hierarchical task decomposition (P. Johnson, H. Johnson, Waddington, & Shouls, 1988), which are influenced by the human factors evaluation view of information-processing psychology, and others, such as goals, operators, methods, and selection rules—GOMS (Card, P. T. Moran, & Newell, 1983), task-action grammar—TAG (Payne, 1984), the user-action notation—UAN (Hartson, Siochi, & Hix, 1990), and so on, that adopt the methodological perspectives and assumptions of cognitive science. More recently, less formal techniques, such as inspection methods (Nielsen & Mack, 1994), have been introduced to provide more timely input and feedback to design activities.

Regarding design outcomes, the vast majority of existing techniques are artifact oriented. They do not explicitly record and document design rationale, thus providing limited (if any) account of the rationale underpinning the various design options, or the decision-making points that have shaped a particular design effort. More recently, there have been developments offering insights into the process, notations, and tools for capturing, encoding, and articulating design rationale (T. P. Moran & Carroll, 1996). These techniques focus explicitly on the process and argumentation through which design artifacts are generated.

Unified user interfaces encapsulate automatically adapted behaviors and provide end-users with appropriately individualized interaction facilities. Hence, the process of designing unified user interfaces does not lead to a single design outcome (i.e., a particular design instance for a particular end-user). Rather, it collects and appropriately represents, alternative design "facets," as well as the conditions under which each of these should be instantiated (i.e., a kind of design rationale). Two major challenges that can be identified in this respect concern (a) the process for the production of the various alternative designs and (b) the organization of all potential design instances into a single design structure.

Clearly, producing and enumerating distinct designs through the execution of multiple design processes is not a practical solution. Ideally, a single design process is desirable, leading to a design outcome that may directly be mapped to a single (i.e., unified) software system implementation. This, however, introduces two important requirements for a suitable design method. The first is that such a method should offer the capability to associate alternative artifacts (depicting different contexts of use) to a single design problem. The second requirement is that the method should preserve the *hierarchical discipline* of the HCI design process, by means of a "divide and conquer" strategy, in which design problems are incrementally broken down and systematically addressed.

The first requirement leads to the definition of a *polymorphic design artifact,* as a collection of alternative solutions for a single design problem, where each alternative addresses different problem parameters. In this context, the problem can be in general identified as optimally designing artifacts for end-users and contexts of use, whereas the problem parameters are the various attributes characterizing the users, or the contexts of use. The second requirement points to the fact that polymorphic design artifacts should be hierarchically structured, thus giving rise to a *polymorphic task hierarchy.*

THE UNIFIED USER INTERFACE DESIGN METHOD

On the grounds of the preceding discussion, the *unified user interface design method* has been defined to address two objectives: (a) enable the "fusion" of all potentially distinct design alternatives into a single unified form, without, however, requiring multiple design phases, and (b) produce a design structure that can be easily translated by user interface developers into an implementation form.

Some of the distinctive properties of this method are elaborated next by addressing its links with HCI design and by providing an overview of what the outcomes and design deliverables are.

Links With HCI Design

The unified user interface design method is characterized by two properties that distinguish both the conduct of the method and its respective outcomes. First, the method adopts an analytical design perspective, in the sense that it requires an insight into how users perform tasks in existing task models, as well as what design alternatives and underpinning rationale should be embedded in the envisioned and reengineered task models. In this context, the method links with other analytical perspectives into HCI design, such as design rationale and ethnography, to obtain the real-world insight that is required, while it extends the traditional design inquiry by focusing explicitly on *polymorphism* as an aid to designing and implementing user and use-adapted behaviors.

Second, the method supports a disciplined hierarchical approach to populating and articulating rationalized design spaces. This entails a middle-out design perspective, whereby enumerated design alternatives are fused into design abstractions and subsequently polymorphosed in a rational manner to facilitate automatic realization of alternative interactive behavior. In terms of conduct, the method is related to hierarchical task analysis, with the distinction that alternative decomposition schemes can be employed (at any point in the hierarchical task analysis process), where each decomposition seeks to address different values of the driving design parameters. This approach leads to the notion of design polymorphism, which is characterized by the pluralism of plausible design options consolidated in the resulting task hierarchy.

Overall, the method introduces the notion of *polymorphic task decomposition,* through which any task (or subtask) may be decomposed in an arbitrary number of alternative subhierarchies (Savidis, Paramythis, Akoumianakis, & Stephanidis, 1997). The design process realizes an exhaustive hierarchical decomposition of various task categories, starting from the abstract level, by incrementally specializing in a polymorphic fashion (because different design alternatives are likely to be associated with differing user and usage-context attribute values), toward the physical level of interaction. The outcomes of the method include (a) the design space that is populated by collecting and enumerating design alternatives, (b) the polymorphic task hierarchy that comprises alternative concrete artifacts, and (c) the recorded design rationale, for each design artifact produced, that has led to the introduction of this particular design artifact.

The Design Space

Design alternatives are necessitated by the different contexts of use and provide a global view of task execution. This is to say that design alternatives offer rich insight into how a particular task may be accomplished by different users in different contexts of use. Because users differ with regards to their abilities, skills, requirements, and preferences, tentative designs should aim to accommodate the broadest possible range of capabilities across different contexts of use. Thus, instead of restricting the design activity to producing a single outcome, designers should strive to compile design spaces containing plausible alternatives.

As an example, consider the primitive interaction task of selection. A selection may be made either by choosing an option from a menu—see Options (a) and (c) in Fig. 21.1—or by issuing a command, like Option (b) in Fig. 21.1. Moreover, as illustrated by options (a) and (c) in Fig. 21.1, the menu may be conveyed in different design languages.[1] For example, the use of the word *menu* in Option (a) is borrowed from the "restaurant" domain of discourse; the command in Option (b) follows the typewriter's metaphor; and the circular clock in Option (c) resembles the operation of an electric device (e.g., a potentiometer). What is important to note, however, is that none of the aforementioned alternatives, or any other visual option that one may come up with, would be suitable for a blind user who lacks the capability to attain information conveyed in the visual modality. Instead, (physically or situationally) blind people would be more comfortable using alternative manifestations conveyed either through the audio or tactile modalities (see Fig. 21.2).

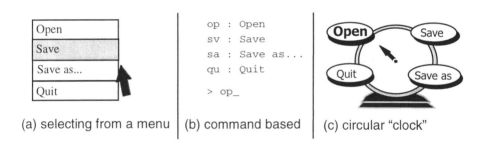

(a) selecting from a menu | (b) command based | (c) circular "clock"

FIG. 21.1   Alternative embodiments of selection in different design languages.

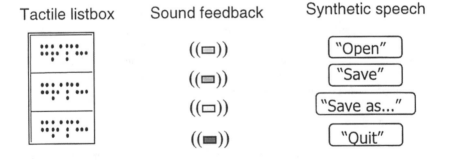

FIG. 21.2.   Nonvisual alternatives for selection.

---

[1]A design language is defined as a mechanism mediating the mapping of concepts in a source domain (e.g., restaurant, typewriter, electric device) to symbols in a presentation domain (e.g., interaction elements offered by a particular toolkit).

Polymorphic Task Hierarchies

A polymorphic task hierarchy combines three fundamental properties: (a) *hierarchical organization,* (b) *polymorphism,* and (c) *task operators.* The hierarchical decomposition adopts the original properties of hierarchical task analysis (P. Johnson et al., 1988) for incremental decomposition of user tasks to lower level actions. The polymorphism property provides the design differentiation capability at any level of the task hierarchy, according to particular user and usage-context attribute values. Finally, task operators, which are based on the powerful CSP (Communicating Sequential Processes) language for describing the behavior of reactive systems (Hoare, 1978), enable the expression of dialogue control flow formulae for task accomplishment. Figure 21.3 illustrates the basic set of operators provided; designers may freely employ additional operators as needed.

The concept of polymorphic task hierarchies is illustrated in Fig. 21.4. Each alternative task decomposition is called a decomposition style, or simply a style, and is given an arbitrary name; the alternative task subhierarchies are attached to their respective styles. The example polymorphic task hierarchy of Fig. 21.4 shows how two alternative dialogue styles for a "Delete File" task can be designed, one exhibiting direct manipulation properties with object-function syntax (i.e., the file object is selected prior to operation to be applied) with no confirmation, and another realizing modal dialogue with a function-object syntax (i.e., the delete function is selected, followed by the identification of the target file) and confirmation.

Additionally, the example demonstrates the case of physical specialization. Because "selection" is an abstract task, it is possible to design alternative ways for physically instantiating the selection dialogue (see Fig. 21.4, lower-part): via scanning techniques for motor-impaired users, via three-dimensional hand pointing on three-dimensional auditory cues for blind people, via enclosing areas (e.g., irregular "rubber banding") for sighted users, and via Braille output and keyboard input for deaf-blind users. The unified user interface design method does not require the designer to follow the polymorphic task decomposition all the way down the user-task hierarchy, until primitive actions are met. A nonpolymorphic task can be specialized at any level, following any design

Operator	Explanation
before	*sequencing*
or	*parallelism*
xor	*exclusive completion*
*	*simple repetition*
+	*absolute repetition*

FIG. 21.3. Basic task operators in the unified user interface design method.

method chosen by the interface designer. For instance, in Fig. 21.4 (lower part) graphical illustrations are used to describe each of the alternative physical instantiations of the abstract selection task.

It should be noted that the interface designer is not constrained to using a particular model, such as CSP operators, for describing user actions for device-level interaction (e.g., drawing, drag-and-drop, concurrent input). Instead, an alternative may be preferred, such as an event-based representation, for example, ERL (Hill, 1986) or UAN (Hartson & Hix, 1989).

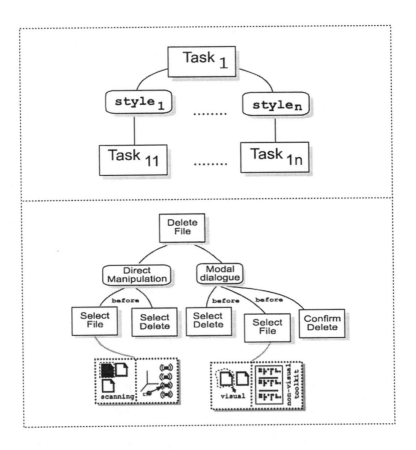

FIG. 21.4. The polymorphic task hierarchy concept, where alternative decomposition "styles" are supported (upper part), and an exemplary polymorphic decomposition, for two different user groups, namely blind and motor-impaired users (lower part).

The most common question regarding the need for a polymorphic task decomposition approach in the unified user interface design method challenges the argument that it is not sufficient to represent alternative task hierarchies by means of the traditional task model, employing the *xor* operator among alternatives.

The answer to this question requires some elaboration (see also Fig. 21.5). First, the *xor* operator in the traditional task model is interpreted as "the user is allowed to perform any, but only one, of the *N* subtasks." This means that the physical interaction context (i.e., interface components) for performing any of the subtasks is made available to the user, whereas the user is required to accomplish only one of those subtasks. However, if alternative subhierarchies are related via polymorphism, it is implied that "a particular end-user will be provided with the design (out of the *N* alternative ones) that maximally matches the specific user's characteristics." Clearly, it is meaningless to provide all designed artifacts (which are likely to address diverse user characteristics) concurrently to end-users, and force users to work with only one of those. Hence, the *xor* operator is not the appropriate way for organizing alternative dialogue patterns.

As discussed in more detail later on, design polymorphism entails a decision-making capability for context-sensitive selection among alternative artifacts, so as to assemble a suitable interface instance, whereas task operators support temporal relationships and access restrictions applied to the interactive facilities of a particular interface instance.

Adaptation Rationale

When a particular task is subject to polymorphism, alternative subhierarchies are designed, each associated with different user and usage-context parameter values. A running interface, implementing such alternative artifacts, should encompass decision-making capability, so that, before initiating interaction with a particular

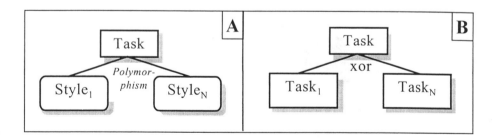

FIG. 21.5.   (a) The use of polymorphism for alternative styles (dialogue patterns), when those address different user and usage-context attributes; and (b) the *xor* operator for various tasks, when exclusive completion has to be imposed.

end-user, the most appropriate of those artifacts are activated for all polymorphic tasks. Hence, polymorphism can be seen as a technique potentially increasing the number of alternative interface instances represented by a typical hierarchical task model. If polymorphism is not applied, a task model merely represents a single interface design instance, on which further run-time adaptation is restricted; in other words, *there is a fundamental link between adaptation capability and polymorphic design artifacts.*

This issue is further clarified with the use of an example. Consider the case where the design process reveals the necessity of having multiple alternative subdialogues available concurrently to the user for performing a particular task. This scenario is related to the notion of multimodality, which can be more specifically called *task-level multimodality,* in analogy to the notion of *multimodal input,* which emphasizes pluralism at the input-device level. We use the physical design artifact of Fig. 21.6, which depicts two alternative dialogue patterns for file management: one providing direct manipulation facilities, and another employing command-based dialogue. Both artifacts can be represented as part of the task-based design, in two ways:

1.   Through polymorphism, where each of the two dialogue artifacts is defined as a distinct style; the two resulting styles are defined as being *compatible,* which implies that they may coexist at run time (i.e., the end-user may freely use the command line or the interactive file manager interchangeably).

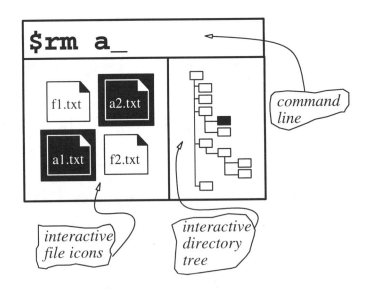

FIG. 21.6.   A design scenario for alternative concurrent subdialogues, in order to perform a single task (i.e., task multimodality).

2.   Via decomposition, where the two artifacts are defined to be concurrently available to the user, *within the same interface instance,* via the *or* operator; in this case, the interface design is "hard-coded," representing a single-interface instance, without needing further decision making.

These two alternative approaches are illustrated in Fig. 21.7.

The advantages of the polymorphic approach are: (a) it is possible to make only one of the two artifacts available to the user, depending on user parameters; (b) even if, initially, both artifacts are provided to end-users, when a particular preference is dynamically detected for one of those, the alternative artifact can be dynamically disabled; and (c) if more alternative artifacts are designed for the same task, the polymorphic design is directly extensible, whereas the decomposition-based design would need to be turned into a polymorphic one (except in the unlikely case where it is still desirable to provide all defined subdialogues concurrently to the user).

CONDUCTING POLYMORPHIC TASK DECOMPOSITION

In this section, we provide a consolidated account of how the unified interface design method can be practiced.

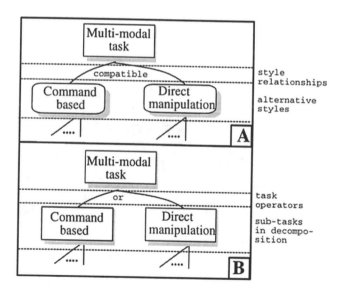

FIG. 21.7.   Two ways for representing design alternatives when designing for task-level multimodality: (A) via polymorphism, adding run-time control on pattern activation; and (B) via the *or* operator, hard-coding the two alternatives in a single task implementation.

Categories of Polymorphic Artifacts

In the unified user interface design method there are three categories of design artifacts, all of which are subject to polymorphism on the basis of varying user and usage-context parameter values. These three categories are (see Fig. 21.8):

1. *User tasks,* relating to what the user has to do; user tasks are the center of the polymorphic task decomposition process.
2. *System tasks,* representing what the system has to do, or how it responds to particular user actions (e.g., feedback); in the polymorphic task decomposition process, they are treated in the same manner as user tasks.
3. *Physical design,* which concerns the various interface components on which user actions are to be performed; physical structure may also be subject to polymorphism.

System tasks and user tasks may be freely combined within task "formulae," defining how sequences of user-initiated actions and system-driven actions interrelate. The physical design, providing the interaction context, is always associated with a particular user task. It provides the physical dialogue pattern associated to a task-structure definition. Hence, it simply plays the role of annotating the task hierarchy with physical design information. An example of such annotation is shown in Fig. 21.4, where the physical designs for the "Select Delete" task are explicitly depicted.

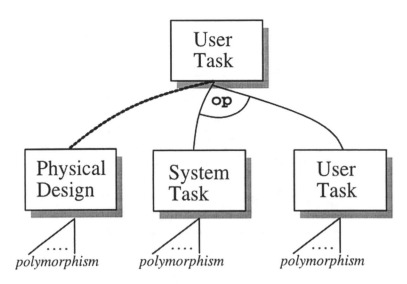

FIG. 21.8.   The three artifact categories in the unified user interface design method, for which polymorphism may be applied, and how they relate to each other.

In some cases, given a particular user task, there is a need for differentiated physical interaction contexts, depending on user and usage-context parameter values. Hence, even though the task decomposition is not affected (i.e., the same user actions are to be performed), the physical design may have to be altered. One such representative example is relevant to changing particular graphical attributes, on the basis of ethnographic user attributes. For instance, Marcus (1996) discussed the choice of different iconic representations, background patterns, visual message structure, and so on, on the basis of cultural background (see also chap. 3, this volume).

However, there are also cases in which the alternative physical designs are dictated due to alternative task structures (i.e., polymorphic tasks). In such situations, each alternative physical design is directly attached to its respective alternative *style* (i.e., subhierarchy).

In summary, the rule for identifying polymorphic artifacts is: *If alternative designs are assigned to the same task, then attach a polymorphic physical design artifact to this task; the various alternative designs depict the styles of this polymorphic artifact* (see Fig. 21.9, Part A). *If alternative designs are needed due to alternative task structures (i.e., task-level polymorphism), then each alternative physical design should be assigned to its respective style* (see Fig. 21.9, Part B).

Steps in Polymorphic Task Decomposition

User tasks and, in certain cases, system tasks need not always be related to physical interaction, but may represent *abstraction* on either user or system actions. For in-

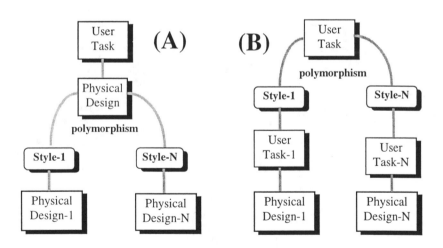

FIG. 21.9.   Representation of alternative physical artifacts: (A) in the case of the same nonpolymorphic task; and (B) in the case where polymorphic task decomposiiton is needed.

stance, if the user has to perform selection, then, clearly, the physical means of performing the selection are not explicitly defined, unless the dialogue steps to perform selection are further decomposed. This notion of continuous refinement and hierarchical analysis, starting from higher level abstract artifacts, and incrementally specializing toward the physical level of interaction, is fundamental in the context of hierarchical behavior analysis, either regarding tasks that humans have to perform (P. Johnson et al., 1988), or when it concerns functional system design (Saldarini, 1989). At the core of the unified user interface design method lies the *polymorphic* task decomposition process, which follows the methodology of abstract task definition and incremental specialization, where tasks may be hierarchically analyzed through various alternative schemes.

In such a recursive process, involving tasks ranging from the abstract task level to specific physical actions, decomposition is applied either in a traditional *unimorphic* fashion or by means of alternative styles. The overall process is illustrated in Fig. 21.10; the decomposition starts from abstract or physical task design, depending on whether top-level user tasks can be defined as being abstract or not. Next follows the description of the various transitions (i.e., design specialization steps), from each of the four states illustrated in the process state diagram of Fig. 21.10.

Transitions From the "Abstract Task Design" State.   An abstract task can be decomposed either in a polymorphic fashion, if user and usage-context attribute

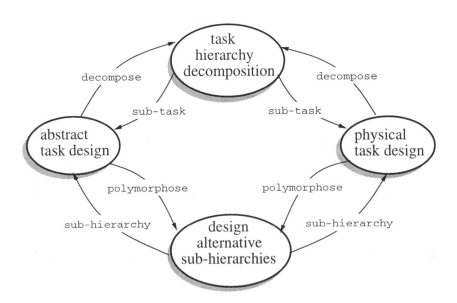

FIG. 21.10.   The polymorphic task decomposition process in the unified user interface design method.

values pose the necessity for alternative dialogue patterns, or in a traditional manner, following a unimorphic decomposition scheme. In the case of a unimorphic decomposition scheme, the transition is realized via a *decomposition* action, leading to the *task hierarchy decomposition* state. In the case of a polymorphic decomposition, the transition is realized via a *polymorphose* action, leading to the *design alternative subhierarchies* state.

Transitions From the "Design Alternative Subhierarchies" State.

Reaching this state means that the required alternative dialogue styles have been identified, each initiating a distinct subhierarchy decomposition process. Hence, each such subhierarchy initiates its own instance of polymorphic task decomposition process. While initiating each distinct process, the designer may start either from the *abstract task design* state, or from the *physical task design* state. The former is pursued if the top-level task of the particular subhierarchy is an abstract one. In contrast, the latter option is relevant in case that the top-level task explicitly engages physical interaction issues.

Transitions From the "Task Hierarchy Decomposition" State.   From

this state, the subtasks identified need to be further decomposed. For each subtask at the abstract level, there is a *subtask* transition to the *abstract task design* state. Otherwise, if the subtask explicitly engages physical interaction means, a *subtask* transition is taken to the *physical task design* state.

Transitions From the "Physical Task Design" State.   Physical tasks may

be further decomposed either in a unimorphic fashion, or in a polymorphic fashion. These two alternative design possibilities are indicated by the *decompose* and *polymorphose* transitions respectively.

An Example of Polymorphic Task Decomposition

To illustrate the process of polymorphic task decomposition (see Fig. 21.10), we refer to an example that is depicted in Fig. 21.4 (lower part). The sequence of steps is illustrated in Fig. 21.11 (states are mentioned with brief names).

The initial state is *abstract task* (Step 1), because "delete file" can be defined as an abstract task. Through the *polymorphose* transition, two alternative styles are defined, resulting in two distinct subhierarchies (Step 2). Each such subhierarchy is further decomposed, by first deciding whether the top-level task is an abstract or a physical task, so as to continue the process (Step 3). For instance, both the "select file" and the "select delete" tasks are abstract (the rest of the tasks are not shown for clarity). Then, the steps for the "select delete" task are shown; this task is polymorphosed (Step 4), resulting in two alternative subhierarchies (one for visual dialogue and another for nonvisual dialogue). The top-level tasks for each of the two subhierarchies are in this case physical (Step 5), as opposed to Step 3, where all tasks are abstract. The "visual rubber-banding" task is subsequently decomposed (Step 6) to a unimorphic task hierarchy. It should be noted that, instead of pursuing the polymorphic task decomposition approach, we could alternatively continue from Step 6, via any other appropriate design practice, such as, for example, event modeling.

Designing Alternative Styles

The polymorphic task model provides the design structure for organizing the various alternative dialogue patterns of automatically adapted interfaces into a unified form. Such a hierarchical structure realizes the fusion of all potential distinct designs that may be explicitly enumerated given a particular unified user interface. Apart from the polymorphic organization model, the following primary issues need to be also addressed: (a) when polymorphism should be applied, (b) which are the user and usage-context attributes that need to be considered, (c) which are the run-time relationships among alternative styles, and (d) how the adaptation ra-

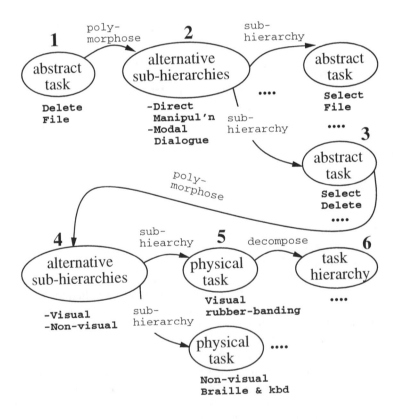

FIG. 21.11.   An example of a polymorphic task decomposition process diagram.

tionale, connecting the designed styles with particular user and usage-context attribute values, is documented.

Identifying Levels of Potential Polymorphism.   In the context of the unified user interface design method, and as part of the polymorphic task decomposition process, designers should always assert that every decomposition step (i.e., those realized either via the *polymorphose* or through the *decompose* transitions of Fig. 21.8) satisfies all constraints imposed by the combination of target user and usage-context attribute values. These two classes of parameters are referred to collectively as *decision parameters/attributes*. An "accessibility gap" is usually encountered when there is a particular decomposition (for user or system tasks, as well as for physical design) that does not address some combination(s) of the decision attribute values. Such a design gap can be remedied by constructing the necessary alternative subhierarchy(ies) addressing the excluded decision attribute values.

Constructing the Space of Decision Parameters.   We discuss the definition of user attributes, which are of primary importance, whereas the construction of context attributes may follow the same representation approach. In the unified user interface design method, end-user representations may be developed using any suitable formalism that can encapsulate user characteristics in terms of attribute-value pairs. There is no predefined/fixed set of attribute categories. Some examples of attribute classes are: general computer-use expertise, domain-specific knowledge, role in an organizational context, motor abilities, sensory abilities, mental abilities.

The value domains for each attribute class are chosen as part of the design process (e.g., by interface designers, or human factors experts), whereas the value sets need not be finite. The broader the set of values, the higher the differentiation capability among various individual end-users. For instance, commercial systems realizing a single design for an "average" user have no differentiation capability at all. The unified user interface design method does not pose any restrictions as to the attribute categories considered relevant, or the target value domains of such attributes. Instead, it seeks to provide only the framework in which the role of user and usage-context attributes constitute an explicit part of the design process. It is the responsibility of interface designers to choose appropriate attributes and corresponding value ranges, as well as to define appropriate design alternatives. A simple example of an individual user profile, complying with the attribute/value scheme, is shown in Fig. 21.12. For simplicity, designers may choose to elicit only those attributes from which differentiated design decisions are likely to emerge.

Relationships Among Alternative Styles.   The need for alternative styles emerges during the design process, when it is identified that some particular user and/or usage-context attribute values are not addressed by the various dialogue artifacts that have already been designed. Starting from this observation, one could argue that "all alternative styles, for a particular polymorphic artifact, are mutually

Computer Knowledge	expert	frequent	average	casual	naive
Web Knowledge	very good	good	average	some	limited	none
Ability to Use Left Hand	perfect	good	some	limited	none

FIG. 21.12. An example of a user profile, as a collection of values from the value domains of user attributes. From Savidis, Akoumianakis, and Stephanidis (1997). Adapted by permission.

exclusive to each other" (in this context, exclusion means that, at run time, only one of those styles may be "active").

However, there exist cases in which it is meaningful to make artifacts belonging to alternative styles concurrently available in a single adapted interface instance. For example, in Fig. 21.6 we have discussed how two alternative artifacts for file management tasks, a direct-manipulation one and a command-based one, can both be present at run time. In the unified user interface design method, four design relationships between alternative styles are distinguished (see Table 21.1), defining whether alternative styles may be concurrently present at run-time. We now show how these four fundamental relationships reflect pragmatic, real-world design scenarios:

- The *exclusion* relationship is applied when the various alternative styles are deemed to be usable only within the space of their target user and usage-context attribute values. For instance, assume that two alternative artifacts for a particular subtask are being designed, aiming to address the "user expertise" attribute: one targeted to users qualified as "novice" and the other targeted to "expert" users. Then, these two are defined to be mutually exclusive to each other, because it is probably meaningless to concurrently activate both dialogue patterns. For example, at run time a novice user might be offered a functionally "simple" alternative of a task, where an "expert" user would be provided with additional functionality and greater "freedom" in selecting different ways in which to accomplish the same task.

TABLE 21.1.

Design Relationships Among Alternative Styles and Their Run-Time Interpretation

Exclusion	Relates many styles. Only one from the alternative styles may be present.
Compatibility	Relates many styles. Any of the alternative styles may be present.
Substitution	Relates two groups of styles together. When the second is made "active" at run time, the first should be "deactivated."
Augmentation	Relates one style with a group of styles. On the presence of any style from the group at run time, the single style may be also "activated."

- *Compatibility* is useful among alternative styles for which the concurrent presence during interaction allows the user to perform certain actions in alternative ways, without introducing usability problems. The most important application of compatibility is for *task multimodality,* as it has been previously discussed (see Fig. 21.6 where the design artifact provides two alternative styles for interactive file management).

- *Substitution* has a very strong connection with adaptivity techniques. It is applied in cases where, during interaction, it is decided that some dialogue patterns need to be substituted by others. For instance, the ordering and the arrangement of certain operations may change on the basis of monitoring data collected during interaction, through which information such as frequency of use and repeated usage patterns can be extracted. Hence, particular physical design styles would need to be "canceled," whereas appropriate alternatives would need to be "activated." This sequence of actions, that is, "cancellation" followed by "activation," is the realization of substitution. Thus, in the general case, substitution involves two groups of styles: Some styles are canceled and substituted by other styles that are "activated" afterward.

- *Augmentation* aims to enhance the interaction with a particular style that is found to be valid, but not sufficient to facilitate the user's task. To illustrate this point, let us assume that during interaction, the user interface detects that the user is unable to perform a certain task. This would trigger an adaptation (in the form of adaptive action) aiming to provide task-sensitive guidance to the user. Such an action should not aim to invalidate the active style (by means of style substitution), but rather to augment the user's capability to accomplish the task more effectively, by providing informative feedback. Such feedback can be realized through a separate, but compatible style. It follows, therefore, that the augmentation relationship can be assigned to two styles when one can be used to enhance the interaction while the other is active. Thus, for instance, the adaptive prompting dialogue pattern, which provides task-oriented help, may be related via an augmentation relationship with all alternative styles (of a specific task), provided that it is compatible with them.

Engaging Abstract Interaction Objects

During the task decomposition process, some subtasks can be directly related to user-input actions that can be managed via interaction objects. For instance, selecting from a list of options, interactively changing the state of a Boolean parameter, providing an arithmetic value, and so on, are all typical examples of input tasks that can be realized via the predefined dialogues implemented by various interaction objects. In such cases, it is desirable to employ general/abstract object classes, in order to enable alternative physical object classes to be selected, reflecting different user, usage-context, and domain properties.

It is argued that designers primarily think in terms of specific instances and physical interface scenarios—especially if the task analysis and graphic design processes are carried out by different teams—rather than composing interface components via abstract behaviors and objects. In this context, we have defined a role-based model (see Fig. 21.13) for "filtering" already made design decisions in order to identify "points" in which abstract interaction objects can be employed in the design representation. Three role categories for interaction objects are identified, namely lexical, syntactic, and semantic; the description of each role follows.

Lexical Role.   In this case, the interaction object is employed for appearance/presentation needs. If such a role can be applied independently of physical realization, then an abstraction can be identified. For example, assume a "message" interaction object, which has only one attribute defining the message content (e.g., a string). The content could be verbal (i.e., the string is a phrase), if the user understands natural language, or symbolic (i.e., the string is a file name where a symbolic sequence is stored), if the user understands symbolic languages. The presentation properties (e.g., emphasizing with an icon, or other visual/auditory effects) concern the physical implementation that may have alternative realizations.

Syntactic Role.   The interaction object serves a specific purpose in the design of dialogue sequencing. If the role can be applied independently of physical re-

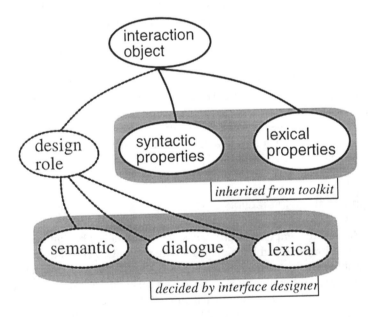

FIG. 21.13.   Role-based model of interaction objects.

alization, then an abstraction can be identified. For example, a "continue" button, a "confirm" button, or a button to initiate an operation, all play the role of a "command" given by the user, in the particular dialogue context. Such a "command" class may be used to support, for instance, execution, confirmation, or cancellation tasks, and may be applicable under various interaction metaphors. It could be physically realized as a conventional push-button for the desktop windowing interaction metaphor, as a voice-input command object for nonvisual interaction, and as a particular symbol structure for language-impaired users. The abstract interaction object "command" may have only one Boolean attribute to control, whether it is accessible or not, whereas the presentation feedback for indicating accessibility status could be different, depending on its physical realization.

Semantic Role.   In this case, an interaction object interactively realizes a domain object. For instance, an interaction object may present a domain object's content, or provide the means to enable "editing" of the content by the user. In such cases, it is always possible to transform the role into a proper abstract class. A typical example is the provision of a numeric value by the user. A "valuator" abstract object could be defined for this purpose, having various properties related to the type of numeric value required (e.g., range, discrete, or real).

Reengineering Designs Through the Role-Based Model

The role-based model can be applied on an existing physical design, in order to produce a higher level design scenario. Such a scenario will serve as an abstract design representation, which may form the basis for deriving further alternative physical design scenarios. This notion of "filtering" physical scenarios via the role-based model, so as to subsequently construct a higher level design representation, is illustrated in Fig. 21.14. We demonstrate the power of such a design reengineering process through an example.

Figure 21.15 depicts a form-based dialogue, typically found in Web documents, for providing credit card information. In Fig. 21.16, the physical design scenario is analyzed in order to identify object roles. The resulting higher object model is depicted in the diagram of Fig. 21.17. Figures 21.18 and 21.19 illustrate two alternative physical realizations that can be derived and that comply with the abstract scenario of Fig. 21.16. The realization depicted in Fig. 21.17 indicates one potential option for graphical interaction, whereas the corresponding depiction of Fig. 21.18 presents a nonvisual scenario that conveys the same information in an alternative metaphoric representation, namely that of the room.

DISCUSSION AND CONCLUSIONS

This chapter has presented the unified user interface design method in terms of primary objective, underlying process, representation, and design outcomes. Unified

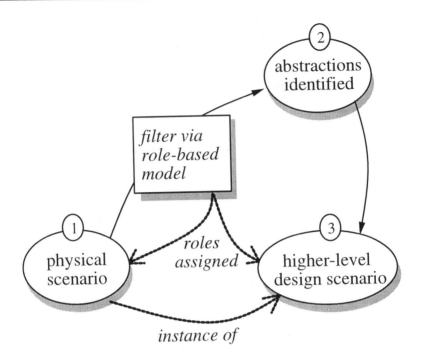

FIG. 21.14.   Producing higher level design scenarios through the role-based model.

user interface design is intended to enable the "fusion" of potentially distinct design alternatives, suitable for different user groups, into a single unified form, as well as to provide a design structure that can be easily translated into a target implementation. By this account, the method is considered to be especially relevant for the design of systems that are required to exhibit adaptable and adaptive behavior, in order to support individualization to different target user groups. In such interactive applications, the design of alternative dialogue patterns is necessitated due to the varying requirements and characteristics of end-users.

In terms of process, the method postulates polymorphic task decomposition as an iterative engagement through which abstract design patterns become specialized to depict concrete alternatives suitable for the designated situations of use. Through polymorphic task decomposition, the unified user interface design method enables designers to investigate and encapsulate adaptation-oriented interactive behaviors into a single design construction. To this effect, polymorphic task decomposition is a prescriptive guide of what is to be attained, rather than how it is to be attained, and thus it is orthogonal to many existing design instruments.

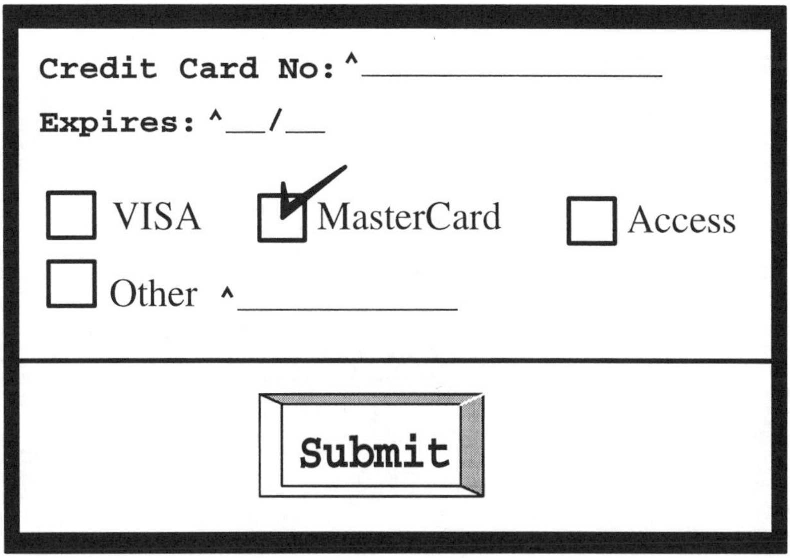

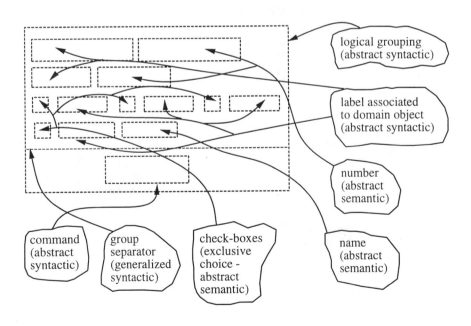

Credit Card No: ^_____

Expires: ^__/__

☐ VISA ☑MasterCard ☐ Access

☐ Other ^_____

Submit

FIG. 21.15.   The physical design scenario that will be reengineered.

logical grouping (abstract syntactic)

label associated to domain object (abstract syntactic)

number (abstract semantic)

name (abstract semantic)

command (abstract syntactic)

group separator (generalized syntactic)

check-boxes (exclusive choice - abstract semantic)

FIG. 21.16.   Assigning roles to physical interaction objects.

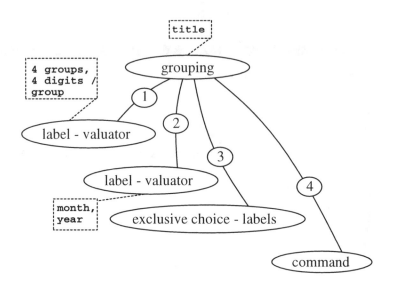

FIG. 21.17.   The resulting higher level object model.

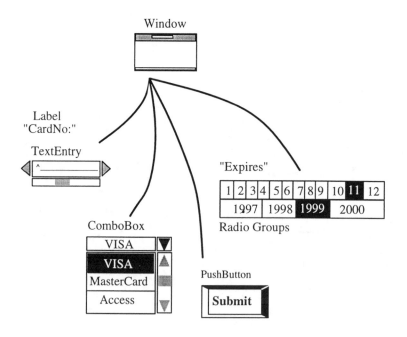

FIG. 21.18.   An alternative graphical design derived on the basis of the abstract object model.

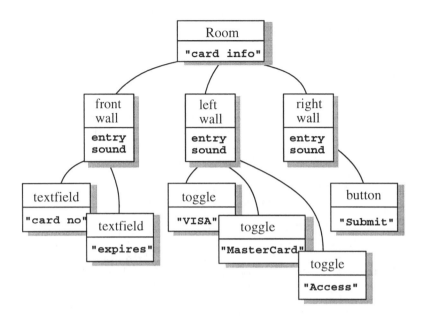

FIG. 21.19.   An alternative nonvisual rooms design, derived from the higher level object model.

The outcomes of unified user interface design include the polymorphic task hierarchy and a rich design space that provides the rationale underpinning the context-sensitive selection amongst design alternatives. A distinctive property of the polymorphic task hierarchy is that it can be mapped into a corresponding set of specifications from which interactive behaviors can be generated. This is an important contribution of the method to HCI design, because it bridges the gap between design and implementation, which has traditionally challenged user interface engineering.

The main conclusion from this chapter is that interaction design becomes increasingly a knowledge-intensive endeavor. Designers should, therefore, be prepared to cope with large design spaces to accommodate design constraints posed by diversity in the target user population and the emerging contexts of use. To this end, analytical design methods, such as unified user interface design, will become necessary tools for capturing and representing the global execution context of interactive products and services in the emerging information society. Moreover, adaptation is likely to predominate as a technique for addressing the compelling requirements for customization, accessibility, and high quality of interaction. Thus, it must be carefully planned, designed, and accommodated into the life-cycle of an interactive system, from the early exploratory phases of design, through to evaluation, implementation, and deployment.

REFERENCES

Card, S. K., Moran, T. P., & Newell, A. (1983). *The psychology of human–computer interaction.* Hillsdale, NJ: Lawrence Erlbaum Associates.

Hartson, H. R., & Hix, D. (1989). Human–computer interface development: Concepts and systems for its management. *ACM Computing Surveys, 21*(1), 241–247.

Hartson, H. R., Siochi, A. C., & Hix, D. (1990). The UAN: A user-oriented representation for direct manipulation interface design. *ACM Transactions on Information Systems, 8*(3), 181–203.

Hill, R. (1986). Supporting concurrency, communication and synchronisation in human–computer interaction—The Sassafras UIMS. *ACM Transactions on Graphics, 5*(3), 179–210.

Hoare, C. A. R. (1978). Communicating sequential processes. *Communications of the ACM, 21*(8), 666–677.

Johnson, P., Johnson, H., Waddington, P., & Shouls, A. (1988). Task-related knowledge structures: Analysis, modeling, and applications. In D. M. Jones & R. Winder (Eds.), *People and computers: From research to implementation—Proceedings of HCI '88* (pp. 35–62). Cambridge, England: Cambridge University Press.

Marcus, A. (1996). Icon design and symbol design issues for graphical interfaces. In E. Del Galdo & J. Nielsen (Eds.), *International user interfaces* (pp. 257–270). New York: Wiley.

Moran, T. P. & Carroll, J. M. (1996). *Design rationale: Concepts, techniques, and use.* Mahwah, NJ: Lawrence Erlbaum Associates.

Nielsen, J. & Mack, R. L. (Eds.). (1994). *Usability inspection methods.* New York: Wiley.

Payne, S. (1984). Task-action grammars. In *Proceedings of IFIP Conference on Human–Computer Interaction: INTERACT '84* (Vol. 1, pp. 139–144). Amsterdam: North-Holland, Elsevier Science.

Saldarini, R. (1989). Analysis and design of business information systems. In *Structured systems analysis* (pp. 22–23). New York: Macmillan.

Savidis, A., Akoumianakis, D., & Stephanidis, C. (1997). Software architectures for transformable interface implementations: Building user-adapted interactions. In *Proceedings of HCI International '97* (pp. 453–456). Amsterdam: Elsevier, Elsevier Science.

Savidis, A., Paramythis, A., Akoumianakis, D., & Stephanidis, C. (1997). Designing user-adapted interfaces: The unified design method for transformable interactions. In *Proceedings of the ACM Conference on Designing Interactive Systems: Processes, Methods and Techniques* (pp. 323–334). New York: ACM Press.

22 Development Requirements for Implementing Unified User Interfaces

Anthony Savidis
Constantine Stephanidis

This chapter describes a set of requirements for the development of Unified User Interfaces. The requirements fall into two main clusters: (i) supporting new interaction metaphors; and (ii) manipulating interaction primitives and alternative dialogue components. These clusters are subsequently mapped onto a set of implementation mechanisms and related functional properties, which should be embedded within user interface tools in order to facilitate the development of Unified User Interfaces. For each mechanism, we provide a review of the support offered by existing user interface software tools. The analysis identifies strong and weak points in existing user interface development tools and points out several future research areas, where additional findings are required in order to provide comprehensive support for the development of Unified User Interfaces.

In chapter 19 of this volume, a unified user interface has been defined as an interactive system, which encapsulates alternative (user- and use-adapted) interactive behaviors into a single software implementation. By implication, a unified user interface exhibits the following properties: (a) it encompasses alternative implemented dialogue patterns, depicting alternatives within the design space; (b) it possesses the capability to automatically select the most suitable dialogue pattern based on user- and context-oriented information (i.e., task-, user-, and use-specific); and (c) it realizes physical dialogue patterns in ways that are not bound to particular interaction platforms (e.g., by "linking" to, rather than directly "calling" an interaction platform). Moreover, as already pointed out, these properties depict functionality implemented into a single software system. This is very important, because developing multiple interface instances corresponding to the plausible design alternatives is practically unacceptable, due to the inherently large development and maintenance costs.

It follows, therefore, that the implementation of unified user interfaces is a very demanding task, because it requires the manipulation of diverse categories of interaction elements to construct dialogue components, which may frequently need to comply to different interaction metaphors. At the same time, key architectural qualities such as openness, extensibility, and portability need to be preserved. In this chapter, the focus is on the implementation of the various dialogue patterns corresponding to outcomes of a unified design process, as elaborated in chapter 21 of this volume. The primary objective is to identify some of the important implementation mechanisms and related functional properties, which are required when developing such dialogue artifacts exhibiting the property of polymorphism.

REQUIREMENTS FOR UNIFIED USER INTERFACE DEVELOPMENT

Unified user interfaces pose several challenges to the overall development process, with the following prominent requirements: (a) supporting new interaction metaphors and (b) manipulating alternative dialogue patterns and interaction objects.

Supporting New Interaction Metaphors

The proliferation of advanced interaction technologies and the diffusion of the Internet have enabled the construction of new virtualities in novel application domains. Typically, these technologies extend the scope of user tasks beyond the traditional business domain. Specifically, multimedia have augmented the capabilities of computers not only with regard to what can be embodied, but also with regard to the range and type of interactions that users can experience. Thus, for example, educational software titles provide new interactive computer embodiments based on themes that children are familiar with, such as the playground or interactive books. The Internet, on the other hand, has enabled new types of interpersonal communication, and new forms of knowledge exchange and experience sharing. The Web infrastructure has enriched the desktop with novel navigation concepts. Advanced collaboration technologies have provided powerful tools for group-centered activities. Mobile devices have extended the conception of computer-mediated human activities beyond the traditional desktop.

It can, therefore, be concluded that with the emergence of the information society, new interactive metaphors will be introduced to facilitate the broad range of computer-mediated human activities likely to evolve. These metaphors will provide users with more natural means to attain a wider range of tasks in a manner that is effective, efficient, and satisfactory, by providing a better cognitive fit between the interactive embodiment of the computer and real-world analogies with which users are already familiar. It is expected that such metaphors will depart significantly from graphical window-based interaction, which, inspired from the Star interface (Canfield Smith, Irby, Kimball, Verplank, & Harlsem, 1982), was intended to meet the demands of able-bodied users working in an office environment.

Along these lines, metaphors have the potential of improving learnability, ease of use, and user satisfaction of user interfaces (Carroll, Mack, & Kellogg, 1988). For that to be attained, the design of interaction metaphors requires in-depth consideration of (a) the target user group, (b) the application domain to be represented metaphorically, and (c) the properties of real-world analogies to be embodied in the computer. Furthermore, the design and implementation of metaphors need to be supported by means of appropriate methodologies and tools.

Currently, the design and realization of new, specialized interaction metaphors is not supported by corresponding reusable software libraries and toolkits; rather, it is hard-coded into the user interface software of computer-based applications. Existing multimedia application construction libraries are too low-level, requiring that developers undertake the complex task of building all metaphoric interaction features from primitive interaction elements. For instance, even though various educational applications provide virtual worlds familiar to children as an interaction context, the required "world" construction kits for such specialized domains and metaphors are lacking.

This is analogous to the early period of graphical user interfaces (GUIs) where developers used a basic graphics package for building window-based interactive applications. However, the evolution of GUIs into a de facto industry standard did not take place until tools for developing such user interfaces became widely available. Similarly, it can be argued that the evolution of new metaphors to facilitate the commercial development of novel applications and services targeted to the population at large will require the provision of the necessary implementation support within interface tools.

Manipulating Alternative Interaction Objects and Dialogue Patterns

The conclusion from the previous section is that the effective deployment of new interaction metaphors is closely related to the availability of suitable tools to support the implementation of alternative dialogue patterns and artifacts (e.g., visual windowing-, three-dimensional auditory-, tactile-, and switch-based dialogues). As in the case of GUIs, the primary means to construct metaphoric interactions are likely to be in the form of implemented reusable interaction elements provided by software libraries called *toolkits* (e.g., OSF/Motif, WINDOWS Object Library, InterViews, Xaw/Athena widget set). Such tools provide development facilities to manipulate (a) interaction objects/interaction techniques/object hierarchies, (b) input events, (c) graphic primitives, and (d) callbacks/methods/notifications.

In the past, interaction objects (e.g., windows, buttons, check boxes) have been the most important category of interaction element, because the largest part of existing interface development toolkits is devoted to providing rich sets of interaction objects, accompanied by all the necessary functionality. Moreover, interaction objects constitute a common vocabulary for both designers and user interface programmers, even though the type of knowledge being possessed by each group is rather different. Thus, designers have more detailed knowledge regarding the appropriateness of different interaction objects for particular user tasks, whereas programmers have primarily implementation-oriented knowledge. In any case, a "button," or a "window" has the same meaning for both designers and programmers when it comes to the physical entity being represented. Interface toolkits have traditionally provided a means for bridging the gap between lexical-level design and implementation. In other words, they provide a vehicle for mapping lexical-level user tasks, revealed through a design activity, to implementation constructs available in a target toolkit.

Recently, we have experienced the development of new, functionally extended forms of interaction objects (as compared to traditional graphical toolkit elements), which have been necessitated by changes in the development paradigm, introduced by innovations, such as, for example, the Web infrastructure for distributed interactive hypermedia documents. Thus, in HTML, "Form" elements support a comprehensive collection of interaction objects; in the JAVA language, the Abstract Window Toolkit (AWT) library provides a rich set of interaction objects supporting *retargetability* (i.e., mapping to multiple graphical platforms). The functional needs for manipulating interaction objects have been studied in the context of real-life application development (e.g., dual hypermedia electronic book [Petrie, Morley, McNally, O'Neill, & Majoe, 1997], interpersonal communicator [Kouroupetroglou, Viglas, Anagnostopoulos, Stamatis, & Pentaris, 1996], unified Web browser [Stephanidis et al., 1998]), targeted to diverse user groups, differing with respect to characteristics such as physical/sensory/motor abilities, preferences, domain-specific knowledge, role in organizational context, and so on. In this context, various interaction technologies and interaction metaphors had to be employed, including graphical windowing environments, auditory/tactile interaction, rooms-based interaction metaphors, and the like.

In the aforementioned developments, new functional requirements emerged regarding the manipulation of interaction objects in user interface tools, which can be summarized as follows:

- *Integration* of additional toolkit(s) providing software libraries for implementing new interaction facilities, possibly complying to alternative interaction metaphors.
- *Augmentation,* that is, providing additional interaction techniques when the dialogue for objects offered by the available toolkit(s) is not adequate.
- *Expansion* of the particular toolkit, when interaction objects considered necessary are not provided by the toolkit(s) being utilized, and/or new custom-made interaction objects are to be designed.
- *Abstraction* applied on interaction objects, to allow the manipulation of objects at a level "higher" than the typical implementation layer of toolkits, in order to make the dialogue design applicable to multiple user groups and different target toolkits.

MAPPING REQUIREMENTS TO DEVELOPMENT MECHANISMS

The preceding requirements translate to several mechanisms that are essential to facilitate unified user interface development. In this section, we describe each of those mechanisms in terms of underlying functional properties that user interface development tools should exhibit in order to provide adequate support for the identified mechanisms. Table 22.1 summarizes and provides a roadmap to the mechanisms and respective development tool properties, separating the latter into required and recommended ones (a separation follows in the forthcoming analysis as well).

Metaphor Development

In the unified development paradigm, the metaphor development process is split in three distinct phases (see Fig. 22.1): (a) *design* of the required metaphoric representation that entails both the selection of suitable metaphoric entities and the definition of their computer equivalents in terms of presentation properties, interactive behaviors, and relationships, (b) *realization* of the interactive embodiment of the metaphor through the selection of media and modalities, interaction object classes, and associated attributes, and (c) *implementation* of a metaphor realization, through the provision of interface development software libraries, which comprise dialogue elements that comply with that particular metaphor realization.

It should be noted that we distinguish between metaphor realization and metaphor implementation to account for the fact that there may be many realizations of a real-world metaphor and many implementations for a particular realization of a metaphor. For instance, a room can be realized visually as a graphical entity, as in Card and Henderson (1987), but also nonvisually as in COMONKIT (Savidis & Stephanidis, 1995a) and AudioRooms (Mynatt & Edwards, 1995).

Additionally, various implementations can be built for a particular metaphor realization. This is the case with the various existing window-based toolkits, which may differ with respect to their software implementation and programming models, implementing, however, a common set of dialogue techniques corresponding to the visual realization of the desktop metaphor. It follows, therefore, that such a distinction between metaphor realization and metaphor implementation is important, be-

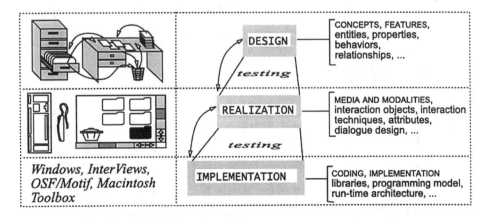

FIG. 22.1. Metaphor development stages in the unified user interface engineering paradigm.

TABLE 22.1.
Overview of Mechanisms and Related Functional Properties

Mechanisms	Required Tool Properties	Recommended Tool Properties
Toolkit integration The capability to import any particular toolkit, so that all interaction elements of the imported toolkit(s) can be made available through the original interaction building techniques of the "host" tool.	• ability to link/mix code at the software library level • documented "hooks," to support interconnections at the source code level	• "well-behaved" and well-documented compilation/translation and linking cycles • single implementation model for all the integrated toolkits • ability to change aspects of the programming interface of each of the imported toolkits • support for resolving language, compilation, linking, and execution conflicts • ability to import any type of interface toolkit, irrespective of the style of interaction supported • ability to combine toolkits for creating cross-toolkit object hierarchies (toolkit interoperability)
Toolkit augmentation The process through which additional interaction techniques are "injected" into the original (native) interaction elements supplied by a particular toolkit.	• support for device integration • programmatic manipulation of the "focus object" • programmatic manipulation of the interaction object hierarchy	• expansion of object attributes and methods, without affecting their original programming interface of the original toolkit classes • "visible" and extensible constructor, where additional interactive behavior can be added • modular device installation/integration layer
Toolkit expansion The process through which toolkit users (i.e., user interface developers) introduce new interaction objects, not originally supported by that toolkit.	• presence of an object expansion framework	• Implementation of the dialogue for new interaction objects, via the native dialogue construction facilities of the development tool (closure property)

Toolkit abstraction

The ability of the interface development tool to support manipulation of interaction objects, which are entirely decoupled from physical interaction properties.

- a predefined collection of abstract interaction object classes (closed set of abstractions)
- support for a predefined mapping scheme, for each abstract object class, to various alternative physical object classes (bounded polymorphism)
- ability to select which of the alternative physical classes (in the mapping scheme) will be instantiated, for each abstract object instance (controllable instantiation)
- support more than one active physical instances, for a particular abstract object instance (plural instantiation)

- capability to define new abstract interaction object classes (open abstraction set)
- support for alternative schemes for mapping abstract object classes to physical object classes (open polymorphism)
- capability to define run-time relationships between an abstract instance and its various concurrent physical instances (physical mapping logic)
- direct programming access, through the abstract object instance, to all associated (concurrent) physical instances (physical instance resolution)

Coordination and manipulation of alternative dialogue patterns

Facilities of the interface development tool that support the implementation of alternative dialogue patterns, as well as their manipulation in the context of the adaptation process.

- explicit component class model for dialogue control, supporting dynamic instantiation or destruction of components (based on the evaluation of run-time conditions)
- programmatic component instantiation or destruction, in a synchronous or asynchronous manner
- dialogue control constructs for asynchronously receiving and applying adaptation decisions originating from external decision-making modules
- support for hierarchical component instantiation or destruction relationships, reflecting hierarchical polymorphic task models
- support for orthogonal expansion of interface components
  —addition of new dialogue patterns
  —addition of monitoring code

447

cause it allows modifications to be introduced at a particular level without necessarily affecting the levels above it.

Interaction metaphors can be realized either by developing new dedicated toolkits, such as in the case of the nonvisual implementation of the room metaphor in COMONKIT, or by augmenting an existing toolkit, such as in the case of the book metaphor in (Moll-Carrillo, Salomon, March, Fulton Suri, & Spreenber, 1995). Because we require that a unified user interface should be able to encapsulate alternative interactive embodiments, complying, perhaps, to different real-world metaphors, it follows that interface tools for building such interfaces need to provide support for effective manipulation of the potentially different implemented versions of the same or alternative metaphors. In other words, tools for unified user interface development should provide support for integrating, as well as augmenting toolkits (e.g., toolkit integration and toolkit augmentation, respectively).

Toolkit Integration

A given interface development tool is considered to support toolkit integration if it can import *any* particular toolkit, so that *all* interaction elements of the imported toolkit(-s) can be subsequently "exposed" (i.e., made available) through the original interaction building techniques of the "host" tool. For instance, if a user interface builder providing graphical construction techniques was to support toolkit integration, then it should be possible to use the tool's native construction techniques to manipulate additional object classes, offered by an imported toolkit. It should be noted that this definition does not assume any particular interface construction method for interface tools. Hence, toolkit integration could be supported by programming-based tools, interactive user interface builders, state-based tools, event-based tools, demonstration-based tools, fourth-generation languages (4GLs) for user interface implementation, and so on.

At this point, an explicit distinction needs to be made between the toolkit integration requirement and the multiplatform capability of certain toolkits. In the latter case, a single toolkit is provided with multiple (hard-coded) implementations across different operating systems, available when the toolkit product is released, that is, multiplatform toolkits like YACL, Amulet, XVT, and the JAVA AWT library (a review can be found in Guinan, 1997). In the former case, a tool is made open, so that its users can take advantage of a well-documented functionality for connecting to arbitrary toolkits.

The need for importing toolkits is evident in cases where the interaction elements originally supported by a particular interface development tool do not suffice. This is a possible scenario if interface development for diverse user groups needs to be addressed. For instance, in the context of dual interface development (Savidis & Stephanidis, 1995b), where it is necessary to construct interfaces concurrently accessible by sighted and blind users, nonvisual interaction techniques are required, in tandem with typical graphical interaction elements. Existing windowing toolkits do not supply such interaction techniques; hence, integration of special-purpose, nonvisual interaction toolkits, such as COMONKIT (Savidis & Stephanidis, 1995a) or HAWK (Savidis, Stergiou, & Stephanidis, 1997) is necessitated. In the more general case of unified user interfaces, it can be argued that scenarios necessitating toolkit integration are likely to emerge, not only as a result of user diversity, but also as a consequence of proliferating interaction technologies and the requirement for portable and platform-independent user interface software.

The toolkit integration capability implies that interface tools supply mechanisms that are made available to developers and can be used *after* a particular user interface development toolkit is launched. Currently, a very small number of interface tools support this notion of *platform connectivity,* that is, the capability to implementationally "connect" to any particular target toolkit platform (e.g., OSF/Motif, MS Windows object library), enabling developers to manipulate its interaction elements as if they are an integral part of the interface tool. The first tool to provide comprehensive support for toolkit integration was the SERPENT UIMS[1] (Bass, Hardy, Little, & Seacord, 1990), where the toolkit layer was termed *lexical technology layer.* The architectural approach developed in the SERPENT UIMS revealed key issues related to the programmatic interfacing of toolkits. Toolkit integration has also been supported by the HOMER UIMS (Savidis & Stephanidis, 1998), developed to facilitate the construction of dual user interfaces. The HOMER UIMS provides a powerful integration model, which is general enough to enable integration of nonvisual interaction libraries, as well as traditional visual windowing toolkits.

Required Tool Properties for Toolkit Integration.   The required development tool properties for toolkit integration are intended to characterize a particular tool with respect to whether it supports some degree of openness, so that interaction elements from external (to the given interface tool) toolkits can be utilized (subject to some particular implementation restrictions). For a user interface tool to support toolkit integration, the required properties are twofold, namely: (a) ability to link/mix code at the software library level (e.g., combining object files, linking libraries together) and (b) support for documented hooks, in order to support interconnections at the source code level (e.g., calling conventions, type conversions, common errors and compile conflicts, linking barriers). If the required properties are present, it is possible for the developer to import and combine software modules that utilize interaction elements from different toolkits.

The critical issue, which determines the capability of a particular development tool to exhibit the functional properties required for toolkit integration, relates to the type of interaction elements offered by the target toolkit (i.e., the toolkit to be integrated). In particular, the properties are satisfied in those cases where the target toolkit provides different categories of interaction elements than the interface tool being used (i.e., the tool providing the integration service). This is, for instance, the case when the interface tool is a programming-based library of windowing in-

---

[1]UIMS: user interface management system.

teraction elements and the target toolkit offers audio-processing functionality for auditory interaction. In such cases, potential conflicts can be easily resolved and elements from the two toolkits can be combined.

In contrast to the aforementioned scenario, the required tool properties are not met when the target toolkit supplies similar categories of interaction elements with the interface tool being used. For example, when trying to combine various libraries of windowing interaction elements (e.g., MS Windows object libraries from different vendors, Xt-based toolkits like OSF/Motif and the Athena widget set), serious conflicts are encountered. The primary source of such conflicts is name collision, at the binary library level, due to the common constructs (e.g., "object," "window," "button," "initialize_toolkit," "close_toolkit," "event_handler," etc.) employed by these toolkits. In all such cases, the interface toolkits reserve exclusively some names (such as those previously mentioned) for particular software library entities, and they do not allow other implemented entities to be registered with those specific identifiers.

Additionally, in cases where the various toolkits to be concurrently utilized share common libraries (e.g., like Xt-/Xlib-based toolkits), it is not allowed to call particular services (e.g., functions) more than once. Typically, such "problematic" services provide some start-up registration facilities, like establishing a connection with a window manager. In all cases studied, multiple calls to such services caused run-time errors, thus practically disabling initialization sequences to be executed for more than one toolkit (in a client utilizing more than one toolkit). Due to their nature, we refer to this category of conflicts as the *exclusive registration* phenomenon.

Recommended Tool Properties for Toolkit Integration.   Ideally, a development tool should offer additional means to effectively support toolkit integration. These additional features constitute the comprehensive set of recommended tool properties for toolkit integration. First, the tool should supply well-behaved (i.e., functionally robust/reliable) and well-documented (i.e., developers should be able to predict functional behavior from comprehensive documentation resources) compilation/translation and linking cycles for interfaces utilizing the integrated toolkit(s); this applies to all types of interface building methods, not only to methods supported by programming-oriented tools.

Second, it should be possible to adopt a single implementation model for all the integrated toolkits. Thus, when the development tool provides visual construction methods, then the same facilities should allow manipulation of interface elements from all the integrated toolkits.

Third, the development tool should provide the ability to change aspects of the programming interface (i.e., the programmable "view") of each of the imported toolkits. This would minimize the effort required for programmers to become familiar with the programming style of the integrated toolkits.

Fourth, the development tool should provide means for resolving problems arising when trying to combine multiple toolkits. These problems include (a) *language conflicts,* because not all software libraries are made available through the same programming language, (b) *compilation conflicts,* because collision of data types, variables, constants, and so on, is a common phenomenon, even when the same (or compatible) programming languages are supported among toolkits, (c) *linking conflicts,* mainly due to name collision at the binary library levels, and (d) *execution conflicts,* due to the exclusive registration phenomenon.

Fifth, the development tool should be able to import any type of interface toolkit, irrespective of the style of interaction supported (e.g., windowing toolkits, auditory/tactile toolkits, virtual-reality-based interaction toolkits).

Finally, the development tool should allow programmers to combine toolkits for creating cross-toolkit object hierarchies; the latter is defined as the *toolkit interoperability* property.

To present, there has been no single interface tool reported in the literature that exhibits the comprehensive set of recommended properties introduced in this chapter. Regarding the notion of a single programming interface, multiplatform toolkits already provide adequate support, by means of a fixed programming layer. Only UIMS tools like SERPENT (Bass et al., 1990), HOMER (Savidis & Stephanidis, 1995b), and I-GET (see chap. 24, this volume) supply adequate support for toolkit integration, by enabling the establishment of developer-defined programming interfaces on top of software toolkits. When a single programming interface is supported for all platforms, one important question concerns the "look and feel" of supported interaction objects across those platforms. There are three alternative approaches in this case: (a) *employing the native interaction controls of the platform,* that is, making direct use of the platform's controls, along with their presentational and behavioral attributes (e.g., as in Java's AWT library), (b) *mimicking the native controls of the platform,* that is, providing controls that are capable of altering their presentation and behavior to match the respective attributes of the target platform's native controls (e.g., as in Java's Swing library), and (c) *providing custom interaction elements across all platforms,* that is, providing custom controls that "look" and behave the same across platforms, independently of the platforms' native controls (e.g., Tcl/Tk; Ousterhout, 1994).

Regarding toolkit interoperability (see Fig. 22.2), only the Fresco User Interface System (X Consortium, 1994) is known to support cross-toolkit hierarchies. In particular, Fresco facilitates the mixing of InterViews-originated objects with Motif-like widgets.

It should be noted that, even though elements from different toolkits may be combined, possibly employing a different "look and feel," consistency is not necessarily compromised. Figure 22.3 outlines a scenario depicting the combination of a windowing toolkit with a toolkit implementing the "book" metaphor, exemplifying support for cross-toolkit object hierarchies at the implementation level. The advantages of mixing multiple toolkits are more evident in those cases that the combined toolkits offer container objects with different metaphoric representations (like the example of Fig. 21.3), thus practically leading to the fusion of alternative metaphors.

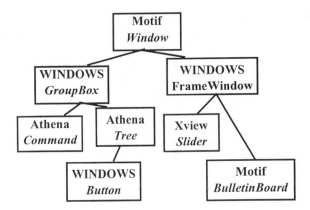

FIG. 22.2.   An example of a cross-toolkit object hierarchy (WINDOWS, Motif, Athena, and X view objects are mixed).

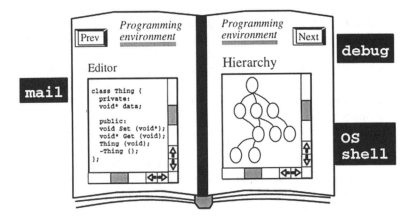

FIG. 22.3.   A dialogue artifact for mixing container objects (coming from different toolkits) that comply to different interaction metaphors.

Toolkit Augmentation

Augmentation is defined as the design and implementation process through which additional interaction techniques are "injected" into the original (native) interaction elements supplied by a particular toolkit, thus leading to improved accessibility or enhanced interaction quality for specific user categories. Newly introduced interaction techniques become an integral part of the existing interaction elements, whereas existing applications that make use of the toolkit "inherit" the extra dialogue features, without requiring revisitation of their implementation (e.g., through recompilation, or relinking, in the case of programming-based implementation approaches).

The need for toolkit augmentation arises mainly from shortcomings, or design deficiencies regarding the supplied interaction entities of existing user interface development toolkits. Because the majority of interactive software products are built by utilizing such toolkits, these shortcomings and deficiencies are propagated to the resulting applications. For instance, graphical interaction libraries do not support voice-controlled interaction. Thus, nonvisual access to the interface, in a situation where direct visual attention is not possible (e.g., while driving) cannot be supported. Another typical example where augmentation is required is the case of accessibility of window-based interaction by motor-impaired users. In this case, additional interaction techniques are needed, which will enable a motor-impaired user to access the user interface through specialized input devices (e.g., binary switches). In both cases, augmentation implies the development of new interaction techniques, as well as the integration of (support for) special-purpose I/O (input/output) devices (e.g., voice I/O hardware, binary switches).

Required Tool Properties for Toolkit Augmentation.   The required user interface development tool properties for toolkit augmentation depict the set of functional capabilities that enable augmented object classes to be introduced into the original software libraries of a toolkit. The set of these properties is not bound to any particular programming language, or even to any category of programming languages (e.g., procedural, object oriented, scripting). However, it is assumed that typical development functionality, such as creating object hierarchies, accessing or modifying object attributes, defining callbacks, and implementing event handlers, is supported.

The first required property for toolkit augmentation is the capability to support device integration. For example, when one considers input devices, this may require low-level software to be written and implemented either through a *polling-based* scheme (i.e., continuously checking device status), or through a *notification-based* scheme (i.e., device-level software may asynchronously send notifications, when device input is detected).

Second, the development tool should allow programmers to manipulate the "focus object" (i.e., the object to which input from devices will be directed). During interactive episodes, different interaction objects will normally gain and lose the input focus, via user control (e.g., through mouse or keyboard actions in windowing environments). In order to augment interaction, it is necessary to have programmatic control of the focus object, because any device input originating from the additional peripheral devices will need to be relayed to that particular object.

Finally, the development tool should allow for programmatic manipulation of the object hierarchy. When implementing augmented interaction, it is necessary to provide augmented analogies of the user's control actions. This is necessary because the user must be enabled to "navigate" within the interface. Hence, the hierarchical structure of the interface objects must be accessed in a programmatic manner, as part of the navigation dialogue implementation.

The relationship between the native object classes and the augmented object classes is a typical is-a relationship, in the sense that augmented toolkit object classes inherit all the features of the original classes (see Fig. 22.4). This scheme can be implemented either via subclassing, if the toolkit is provided in an object-oriented programming (OOP) framework, or via composition, in the case of non-OOP frameworks. In the latter case, composition is achieved through the definition of an augmented object structure that comprises the original object features (by directly incorporating an instance of the original object class), as well as the augmented ones. Such an explicit instantiation is necessary to physically realize the toolkit object, which, in the case of subclassing, would be automatically carried out. Additionally, the interface tool should provide facilities for specifying the mapping between the attributes of the toolkit object instance and its corresponding features defined as part of the new object structure (e.g., procedural programming, constraints, monitors).

Recommended Tool Properties for Toolkit Augmentation.   The comprehensive set of recommended tool properties for full support of toolkit augmentation introduces additional functional properties that need to be present in the development tool. These properties are primarily targeted toward enabling an easier and more modular implementation of the augmented object classes. First, the tool should support the expansion of object attributes and methods "on top" of the original toolkit classes, that is, without affecting their original programming interface (see Fig. 22.4). This alleviates the problem of defining new object classes, while it enables old applications to directly make use of the augmented dialogue features without modifications at the source code level.

Second, the interface tool that supplies the toolkit integration facility should provide a "visible" (i.e., accessible to programming entities outside the scope of

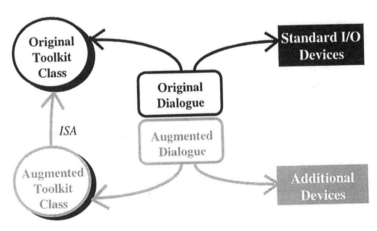

FIG. 22.4.   Relationship between original and augmented toolkit classes in toolkit augmentation.

the object class) and extensible constructor, where additional interactive behavior can be added. This allows for the installation of new event handlers and the performance of all necessary initializations directly at the original class level. The notion of a constructor (in its object-oriented sense) may also be supported by nonprogramming-oriented interface tools, by means of user-defined initialization scripts.

Finally, a modular device installation/integration layer (see Fig. 22.5) is necessary, so that new device input can be attached to the toolkit input-event level. This facilitates the management of additional peripheral devices, through the original event-management layer of the given interface development tool.

The most important advantage to be gained from a tool exhibiting the comprehensive set of recommended properties for toolkit augmentation is the elimination of the need to introduce new specialized object classes, encompassing the augmented interaction features. As a result, it is not necessary to make changes to existing applications (if only the required properties were supported, the augmented capabilities would most probably be conveyed by newly defined object classes, not originally referenced within applications developed prior to the augmentation process). If developers desire to introduce some additional dialogue control logic within existing applications, so as to take advantage of the augmented object attributes and methods, modifications, recompilation, and relinking are evidently required in all cases. Even if no changes are needed in existing applications, recompilation or relinking may still be necessary (especially in the case of purely compiled languages and static linkage to the toolkit libraries).

Toolkit Expansion

Expansion over a particular toolkit is defined as the process through which toolkit users (i.e., user interface developers) introduce new interaction objects, not originally supported by that toolkit. An important requirement of toolkit expansion is

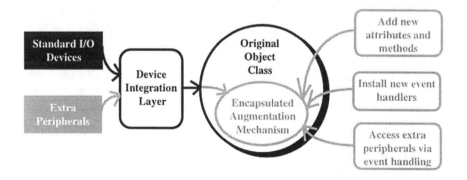

FIG. 22.5. Toolkit augmentation based on the related recommended functional properties.

that all newly introduced interaction objects are made available, in terms of the manipulation facilities offered, in exactly the same manner as original interaction objects—in effect rendering "add-on" objects indistinguishable from original objects, from the toolkit user's point of view.

A typical case for toolkit expansion is the introduction of new interaction object classes, built "on top" of existing interaction facilities, to embody an alternative interactive metaphor (cf. the case of a booklike metaphor in Moll-Carrillo et al., 1995). Toolkit expansion may also be necessitated by domain-specific, or application-specific functionality that needs to be presented as a separate self-contained interactive entity (e.g., consider a temperature-pressure graphical interaction object, to be employed in the implementation of a factory process control system, used for temperature and/or pressure visualization and/or control).

Currently, toolkit expansion facilities are supplied within various categories of commercially available interface development tools. There exists a considerable variety with respect to the way expansion is supported, the process developers have to follow to define new interaction objects, and the method for employing newly introduced objects within interface development. In the present context, we put particular emphasis on the easiness of the approach, because the complexity of the mechanism may pose practical barriers, even though powerful expansion features may be supported.

One of the early toolkits providing expansion support has been the generic Xt toolkit, built on top of the Xlib library for the X Windowing System. The Xt mechanism provides a template widget structure, where the developer has to provide some implemented constructs. The mechanism of Xt is complicated enough to turn expansion to an expert's programming task. Other approaches to expansion concern toolkit frameworks supported by OOP languages, such as C++ or JAVA. If key superclasses are distinguished, with well-documented members providing the basic interaction object functionality, then expansion becomes a straightforward subclassing task. This is the typical case with OOP toolkits like the MS Windows object library or InterViews.

Apart from user interface programming toolkits, the expansion mechanism is also supported in some higher level development tools, such as UIMSs. The demonstration-based method for defining interactive behaviors in Peridot (Myers, 1988) leads to the introduction of interaction objects that can be subsequently recalled and employed in interface construction. This capability can be viewed as expansion functionality. The Microsoft Visual Basic development environment for interface construction and scripting is currently supported with various peripheral tools from third-party vendors; one such tool, called VBXpress, introduces expansion capabilities by supporting the interactive construction of new VBX interaction controls.

Finally, a more advanced approach to toolkit expansion concerns distributed object technologies for interoperability and component-based development. New interaction objects can be introduced through the utilization of a particular tool, while being employed by another. This functionality is accomplished on the basis of generic protocols for remote access to various software resources, supporting distribution, sharing, functionality exposure, embedding, and dynamic invocation.

Microsoft has been the first to allow ActiveX controls[2] to be embedded in JavaBeans[3] containers. JavaSoft's Migration Assistant[4] accomplishes exactly the opposite (or symmetric) link, thus enabling JavaBeans to work inside ActiveX containers. The result is that today we have software that enables the interoperation of ActiveX and JavaBeans components in both directions. For programmers using one of these component categories, this capability is an expansion of the set of available interaction controls; for instance, ActiveX programmers are enabled to use directly JavaBeans objects within ActiveX containers.

Required Tool Properties for Toolkit Expansion.   The required development tool properties relate to the presence of an appropriate object expansion framework supported by the user interface development tool. The most typical forms of such an expandable object framework are:

- *Super class.* Expansion is achieved by taking advantage of the inheritance mechanism in OOP languages, whereas expanded objects are defined as classes derived from existing interaction object classes. Examples of such an approach are the MS Windows object library and the InterViews toolkit.
- *Template structure.* In this case, an object implementation framework is provided, requiring developers to fill in appropriate implementation "gaps" (i.e., supply code), mainly relevant to dialogue properties, such as visual attributes, display structure, and event handling. The most representative example of this approach is the Xlib/Xt widget expansion model of the X Windowing System. The JavaBeans approach is a more advanced version of an object implementation framework.
- *Application programming interface (API).* In this case, resource manipulation and event propagation corresponds to services and event notifications, realizing object management APIs that are built on top of standardized communication protocols. This approach in usually blended with an object-oriented implementation framework, providing a way to combine objects irrespective of their binary format, thus achieving open, component-based development. The ActiveX model is the most typical example in this case. The OMG CORBA model, though not providing a standardized API, allows customized APIs to be built, for example, the Fresco User Interface System (X Consortium, 1994).
- *Physical pattern.* In this case, newly introduced object classes are built via interactive construction methods. For example, construction could start from basic physical structures (e.g., rectangular regions), adding various

---

[2]For more information, see http://microsoft.com/com/activex.asp

[3]For more information, see http://java.sun.com/beans/

[4]For more information, see http://java.sun.com/beans/docs/jbmigratex/html and http://www.javasoft.com/beans/faq/faq.migration.html

physical attributes (e.g., textual items, colors, borders, icons), defining logical event categories (e.g., "selected"), and implementing behavior via event handlers (e.g., highlighting on gaining focus, returning to normal state upon losing focus). The way in which physical patterns are supported varies depending on the tool. For instance, Microsoft Visual Basic provides an "exhaustive" definition and scripting approach, whereas Peridot (Myers, 1988) offers a demonstration-based approach.

- *4GL model.* Fourth-generation interface development languages allow the combination of their interaction object model with the dialogue construction methods of the language, allowing new object classes to be built. These dialogue implementation methods are to be utilized for implementing the interactive behavior of the new objects. The I-GET UIMS is a typical example of an interface tool supporting a 4GL expansion model (see chap. 24, this volume).

Recommended Tool Properties for Toolkit Expansion.   The comprehensive set of development tool properties for toolkit expansion includes one additional recommended functional property: If an interface tool is to fully support object expansion, then it should empower its users to implement the dialogue for new interaction objects via its native dialogue construction facilities (closure property, see Fig. 22.6). In other words, developers should be allowed to define dialogues for new interaction objects via the facilities they have already been using for implementing conventional interfaces. For instance, in an interactive construction tool, the full functionality for expansion is only considered to be available when interactive object design and implementation is enabled.

Toolkit Abstraction

Toolkit abstraction is defined as the ability of the interface tool to support manipulation of interaction objects that are entirely decoupled from physical interaction properties. Abstract interaction objects are high-level interactive entities reflecting generic behavioral properties having no input syntax, interaction dialogue, and physical structure. An example of an abstract interaction object is provided in Fig.

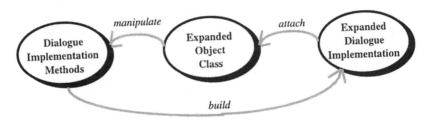

FIG. 22.6. *Closure* property in maximal expansion—the resulting expanded objects are constructed through the original dialogue implementation facilities.

22.7, where an abstract *selector* object is illustrated. Such an abstract interaction object has only two properties: the number of options and the selection (as an index) made by the user. Additionally, it may encompass various other programming attributes, such as a callback list (i.e., reflecting the *select* method), and a Boolean variable to distinguish among multiple-choice and single-choice logical behaviors.

As illustrated in Fig. 22.7, multiple physical interaction styles, possibly corresponding to different interaction metaphors, may be defined as physical instantiations of the abstract selector object class. When designing and implementing interfaces for diverse user groups, even though considerable structural and behavioral differences are naturally expected, it is still possible to capture various commonalities in interaction syntax, by analyzing the structure of subdialogues at various levels of the task hierarchy.

In order to promote effective and efficient design, implementation, and refinement cycles, it is crucial to express such shared patterns at various levels of abstraction, in order to support modification at only a single level, that is, the abstract level. Such a scenario requires implementation support for (a) organizing interac-

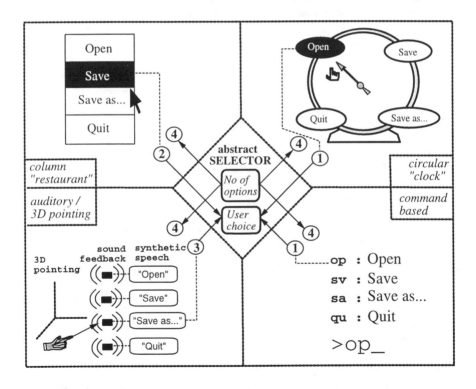

FIG. 22.7. An abstract SELECTOR interaction object class, having only two abstract attributes (central diamond); four alternative physical realizations are shown.

tion objects at various levels of abstraction, (b) enabling developers to define the way in which abstract objects may be mapped (i.e., physically instantiated) to appropriate physical artifacts, and (c) providing the means to construct dialogues composed of dialogue objects.

Abstract interaction objects can be employed for the design and implementation of dialogue patterns that do not need to have physical interaction properties. In this sense, such dialogue patterns are not restricted to any particular user group, or interaction style. The introduction of the intermediate physical instantiation levels is also required, so that abstract forms can be mapped to concrete physical structures. By automating such an instantiation mechanism, development for diverse users is facilitated in a unified fashion (at an abstract layer), whereas the physical realization is automated on the basis of an appropriate object instantiation mechanism.

The notion of abstraction has gained increasing interest in software engineering as a solution to recurring development problems. The basic idea has been the establishment of software frameworks that clearly separate the implementation layers relevant only to the nature of the problem, from the engineering issues, which emerge when the problem class is instantiated in different forms. The same philosophy applies to the development of interactive systems, to allow a dialogue structure composed of abstract objects to be retargeted to various alternative physical forms, through an automation process controlled by the developer.

Before attempting to define the functional properties for supporting toolkit abstraction, it is important to briefly investigate the properties of abstraction in the context of interaction objects. In doing so, we briefly review previous research efforts that can be associated to the notion of toolkit abstraction. These mainly fall within the following four categories:

- *Generalizing and reusing primitive behaviors.* The common perspective in such research efforts is that various input behavioral aspects of interaction in direct manipulation graphical interfaces are captured, generalized, and expressed via a set of generic, highly parameterized primitives, through which complex behaviors can be assembled in a modular manner. Initial work in this field is mainly concerned with *interaction tasks* (Foley, Wallace, & Chan, 1984), providing a comprehensive theoretical framework. *Interactors,* a subsequent effort by Myers (1990), provided an implementation-based framework for combining input behaviors to create complex interactions. Both efforts addressed the desktop metaphor for graphical interaction and proposed an alternative structural and behavioral "view" of the lexical interaction level; hence, they offer only a very limited view into abstraction.
- *Visual construction with behavior abstraction.* This approach has been realized initially in the ADG (application display generator) system (Desoi, Lively, & Sheppard, 1989) and is based on the visual construction of interfaces, on the basis of behavioral abstraction of interaction objects. The developer may choose among different physical realizations of such abstract

behaviors. Through such different choices for physical realization, alternative versions of the lexical structure of the developed interface are produced. The abstractions supported in the ADG system are visual objects, such as "bar gauges," "scrollbars," and so on, that can have alternative representations. The number of such abstract objects, as well as the links between abstract objects and specific physical alternatives, are fixed in the ADG system. This type of "abstraction" is better expressed by the term *generalization,* because the resulting object classes are still bound to the desktop metaphor, merely providing a collection of generically applicable desktop behaviors.

- *Formal models of interaction objects.* Such models have been strongly related to methods for automatic verification of interactive systems in which formal specification of the user interface is employed to assert certain properties. The theoretical background of such approaches is related to reactive systems from the distributed computation theory. The most representative example is the interactor model (Duke, Faconti, Harrison, & Paternó, 1993; Duke & Harrison, 1993). Interactors convey both state and behavior, and communicate with each other, as well as with the user and the underlying application. Such an entity category is computationally very primitive (and as a result, highly generic), and is appropriate for detailed modeling of the dialogue of objects from "within" (i.e., the physical dialogue of an object as such). The interactor model itself constitutes an algorithmic computation structure, like state automata, and as a result is far from the typical programming models of interactive entities, such as interaction objects in toolkits. Hence, interactors can be highly expressive, enabling interaction objects to be formally represented as collections of communicating interactor instances. However, they are less convenient as a development entity in an interface tool (in the same manner that the Turing machine, though the most generic algorithmic structure, is not employed as an explicit programming model in programming languages). In conclusion, the interactor model is better suited for formal verification approaches (and has been very successful in that respect), rather than for interface development.
- *Metawidgets and virtual interaction objects.* The concept of metawidgets is based on the abstraction of interaction objects above physical platforms (Blattner, Glinert, Jorge, & Ormsby, 1992). A metawidget is free of physical attributes and presentation issues and is potentially capable of modeling interaction objects above metaphors. Currently, meta-widgets have been provided with a fixed implementation (Wise & Glinert, 1995), as far as their classes and physical instantiation are concerned. The notion of *virtual objects,* originally referring to multiplatform objects (Myers, 1995), has been revisited and enhanced in Savidis and Stephanidis (1995b) to address abstraction above toolkits and interaction metaphors. Implementation support for this notion of virtual objects has been origi-

nally provided in the context of the HOMER UIMS (Savidis & Stephanidis, 1998), in which dual physical instantiations for virtual objects have been supported. This basic form of abstract objects has been enhanced in the I-GET UIMS, which supports virtual objects with plural instantiations, realizing the physical mapping on multiple imported toolkits (see chap. 24, this volume). The notions of metawidgets and virtual objects have many high-level similarities, though virtual objects have been accompanied, in the context of the HOMER and the I-GET UIMSs, with more concrete engineering models.

In conclusion, existing work has not addressed the provision of interface development techniques for (a) creating abstract interaction objects, (b) defining alternative schemes for mapping abstractions to physical entities (i.e., polymorphism), and (c) selecting the desirable active mapping schemes for abstract object instances, in the context of interface implementation. Additionally, the ability to handle multiple toolkits, which has already been introduced as a fundamental requirement for the development of unified user interfaces, is not supported. The absence of such development methods from existing tools is due to the different development needs of typical graphical interactive applications, as compared to unified user interfaces.

Required Tool Properties for Toolkit Abstraction. The required development tool properties for toolkit abstraction define the basic set of functionalities through which a user interface development tool facilitates construction based on abstract objects. Additionally, we discuss some high-level implementation issues, revealing the complexity of explicitly programming abstract objects, if the interface development tool does not support them inherently. The tool properties are as follows:

- The first required property is that the user interface tool supplies a predefined collection of abstract interaction object classes *(closed set of abstractions)*.
- For each abstract object class, the tool should support a predefined mapping scheme to various alternative physical object classes *(bounded polymorphism)*.
- Moreover, for each abstract object instance, the developer should be able to select which of the alternative physical classes (in the mapping scheme) will be instantiated *(controllable instantiation)*.
- It may also be the case that more than a single physical instance might need to be active for a particular abstract object instance *(plural instantiation)*.

The aforementioned properties enable the developer to instantiate abstract objects, while having control over the physical mapping schemes that will be active for each abstract object instance. Mapping schemes define the candidate classes for physically realizing an abstract object class. The need for having multiple physical instances active, all attached to the same abstract object instance (i.e., plural instantiation), is necessitated in those cases that a running interactive application maps to multiple, concurrent physical interfaces. This is a typical case for applications in the field of computer-supported cooperative work (CSCW), and it is also required in the context of dual interface development (Savidis & Stephanidis, 1995b), where two instances of the interface (a visual and a nonvisual one) are concurrently active, for each abstract object instance. It should be noted at this point that the notions of polymorphic physical mapping and of plural instantiation have fundamentally different functional requirements, with respect to polymorphism in OOP languages. The key differences are outlined in Table 22.2.

Clearly, the traditional schema of abstract/physical class separation in OOP languages, by means of class hierarchies and ISA relationships, cannot be directly applied for implementing the abstract/physical class schema, as needed in unified user interface development. An explicit run-time architecture is necessary, where connections among abstract and physical instances are explicit programming references, beyond the typical instance of run-time links of ISA hierarchies.

Recommended Tool Properties for Toolkit Abstraction.   The comprehensive set of recommended properties introduced next can be used to judge whether an interface tool provides powerful methods for manipulating abstractions, such as defining, instantiating, polymorphosing, and extending abstract interaction object classes. Support for such facilities entails, in addition to controllable instantiation and plural instantiation (as these were defined previously), the following:

- Facilities to define new abstract interaction object classes *(open abstraction set)*.

TABLE 22.2.
Key Differences, With Respect to Functional Properties Related to Polymorphism, Between OOP Development Languages and Unified User Interface Support

Polymorphism in Unified Development	*Polymorphism in OOP Languages*
• Aims to support various dialogue faces (i.e., polymorphism with its direct physical meaning).	• Aims to primarily support reuse and implementation independence (i.e., polymorphism with its metaphoric meaning).
• Instantiation must be applied on abstract classes.	• Instantiation is always applied on derived nonabstract classes.
• Multiple physical instances, manipulated via the same abstract object instance, may be active in parallel.	• References to an abstract class may be mapped only to a single derived object instance.

- Methods to define alternative schemes for mapping abstract object classes to physical object classes, so that, for example, an abstract "selector" may be mapped to a visual "column menu" and a nonvisual "list-box" *(open polymorphism)*.
- Facilities for defining run-time relationships between an abstract instance and its various concurrent physical instances. This may require the definition of attribute dependencies and propagation of callback notifications; that is, if a particular physical instance is manipulated by the user, the abstract instance must be appropriately notified *(physical mapping logic)*.
- Enabling direct programming access, through the abstract object instance, to all associated (concurrent) physical instances *(physical instance resolution)*.

Coordination and Manipulation of Alternative Dialogue Patterns

The various design artifacts that are the outcome of the unified user interface design process need to be mapped to implemented dialogue patterns. Developers will manipulate such implementation components in order to accomplish the desired, automatically adapted, interactive behavior. The properties that are formulated next are based on the goal of facilitating the implementation and manipulation of such alternative interaction patterns.

The functional properties are not distinguished into required and recommended ones, as with previously discussed development mechanisms. Instead, we define a single set of properties, all of which must be exhibited by interface tools if the implementation and manipulation of alternative dialogue patterns is to be effectively supported. These properties are:

- The presence of an explicit component class model for dialogue control, supporting dynamic instantiation or destruction of components (based on the evaluation of run-time conditions). This capability is required to efficiently support *activation* or *cancellation* decisions of alternative dialogue components. Such a requirement is met by OOP languages, where components refer to object classes. It is also met by interface builders that support some scripting facilities and attach programmable identifiers to interactively constructed components.
- Support for programming component instantiation or destruction, in a synchronous or asynchronous manner. In this context, *synchronous* means that developers add instantiation or destruction statements as part of a typical program control flow (i.e., via statements or calling conventions). *Asynchronous* means that the instantiation or destruction events are associated to declarative constructs, such as preconditions or notifications. Normally, instantiation or destruction of components will be "coded" by developers in those points within the implementation that certain conditions dictating those events are satisfied. For this purpose, the asynchronous approach of-

fers the significant advantage of relieving developers from the burden of algorithmically and continuously testing those conditions during execution, for each component class.

- Provision of dialogue control constructs for asynchronously *receiving* and locally *applying* adaptation decisions originating from external decision making modules (see chap. 20, this volume, for more details). In programming languages, this may be carried out via lower level communication functions, whereas in UIMS tools, the API capability should offer the necessary support.

- Support for hierarchical component instantiation or destruction relationships, reflecting hierarchical polymorphic task models (for a definition of, and a discussion on, polymorphic task hierarchies please see chap. 21, this volume). This implies that some components may be dependent on the presence of other, hierarchically higher, components. This reflects the need to make the interface context for particular subtasks available (to end-users of the interface), if and only if the interface context for *ancestor* tasks is already available. For instance, the "Save file as" dialogue box may appear only if the "Editing file" interface is already available to the user.

- Support for orthogonal expansion of interface components. This implies that when adding new implemented dialogue components, or even interaction monitoring components, the particular structure of the control, coordination, and communication software should not be affected. For instance, in conventional programming languages, which do not support declarative constructs (like preconditions, monitors, and constraints), when adding interaction monitoring code to interface components, the following two steps generally need to be taken: (a) add a new monitoring activation type in the dispatcher of externally received monitoring control requests and (b) make a call for the instantiation or installation of the monitoring software, directly accessing the interaction objects of interest.

DISCUSSION AND CONCLUSIONS

Interaction objects play a central role in interface development; consequently, a large part of the software for implementing commercially available interface tools is dedicated to the provision of comprehensive collections of graphical interaction objects. The basic layer providing the implementation of interaction objects is the *toolkit layer,* whereas interface tools typically provide additional layers on top of that. We have studied object-based interface development under the perspective of unified user interfaces, and identified four key mechanisms for manipulating interaction objects: integration, augmentation, expansion, and abstraction. The relationships among these fundamental mechanisms are illustrated in Fig. 22.8.

The support for each of the basic mechanisms varies today. Regarding toolkit integration, the vast majority of commercial tools is targeted toward multiplatform

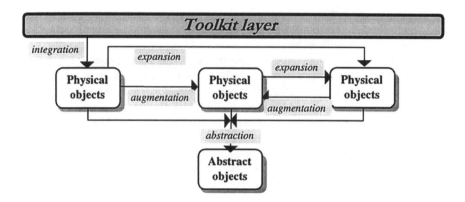

FIG. 22.8.   Relationships among the four fundamental mechanisms for object manipulation.

support in a hard-coded manner, rather than providing open mechanisms for connecting to arbitrary toolkits. Toolkit augmentation is supported in most programming-based interface tools, whereas higher level development tools are very weak in this perspective. Toolkit expansion is also supported in most programming-oriented interface tools, but the considerable overhead required, as well as the inherent implementation complexity, turns the expansion task to an activity primarily targeted to expert programmers. Regarding higher level development tools, there is an increasing number of interactive construction tools supporting expansion, whereas there are currently only two 4GL-based interface tools supporting expansion. Finally, toolkit abstraction, although the most important mechanism in terms of its practical contribution to the unification concept, is also the least supported mechanism in existing interface tools; the I-GET 4GL-based UIMS (see chap. 24, this volume) is the only tool reported in the literature to exhibit the full set of properties required for supporting abstraction.

Regarding the manipulation of alternative dialogue patterns, most programming toolkits provide adequate support to fulfil the various coordination and management requirements. In such tools, there is no built-in construct for dialogue patterns; instead, it is expected that programmers will build object classes, or even component classes (when a component technology is present). However, it is argued that, in order to facilitate the work of developers, a concrete implementation model should be supplied a priori, allowing for some typical interface construction and interaction management issues to be easily addressed.

Currently, there is no commercially available tool that provides adequate support for all the mechanisms. Because distinct tools provide different degrees of support for each mechanism, it becomes necessary to enable developers to employ multiple tools, taking advantage of the distinctive features of each. This paradigm shift from monolithic tools/environments, toward the employment of multiple servers and tools, providing a wide range of implementation services, resources, and functionality (i.e., *multitoolkit platforms*), requires specific advances in the area of interface development tools. In particular, interoperability, distribution, and component-based development are the key properties required. In the past few years, commercial support in this context has exploded, leading to new paradigms in the use of component technologies and multiple-server-based interface development.

The Web, starting from early support for interactive distributed hypertext documents, is evolving to an application development layer, supporting remote downloading of application components for highly interactive software products, assembled "on the fly." The fundamental software layers to accomplish this goal fall in the domain of distributed software technologies such as CORBA (by the Object Management Group) and DCOM (by Microsoft), and software component technologies like JavaBeans (by JavaSoft) and ActiveX (by Microsoft). By blending such technologies with typical toolkit architectures, and by enhancing the present facilities regarding the identified four fundamental mechanisms, the manipulation capabilities for objects are considerably enhanced, bridging the gap between the present software interface technology and the objective of offering practical development support for the construction of *user interfaces for all*.

REFERENCES

Bass, L., Hardy, E., Little, R., & Seacord, R. (1990). Incremental development of user interfaces. In G. Cockton (Ed.), *Engineering for Human–Computer Interaction* (pp. 155–173). Amsterdam: North-Holland, Elsevier Science.

Blattner, M., Glinert, E., Jorge, J., & Ormsby, G. (1992). Metawidgets: Towards a theory of multimodal interface design. In *Proceedings of COMPSAC '92 Conference* (pp. 115–120). New York: IEEE Computer Society Press.

Canfield Smith, D., Irby, C, Kimball, R., Verplank, B., & Harlsem, E. (1982). Designing the star user interface. *Byte Magazine, 7*(4), 242–282.

Card, S. K., & Henderson, D. (1987). A multiple, virtual-workspace interface to support user task switching. In J. Carroll & P. Tanner (Eds.), *Proceedings of the ACM Conference on Human Factors in Computing Systems and Graphics Interfaces* (pp. 53–59). New York: ACM Press.

Carroll, J., Mack, R. L., & Kellogg, W. A. (1988). Interface metaphors and user interface design. In M. Helander (Ed.), *Handbook of human–computer interaction* (pp. 67–85). Amsterdam: North-Holland, Elsevier Science.

Desoi, J., Lively, W., & Sheppard, S. (1989). Graphical specification of user interfaces with behavior abstraction. In *Proceedings of the ACM Conference on Human Factors in Computing Systems* (pp. 139–144). New York: ACM Press.

Duke, D., Faconti, G., Harrison, M., & Paternò, F. (1994). *Unifying view of interactors* (Amodeus Project Document No. SM/WP18).

Duke, D., & Harrison, M. (1993). Abstract interaction objects. *Computer Graphics Forum, 12*(3), 25–36.

Foley, J. D., Wallace, V. L., & Chan, P. (1984). The human factors of computer graphics interaction techniques. *IEEE ComputerComputer Graphics & Applications, 4*(11), 13–48.

Guinan, J. (1997). Platform-independent C++ GUI toolkits. *C/C++ Users Journal (CUJ)*, *15*(1), 19–26.

Kouroupetroglou, G., Viglas, C., Anagnostopoulos, A., Stamatis, C., & Pentaris, F. (1996). A novel software architecture for computer-based interpersonal communication aids. In J. Klauss, E. Auff, W. Kremser, & W. Zagler (Eds.), *Interdiscplinary aspects on computers helping people with special needs, Proceedings of 5th International Conference on Computers Helping People with Special Needs* (pp. 715–720). Wien, Austria: Oldenberg.

Moll-Carrillo, H. J., Salomon, G., March, M., Fulton Suri, J., & Spreenber, P. (1995). Articulating a metaphor through user-centred design. In *Proceedings of the Conference on Human Factors in Computing Systems* (pp. 566–572). New York: ACM Press.

Myers, B. (1988). *Creating user interfaces by demonstration*. Boston: Academic Press.

Myers, B. (1990). A new model for handling input. *ACM Transactions on Information Systems, 8*(3), 289–320.

Myers, B. (1995). User interface software tools. *ACM Transactions on Computer–Human Interaction, 12*(1), 64–103.

Mynatt, E., & Edwards, W. (1995). Metaphors for non-visual computing. In A. Edwards (Ed.), *Extra-ordinary human–computer interaction—Interfaces for users with disabilities* (pp. 201–220). New York: Cambridge University Press.

Ousterhout, J. (1994). *Tcl and the Tk toolkit*. Reading, MA: Addison-Wesley Professional Computing Series.

Petrie, H., Morley, S., McNally, P., O'Neill, A-M., & Majoe, D. (1997). Initial design and evaluation of an interface to hypermedia systems for blind users. In *Proceedings of Hypertext '97* (pp. 48–56). New York: ACM Press.

Savidis, A., & Stephanidis, C. (1995a). Building non-visual interaction through the development of the rooms metaphor. In *Companion Proceedings of ACM Conference on Human Factors in Computing Systems* (pp. 244–245). New York: ACM Press.

Savidis, A., & Stephanidis, C. (1995b). Developing dual interfaces for integrating blind and sighted users: The HOMER UIMS. In *Proceedings of the ACM Conference on Human Factors in Computing Systems* (pp. 106–113). New York: ACM Press.

Savidis, A., & Stephanidis, C. (1998). The HOMER UIMS for dual interface development: Fusing visual and non-visual interactions. *Interacting With Computers, 11*, 173–209.

Savidis, A., Stergiou, A., & Stephanidis, C. (1997). Generic containers for metaphor fusion in non-visual interaction: The HAWK interface toolkit. In *Proceedings of INTERFACES '97 Conference* (pp. 194–196). EC2 & Development (ISBN 2-910085-21X)

Stephanidis, C., Paramythis, A., Sfyrakis, M., Stergiou, A., Maou, N., Leventis, A., Paparoulis, G., & Karagiannidis, C. (1998). Adaptable and adaptive user interfaces for disabled users in the AVANTI project. In S. Trigila, A. Mullery, M. Campolargo, H. Vanderstraeten, & M. Mampaey (Eds.), *Proceedings of the 5th International Conference on Intelligence in Services and Networks (IS&N '98), "Technology for Ubiquitous Telecommunication Services"* (pp. 153–166) (Lecture Notes in Computer Science, Vol. 1430). Berlin, Germany: Springer-Verlag.

Wise, G. B., & Glinert, E. P. (1995). Metawidgets for multimodal applications. In *Proceedings of the RESNA '95 Conference* (pp. 455–457). Washington, DC: RESNA Press.

X Consortium. (1994). *FRESCO™ Sample Implementation Reference Manual* (Version 0.7). Working Group Draft. Mountain View, CA: Silicon Graphics & Japan: Fujitsu.

23

USE-IT: A Tool for Lexical Design Assistance

Demosthenes Akoumianakis
Constantine Stephanidis

This chapter presents USE-IT, a design tool which is a component of the Unified User Interface development platform, responsible for the derivation of recommendations for the lexical level of interaction. Such recommendations are intended to ensure the accessibility of the user interface by the target user group(s). They depict maximally preferred assignments to attributes of abstract interaction object classes, given a set of design constraints. This chapter describes the formulation of the problem in terms of knowledge representation, as well as how the resulting recommendations can be appropriated during user interface development.

As already discussed in chapter 19 of this volume, adaptation in unified user interface development aims to (a) ensure the accessibility of the user interface and (b) enhance the interactive experience of the user through adaptive behavior. This chapter is concerned with adaptations intended to address the former objective, as this has been relatively unexplored by recent efforts addressing user interface adaptations. In particular, recent research on user interface adaptation has concentrated either on the dynamic modification of dialogue characteristics during user–computer interaction (adaptivity), or on the provision of services through which the user can customize the interactive application while using it (adaptability[1]). Adaptive system capabilities are typically supported by dedicated tools referred to as user-modeling components (see chap. 14, this volume), such as BGP-MS (Kobsa & Pohl, 1995), the um toolkit (Kay, 1995), and so on, or user-modeling servers (Orwant, 1995), which maintain assumptions about the user and identify and initiate adaptations accordingly. On the other hand, adaptable systems, such as OBJECTLENS (Lai & Malone, 1988), BUTTONS (MacLean, Carter, Lovstrand, & Moran, 1990) and Xbuttons (Robertson, Henderson, & Card, 1991), provide the user with facilities to tailor certain aspects of the interaction while working with the system.

---

[1]This is an early definition of adaptability and it should not be confused with the working definition of adaptability as introduced in chapter 20 of this volumne.

In all recent efforts addressing user interface adaptation, the assumption is that users can at least initiate interaction with the system, which, in turn, will strive to enhance the dialogue with the user through customization services or adaptive behavior. In other words, systems are assumed to be *a priori accessible,* thus reducing adaptability to the customization of some of the system attributes. In the context of unified user interface development, such an assumption is no longer valid as the target user groups might have radically different interaction capabilities, thus requiring accessibility to be built into the system through adaptation, rather than assumed as a by-product of using a particular interaction platform.

This chapter briefly describes USE-IT, a design tool that implements a knowledge-based approach to automatically deriving design recommendations regarding the accessibility of a user interface. The recommendations cover lexical attributes of the user interface, as these determine the accessibility of a system. The capability to automatically derive such recommendations introduces another dimension to the study of user interface adaptation in general, and to computer-aided user interface adaptation in particular. We refer to this type of adaptation as *lexical adaptability,* because it takes place during the development phase of the system and seeks to ensure its accessibility. The appropriation of the benefits of this type of adaptation has only recently been realized (Stephanidis, 1995) in the context of the ACCESS project (see Acknowledgments). Figure 23.1 highlights the attributes of lexical adaptability that differentiate it from other forms of user interface adaptation.

The value of such user interface *adaptability* is evident when considering interface design for users with diverse abilities, skills, requirements, and preferences. In these cases, it is argued that supporting the run-time adaptation of the user interface of an interactive software system is not sufficient (and sometimes not appropriate) for addressing the problem of user interface accessibility by different user groups. For instance, a totally inaccessible user interface cannot be altered during user–computer interaction, on the basis of dynamically constructed assumptions

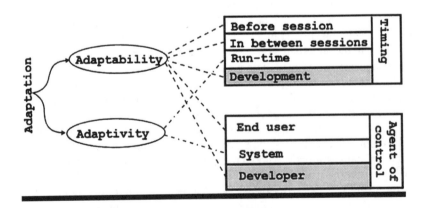

FIG. 23.1.   Dimensions in the study of adaptation.

about the user, because no interaction can be actively accommodated. It follows, therefore, that lexical adaptability complements both the traditional notion of customization and of adaptivity.

In what follows, we describe the way in which USE-IT manipulates a broad range of design knowledge (e.g., user-related, task-context-oriented, and platform-specific attributes) to automatically construct plausible adaptability scenarios and determine the maximally preferred ones, in order to decide on the various characteristics of the lexical level user–computer interaction. The USE-IT tool is embedded in a powerful user interface development platform, which supports unified user interface development (see chap. 22, this volume) with respect to the various user groups (including people with disabilities), lexical technologies (i.e., toolkits), and interaction metaphors.

ADAPTING USER INTERFACES DURING THE EARLY DESIGN AND DEVELOPMENT PHASES

To accomplish its intended objective, USE-IT combines a user-modeling component (for acquiring and maintaining user models), a declarative task description (depicting task requirements and interaction/dialogue constraints), a representation of lexical-level adaptable constituents (i.e., attributes of abstract interaction objects), and algorithms that reason about plausible alternatives and generate a collection of maximally preferred lexical adaptability recommendations (or rules). As shown in Fig. 23.2, such recommendations are encoded in a representation that can subsequently be interpreted by the run-time libraries of a user interface development toolkit to realize the derived recommendations into a user interface that is accessible and adapted to the target end-user (group).

User interface adaptation during the design phase may relate to the lexical, syntactic, or semantic layer of a user interface. In the context of the ACCESS project, we have tackled the lexical level, as this constitutes a prerequisite for accessibility of a user interface by different target user groups with varying abilities, skills, requirements, and preferences, including people with disabilities.

Binding Adaptations to Context

Adaptability recommendations are generated in USE-IT through reasoning about context-oriented parameters to derive maximally preferred assignments to lexical attributes of abstract interaction objects. The notion of context is determined by three first-class design primitives, namely the *interaction metaphor,* the application-specific *task contexts,* and the range of *interaction elements* (i.e., object classes and their attributes). This means that the designer, when considering adaptation of lexical elements of a user interface, explicitly defines the scope of the adaptation by binding it to a particular interaction metaphor, task context, and abstract object class.

Metaphors either may be embedded in the user interface (e.g., menus as interaction objects follow the "restaurant" metaphor) or may characterize the properties

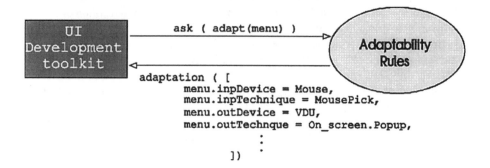

FIG. 23.2.   Interoperation between USE-IT and user interface development toolkits.

and the attitude of the overall interaction environment (e.g., the desktop metaphor presents the user with an interaction environment based on sheets of papers, called windows). The interactive environment of a metaphor is realized by specific user interface development toolkits. Thus, for example, OSF/Motif and MS-Windows support a particular embodiment of the visual desktop metaphor. Different interaction metaphors may be facilitated either through the enhancement of existing development toolkits or by developing new ones. For instance, an enhancement of the interactive environment may be facilitated by embedding in the toolkit automatic scanning facilities for interaction object classes (Akoumianakis, Savidis, & Stephanidis, 1996). Alternatively, new interaction metaphors may be realized through novel toolkits. An example of the latter case is reported in Savidis and Stephanidis (1995b), where Commonkit is used to support nonvisual interaction based on a nonvisual embodiment of the rooms metaphor. Depending on the choice of metaphor, the designer will typically have to deal with different object classes and alternative interactive behavior. Consequently, an interface may be adapted with regard to a particular metaphor so that it exhibits the attitude and characteristics of that metaphor. This entails the representation of knowledge determining the choice of the metaphor and, by implication, the choice of the toolkit for which recommendations are to be derived.

The second first-class design primitive is the application-specific task context. The explicit declaration of task contexts facilitates task-oriented differentiation of lexical and syntactic properties of a user interface depending on what the user is trying to accomplish, or what the interface is attempting to convey. Task contexts are identified and explicitly modeled as part of the design activity. In other words, the interaction designer decides the dialogue states in which the interface should exhibit differentiated behavior. Subsequently, these task contexts are explicitly accounted for during interface development, as is further elaborated later on in this chapter. There are two important requirements related to the notion of application-specific task contexts: (a) a design representation requirement entailing the provision of suitable constructs to enable the designer to build a design representation around the identified task contexts and (b) a development requirement reflecting the need for binding interaction objects to task contexts or dialogue states, during user interface development. Through such a localization of the interactive behavior of an object, it is possible to practically support syntactic and lexical differentiation across different task contexts of the same user interface.

The third binding component of adaptation is the notion of abstract physical interaction objects (Savidis & Stephanidis, 1995a; see also chap. 22, this volume). These are instances of abstract interaction object classes bound to a particular interaction toolkit. In recent bibliographies (Bodart, Hennebert, Leheureux, Provot. & Vanderdonckt, 1994; Myers, 1990), the term *abstract interaction object* (AIO) has been associated with several properties, such as (a) application domain independence, (b) encapsulation of all the necessary interaction properties (i.e., appearance, placement, behavior, state, etc.) by means of attributes (i.e., size, width, color, and methods such as selection, activation, state change, etc.), and (c) independence from particular windowing systems and environments (i.e., platform independent). In the context of the present work, the term is used in a broader sense to include additional properties, such as the following: (a) AIOs are adaptable to the end-user (i.e., their attributes can be adapted through reasoning), and (b) AIOs are metaphor independent (e.g., an AIO, such as the button object, can be applicable for both the desktop and rooms metaphors [Savidis & Stephanidis, 1998], through perhaps different realizations). An abstract interaction object, when bound to a particular interaction metaphor or lexical technology, inherits additional attributes specific to that interaction metaphor (see next section).

Adaptation Decisions

The preceding design primitives give rise to an adaptation space that is conceptually depicted in the relational view of Fig. 23.3. Normalizing the data model of Fig. 23.3, one arrives at the diagram of Fig. 23.4. It follows that an adaptation may be contextually defined by a five-tuple relation, such as <*Metaphor, State, Object, Attribute, Assignment*> where Metaphor is the interaction metaphor (as embedded in a particular user interface development toolkit), State is a dialogue state or task context (identified and specified during the design phase), Object is the object class, Attribute is the object attribute being adapted, and Assignment is the proposed adaptation.

Thus, an adaptation decision is bound by the lexical attribute to which it relates, the interaction object class possessing the attribute being adapted, the dialogue state within which the object exists, and the interaction metaphor.

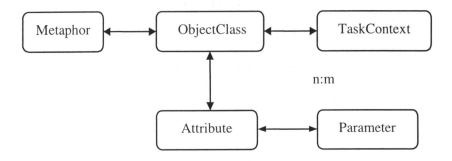

FIG. 23.3.   Structure of object classes.

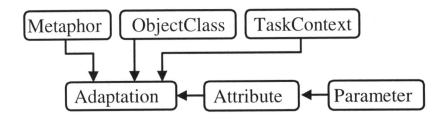

FIG. 23.4.   The normalized data model.

An Illustrative Example

To demonstrate the principles just discussed, let us assume the visual desktop as supported by MS-WINDOWS™ and a hypothetical nonvisual interaction metaphor. An abstract interaction object class, called Button, is defined as possessing one attribute, namely Button.Label, of type *string* and one method, namely Button.Selected, of type *boolean*. When bound to the first interaction metaphor (i.e., the visual desktop), Button inherits additional lexical attributes specific to that metaphor, whereas, when bound to the nonvisual metaphor, Button inherits other attributes, such as language and voice. Consequently, when Button is adapted for the visual desktop, the adaptation constituents are the lexical attributes of the abstract interaction object class (i.e., label and selected), as well as the metaphor-specific attributes (e.g., font.Family, font.Size, color.Bkground). Similarly, when adapted for the nonvisual interaction metaphor, the adaptation constituents will reflect the details of that metaphor. It should be noted that certain attributes may be aggregate types, thus requiring specific parameters in order to be fully defined. For instance, let us assume that there exists an attribute called outputTechnique and that this attribute is to be adapted for the nonvisual interaction metaphor mentioned previously. Let us further assume that the user interface development toolkit supporting this interaction metaphor provides two nonvisual modalities, namely {tactile, auditory}, where the tactile version is implemented using Braille, whereas the auditory version is implemented using a speech synthesizer. Then, depending on the supported interaction techniques, the attribute outputTechnique requires additional parameters to be fully defined. For instance, if its value is set to tactile, then the parameters NumberOfBrailleCells and NumberOfBrailleLines are also lexical constituents to be assigned. From the preceding, it becomes evident that the intention of the relation adaptation of Fig. 23.4 depends on the underlying user interface development system, the interaction objects it offers, and their attributes (i.e., presentation structure, interaction techniques, input/output [I/O] devices, feedback, etc.).

Another important consideration regarding the underlying user interface development system relates to the issue of a task context or dialogue state and the way in which this can be supported. More specifically, the foregoing discussion implies that any adaptation decision is to be interpreted and applied (by the user interface development system) in the context of particular dialogue states (*locality* of decision). This requires that when an interaction object in constructed using the user interface development system, it is assigned to a particular dialogue state, or application-specific task context. Moreover, any adaptation that is to be applied to this object should cover the interaction with this object only during the assigned dialogue state. Although the details in which this is achieved are beyond the scope of the present chapter, it should be mentioned that such toolkits have been developed (Savidis, Stergiou, & Stephanidis, 1997; Stephanidis & Savidis, 1995) and used by the ACCESS consortium partners to develop user interfaces in selected application domains (i.e., nonvisual hypermedia accessible by blind people and interpersonal communication aids for language-cognitive- and speech-motor-impaired users).

AUTOMATING THE GENERATION OF LEXICAL ADAPTATIONS

To support the contextual binding of adaptation decisions to specific interaction metaphors and contexts of use, and the generation of corresponding lexical adaptability rules, USE-IT automates the derivation of lexical adaptation decisions (Akoumianakis & Stephanidis, 1997a, 1997b). USE-IT provides the designer with the interactive means that enable the elicitation and articulation of design knowledge into declarative models from which an adaptation scenario (i.e., a collection of adaptation decisions) is automatically generated. Moreover, such decisions are subsequently interpreted and applied by the target user interface development system to construct a user-adapted interface (see Fig. 23.5), or alternatively, provide a reusable repository of design recommendations.

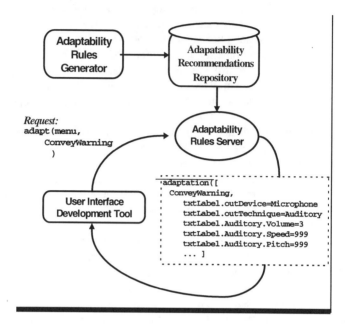

FIG. 23.5.   Applying adaptability rules during user interface development.

The process of deriving adaptability decisions follows two steps. The first step is concerned with the development of a design representation. The second step entails the assessment of this design representation by an adaptation engine that is responsible for the derivation of the lexical specification. The development of a design representation is an iterative process involving the consolidation of design knowledge into the knowledge representation language supported by this tool. USE-IT implements an adaptation decision engine that entails three stages, namely, reasoning about design alternatives, selecting plausible adaptations, and deciding on maximally preferred assignment for adaptation constituents. Each one of these steps is briefly reviewed in the following subsections.

Design Primitives and Knowledge Representation

Reasoning about design alternatives is based on the notion of determining adaptations given three sets of design constraints: (a) constraints related to the user of the user interface, (b) constraints related to application-specific task context requirements, and (c) constraints related to the platform (e.g., range of interaction objects, their attributes and I/O devices). Each class of constraints is generated automatically by appropriate inference mechanisms.

User Modeling.   In particular, constraints related to the user deal with the capabilities of the target user that are modeled in a user-modeling language (UmoL) defined by the triad UMoL = $\langle C,Pr,\Re \rangle$ where:

- C is the set of UMoL constant symbols, defined as the union of the parameter constants, their values, and adaptation values.
- Pr = {x:x = evalue(Parameter,Value) ∨ control_act (Scalar_Value) ∨ contact_site(Scalar_Value) ∨ constraint(user,Attr,Value)}
- $\Re$ is a rule set that is used to generate user-centered constraints from a user model expressed in UMoL.

A typical user model is a collection of formulae in UMoL. User-centered design constraints are declared by a three-argument predicate:

```
constraint(user,Constituent,Assignment)
```

Such constraints are derived automatically by interpreting the contents of a user model against a selected device model. A device model declares the availability of I/O devices, as well as the requirements for operating the device (e.g., alternative control acts, contact sites, and physical abilities required for the operation of the device).

The interpreter is a routine that translates the disjunctive semantics of a device model into a set of rules, and subsequently runs these rules against the current user model. Disjunction in the device model is due to the fact that a device may be operated with more than one control act and for each control act more than one contact site may be used. This gives rise to a disjunctive problem description which is translated into a conjunctive formulation by means of compiling rules. Such rules are stored in a file, which is subsequently run against the user model.

Task Context Schema.   In addition to user-oriented design constraints, the designer of the interface is provided with a tool that enables the declaration of syntactic knowledge and thereby the elicitation of task-oriented design constraints. Syntactic knowledge is based on the notion of a *task context*. Knowledge about each task context of a given application is collected and represented in a *task context schema language*. This is a representation tool that combines three types of characterizations.

First, it comprises a set of attributes that (a) identify the type of task context (i.e., input or output), (b) classify the task context according to the primary interaction task that is performed, and (c) declare the application requirements during the task context, depending on the previous two characterizations.

Second, the designer is allowed to specify aggregation policies that should determine the adaptation of lexical user interface constituents during a particular task context and a set of initial preference expressions. Preference expressions are primarily aimed at capturing intentions, or general ergonomic guidelines, and experimentally justified results that a designer may wish to convey during a particular task context. An aggregation policy is defined by declaring a *design objective* (e.g., speed of cursor movement, precision in positioning, frequency), the relevant *adaptation constituent,* and the associated task context. A heuristic rule derives a partial ordering of design alternatives based on the assigned design objective (e.g., if `speed_of_cursor_movement(true)`, and the adaptation constituent is `inputDevice`, then prefer `continuous` to `discrete` devices). A preference ordering is a ranking of equivalent classes of alternatives with respect to a criterion C that is suggested by the design objective. These equivalent classes are also called indifference classes. We adopt the notation $\{x, \dots, y\}$ to represent indifference classes. Thus, $\{x,y\}$ means that x and y are of the same equivalence class with respect to a criterion C.

The general representation scheme, which is used to declare an aggregation policy, is as follows:

```
policy(taskContextName,Constituent,booleanExpression)

heuristic(taskContextName,Constituent,Preferred,Option)
```

Finally, the designer is allowed to explicitly declare a set of preference expressions that are to be used when aggregating toward task-context-oriented design constraints. A preference expression is a four-argument predicate such as:

```
E(taskContextName,lexicalAttribute,booleanExpression,
Criterion)
```

The predicate E may denote either a *preference* (represented by the predicate p) or an *indifference* (represented by the predicate i) relation. It is important to note that the designer is free to declare as many (if any) preference and indifference relations as it may be appropriate or desirable. Moreover, it is possible to declare aggregation policies and/or preference and indifference relations applicable to all task contexts. To facilitate the elicitation of task-oriented design constraints, based on syntactic knowledge such as the preceding, an inference engine has been designed that consults the task context schema and computes *indifference classes* per adaptation constituent based on the aggregation policy specified and the set of initial preference and indifference relations.

In the present version of the prototype, the inference engine assumes equal voting power for all preference expressions (i.e., simple majority rule for aggregation). The inference engine comprises a set of preference constraints that take the form of general derivation rules or integrity constraints. For purposes of accuracy, the two sets of rules are presented using first-order predicate logic. The minimal set of general derivation rules are as follows:

For any X, Y, Z that belongs to the set S of alternatives for a problem, and any task context T, attribute A, and criterion C, the following properties hold:

Reflexivity and Symmetry

$i(T,C,A,X,X)$

$i(T,C,A,X,Y) \rightarrow i(T,C,A,Y,X)$

Connectivity

$p(T,C,A,X,Y) \lor p(T,C,A,Y,X) \lor i(T,C,A,Y,X)$

Transitivity

$i(T,C,A,X,Z) \rightarrow i(T,C,A,X,Y) \land i(T,C,A,Y,Z)$

$p(T,C,A,X,Z) \rightarrow p(T,C,A,X,Z) \land p(T,C,A,Y,Z)$

$p(T,C,A,X,Z) \rightarrow p(T,C,A,X,Y) \land i(T,C,A,Y,Z)$

$p(T,C,A,X,Z) \rightarrow i(T,C,A,X,Y) \land p(T,C,A,Y,Z)$

Asymmetry

$p(T,C,A,X,Y) \rightarrow not(p(T,C,A,Y,X))$

$p(T,C,A,X,Y) \rightarrow not(i(T,C,A,X,Y))$

$i(T,C,A,X,Y) \rightarrow not(p(T,C,A,X,Y))$

$i(T,C,A,X,Y) \rightarrow not(p(T,C,A,Y,X))$

The preceding constraints have the following meaning: (a) preference is reflexive, asymmetric, and transitive; (b) indifference is reflexive, symmetric, and transitive; (c) preference and indifference are mutually exclusive. The connectivity constraint states that any two competing alternatives must be related in one preference expression, but not both. Transitivity holds for all competing alternatives. For example, when visual feedback is preferred to auditory feedback, and auditory feedback to tactile feedback, then visual feedback is preferred to tactile feedback. It is important to mention that transitivity characterizes the preferences of a rational agent; thus, it is desirable in the context of the proposed framework as transitive preference orderings organize preferences into a simple and tractable structure.

The foregoing rules facilitate the construction of a preference ordering of design alternatives for the specified attributes and task context. In order to demonstrate the notion of an indifference class and the way in which it may be constructed using the representational tool just discussed, we consider another example that demonstrates how task-oriented design constraints can be computed using the aforementioned preference scheme and the design knowledge described earlier.

Let T be the task context "Link Selection." For the attribute `inputDevice`, the designer assigns the criterion `speedCursorMovement(true)`, which implies that indirect devices should be preferred to direct devices. Additionally, the designer declares that the available indirect devices are related by the following preference expressions: the keyboard is preferred to mouse, the mouse is indifferent to trackball, the trackball is preferred to data tablet, which in turn is indifferent to joystick. From such a description and given the aforementioned preference constraints, the indifference classes for the problem at hand are:

1st indifference class:<keyboard>;

2nd indifference class:<mouse,trackball>;

3rd indifference class:<data_tablet,joystick>

Platform Constraints.   Finally, platform-oriented constraints can be built into the system by explicitly naming the I/O devices available and their relationship with the high-level design criteria (i.e., braille is a device that supports nonVisual interaction, mouse is a continuous and relative device, etc.). The resulting representation is a semantic network.

Selection of Design Alternatives: Using Model Trees to Represent Plausible Adaptations

To support the lexical level design of a user interface, a mechanism is required to aggregate the three sets of design constraints reviewed in the previous section into an unambiguous statement regarding the plausible values of adaptable constituents (i.e., object attributes). The solution we have given to the preceding problem comprises the design of a sufficiently expressive data structure for representing the problem, and the development of an algorithm for deciding on the most preferable lexical adaptations. Here, we provide a brief account of the underlying notions, as a comprehensive description is reported in Akoumianakis and Stephanidis (1996, 1997a). The representation of all plausible assignments to a lexical attribute follows a treelike structure, called the *adaptability model tree.*

To illustrate the concept, let us consider the attribute inputDevice of a physical interaction object class. Let us further assume that from the user model it is possible to infer that the user has sufficient abilities to use the keyboard, the mouse, and the joystick. In this case, the user-centered design constraint set for the attribute inputDevice will include:

```
U = {input_device(Keyboard), input_device(Mouse),
            input_device(Joystick)}
```

Furthermore, from the task schema description, it is possible to infer that the task requires two-dimensional positioning, which can be achieved with the mouse or the joystick. Thus, correspondingly, we have the following constraint set for the same attribute:

```
T = {input_device(Mouse),input_device(Joystick)}
```

From the point of view of device availability, we assume the following clause:

```
D = {input_device(Keyboard), input_device(Speech),
     input_device(Mouse),input_device(TrackBall),
            input_device(Joystick)}
```

The adaptability model tree of the previous example is depicted in Fig. 23.6.

Ambiguity occurs when there is more than one plausible assignment for a particular lexical attribute. For instance, the adaptability model tree of inputDevice attribute may suggest that mouse, keyboard, and eye gaze are equally valid assignments. Proper adaptation of the inputDevice attribute, therefore, requires some kind of knowledge that will enable the tool to reason toward a resolution of the ambiguity that arises from the multiplicity of plausible devices that could be used for input. This problem is addressed by compiling the *minimal model tree,* which is defined by the intersection of the branches of the adaptability model tree. For example, the intersection of the three branches of Fig. 23.6 defines the minimal model tree that satisfies all design constraints. Thus, for the situation described in the diagram of Fig. 23.6, the minimal model tree is defined by the set:

$$\mathtt{MIN_{model}} = \{((\mathtt{input\_device(mouse),\ input\_device(joystick)))}\}$$

Any one of the elements of this set could be a plausible adaptation for the attribute input_device. However, for the purposes of the present work, the USE-IT system decides in favor of the solution that preserves maximal multimodality. Thus, the maximally preferred option is defined by the expression:

$$\mathtt{input\_device(joystick)\ \land\ input\_device(mouse)}$$

Consequently, the adaptability decision compiled for this attribute is as follows:

$$\mathtt{Metaphor.Task.Object.\textit{input\ device}} = \mathtt{[joystick,mouse]}$$

The procedure just discussed is applied for the adaptation of general attributes of abstract interaction objects such as inputDevice, inputTechnique, outputDevice, outputTechnique, initiationFeedack, interimFeedbackall, completion Feedback, and additionally, accessPolicy, topologyPolicy and navigationPolicy, for container objects.

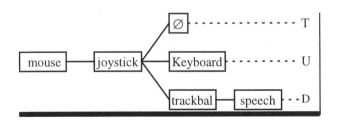

FIG. 23.6.   Example of an adaptability model tree.

IMPLEMENTATION

The USE-IT system has been fully implemented under MS Windows 95 and used by the partners of the ACCESS consortium (see Acknowledgments section) to develop design representations and corresponding adaptation decisions for two application domains, namely nonvisual hypermedia for blind users (Petrie, Morley, McNally, O'Neill, & Majoe, 1997) and interpersonal communications aids for speech-motor- and language-cognitive-impaired users (Kouroupetroglou, Viglas, Anagnostopoulos, Stamatis, & Pentaris, 1997).

USE-IT provides the designer with tools and templates to develop a fully operational design representation as described in the previous sections and, in addition, offers facilities for defining the range of interaction object classes subject to adaptation, their lexical attributes, acquiring conditional rules, defaults, task-oriented and global preferences, as well as limited critiquing of tentative designs. Figure 23.7 depicts the tools that may be used to define the range of interaction elements available in a particular platform, as well as those that the designer deems appropriate to be considered for adaptation.

The example in Fig. 23.7 depicts the interaction elements of the HKTOOL toolkit (Savidis et al., 1997) of the unified user interface development platform, used to develop the nonvisual application. Figure 23.8 illustrates the dialogues for populating task contexts and constructing task-context-oriented design representations (i.e., assigning policies, preference expressions, checking indifference classes—Fig. 23.9, etc.). It should also be mentioned that incremental design is supported by maintaining a default task context called Any_other, which captures global default behavior for an interface. This means that any time a new task context is introduced, it is assigned the required properties without affecting the default behavior of the interface specified in

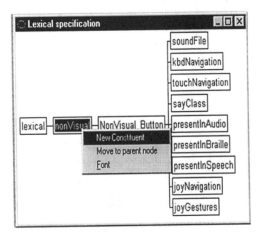

FIG. 23.7.   Defining lexical interaction elements.

the Any_other task context schema. Figures 23.10 and 23.11 depict sample adaptability decisions generated by USE-IT and provided to the designer either in the format required by the user interface development tools (Fig. 23.10) or as a reusable design server (an example of the interrogation of the server is depicted in Fig. 23.11).

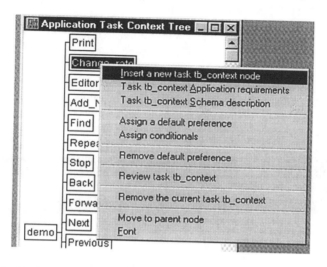

FIG. 23.8.   Task-context-building tools.

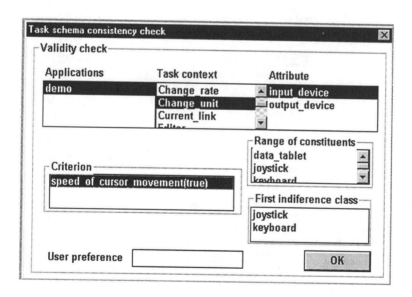

FIG. 23.9.   Querying task contexts schema.

Additional facilities include full support for UMoL, critiquing of tentative designs, as well as knowledge-base editing (i.e., introducing new aggregation policies, heuristic rules and preference profiles). A more detailed account of these features, as well as the usability evaluation of the overall system can be found in Akoumianakis and Stephanidis (1997a, 1997b).

DISCUSSION AND CONCLUSIONS

This chapter has described a technique and a supporting tool environment for generating task-context-adapted lexical specifications of a user interface, through reasoning toward maximally preferred assignments of values to attributes of abstract

FIG. 23.10.   Extract from the derived specification.

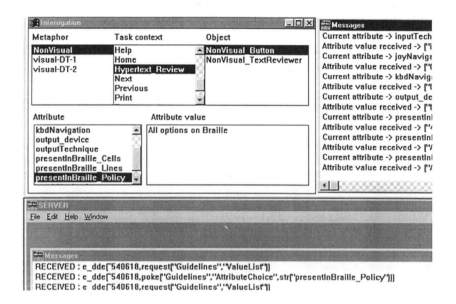

FIG. 23.11. Client server collaboration.

interaction object classes. The main contribution of this work to the study of user interface adaptations is that it covers a dimension that has not been addressed by previous efforts. In particular, it is claimed that the way in which user interface adaptation has been conceived and supported by recent research efforts does not provide the means for addressing radically different interaction requirements.

The main reason for this is that, at present, the techniques of user interface adaptability and adaptivity are implemented on top of a class of user interface development tools, which do not support the development of high-quality user interfaces suitable for diverse user groups, that is, users with different abilities, skills, requirements, and preferences.

This implies that the supported adaptations, irrespective of their timing or agent of initiation, are bound by the limitations of the currently prevailing visual desktop embodiment of the computer and the way that this is implemented in the various user interface development tools and systems (e.g., range of interaction objects, attributes, interaction techniques offered by these development tools). Additionally, the range of adaptation constituents considered is inherently limited to either predetermined dialogue syntax or some presentation attributes, thus failing to account for basic design concerns, such as the details of the way in which interaction is managed at the physical level.

In order to eliminate some of these shortcomings, the work presented in this chapter provides the rationale and technical ground for an alternative approach to user interface adaptation, which is initiated by the developer of the user interface and takes place during the early design and development phases of an interactive application. This type of adaptation is currently not supported by any of the existing tools, systems, or environments for user interface development, as it requires certain qualities that are missing from the present generation of user interface development tools and the presently adopted paradigm. The approach described in this chapter was developed in the context of the ACCESS project, and is part of a novel set of tools for unified user interface development. USE-IT is the tool that automates the derivation of lexical adaptability recommendations to ensure the accessibility of the target user interface implementation. As pointed out, the derived recommendations are interpreted and applied by the run-time libraries of user interface development toolkits so as to realize the designated interactive behaviors.

REFERENCES

Akoumianakis, D., Savidis, A., & Stephanidis, C. (1996). An expert user interface design assistant for deriving maximally preferred lexical adaptability rules. In *Proceedings of 3rd World Congress on Expert Systems* (pp. 1298–1315). Cognizant Communication Corporation (ISBN 1-882345-04-5)

Akoumianakis, D., & Stephanidis, C. (1997a). Knowledge-based support for user-adapted interaction design. *Expert Systems With Applications, 12*(2), 225–245.

Akoumianakis, D., & Stephanidis, C. (1997b). Supporting user-adapted interface design: The USE-IT system. *Interacting With Computers, 9*(1), 73-104.

Bodart, F., Hennebert, A-M., Leheureux, J. M., Provot, I., & Vanderdonckt, J. (1994). A model-based approach to presentation: A continuum from task analysis to prototype. In F. Paternò (Ed.), *Proceedings of Eurographics Workshop on Design, Specification and Verification of Interactive Systems* (pp. 25–39). Berlin: Springer-Verlag.

Kay, J. (1995). The um toolkit for reusable, long-term user models. *User Modelling and User-Adapted Interaction, 4*(3), 149–196.

Kobsa, A., & Pohl, W. (1995). The user modelling shell system BGP-MS. *User Modelling and User-Adapted Interaction, 4*(2), 59–106.

Kouroupetroglou, G., Viglas, C., Anagnostopoulos, A., Stamatis, C., & Pentaris, F. (1996). A novel software architecture for computer-based interpersonal communication aids. In J. Klaus, E. Auff, W. Kresmer, & W. Zagler (Eds.), *Interdisciplinary aspects on computers helping people with special needs, Proceedings of 5th International Conference on Computers Helping People with Special Needs* (pp. 715–720). Wien, Austria: Oldenbourg.

Lai, K., & Malone, T. (1988). Object Lens: A spreadsheet for cooperative work. In *Proceedings of the Conference on Computer-Supported Cooperative Work* (pp. 115–124). New York: ACM Press.

MacLean, A., Carter, K., Lovstrand, L., & Moran, T. (1990). User-tailorable systems: Pressing the issues with buttons. In *Conference Proceedings on Empowering People: Human Factors in Computing System: Special Issue of the SIGCHI Bulletin* (pp. 175–182). New York: ACM Press.

Myers, B. (1990). A new model for handling input. *ACM Transactions on Information Systems, 8*(3), 289–320.

Orwant, J. (1995). Heterogeneous learning in the doppelganger user modelling system. *User Modelling and User Adapted Interaction , 4*(2), 107–130.

Petrie, H., Morley, S., McNally, P., O'Neill, A-M., & Majoe, D. (1997). Initial design and evaluation of an interface to hypermedia systems for blind users. In *Proceedings of Hypertext '97* (pp. 48–56). New York: ACM Press.

Robertson, G., Henderson, D., & Card, S. (1991). Buttons as first class objects on an XDesktop. In *Proceedings of ACM Symposium on User Interface Software and Technology* (pp. 35–44). New York: ACM Press.

Savidis, A., & Stephanidis, C. (1995a). Building non-visual interaction through the development of the rooms metaphor. In *Companion Proceedings of Human Factors in Computing Systems* (pp. 244–245). New York: ACM Press.

Savidis, A., & Stephanidis, C. (1995b). Developing dual user interfaces for integrating blind and sighted users: The HOMER UIMS. In *Proceedings of the ACM Conference on Human Factors in Computing Systems* (pp. 106–113). New York: ACM Press.

Savidis, A., & Stephanidis, C. (1998). The HOMER UIMS for dual user interface development: Fusing visual and non-visual interactions. *Interacting With Computers, 11*(2), 173–209.

Savidis, A., Stergiou, A., & Stephanidis, C. (1997). Generic containers for metaphor fusion in non-visual interaction: The HAWK interface toolkit. In *Proceedings of Interfaces '97 Conference* (pp. 194–196). EC2 & Development (ISBN 2-910085-21X)

Stephanidis, C. (1995). Towards user interfaces for all: Some critical issues. In *Conference Proceedings of HCI International '95* (pp. 137–143). Amsterdam: Elsevier, Elsevier Science.

Stephanidis, C., & Savidis, A. (1995). *Progress Report n.3 on the development of use interface development tools.* TIDE ACCESS TP 1001, Project Report. ACCESS Consortium (available from the authors, ICS-FORTH, Heraklion, Crete, Greece.

24 The I-GET UIMS for Unified User Interface Implementation

Anthony Savidis
Constantine Stephanidis

This chapter describes the I-GET User Interface Management System and outlines the mechanisms it offers in order to support the full set of recommended development-tool properties for Unified User Interface implementation. I-GET follows the tradition of language-based User Interface Management Systems and implements an extended Arch model as the architectural abstraction for Unified User Interface implementation. With the exception of toolkit interoperability, all other mechanisms for Unified User Interface development have been tested in the course of prototypical developments carried out with the I-GET User Interface Management System. As a result, the system can be considered as an appropriate tool for implementing Unified User Interfaces. Possible extensions that would enable I-GET to take advantage of emerging distributed and component-based technologies are also briefly discussed.

Chapter 22 of this volume has identified the desirable functional properties of interface development tools, which would enable them to adequately support the implementation of unified user interfaces. These requirements have emerged primarily due to the need for unifying and concurrently managing implemented dialogue patterns, which address diverse user and usage-context characteristics. Thus, a unified implementation strategy should provide the means for *managing diversity* in human–computer interaction (HCI). Additionally, if we study the development intrinsics of the various distinct dialogue patterns, isolated from the unification perspective (i.e., *implementing diversity*), it is expected that typical functional requirements related to dialogue construction will apply. It is evident that both managing and implementing diversity will have to be substantiated in the context of the unified user interface development process.

The relationship between the aforementioned two categories is illustrated in Fig. 24.1, with an example involving geometrical shapes. In this example, managing diversity implies the capability to unify the physical properties of visually diverse structures into a single unified representation involving abstract geometric properties. Implementing diversity is an important task, because different techniques have to be applied for drawing each distinct geometric shape, even though all the unified artifacts are well represented by the abstract relationships.

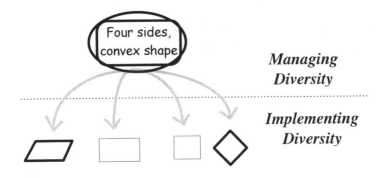

FIG. 24.1. Managing diversity versus implementing diversity in a simple example taken from geometry.

In this chapter, we discuss the I-GET language-based user interface management system (UIMS), which is particularly suited to the *management and implementation of diversity,* in the context of unified user interface development. The I-GET UIMS provides specific development mechanisms to address the fundamental implementation requirements of unified user interfaces, as they have been introduced and discussed in chapter 22, this volume. This chapter aims to demonstrate that developers using the I-GET UIMS may directly employ the advanced mechanisms available, in order to address the demanding functional requirements for:

- Toolkit integration, augmentation, expansion, and abstraction, being the four key mechanisms for effective manipulation of interaction objects in building unified user interfaces.
- Coordination and manipulation of alternative dialogue components, reflecting diverse user and usage-context design dimensions.

Regarding metaphor development, it is argued that the major development overhead is associated with the design and evaluation phases. This is important to ensure that any new interaction metaphor supports accessibility and high quality of interaction for its intended users, usage contexts, and application domains. From the implementation point of view, the overhead and complexity of metaphor development is similar to the development of an interaction toolkit "from scratch." Consequently, the implementation demands for metaphor development are the same as those entailed by "toolkit implementation." In this context, the I-GET UIMS supplies no specific implementation facilities for building toolkits from scratch, but it provides mechanisms to apply integration, augmentation, expansion, and abstraction on existing toolkits.

The chapter is structured as follows. First, we provide an analysis of related work on prevalent interface implementation practices, by assessing existing methods mainly in terms of their appropriateness for managing diversity (as this is the key ingredient of unified user interface implementation). Then, we present an overview of the I-GET UIMS, discussing general characteristics, its run-time model, particular development roles supported, and the various layers of the I-GET language (the language of the I-GET UIMS for building interaction dialogues). Subsequently, we give a detailed presentation of the way in which I-GET facilitates the manipulation of interaction objects, as well as the coordination and manipulation of dialogue components. Finally, we briefly discuss possible future extensions of the I-GET UIMS.

RELATED WORK

There are numerous dialogue-building techniques, as well as categories of interface tools; a comprehensive survey can be found in Myers (1995). For the purposes of tool assessment, we have compiled a list of the most prevalent practices for building and manipulating dialogue artifacts, according to the type of implementation facilities offered to interface developers (see Table 24.1).

It is argued that the key ingredient in managing diverse dialogue patterns is the capability to handle abstraction. In what follows, we briefly review the techniques discussed earlier with the intention to assess the possibility and availability of constructs to handle abstractions.

Presentation-based techniques entail the graphical construction of interface instances based on the range of interaction primitives offered by a particular toolkit. Examples of tools supporting presentation-based techniques are: MENULAY (Buxton, Lamp, Sherman, & Smith, 1983), one of the first interface builders; LUIS (Manheimer, Burnett, & Wallers, 1989); TAE Plus™ (TAE Plus, 1998[1]); and Microsoft Visual Basic™. Being tightly coupled to a particular presentation medium, namely the screen, these techniques do not offer the type of abstraction that is needed to handle alternative dialogue patterns (e.g., nonvisual artifacts).

TABLE 24.1
Categories of Interface Development Practices

Presentation based	Constructing the appearance of the artifact
Physical task based	Decomposing physical user actions
Demonstration based	Demonstrating physical user actions
State based	Implementing dialogue-state transition logic
Event based	Implementing input-event reaction logic
Declarative 4GL methods	Implementing around "when," as opposed to "how"
Model based	Defining various models (e.g., dialogue, data)
Abstract objects/components	Manipulating artifacts "relieved" of physical issues
Abstract task based	Decomposing "what" the user will do, but now "how"
Semantic based	Defining only the semantic information and services

[1]Online. Available: http://www.cen.com/tae

Physical-task-based techniques include TAG (Reisner, 1981), one of the first task-based dialogue specification methods, and UAN (Hartson, Siochi, & Hix, 1990). Typically, such techniques lead to one particular decomposition of user actions (i.e., unimorphic decomposition), thus impeding the design and implementation of alternative task structures (i.e., polymorphic decomposition), as is potentially required when designing for diverse users and contexts of use.

Demonstration-based techniques allow the interactive definition of a target physical interface, in an example-based fashion. Examples of tools supporting demonstration-based techniques are: PERIDOT (Myers, 1988), the first system to support example-based methods; Pavlov (Wolber, 1996); TRIP3 (Miyashita, Matsuoka, Takahashi, Yonezawa, & Kamada, 1992); and DEMOII (Fischer, Busse, & Wolber, 1992). In essence, these techniques still fall in the category of graphical interface construction methods, because the end result is one particular physical interface instance tight to the presentation medium.

State-based techniques are not necessarily bound to any particular development style, because they constitute generic computable models. However, state models do not incorporate any abstraction facilities as such. They only provide an implementation framework for algorithmically representing dialogue control logic. Such implementation models are less popular today than in the past, due to the large development overhead resulting from the fundamentally primitive nature of the basic implementation constructs (i.e., states, transitions, and basic input/output). Examples of state-based techniques are: StateCharts (Wellner, 1989) and STN (Jacob, 1988).

Event-based techniques are the interactive system equivalent of state-based techniques for general computation systems. Even though it is possible to implement abstraction via events (such as logical input), the technique seems to be more appropriate for addressing intracomponent dialogue control requirements (such as handling physical user input within a specific dialogue context). These are in contrast with intercomponent dialogue control requirements, which apply in the coordination of alternative implemented interface components. The Sassafras UIMS (Hill, 1986) was an event-based UIMS. Today, the event model is supported by all known interface toolkits.

Development techniques, which are further from the physical level of interaction, focusing on higher level dialogue properties, are generally more appropriate for addressing dialogue diversity needs. For instance, *declarative fourth-generation language (4GL) methods,* supporting precondition-/notification-based schemes for activating subdialogues, are considered good implementation models for coordinating alternative dialogue patterns. Examples of tools supporting declarative 4GL methods are: ViewControllers in the SERPENT UIMS (Bass, Hardy, Little, & Seacord, 1990); dual dialogue agents in the HOMER UIMS (Savidis & Stephanidis, 1995); and agent classes in the I-GET UIMS (see later section, Support for Coordination and Manipulation of Alternative Dialogue Components).

Development approaches that rely on *abstract interaction objects* or *dialogue components,* by supporting alternative physical realizations, are very promising in the context of unified user interface development, because of their built-in implementation support for polymorphism at the physical interaction level. Examples of methods supporting abstractions of interaction objects are metawidgets (Blattner, Glinert, Jorge, & Ormsby, 1992) and virtual objects (Savidis & Stephanidis, 1995).

Similarly, techniques supporting *abstract tasks* enable the construction of user dialogues not being bound to a single, "hard-coded" task model, as in the case of physical task decomposition. Abstract task models will normally have to be instantiated as concrete physical task structures. Embedding alternative instantiations within the implementation, so as to facilitate run-time coordination and selection, constitutes a big step toward the unified user interface development paradigm. To our knowledge, there are no tools available that support the notion of user-task abstraction.

Semantic-based techniques are very promising, even though the level of support presently available demonstrates a very narrow view of the capabilities that could be integrated within a semantic-based development process. Today, semantic-based development tools directly produce a single-minded physical design from semantic specifications, mainly attributing to the type of functional services/information intended to be interactively supplied to the user. There is a need for generating intermediate dialogue layers, supporting design parameters that the developer may control in order to affect the type of dialogue artifacts produced. In this way, semantic-based methods may serve as design generators, still requiring methods for coordination and manipulation of the aforementioned dialogue artifacts.

In Fig. 24.2, the distinctive role of semantic-based methods in automated artifact production is indicated, along with the various levels of abstraction (or specialization, when seen from the opposite perspective) needed, according to various design parameters controlled by developers. Following this perspective, semantic-based methods are considered as a very promising method for unified development (even though they do not directly contribute to managing diversity), due to their fundamental design-oriented interface engineering policy. Examples of tools supporting semantic-based development are Mickey (Olsen, 1989) and UofA (Singh & Green, 1989).

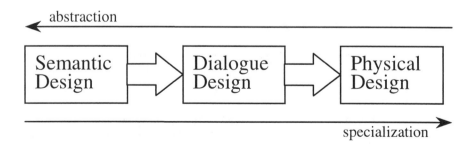

FIG. 24.2.   Generating intermediate design layers in semantic-based methods.

Finally, *model-based techniques* are, in principle, the most promising category of interface tools because they potentially incorporate computable user and design models. However, in most existing model-based tools the emphasis has been shifted from user modeling toward application and dialogue modeling. As a result, the interface generation policies have been centered around the problem of "closing the gap between the application model and the dialogue model," rather than "closing the gap between the user model and the dialogue model." Some of the model-based tools do include various heuristic design rules on the basis of which the dialogue model is incrementally constructed according to hard-coded design rules, mapping specific categories of application structures to particular classes of dialogue objects. This type of functional behavior falls in the premises of design-oriented interface engineering. However, it is far away from adaptable interface generation, engaging primarily user and usage-context design parameters. Another disadvantage of today's model-based tools is their monolithic nature. It is believed that the emphasis should be shifted toward model-based development in an interdisciplinary fashion, allowing the employment of diverse tools, each serving a specific distinctive purpose in interface development. Furthermore, tools should be connected on the basis of a common architectural vision, a concrete development model, and protocols for information exchange. Examples of model-based tools are UIDE (Foley, Wallace, & Chan, 1984) and Mobi-D.[2]

The preceding review of the various interface development methods leads to the general conclusion that techniques targeted to the production of artifacts, which are relatively close to the physical level of interaction, are generally less appropriate for managing diverse dialogue patterns. In contrast, development methods that support manipulation of artifacts that do not directly engage lexical interaction issues are in an advantageous position for addressing the needs of unified development. These observations lead to the separation of the presented development practices in two categories, as indicated in Table 24.2.

THE I-GET UIMS

The I-GET UIMS has been developed in the context of the ACCESS Project (see Acknowledgments), in order to support the implementation of unified user interfaces. With regard to the unified user interface architecture, I-GET addresses the implementation of the dialogue patterns component (see chap. 20, this volume). Its run-time architecture is Arch-compliant and, thus, fully orthogonal to the unified user interface software architecture, enabling orthogonal expansion for connecting with the user information server, the context parameters server, and the decision-making component (see chap. 20 for details on the architecture and the role of the individual components).

The I-GET language borrows elements from both interface languages and 4GLs. Interface languages are different from mainstream programming lan-

[2]Stanford University, Knowledge Modeling Group. *The MOBI-D Interface Development Environment* [Online]. Available: http://smi-web.stanford.edu/project/mecano/mobi-d.htm

TABLE 24.2
Classifying Development Practices According to Their Appropriateness
for Managing Diversity

Promising for Managing Diversity	Not Appropriate for Managing Diversity
• Declarative 4GL methods	• Presentation based
• Abstract objects/components	• Physical task based
• Model based	• Demonstration based
• Abstract task based	• State based
• Semantic based	• Event based

guages, like C++ and Java, because the former have built-in language constructs to support interaction-specific concepts (e.g., interaction object, attribute, method, event, notification, event handler, object hierarchy). 4GLs exist for various programming paradigms, such as procedural, object-oriented (OO), functional, formal, and logic techniques, and are characterized by the introduction of declarative algorithmic constructs. The I-GET language is closer to the procedural and OO paradigms, though it introduces various declarative program control constructs such as preconditions, constraints, and monitors.

The choice of providing a language-based development method has been dictated by the need to support unified user interface development. Interactive construction techniques were found to be inappropriate in this context, because they cannot support the manipulation of diverse interaction elements, while remaining open with respect to integrated interaction toolkits. The primary reasons for this are two-fold; on the one hand, interactive building tools "presuppose" a particular interaction technology (e.g., OSF/Motif, Windows, X/Athena), and, on the other hand, interactive construction cannot support facilities such as abstraction.

Along the same lines, syntax-oriented methods, such as task notations (Hartson et al., 1990) and action grammars (Reisner, 1981), suffer from the shortcoming that they do not allow potentially (different) alternative syntactic designs to coexist in a single running interface. Finally, other techniques, such as event-based models (Hill, 1986) and state-based methods (Jacob, 1988), have been rejected due to their fundamentally low-level dialogue control approach (which is close to programming), as well as their lack of support for abstraction and polymorphism (both of which have been put forward as critical functional features for unified user interface development).

Run-Time Architecture and Development Roles Supported

The run-time architecture of interfaces developed through the I-GET UIMS is illustrated in Fig. 24.3. This architecture extends the Arch UIMS model (UIMS Developers Workshop, 1992) by introducing explicit communication interfaces among the various components, the most important of which is the generic toolkit interfacing protocol (GTIP) (Savidis, Stephanidis, & Akoumianakis, 1997). As it has been discussed in chap. 20 of this volume, the unified user interface architectural framework can be appropriately combined with these UIMS architectural paradigms. Thus, I-GET can be directly employed for the development of the dialogue patterns component in a unified system architecture. The detailed explanation of the various architectural components shown in Fig. 24.3 is as follows.[3]

- A *Toolkit server* plays the role of an intermediate translator between the dialogue control components' "view" of a particular toolkit and the toolkit's programming interface. In terms of implementation, the "real" physical interaction elements will be managed and maintained locally at the toolkit server side, whereas the dialogue control components' view comprises lightweight software elements (which are linked to their physical counterparts through the toolkit servers' translation "filter").
- The *toolkit interfacing* mechanism of I-GET is based on a special-purpose communication protocol, the GTIP. This protocol has been designed in order to enable physical separation between the programming interface of

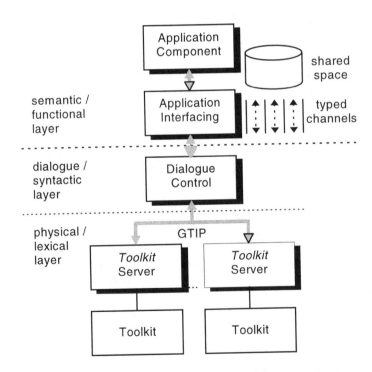

FIG. 24.3.   Run-time architecture of interfaces developed through the I-GET UIMS.

_____

[3]It should be noted that each architectural component is, at run time, a distinct system process.

a toolkit (e.g., object classes, data types, and procedures) from the actual implementation of physical interaction elements (e.g., rendering, device handling, display algorithms).

- The implementation of the *dialogue control* component is realized through the I-GET language, and encompasses all the various interactive modules required by the dialogue patterns component of the unified software architecture.
- The role of the *application interfacing* software component is to coordinate and mediate communication between the application component and the dialogue control component. In this context, it maintains a space where objects can be shared among the application component and the dialogue control, while it also controls the "flow" of messages between them, through the support of typed messaged channels (i.e., message channels carrying only a specific data type of information entities).
- The *application component* comprises the various modules supplying noninteractive ("application") functionality to a software system. An application component can be integrated into the I-GET run-time architecture by implementing a small set of special-purpose C++ functions, as part of the instantiation of a specialized "template" structure provided by I-GET. An application component never communicates directly with the dialogue control component. Instead, it may either "access" the shared space directly, or post messages in a message channel. In the same manner, an application component is implemented by utilizing I-GET library functions for receiving notifications about events occurring in the shared communication space (which is maintained by the application interfacing process).

I-GET supports different *development tasks* that logically correspond to distinct implementation layers (see Fig. 24.4): (a) development of the *functional layer,* which involves the development of the application component and the design and specification of the application interfacing space, (b) development of the *dialogue layer,* which entails the design and implementation of the user interface, and (c) development of the *physical layer,* which involves the integration of toolkits, the specification of the programming interface for each toolkit, as well as the development of the respective toolkit server.

Layers of Implementation Constructs in the I-GET Language

The I-GET language consists of four logical layers of constructs, as illustrated in Fig. 24.5:

1.   The *application programming interface (API) layer,* comprising constructs for the specification of the application interfacing space.
2.   The *agent layer,* containing constructs that address the specification of agent classes (these play the role of dialogue control component classes and

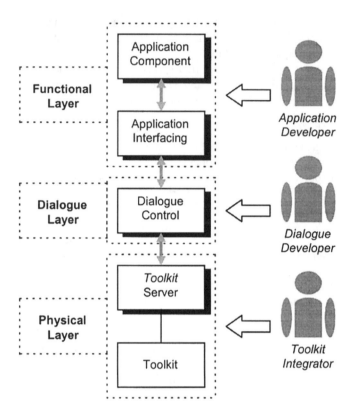

FIG. 24.4.   Development roles in I-GET.

should not be confused with the preserved notion of intelligent/autonomous software agents).

3.   The *objects layer,* which is related to the levels of interaction objects supported in the I-GET language. Physical objects and virtual objects are both supported.

4.   The *common layer,* which provides language constructs that can be employed in any of the previous three layers: (a) programming kernel, (b) constraints and monitors, (c) hooks and bridges, and (d) prototype/implementation facilities, which support separate compilation.

MECHANISMS FOR UNIFIED USER INTERFACE DEVELOPMENT IN THE I-GET UIMS

Having briefly outlined some of the features and the architectural abstractions of the I-GET UIMS, we now discuss some techniques that have been integrated into the

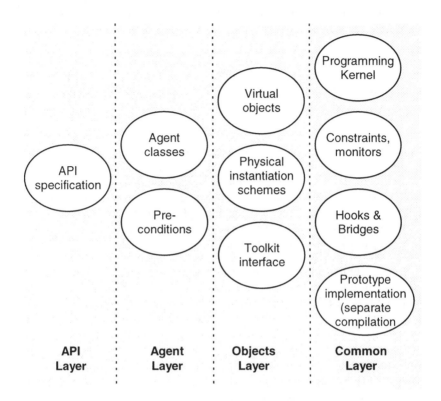

FIG. 24.5. The four layers of implementation constructs in the I-GET language.

I-GET UIMS, in order to address the comprehensive set of implementation requirements posed by the unified user interface development (see chap. 22, this volume).

Support for Toolkit Integration

The I-GET UIMS provides toolkit integration facilities, which satisfy the comprehensive set of related implementation requirements, as the latter have been defined
in chapter 22 of this volume. The design of the I-GET toolkit integration mechanism addresses two key issues: (a) how developers view and manipulate, through
the I-GET language, the imported interaction elements *(toolkit interface specification)* and (b) how the run-time connection with the underlying toolkit software
library(ies) is established *(toolkit server development)*. In the following subsections, we briefly discuss how these two high-level issues have been addressed,
while outlining the toolkit integration strategy of I-GET.

Toolkit Interface Specification.   The first step in toolkit integration is toolkit interface specification. In this stage, the structure and type of the various interaction elements to be integrated is specified. Such a specification is characterized as an "interface" specification (in a programming sense of the term *interface*), because it does not encompass any type of executable statements or constructs (analogous to the CORBA Interface Definition Language). The role of the resulting specification is twofold:

1. It provides the programming interface to imported interaction elements in the I-GET language, encompassing the necessary information for developers to effectively utilize those elements (in the same sense that function prototypes include enough information for someone to call functions in C / C++).
2. It is parsed by the I-GET compiler, to generate special-purpose C++ software modules, which are subsequently employed in the development of a corresponding toolkit server (the latter acting as a run-time translator between the developed I-GET interfaces and the imported toolkit library).

With the toolkit interface specification mechanism of the I-GET UIMS, the definition of object properties can be realized through various policies. Toolkit integrators may freely choose how they will name methods and whether they will group together relative attributes (e.g., presentation, behavior, feedback). In this context, and based on our experience in integrating toolkits targeted to diverse user groups with the I-GET UIMS, we have defined a generic programming model that can be adopted during the toolkit interface specification process. This has been termed *minimal object modeling,* and it constitutes a specific programming approach to organizing interaction objects attributes (into appropriate categories), based on their role in terms of interaction (e.g., lexical, syntactic, semantic). Minimal object modeling aims to address two shortcomings of existing toolkits, namely, the differences in programming structures of interaction objects (like dedicated classes or opaque data types) and the "flat" modeling of the various object properties. For instance, widgets, which correspond to interaction object entities in Xt-based toolkits, provide a linear set of attributes, such as "x," "y," "foreground," "background," "accelerators," "font," "sensitive." Such attributes may be further classified as *presentation* attributes (e.g., "x," "y," "foreground," "background," "font"), *input style* attributes (e.g., "accelerators"), and *behavior* attributes (e.g., "sensitive").

Unfortunately, this approach of explicitly supporting property categories is lacking in existing toolkits. Furthermore, in all known toolkits, the available interaction objects typically have hard-coded behavior and presentation structures, and do not provide programmatic control over lexical-level interaction properties. By offering such limited sets of alternatives (e.g., one or two alternative interaction techniques), the programming models of interaction objects supported by existing toolkits cannot offer practical support for well-known powerful models of lexical input behaviors, such as interaction tasks and interaction techniques (Foley et al., 1984).

To address these shortcomings, the I-GET UIMS provides a toolkit interface specification approach that allows for the categorization of the various properties of interaction objects, within appropriately defined property classes, as reflected in the suggested minimal object modeling (e.g., input technique, interim feedback, output structure, behavior attribute, presentation attribute). Some of the advantages offered by this approach include the following. First, it promotes the modular extension of the lexical interaction facilities of interaction objects by providing implemented alternatives for the various property classes (e.g., various input techniques, alternative approaches to interim feedback).[4] Second, it offers interface developers better programming control of interaction properties. Third, it provides a reference model, which can be adopted by user interface design support tools, to derive recommendations regarding the lexical or syntactic level of interaction, on the grounds of complementary design knowledge, such as user and task models (chap. 22, this volume). Such recommendations can be subsequently interpreted and applied by the run-time libraries of toolkits to support user interface adaptation.

From the preceding, it follows that minimal object modeling is particularly suited for toolkit interface specification, because it offers high flexibility in modifying the original programming model of interaction objects. The model, which is illustrated in Table 24.3, provides only a template that has to be adopted, suggesting two main categories of object attributes; it is not prescriptive with respect to the type, number, names, and nature of specific interaction properties. In this sense, it minimally affects the toolkit interface specification process, because it only provides logical grouping of attributes, rather than introducing new attributes not originally supported by the imported toolkits. Specifically, the model proposes grouping (see Table 24.3) according to output techniques (attributes per output technique) and input techniques (attributes per input technique). Minimal object modeling follows the generic toolkit metamodel, the metamodel for toolkits supported by the I-GET language being one particular instantiation of the latter.

Toolkit Server Development.   The toolkit server plays the role of the intermediate "translator" between the dialogue control module of the I-GET run-time architecture and the particular imported toolkit. The toolkit server has a twofold role:

1.   It receives requests from the dialogue control module, corresponding to object instantiations, attribute modifications, and output events, that the interface client program (developed in the I-GET language) will normally perform, and it serves such requests by calling the original toolkit functionality. The "physical" interface is created and maintained locally within the toolkit server.

2.   It sends messages back to the dialogue control module regarding input events, method notification, and attribute modification, as a result of user interaction with the physical interface locally managed by the toolkit server; additionally, return values from the execution of output events are sent back to the dialogue control.

---

[4]It should also be noted that such an approach leads to one possible implementation framework for *toolkit augmentation.*

TABLE 24.3
Minimal Object Modeling—Definition and Explanations

Output Techniques

The different presentation methods supported for a particular object class. For example, a "menu" object in a windowing toolkit might employ the following presentation techniques for the list of displayed options:
- Vertical arrangement.
- Circular arrangement.
- Sequential presentation of options.
- Options presented as stacked cards.

Attributes per Output Technique

There are the various parameters for each particular output technique. For instance, assuming a circular topology of "menu" options, the following attributes may be supported:
- Circle parameters (a and b radius, assuming an ellipse in the general case).
- Flag for auto-splitting the circle in sectors for each option.
- Foreground/background colors.

Input Techniques

The various ways through which the user provides input and interactively manipulates a particular object instance. Input techniques are also related to the notion of interaction tasks, as defined in Foley et al. (1984). Input techniques are in many cases directly dependent on particular output techniques. Hence, the categories of properties in minimal object modeling are not always orthogonal to each other (i.e., combinations may be subject to toolkit implementation restrictions). Examples of input techniques, for a "menu" object are:

- Direct selection,
  —via mouse;
  —via an eye gaze device;
  —through speech-input;
  —through shortcuts.
- Indirect selection,
  —automatic scanning;
  —manual scanning;
  —sequential presentation of options enabling selection;
  —focus on desired option and selection.

Attributes per Input Technique

These are the available parameters for tailoring a particular input technique. Input technique attributes are split into two subcategories: presentation attributes, which concern the various appearance parameters for a particular input technique, and behavior attributes, related to the way in which the behavior associated with an input technique can be customized. For instance, considering the case of indirect selection and automatic scanning, the following attributes are identified:
- Option feedback method
  —Highlight
  —Shape outline
    ☐ Border width   Presentation
    ☐ Line style and color
- Scanning time interval   Behavior
- Scanning direction
  —Always forward and recycle   Behavior
  —Always backward and recycle
  —Start with forward and change direction on recycle

The communication between the dialogue control module and the toolkit server(s) is transparent to I-GET programmers. The backbone of such a communication channel is the GTIP. As it has been illustrated in the run-time architecture of I-GET developed interfaces (Fig. 24.3), the dialogue control module and the toolkit servers are independent, local or remote, processes. In order to run an interface developed with I-GET, the following default steps are taken (these are managed automatically by the I-GET run-time system and are completely transparent to end-users): (a) an instance (i.e., process) of the dialogue control is created locally and (b) the various toolkit servers, which are necessary for the realization of the "running" interface, are also locally executed.

There can be many deviations from the default behavior, which can be defined by the end-user (through special initialization files called "execution plans"). First, it is possible to explicitly define the host machines on which the interface client program (the I-GET interface program after being compiled and linked—that is, the dialogue control module in the I-GET run-time architecture), as well as the host machines on which the toolkit servers will execute. Second, not all the servers need to be running. This may be necessary in the following cases: (a) a toolkit server has not been developed yet or (b) the software/hardware resources required by a toolkit are not present at the target machine (e.g., audio processing capability, video sampling and presentation, nonvisual interaction equipment). The I-GET UIMS supports the development of compact and efficient (in terms of performance) toolkit servers, supporting quick response schemes.

In the case of generalization, toolkit interface specification strategies (i.e., integrating multiple toolkits that implement the same interaction metaphor, into a common toolkit interface specification), resulting in virtual toolkits within the I-GET language layer, multiple servers must be developed, one for each target toolkit to which the virtual toolkit will be mapped. Interfaces utilizing elements of virtual toolkit specifications may run with any of the available servers (but with only one of them for each different execution session); the running interface will have the native "look and feel" of the toolkit associated with the particular running server.

In the context of the ACCESS project (see Acknowledgments), three toolkits were integrated into I-GET. The first toolkit integrated was the standard Xaw/Athena widget set. The second toolkit that was integrated was SCANLIB (Savidis, Vernardos, & Stephanidis, 1997), an augmented version of the basic Microsoft Windows object library. SCANLIB supports one-, two-, and five-switch-based scanning interaction in the Windows environment, with alternative parameterizable dialogue modes. SCANLIB provides facilities for intra- and interapplication navigation, window manipulation (sizing, moving, etc.), and unconstrained interaction with all basic Windows controls (including the capability for text entry, through onscreen keyboards).

The third toolkit integrated into I-GET was HAWK (Savidis, Stergiou, & Stephanidis, 1997), a toolkit intended for the development of purely nonvisual interfaces. HAWK provides a set of standard nonvisual interaction objects and interaction techniques that have been specifically designed to meet the interaction needs of physically or "situationally" blind users. Furthermore, HAWK introduces the concept of a generic nonvisual container that can be parameterized to convey different nonvisual metaphors (through text, speech, and sounds).

Support for Toolkit Augmentation

The I-GET UIMS supports two alternative implementation strategies for toolkit augmentation, reflecting two different technical objectives: (a) ensuring direct presence of augmented dialogue features in existing applications, without imposing recompilation or relinking or (b) ensuring that the implementation of augmented interaction methods is supported to the fullest possible degree.

On the basis of these two alternative technical goals, I-GET supports two augmentation scenarios. The first scenario assumes that augmented dialogue features are "injected" at the toolkit server side, whereas I-GET interactive applications, developed prior to the augmentation process, need no extra recompilation/relinking cycles, due to the fact that they never make direct calls to the original toolkit library. In the second scenario, the augmented interaction techniques are implemented through the I-GET language and "injected" within the toolkit interface specification layer, ensuring upward compatibility. Thus, existing applications utilizing elements from the original toolkit interface specification are not affected.

The first method, relying on the incorporation of augmented dialogue facilities within the toolkit server, is outlined in Fig. 24.6. As shown, the toolkit interface specification does not need to be altered, because the extra dialogue techniques are added at the toolkit server side and do not require any modification in the original

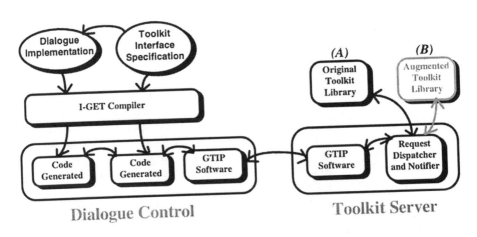

FIG. 24.6.   Implementing augmented interaction software at toolkit server side; the dialogue control software is not affected, because the transition from the original toolkit library, case (A), to the augmented toolkit library, case (B), takes place at the toolkit server.

toolkit interface specification. Even in cases where the toolkit interface specification has to be expanded, so as to include new lexical attributes and/or methods, existing applications will still behave correctly without recompilation/relinking. This is due to the fact that the communication between the dialogue control and the toolkit servers relies on the names of elements and their properties rather than physical resources. As a result, the original naming conventions for imported elements are fully preserved (only some new attributes and methods may be introduced), and the run-time intercomponent communication will still function properly, even with the augmented toolkit server version. This technical approach has been taken for the implementation of the augmented interaction techniques for the MS Windows object library, that is, the SCANLIB library (Savidis, Vernardos et al., 1997).

The module indicated as "Request Dispatcher and Notifier" in Fig. 24.6, is part of a toolkit server implementation and has a twofold role: (a) it serves all requests from the dialogue control, received via the GTIP layer, and (b) it sends back notifications, due to user interaction, or returned parameters, due to dispatching of output events.

The second method, in which all augmented dialogue features are implemented through the I-GET language, is based on the extension of the original toolkit interface specification, to include the augmented dialogue control logic (see Fig. 24.7). This approach is similar to extending C++ classes by adding more member functions, and/or upgrading the implementation of various member functions, without affecting the original class definition (i.e., upward compatibility of class interface). As a result, programs using the old class definition may be safely recompiled with the new version, gaining directly the advantages of the upgraded execution behavior of member functions. In the same manner, interactive components built with I-GET (dialogue implementation in Fig. 24.7) utilize interaction elements based on their definitions within toolkit interface specification. Hence, as the preexisting programming interface of those elements is not changed, their respective client dialogue implementations are not affected.

Support for Toolkit Expansion

The I-GET UIMS is the first known 4GL-based interface development tool that supports the full set of requirements for toolkit expansion, in the context of unified user interfaces.

Table 24.4 presents the two major steps involved in the accomplishment of toolkit expansion through the I-GET language. The first step concerns the definition of the new lexical class. Such a definition is similar to lexical class definitions in a toolkit interface specification and provides the programmable "picture" of the newly introduced object class from the developers' point of view. The second, and normally more resource-demanding step, involves the implementation of dialogue code. This stage will typically require:

- Writing the presentation code for the object (e.g., code that will display the object on the presentation device), which may employ platform-specific primitives (e.g., lines, sounds), as well as other interaction objects (in which case the object would be a composite one).

- Implementing event handlers for input management and feedback control.
- Providing code to map attribute updates to the presentational or behavioral aspects of the real object (e.g., if the "label" attribute of an object is changed, code must be executed to refresh the displayed object and present the new "label").
- Notifying methods at those points of input management code, where the logical actions represented via methods are considered to be accomplished (e.g., a "radio-button" may fire its "state changed" method upon a mouse click occurring within its screen space).

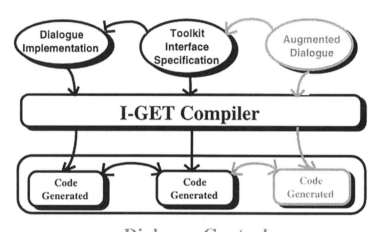

FIG. 24.7.   Injecting augmented dialogue code within toolkit interface specification; dialogue implementations utilizing the old toolkit interface specification version only need recompilation.

TABLE 24.4
The Two Steps in Accomplishing Toolkit Expansion Through the I-GET Language

1. Class definition	• Naming class and associating with proper toolkit name.
	• Defining attributes and their default values.
	• Defining method categories.
2. Dialogue Implementation	• Implementing presentation structure.
	• Implementing input control and feedback.
	• Mapping attribute changes to presentation or behavior.
	• Notifying methods when needed.
	• Ensuring that the object is correctly "hooked" in hierarchies.

Support for Toolkit Abstraction

The I-GET language provides implementation constructs that address the comprehensive set of requirements for toolkit abstraction in unified user interface development. Abstract interaction objects, which are named *virtual objects* in the I-GET language, may be defined through a mechanism called *virtual object genesis*. Virtual objects do not possess physical interaction properties, unless associated with their physical counterparts, through a *lexical instantiation relationship*. Such relationships may encompass multiple alternative *instantiation schemes*, each defining a particular mapping of a virtual object class to a specific *lexical object class*.

The I-GET language supports the parallel management of multiple named toolkits, called *lexical layers*. Therefore, it is possible to define an instantiation relationship, for a given virtual object class for each such imported toolkit. This capability is illustrated in Fig. 24.8, where all the intermediate links for bridging virtual interaction objects to lexical interaction objects are also shown. Interaction objects may either be implemented directly by the underlying toolkit, or result from the application of the augmentation, or expansion mechanisms. Figure 24.9 presents the various implementation levels for interaction objects in the I-GET language, reflecting the various links shown in Fig. 24.8.

In order to demonstrate the maximal abstraction capabilities of the I-GET language, we use an illustrative example taken from the ACCESS project, where the I-GET UIMS has been used for defining and using abstract objects (Stephanidis & Savidis, 1995, 1997). First, we define a virtual "push-button" class, exhibiting no physical interaction properties. Then, we define two instantiation relationships, one for the SCANLIB toolkit (Savidis, Vernardos et al., 1997) and one for the HAWK toolkit (Savidis, Stergiou et al., 1997), both of which have been integrated within the I-GET UIMS (Stephanidis, Savidis, & Gogoulou, 1995), in the context of the ACCESS project. Following the full definition of the physical instantiation logic of the push-button object, we show the declaration of virtual instances, programming access to virtual attributes and implementation of virtual methods, as well as programming access to lexical attributes and implementation of methods for lexical instances.

In Example 1 (left) of Fig. 24.10, the specification of the `Button` virtual class is given. The class definition starts with the `virtual` keyword, followed by the class name and the explicit list of toolkit names, for which the virtual class will be given instantiation relationships, being in our example `Xaw`, `W95`, and `Hawk` (the order of appearance is not important). In practical terms, the purpose of the list is to define the toolkits for which the newly introduced class constitutes an abstraction. Within the virtual class, one virtual attribute, called `Accessible` (to enable/disable dialogue with the user), and one virtual method, called `Pressed`, are locally defined. It is evident that such an object class forms a pure abstraction for all physical interactive behaviors, independently of particular style and/or metaphor, supporting direct user actions (such as a button press). Even though its particular name (i.e., `Button`) is related to a specific real-world metaphor (i.e., electromechanical

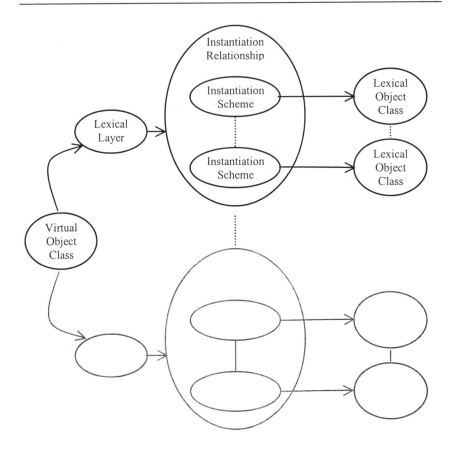

FIG. 24.8.   The links between virtual object classes, lexical layers, instantiation relationships, instantiation schemes, and lexical object classes, in the I-GET language.

devices), in the context of the I-GET language it has a purely notational value (i.e., it is just a class identifier). A better name could be devised that would be more indicative of its generic role, if so desired by the developer.

Similarly, the Selector class encompasses only those attributes that are relieved from physical interaction properties, and a single abstract method called Selected. Apart from the Accessible attribute (which can be seen as a standard property for all virtual classes), it has two additional attributes: NumOfOptions, indicating the total number of items from which the user may select, and UserChoice, which contains the order of the option most recently selected by the user. Note that the actual list of options is not stored within the virtual object class; this has been a design decision based on the fact that options do not always have a specific information type, but may vary for different physical interaction

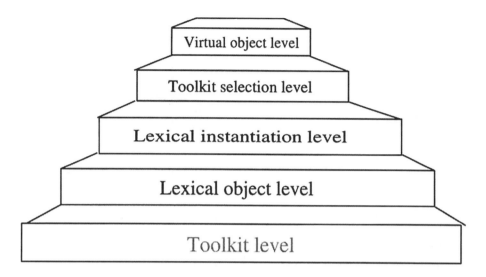

FIG. 24.9.   Implementation layers for interaction objects in the I-GET language.

styles. For instance, options may have textual, iconic, animated, video, or audio content. Hence, the storage of options is moved to the physical interaction object classes. For all virtual classes, a constructor/destructor block is supported, for code that needs to be executed upon object instantiation/destruction.

Examples 2 and 3, of Fig. 24.10, contain the specification of instantiation relationships of the Button virtual class, for the W95 and Hawk toolkits, respectively. An instantiation relationship encompasses the implementation logic for (a) mapping a virtual object class to a lexical object class and (b) defining the run-time relationship among virtual and lexical attributes/methods. An instantiation relationship may encompass an arbitrary number of instantiation schemes, which are defined as follows:

1.   An arbitrary name is given to each instantiation scheme, which has to be unique across its *owner* instantiation relationship block (i.e., defined between an opening and a closing square bracket—" [...] "). For instance, the scheme defined within Example 2 is given the name Button, whereas the only scheme defined as part of the Hawk instantiation relationship of Example 3 is given the name PushButton.

2.   Each scheme maps the virtual object class to a particular lexical class. The name of this lexical class must be provided next to a colon symbol (":"), directly following the scheme name. For example, the Button lexical class for the W95 toolkit is chosen as the target lexical class of the Button scheme in Example 2, whereas the PushButton class is similarly defined for the PushButton

```
virtual Button (Xaw,W95,Hawk) [          virtual Selector (Xaw,W95,Hawk) [
   bool Accessible=true;                    bool Accessible=true;
   method Pressed;                          method Selected;
   constructor []                           int NumOfOptions=0;
   destructor []                            int UserChoice=-1;          ┌───┐
]                                           contructor []               │ 1 │
                                            destructor []               └───┘
                                         ]
```
```
instantiation Button (W95) [ // instantiation relationship for Windows
   Button : Button [
      {me}W95.Accessible := {me}Accessible;
      method {me}W95.Pressed [ {me}->Pressed; ]                         ┌───┐
      constructor []                                                    │ 2 │
      destructor []                                                     └───┘
   ]
   default Button;
]
```
```
instantiation Button (Hawk) [ // instantiation relationship for Hawk
   PushButton : PushButton [
      {me}Hawk.Accessible := {me}Accessible;
      method {me}Hawk.Activated [ {me}->Pressed; ]                      ┌───┐
      constructor []                                                   │ 3 │
      destructor []                                                    └───┘
   ]
   default PushButton;
]
```
```
virtual Container cont :scheme(W95)=FrameWindow;
virtual Button button :parent(W95)={cont}W95
                      : parent(Hawk)={cont}Hawk;                       ┌───┐
virtual Toggle toggle : scheme(W95)=RadioButton                       │ 4 │
                      : parent(W95)={cont}W95                          └───┘
                      : parent(Hawk)={cont}Hawk;
```

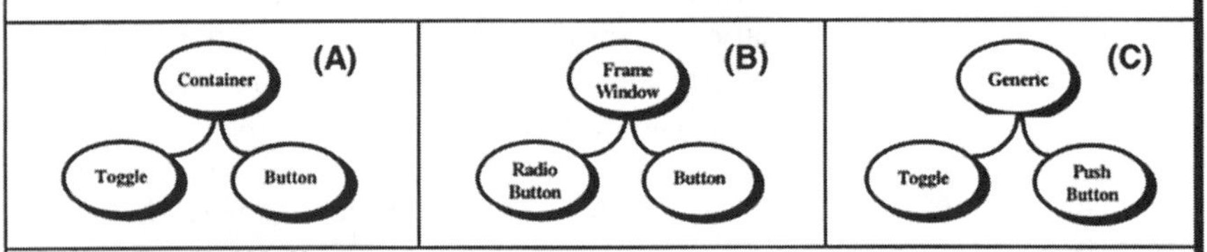

```
Color color=[0,255,0];
{button}W95.bkColor=color;
                        {button}Hawk.soundFile="press.wav";            ┌───┐
{button}Hawk.label={button}W95.label=" Press";                        │ 5 │
{button}.Accessible=false;                                            └───┘
```
```
method {button}.Pressed [ printstr("Pressed") ]
method {button}Hawk.Activated [
   printstr("Blind user action");                                     ┌───┐
]                                                                     │ 6 │
method {button}W95.Pressed [                                          └───┘
   printstr("Sighted user action");
]
```

FIG. 24.10. (1) Virtual class definitions; (2) instantiation relationship for W95; (3) instantiation relationship for Hawk; (4) A, B, and C are resulting hierarchies for the virtual, W95, and Hawk instances, respectively; (5) accessing virtual and lexical attributes; and (6) implementing virtual and lexical methods.

scheme in Example 3. It should be noted that, although we have employed the convention of naming schemes with the names of their respective lexical classes in our examples, the I-GET language does not pose any restrictions in this respect.

3.   In the body of instantiation schemes, the linkage between the virtual and lexical instances is implemented. Normally, constraints and monitors will be employed for bridging together virtual and lexical attributes, whereas artificial method notifications are employed to fire virtual methods, due to lexical method activation. For instance, in Example 2, the lexical attribute `Accessible`, of the `W95 Button` class, syntactically defined as `{me}W95.Accessible`, is constrained by the `Accessible` virtual attribute, syntactically defined as `{me}.Accessible`. Additionally, a local implementation of the Pressed lexical method is supplied, which executes only a single statement, generating a notification of the virtual `Pressed` method; this establishes the link for propagating method notifications from the lexical object instances up to virtual object instances.

4.   The default active instantiation scheme is chosen; in Example 2, the expression `default Button`; defines that `Button` is to be activated as the default lexical instantiation of declared `Button` virtual object instances, for the `W95` toolkit (unless another existing scheme is explicitly chosen, as is explained later on).

At run time, when an instance of a virtual class is declared, the various instantiation relationships, for each toolkit listed in its class definition header, are first identified. Then, for each instantiation relationship, the necessary instantiation scheme is activated; this may be the default scheme, or a scheme explicitly chosen by the interface developer (the details of declaring virtual object instances is discussed next). The scheme activation step results in the construction of an instance of the lexical class that has been associated to that scheme. This lexical instance is attached to the original virtual instance. Finally, the various constructs defined locally, within the scheme body, are activated, establishing the run-time links among the virtual and the physical object instances.

In Example 4 of Fig. 24.10, the declaration of various virtual object instances is illustrated. The three most important properties of virtual instance declarations are: (a) physical instantiations will be made automatically for each target toolkit (i.e., *plural instantiation*); (b) explicit scheme selection is allowed for each toolkit, thus enabling the developer to choose the desirable physical instantiation of a virtual instance for each toolkit (i.e., *controllable instantiation*); and (c) an appropriate parent lexical instance should be explicitly supplied, for each toolkit defined (i.e., *polymorphic object hierarchies*). In Example 4, the expression `:scheme(W95)=FrameWindow` makes an explicit scheme selection for the virtual instance called `cont`. Similarly, the `:parent(Hawk)={cont}Hawk` expression, for the button virtual instance, means that its parent instance for the Hawk toolkit is the Hawk lexical instance, retrieved from the cont virtual instance. (A), (B) and (C) of Fig. 24.10 indicate the virtual-, `W95`-, and `Hawk`-instance hierarchies, all resulting from the declarations of Example 4. In Example 5, access on virtual and lexical attributes is demonstrated. For a virtual instance *<obj>*, the conventions *{<obj>}.<attr>* and *{<obj>}<toolkit>.<attr>* represent a virtual attribute *<attr>* and a lexical attribute *<attr>*, the latter to be found in the active lexical instance of *<toolkit>* toolkit, respectively. Hence, following Example 5, for the button virtual instance, the {button}W95.bkColour is an attribute of the W95 active lexical instance, which, due to the fact that the active W95 scheme for button is Button, must be an attribute of the Button lexical class.

Finally, within Example 6, the implementation of callbacks for the virtual and lexical instances is shown. The {me}.Pressed method is a virtual method, which, according to the specifications in Examples 2 and 3, will be fired if either the Hawk or the W95 instances are notified; hence, a *unified method implementation* is enabled. If, however, further specialization of method behavior is needed for each toolkit, the callbacks of the appropriate lexical instances must be accessed. As shown in Example 6, the {me}Hawk.Activated and the {me}W95.Pressed methods are explicitly implemented, to display an indicative message. According to Example 6, if the blind user presses the "Press" button, the following actions will take place: First, the lexical Activated method implementation, defined locally within the PushButton scheme of the Hawk instantiation relationship, will be executed. As a result, a virtual Pressed method notification will be generated, resulting in a call to the virtual method implementation of Example 6, causing the message *"Pressed"* to be displayed; then, the lexical method registered in Example 6 will be called, resulting in the message *"Blind action"* to be also displayed.

Support for Coordination and Manipulation of Alternative Dialogue Components

The dialogue control component model of the I-GET UIMS is called *dialogue agents* (Savidis & Stephanidis, 1997). Agents in the I-GET language are not related to the concept of intelligent or autonomous agents. They are independent threads of execution performing their own main loop (typical of event-based interactive software), maintaining their own local state, communicating with other agents, managing arbitrary collections of interaction objects (e.g., "windows," "menus," "buttons"), and coordinating/controlling, or being coordinated/controlled by other dialogue agents.

The implementation of dialogue agents is facilitated in the I-GET language via respective agent classes. Agent classes constitute the component class model of the I-GET language and are analogous to object classes in typical object-oriented programming (OOP) languages, in the sense that various instances of an agent class can be created in a dialogue program. Figure 24.11 outlines the agent class structure. Agent classes fall into one of two categories, which are described as follows:

1. *Precondition-based agent classes.* Instances of these classes are created automatically at run time, when their respective creation precondition is satis-

fied. Similarly, created instances are automatically destroyed if the destruction precondition in their respective class (if one exists) is satisfied.

2. *Parameterized agent classes.* These can be instantiated directly via statements, in a way similar to instance creation in typical OOP languages, where agent parameters correspond to class constructor parameters.

Receiving and Applying Adaptation Decisions. In the unified software architecture, adaptation decisions are sent by the decision-making component and received asynchronously by the dialogue patterns component, that is, the dialogue control module of the I-GET run-time system architecture. In the I-GET run-time architecture, the application component provides the link with the decision-making component. As a result, adaptation decisions may be received locally in the dialogue control module via the application interfacing space. The I-GET application interfacing space supports both the shared space model, where arbitrary typed objects can be exposed, and the model of typed communication channels, which supports the establishment of an arbitrary number of strongly typed links for information exchange. The API specification in the I-GET language involves the

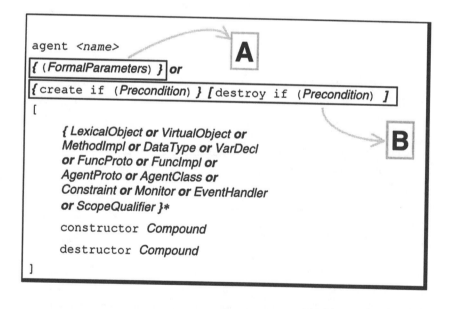

FIG. 24.11.   The agent class structure: (A) the agent class header with its formal parameters, if the class is to be instantiated synchronously (i.e., via a statement), and (B) agent class header with its instantiation precondition, as well as optional destruction precondition, if the class is to be instantiated asynchronously (i.e., via preconditions).

definition of shared data structures as well as the declaration of typed channels; representative examples are provided in 24.12.

In Fig. 24.12, we provide the API definitions, as well as the implementation structure, for locally receiving adaptation decisions. This implementation module does not need to change when the number of adaptation decisions receives changes. More important, it is independent of the number of implemented dialogue components (as agent classes), that reside locally within the dialogue control implementation. As we show in the next section, the locally stored adaptation decisions will be accessed within the preconditions of dialogue components, thus enabling decisions to be automatically applied. This way, components get activated or canceled directly by the precondition management mechanisms, without requiring any extra code to that extent. In Fig. 24.12, Block 1, the definitions providing the API space for communicating decisions are supplied. In Fig. 24.12, Block 2, the local (to the dialogue implementation) data structures and functions for storing adaptation decisions are provided. Finally, Fig. 24.12, Block 3, contains the definition of the agent class that receives and locally stores adaptation decisions (utilizing functions from Block 2).

Orthogonal Expansion of Dialogue Components. Component expansion concerns all cases in which new dialogue components have to be implemented.

```
enum Decision=[Activation, Cancellation];
struct AdaptationDecision [
   Decision decision;
   String task;
   string style;
];
channel Decisions of AdaptationDecision;                          1

local AdaptationDecision *localDecisions=nil,
local int totalDecisions=0;                                       2
AdaptationDecision currDecision;
void AddLocallyDecision(AdaptationDecision decision);

agent ReceiveDecision
create if (message(Decisions decision)) [
   constructor [
      AddLocallyDecision(decision);                               3
      destroy {myagent};
   ]
   destructor []
]
```

FIG. 24.12.   Definitions of (1) a channel to receive adaptation decisions, (2) local structures and functions to store adaptation decisions, and (3) an agent class for receiving and locally storing adaptation decisions.

We demonstrate how the I-GET language supports the implementation of new dialogue components, being part of the dialogue patterns component in a unified user interface architecture. Furthermore, we show that this implementation has minimal effects on the implementation code of old dialogue components. This property is characterized as *orthogonal expansion,* because the software implementation of the various distinct components is made virtually independent (i.e., orthogonal) to each other. This property is also known as the principle of incremental development in the software-engineering field. In the unified user interface development paradigm, new components need to be constructed to address the following scenarios:

- The introduction of additional interactive capabilities (i.e., new tasks or subtasks).
- The implementation of alternative styles for polymorphic tasks (i.e., addressing additional decision parameters).
- The implementation of additional interaction-monitoring components (for dynamic detection of user attribute values).

We now discuss some representative implementation examples (see Fig. 24.13) that clearly show how each of the aforementioned component expansion scenarios is addressed, without, however, affecting the rest of the implementation structure. In Fig. 24.13, Block 1, the addition of an embedded agent class, named File, is employed within the DocumentProcessor agent class, for the implementation of the "File" dialogue box, commonly used in document-centric applications (supporting operations like "Open," "Save," "Save as ...," etc.) This extra facility is also made controllable by the user via a RadioButton object, named file. As shown, additional code is added within the original agent class, without affecting the rest of agent body implementation.

In Fig. 24.13, Block 2, the introduction of an additional implemented style for the "DocumentProcessing" task is shown, as a top-level agent class (DocumentProcessor_AdaptivePrompting) serving adaptive prompting purposes. Its instantiation precondition engages the currDecision global variable, which has been defined within Fig. 24.12, Block 2, for locally storing the most recently received adaptation decision.

In Fig. 24.13, Block 3, the API specifications for receiving monitoring control messages are provided. In the unified software architecture, these messages originate from the user information server. In I-GET, the monitoring messages should be communicated via the API space to the dialogue control module (as in the case of adaptation decisions coming from the decision-making component), because the application component offers the gateway to such external components. The structure of messages reflects the communication protocol between the user information server and the dialogue patterns component. An embedded agent class is defined in order to receive and locally store monitoring messages, named MonitoringControlReceiver; the most recent monitoring control message is stored in a global variable named currMessage.

```
Agent DocumentProcessor ...[
   ...
   lexical(W95) RadioButton file: parent=(win);
   agent File
   create if ({file}.state==RadioOn)
   destroy if ({file}.state=RadioOff) [...]
   ... ]
```
①

```
agent DocumentProcessor_AdaptivePrompting
create if (
currDecision.decision==Activation &&
currDecision.task=="DocumentProcessing" &&
currDecision.style=="AdaptivePrompting") [
... ]
```
②

```
enum MonitoringStatus=[MonitoringOn, MonitoringOff];
enum MonitoringLevel=[
   TaskMonitoring, EventMonitoring, MethodMonitoring
];

struct EventMonitoring [ string objectName, eventCategory; ];
struct TaskMonitoring [ string task; ];
struct MethodMonitoring [ string objectName, methodCategory; ];

struct MonitoringMsg [
   MonitoringStatus status;
   MonitoringLevel level;
   EventMonitoring eventInfo;
   TaskMonitoring taskInfo;
   MethodMonitoring methodInfo;
];
MonitoringMsg currMessage;
channel MonitoringControl of MonitoringMsg;

agent MonitoringControlReceiver
create if (message(MonitoringControl msg)) [
   constructor [ currMessage=msg; destroy {myagent}; ]
   destructor []
]
```
③

```
agent DocumentProcessor ... [
   ...
   agent Monitoring_Object_font_Method
   create if (
   currMessage.status==MonitoringOn &&
   currMessage.level==MethodMonitoring &&
   currMessage.methodInfo.objectName=="font" &&
   currMessage.methodInfo.methodCategory=="StateChanged")
   destroy if (
   currMessage.status==MonitoringOff &&
   currMessage.level==MethodMonitoring &&
   currMessage.methodInfo.objectName=="font" &&
   currMessage.methodInfo.methodCategory=="StateChanged") [...]
   ... ]
```
④

FIG. 24.13. (1) Adding new interactive facilities (child class), (2) adding an adaptive prompting style (top-level class), (3) the API specifications for receiving monitoring control messages, and (4) adding a monitoring component (child class).

Finally, in Fig. 24.13, Block 4, the addition of a monitoring component is shown. Here, we assume that the user information server requests monitoring to be turned on, at the method notification level for the object "font." The monitoring agent class is embedded within the `Document Processor` agent class, because it will have to access the font RadioButton interaction object. Moreover, it is named in a way reflecting its role (i.e., `Monitoring_Object_font_Method`), in the sense that it is responsible for monitoring the object font at the method level. In the instantiation and destruction preconditions of the monitoring agent class access, the `currDecision` global variable is engaged. This variable holds the recent monitoring control message received locally. In the instantiation precondition, it is tested if its content is a "monitoring activation" message, whereas, similarly, in the destruction precondition it is tested if it is a "monitoring cancellation" message.

Mapping the Polymorphic Component Model to Implementation Patterns of the I-GET Language.

Polymorphic task hierarchies are the central outcome of the unified user interface design process (chap. 21, this volume). By minimizing the gap between the resulting polymorphic task design and the necessary target implementation, the overall development overhead is seriously reduced, potentially promoting the quality of the resulting interactive software application. For instance, procedural decomposition could be more easily programmed through a procedure-oriented language (e.g., C or Pascal), whereas an object-oriented design can be more easily transformed into an implementation, if an OOP language is employed (e.g., C++ or Smalltalk). This issue of reducing the gap among the design and implementation worlds has been recently recognized in the interface development tool domain (Graham, 1992), as the necessity of *mapping the behavioral domain* (i.e., task design) *to the constructional domain* (i.e., implementation artifacts), through the provision of comprehensive, well-defined engineering patterns and strategies.

As mentioned earlier, the polymorphic task decomposition process results in a polymorphic hierarchy of dialogue components. In such a component hierarchy, we distinguish the following two general relationship categories that may hold between any pair of components: (a) they both belong to the same set of alternative styles for supporting a particular polymorphic task and (b) the components concern different, either polymorphic or unimorphic, subtasks. For each of these two cases, we identify the set of possible design relationships among the components. These have to be preserved within the target implementation, so that the polymorphic component hierarchy (i.e., the unified user interface design) can be thoroughly mapped into a target implementation form. In this context, we first identify these design relationships, and we subsequently provide the engineering strategy through which they can be easily mapped to implementation patterns in the I-GET language. As a result, we establish a mapping of the polymorphic component hierarchy into appropriate implementation patterns, thus effectively establishing the final link in the chain bridging the unified user interface design "world" and the implementation "world" in I-GET.

Table 24.5 outlines the two categories of relationships among components defined in a polymorphic task hierarchy. The relationships between alternative styles (components) of the same subtask (left part of Table 24.5), have been defined and explained in the discussion of the unified user interface design process (chap. 21, this volume). The definition for the second group of relationships, regarding pairs of components that are associated to different tasks, follows:

- *Containment*. One component is physically contained into the other. For instance, the editing windows of Microsoft Word™ are physically enclosed within a single top-level window of the Microsoft Word™ application.
- *Aggregation*. Both components are direct physical constituents of another component, with which they are both related via a containment relationship. For instance, the "drawing toolbar" and the "formatting toolbar" of Microsoft Word™ are direct constituents of the Microsoft Word™ main application window.
- *Collaboration*. When one component exposes information or services to the other in a "read only" fashion (i.e., "clients" cannot alter the provided information, and services offered do not affect the behavior of the "server"). For instance, monitoring components access the interaction objects residing in the physical components for which they have to collect interaction monitoring information.
- *Control*. One of the components exposes information or services to the other in a "write" fashion. This means that various behavioral or presentational aspects of the "server" component may be changed by a "client" component, by calling those services or by changing the accessed information. For example, in Microsoft Word™, the user is able to turn on or off the appearance of certain toolbars via a pop-up menu displayed when the right mouse button is clicked over any toolbar. Logically, the component responsible for this pop-up dialogue should have "write" access to the "visibility" attribute of those toolbar components.
- *Indifference*. The components are simply not related to each other. For instance, we do not expect any relationship among the "print" dialogue box and the "drawing toolbar" of Microsoft Word™.

TABLE 24.5

Relationships Between Components That Belong in the Same Set of Alternative Styles for a Polymorphic Task (Left Column), and Relationships Among Components that Correspond to Different Tasks (Right Column)

Components Belong in the Same Set of Alternative Styles for a Polymorphic Task	Components Correspond to Different, Polymorphic or Unimorphic, Tasks
• Exclusion	• Containment
• Substitution	• Aggregation
• Augmentation	• Collaboration
• Compatibility	• Control
	• Indifference

The mapping of those relationships to I-GET language constructs is simple and straightforward, similar to typical OOP languages. This implementation facility is offered by the I-GET tool, and is effectively combined with the rest of the mechanisms for implementation, coordination, and manipulation of alternative dialogue patterns (e.g., declarative activation and orthogonal expansion), which are not directly supported in 3GLs. Table 24.6 identifies the corresponding implementation patterns in the I-GET language, for each of the possible component relationships.

SUMMARY, CONCLUSIONS, AND FUTURE WORK

The development of unified user interfaces is a very demanding task. This is mainly due to the following three classes of development requirements: (a) interaction resources from multiple toolkits should be utilized; (b) diverse dialogue patterns need to be implemented, coordinated, and manipulated; and (c) the run-time user- and usage-context-oriented adaptation process has to be realized. From an implementation perspective, specific development mechanisms and architectural disciplines can be identified, which may help to dramatically reduce the overall overhead associated with unified user interface development:

- Toolkit integration, augmentation, expansion, and abstraction, reflecting the fundamental requirements for the manipulation of interaction objects, as these were introduced and discussed in chapter 22 of this volume.
- Implementation, manipulation, and coordination of dialogue components, reflecting the need for mapping the diverse dialogue artifacts produced as part of the unified user interface design process (chap. 21) into an implemented form (see chap. 22 for the relevant discussion).
- The unified user interface software architecture, providing a concrete implementation pattern for structuring the implementation of a unified user

TABLE 24.6

Mapping Relationships Among Components that Correspond to Different Tasks
Into Implementation Patterns of the I-GET Language

Containment	The container supplies an appropriate parent object instance to the contained component, to establish containment appropriately at the object hierarchy level.
Aggregation	The constituent components share the same parent object instance, supplied by their common container component.
Collaboration	Public agent members (i.e., variables and functions), which are made available by collaborating components, not affecting the owner behavior or presentation, are used by their "partner" components.
Control	Public agent members (i.e., variables and functions), which are made available by "controlled" components, affecting the owner's behavior or presentation, are used by the "controller" components.

interface as a set of cooperating components, which communicate in order to provide the end-user with an adapted interface; the unified user interface software architecture is presented in chapter 20, this volume.

In our review of user interface implementation methods, the main issue has been to identify tools that are appropriate for applying these methods in practice. Starting to construct the "profile" of tools that may be employed for unified user interface development, we have classified the required development facilities in two general categories: (a) those for managing diversity and (b) those for implementing diversity. A key ingredient in the successful implementation of unified user interfaces is to drive the development process around the management of diverse dialogue resources, while allowing for their effective implementation in different forms. Hence, "management of diversity" becomes a critical functional property, and interface development practices that support it are likely to be more suitable candidates for unified user interface implementation. From a practical point of view, this implies that development techniques that manipulate resources close to the physical level of interaction inherently become less appropriate for the management of diversity, whereas techniques emphasizing abstraction and syntax-loose control are more appropriate for this purpose. It should be noted that the role of methods bound to the physical level of interaction is not considered less significant, because they are necessary for implementing diversity.

In our effort to establish an implementation platform supporting all the necessary features for unified user interface development, it became evident that most of the existing interface development tools are mainly targeted to the provision of advanced support for implementing physical aspects of the interface, via different techniques (e.g., visual construction, task-based, demonstration-based). Even methods that have long been the subject of research and development efforts in the field, such as abstract objects and toolkit openness, the existing support does not move beyond the GUI.

These observations have led to the development of the I-GET UIMS, a tool that provides all the necessary mechanisms for implementing unified user interfaces. The I-GET UIMS provides the I-GET language for interface implementation, which exhibits a number of powerful new features, when compared to existing development tools, including:

- *Toolkit integration:* I-GET is capable of importing virtually any toolkit, relying on a generic toolkit metamodel, the toolkit interface specification language kernel, and the specifically designed generic toolkit interfacing protocol for communication.
- *Toolkit abstraction:* I-GET makes a step beyond the desk-top metaphor and provides openness and extensibility of virtual objects, by supporting the definition of arbitrary virtual objects, and the specification of the physical instantiation logic, through which alternative ways of mapping abstract objects to physical objects can be defined.

- *Facilities for manipulating diverse dialogue patterns:* I-GET offers a syntax-loose dialogue control method based on component classes called "agents" (exhibiting hierarchical organization and precondition- or call-based instantiation), combined with powerful constructs like monitors, preconditions and constraints for arbitrary variables.

We have integrated the following toolkits in the I-GET UIMS: the Xaw/Athena widget set for X Windowing System (for the Unix version of the I-GET UIMS), the HAWK nonvisual toolkit (for the Windows version), and the SCANLIB toolkits (for the Windows version). Through the abstraction mechanisms, we have defined virtual interaction object classes, as well as the logic of mapping those virtual objects to physical objects from each of the aforementioned imported toolkits.

Looking to the future, we are planning the following two developments regarding the I-GET UIMS:

- To build a CORBA-based toolkit server for the OSF/Motif widget set, and implement the connection of the I-GET dialogue control run-time component with the CORBA-based toolkit server. Such a connection will be handled by "injecting" C++ communication code within the toolkit interface specification of the OSF/Motif imported object classes (a feature supported by the I-GET UIMS). This would effectively allow the development of a distributed interactive application, in which the various components are built via the I-GET UIMS (i.e., I-GET dialogue control running components) and communicate with the same CORBA-based toolkit server running on an end-user machine.
- To provide components that facilitate the development of CSCW applications via the I-GET UIMS. We are planning to provide a central data server application, and implement some specific policies for CSCW management, such as data replication, locking/unlocking methods, and late engagement. Instances of interactive applications, developed in I-GET, and comprising a single computer-supported cooperative network (CSCW) session, will communicate with the same data server. The dialogue implementation will be supported via the I-GET UIMS without introducing any extensions in the I-GET language. We will experiment with multiuser applications in which diverse user groups cooperate from diverse interaction platforms.

REFERENCES

Bass, L., Hardy, E., Little, R., & Seacord, R. (1990). Incremental development of user interfaces. In G. Cockton (Ed.), *Proceedings of Engineering for Human–Computer Interaction* (pp. 155–173). Amsterdam: North-Holland, Elsevier Science.

Blattner, M., Glinert, E., Jorge, J., & Ormsby, G. (1992). Metawidgets: Towards a theory of multimodal interface design. In *Proceedings of the COMPSAC '92 Conference* (pp. 115–120). New York: IEEE Computer Society Press.

Buxton, W., Lamp, M., Sherman, D., & Smith, K. (1983). Towards a comprehensive user interface management system. In *Proceedings of SIGGRAPH '83: Computer Graphics* (pp. 35–42). New York: ACM Press.

Fischer, G., Busse, D., & Wolber, D. (1992). Adding rule-based reasoning to a demonstrational interface builder. In *Proceedings of the ACM Symposium on User Interface Software and Technology* (pp. 89–97). New York: ACM Press.

Foley, J. D., Wallace, V. L., & Chan, P. (1984). The human factors of computer graphics interactive techniques. *IEEE Computer Graphics & Applications, 4*(11), 13–48.

Graham, T. (1992). Future research issues in languages for developing user interfaces. In B. Myers (Ed.), *Languages for developing user interfaces* (pp. 401–418). Boston: Jones & Barlett.

Hartson, H. R., Siochi, A. C., & Hix, D. (1990). The UAN: A user-oriented representation for direct manipulation interface design. *ACM Transactions on Information Systems, 8*(3), 181–203.

Hill, R. (1986). Supporting concurrency, communication and synchronisation in human–computer Interaction—The Sassafras UIMS. *ACM Transactions of Graphics, 5*(3), 179–210.

Jacob, R. (1988). An executable specification technique for describing human–computer interaction. In R. Hartson (Ed.), *Advances in human–computer interaction, 1.* (pp. 211–242). Norwood, NJ: Ablex.

Manheimer, J. M., Burnett, R. C., & Wallers, J. A. (1989). A case study of user interface management system development and application. In *Proceedings of SIGCHI Conference on Wings for the Mind* (pp. 127–132). New York: ACM Press.

Miyashita, K., Matsuoka, S., Takahashi, S., Yonezawa, A., & Kamada, T. (1992). Declarative programming of graphical interfaces by visual examples. In *Proceedings of the ACM Symposium on User Interface Software and Technology* (pp. 107–116). New York: ACM Press.

Myers, B. (1988). *Creating user interfaces by demonstration.* Boston: Academic Press.

Myers, B. (1995). User interface software tools. *ACM Transactions on Computer–Human Interaction, 12*(1), 64–103.

Olsen, D. R. (1989). A programming language basis for user interface. In *Proceedings of SIGCHI Conference on Wings for the Mind* (pp. 171–176). New York: ACM Press.

Reisner, P. (1981). Formal grammar and human factors design of an interactive graphics system. *IEEE Transactions on Software Engineering, 7*(2), 229–240.

Savidis, A., & Stephanidis, C. (1995). Developing dual interfaces for integrating blind and sighted users: The HOMER UIMS. In *Proceedings of the ACM Conference on Human Factors in Computing Systems* (pp. 106–113). New York: ACM Press.

Savidis, A., & Stephanidis, C. (1997). Agent classes for managing dialogue control specification complexity: A declarative language framework. In *Proceedings of HCI International '97* (pp. 461–464). Amsterdam: Elsevier, Elsevier Science.

Savidis, A., Stephanidis, C., & Akoumianakis, D. (1997). Unifying toolkit programming layers: A multi-purpose toolkit integration module. In M. Harrison & J. Torres (Eds.), *Proceedings of the 4th Eurographics Workshop in Design, Specification and Verification of Interactive Systems* (pp. 177–192). Wien, Austria: Springer-Verlag.

Savidis, A., Stergiou, A., & Stephanidis, C. (1997). Generic containers for metaphor fusion in non-visual interaction: The HAWK interface toolkit. In *Proceedings of Interfaces '97 Conference* (pp. 194–196). EC2 & Development (ISBN 2-910085-21X)

Savidis, A., Vernardos, G., & Stephanidis, A. (1997). Embedding scanning techniques accessible to motor-impaired users in the WINDOWS object library. In *Proceedings of HCI International '97* (pp. 429–432) Amsterdam: Elsevier, Elsevier Science.

Singh, G., & Green, M. (1989). A high-level user interface management system. In *Proceedings of SIGCHI Conference on Wings for the Mind* (pp. 133–138). New York: ACM Press.

Stephanidis, C., & Savidis, A. (1995). *Final report on G-DISPEC* (TIDE ACCESS TP1001 Project, Deliverable D.1.3.2). ACCESS Consortium.

Stephanidis, C., & Savidis, A (1997). *Final report on the I-GET tool* (TIDE TP1001 ACCESS Project, Deliverable D.1.4.2). ACCESS Consortium.

Stephanidis, C., Savidis, A., & Gogoulou, R. (1995). *Report on the integration of target platforms* (TIDE ACCESS TP1001 Project, Deliverable D.1.5.1). ACCESS Consortium.

UIMS Developers Workshop. (1992). A meta-model for the run-time architecture of an interactive system. *SIGCHI Bulletin, 24*(1), 32–37.

Wellner, P. D. (1989). Statemaster: A UIMS based on statechart for prototyping and target implementation. In *Proceedings of SIGCHI conference on Wings for the Mind* (pp. 177–182). New York: ACM Press.

Wolber, D. A. (1996). Pavlov: Programming by stimulus-response demonstration. In *Proceedings of the ACM Conference on Human Factors in Computing Systems* (pp. 252–259). New York: ACM Press.

25

A Case Study in Unified User Interface Development: The AVANTI Web Browser

Constantine Stephanidis
Alexandros Paramythis
Michael Sfyrakis
Anthony Savidis

This chapter will present a case study in Unified User Interface development, following the process of constructing a unified interface for a Web browser that provides accessibility and high quality interaction to a wide range of user categories (including disabled and elderly users) in various contexts of use. The design, implementation and evaluation phases of the development process are discussed in turn and important lessons learned along each of the phases are discussed.

The previous chapters in this part of the book have presented the theoretical underpinnings of the unified user interface concept and the practical dimensions of the associated development paradigm. Moreover, chapters 19 through 24 have introduced the concept of, and presented a design method and an architectural framework for the development of, unified user interfaces, as well as tools that facilitate and support the design and implementation phases. This chapter builds on that material, presenting a case study in unified user interface development.

In particular, this chapter reviews and discusses the employment of the unified user interface development paradigm in the design, implementation, and evaluation of a unified interface for a Web browser, developed in the context of the European Commission–funded ACTS AC042 AVANTI project (see Acknowledgments). The rationale that led to the adoption of a Web-browsing interaction paradigm within the project is also outlined. The presentation of the development activities is not exhaustive. Rather, specific aspects of the different development phases are discussed in depth, in order to demonstrate some of the most important practical implications of applying the unification concept in the construction of a user interface. The discussion not only addresses the experience gained within the AVANTI project, but also attempts to provide some general "guidelines" to be followed in the context of unified user interface development.

The chapter is structured as follows. The rest of this section gives a brief overview of the AVANTI system and the related conceptual framework. The second section focuses on the design phase of the AVANTI browser's user interface and details: the design space as it emerged from the project's requirements, and its implications on the overall design of the interface; the task decomposition and the individual stages involved therein, following the unified user interface design method (see chap. 21, this volume); and, the design of the interaction dialogue. The next section presents the architecture of the AVANTI browser, relating it to the unified user interface architecture. Furthermore, it discusses some of the issues involved in the implementation of the unified interface, including the transformation of design rationale into adaptation logic and the different categories of adaptation that emerged during the development of the AVANTI user interface. The section concludes with a series of instances of the interface, adapted for different categories of users and contexts of use. The fourth section presents the expert-based assessment and validation activities that were carried out during the development of the user interface, as well as some of the results of the end-user usability evaluation of the system. The final section discusses the overall development process followed in AVANTI and points out some of the important aspects that differentiate unified user interface development from traditional approaches to developing accessible and high-quality user interfaces.

The AVANTI Information Systems

The AVANTI project, which was concluded in August 1998, aimed to address the interaction requirements of individuals with diverse abilities, skills, requirements, and preferences (including disabled and elderly people), using Web-based multimedia applications and services. AVANTI advocated a new approach to the development of Web-based information systems. In particular, it put forward a conceptual framework for the construction of systems that support adaptability and adaptivity at both the content[1] and the user interface levels. The AVANTI framework comprises five main components[2] (see also Fig. 25.1):

- A collection of multimedia databases, which contain the actual information and are accessed through a common communication interface *(multimedia database interface—MDI)*.

---

[1]For an extensive discussion of content-level adaptation in the AVANTI information systems, please refer to Fink, Kobsa, and Nill (in press).

[2]The authors have been responsible for the design and implementation of the user interface component (Web browser) of the AVANTI information systems. Other partners of the AVANTI consortium (see Acknowledgments) have been responsible for the development of the other modules.

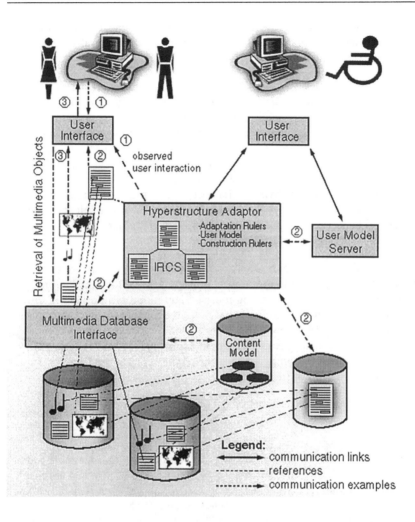

FIG. 25.1.   Overall architecture of the AVANTI system.

- *The user-modeling server (UMS)* (Kobsa & Pohl, 1995; Schreck & Nill, 1998), which maintains and updates individual user profiles, as well as user stereotypes.
- *The content model (CM)* (Fink, 1997), which retains a meta-description of the information available in the system.
- The *hyper-structure adaptor (HSA)* (Fink, Kobsa, & Nill, 1997; Nill, 1998), which adapts the information content, according to user characteristics.
- The *user interface (UI)* component (which is the subject of this chapter).

The cooperation between the main architectural components of the AVANTI system is presented in Fig. 25.1. The following short scenario (also depicted in Fig. 25.1) presents the typical "route" of a request for a hypermedia document in the system:

1.   The user requests a hypermedia document. The user interface forwards this request to the content adaptation component (HSA).

2.   The HSA matches the request to an appropriate hypermedia document "template," assembles the (adapted) document taking into account user- and content-related information that is provided by the user-modeling component (UMS), and propagates the adapted document to the UI.

3.   The UI interprets the hypermedia document, transparently retrieves multimedia objects from the AVANTI databases via the MDI, and finally presents the requested hypermedia page to the user, employing appropriate (accessible) interaction and presentation facilities.

The preceding conceptual framework has been applied in the development of three information systems, in the context of AVANTI: (a) the Siena information system (Del Bianco, 1998), offering touristic and mobility information to residents and visitors of the city of Siena (Italy), (b) the Kuusamo information system (Penttila & Suihko, 1998), providing information on traveling and accommodation in Kuusamo (Finland) and its surroundings, and (c) the Rome information system (Ghetti & Bellini, 1998), aimed at providing cultural and administrative information for the city of Rome (Italy).

DESIGNING FOR MULTIPLE USER GROUPS

Design Space

The front end of the AVANTI system was called to provide an accessible and usable interface to a range of user categories, irrespective of physical abilities or technology expertise. Moreover, it was expected to support various, differing situations of use. The end-user groups targeted in the AVANTI project, in terms of physical abilities, include (a) "able-bodied" people, assumed to have full use of all their sensory and motor communication "channels," (b) blind people, and (c) motor-impaired people, with different forms of impairments in their upper limbs, causing different degrees of difficulty in employing traditional computer input devices. In particular, in the case of motor-impaired people, two coarse levels of impairment were taken into account: "light" motor impairments (i.e., users have limited use of their upper limbs but can operate traditional input devices or equivalents with adequate support) and "severe" motor impairments (i.e., users cannot operate traditional input devices at all).

Furthermore, because the AVANTI system was intended to be used both by professionals (e.g., travel agents), as well as by the general public (e.g., citizens, tourists), the users' experience in the use of, and interaction with, technology was another major parameter that was taken into account in the design of the user interface. Thus, in addition to the conventional requirement of supporting novice and experienced users of the system (Fowler, Macaulay, & Siripoksup, 1987; Perdue-Casali & Chase, 1995), two new requirements were put forward (following the project's requirements analysis phase, and in particular the analysis of user characteristics): (a) supporting users with any level of computer expertise and (b) supporting users with, or without previous experience in the use of Web-based software.

In terms of usage context, the system was intended to be used both by individuals in their personal settings (e.g., within the home or office), and by the population at large through public information terminals (e.g., information kiosks at a railway station, airport). Furthermore, the results of the requirements analysis indicated that, in the case of private use, the front end of AVANTI should be appropriate for general Web browsing, allowing users to make use of the accessibility facilities beyond the context of a particular information system.

Additionally, users were to be continuously supported as their communication and interaction requirements changed over time, due to personal or environmental reasons (e.g., stress, tiredness, system configuration). This entailed the capability, on the part of the system, to detect dynamic changes in the characteristics of the user and the context of use (either of temporary or of permanent nature) and cater for these changes by appropriately modifying itself.

The aforementioned requirements led to the primary adoption of a Web-browser-like approach to the development of the AVANTI front end. Specifically, the user interface component was decoupled from the rest of the modules in the AVANTI system architecture and was given basic Web communication and presentation capabilities, so as to be able to function as a regular Web browser when necessary. As we see later, this also necessitated the introduction of two alternative interaction metaphors, namely the "desktop" metaphor (based on the traditional windowing environment) and the "information kiosk" metaphor (based on interfaces typical of multimedia information systems, usually simplified to address the requirements of inexperienced users).

The decision to support general Web browsing in the AVANTI front end dictated the adoption of a range of tasks that are typical on the Web (see, e.g., Byrne, John, Wehrle, & Crow, 1999; Tauscher & Greenberg, 1997), such as "go to my 'home' document," "select a document from the history list," and so on.

Table 25.1 presents a list of the Web-related interaction tasks that formed the basis of the design for the browser's user interface. As is shown in the forthcoming section, The AVANTI Web Browser Architecture, these tasks have also guided, to a large extent, the functionality built into the user interface component (mainly in terms of communication and presentation facilities).

In practical terms, the interaction requirements posed earlier (especially those concerned with the provision of accessibility to different user groups and the support of run-time adaptations) dictated the development of a new Web-browsing application, which would not be based on preexisting browser technology. The main reasons for that, were the following:

TABLE 25.1

Typical Web-Browsing Tasks

Typical Web-Browsing Interaction Tasks	
• Open location	• Review history
• Go to previous document	—Select page from history
• Go to next document	• Search the Internet
• Go to starting document	• Add bookmark
• Reload document	• Remove bookmark
• Stop loading document	• Review bookmarks
• Select link from document	—Select page from bookmarks list
• Navigate within document	—Remove page from bookmarks list
• Open file	• Search current document
• Save file	• Set browser options
• Print file	• Exit browser
	• Help

1.   Although today's commercially available browsers support customizability (e.g., through "add-on" components), the level of adaptations planned within the project could not be supported using existing approaches (e.g., integrating guidance in system dialogues, dynamically modifying the interaction dialogue).

2.   The accessibility requirements posed by the categories of disability addressed within the project could not be met, either by employing existing browsers, or through the use of third-party assistive technology products (e.g., keyboard enhancements, or screen review software).

In particular, as far as the latter of the two arguments is concerned, it should be noted that there was an explicit commitment in the context of the project not to follow traditional approaches to achieving accessibility, enabling, for example, blind users to access the interface only through a screen reader. Instead, it was desired to fully support the interaction requirements of each individual user category, designing both the task structure and the interaction dialogue to fit their particular needs, instead of "retrofitting" tasks and dialogue to a particular disability.

The preceding brief analysis has demonstrated that the design space for the user interface of the AVANTI Web browser was rather large, covering a range of diverse user requirements, different contexts of use, and dynamically changing interaction situations. Meeting such requirements could not be supported by traditional user interface development methods. This was due to the explicit need for both "horizontal divergence" and "vertical specialization" in the resulting interface (to capture the range of diverse user and usage-context requirements), as well as the need for dynamic instantiation/replacement of interaction patterns, with different design alternatives to support the changing situations of use and the dynamic user states, within or across interaction sessions.

The unified user interface development approach addresses all of the aforementioned requirements, as it provides appropriate methodologies and tools to facilitate the design and implementation of user interfaces that cater for the requirements of multiple, diverse end-user categories and usage contexts. In the next two sections, we outline the process that was followed in the design of the AVANTI user interface and present some exemplary outcomes.

Task Decomposition

Following the unified user interface design method, the design of the user interface follows three main stages: (a) enumeration of different design alternatives to cater to the particular requirements of the users and the specific context of use, (b) encapsulation of the design alternatives into appropriate abstractions and integration into a polymorphic task hierarchy, and (c) development and documentation of the design rationale that will drive the run-time selection between the available alternatives. To demonstrate the process and its outcomes, we follow the design of a Web-browsing user task, namely that of specifying the URL of an arbitrary Web document to be retrieved and presented, through the various stages. We refer to this task as the *open-location* task in the rest of this chapter. Note that, although the following sections present the design activities in a sequential manner, in reality the process is iterative. In fact, iteration (within a stage and between stages) is a vital factor for arriving at coherent results, in the process of developing unified user interfaces.

Enumeration of Alternative Design Artifacts. Having defined the design space as well as the primary user tasks (see 25.1) to be supported by the user interface of the AVANTI Web browser, different dialogue patterns (referred to in the unified user interface design method as interaction styles) were defined/designed for each task, to cater to the various user requirements, abilities, and preferences, as well as to the different contexts of use.

Commencing with our exemplary open-location task, we can observe that it is a typical task in which user expertise is of paramount importance. For instance, an experienced Web "surfer" would probably prefer a "command-based" approach to the accomplishment of the task (which is the approach followed by most modern browsers), whereas a novice user might benefit from a more limited, simplified approach (Fowler et al., 1987), such as choosing from a list of preselected destinations. When the physical abilities of the user are also considered, one can easily understand that the open-location task may need to be differentiated even further, in order to be accessible by a blind person, or a person with severe motor impairment (e.g., through a nonvisual dialogue pattern and a speech command-based one, respectively).

Moreover, there are cases in which the open-location task should not be available at all. This is, for example, the case when the interface is used at a public information point, where it is imperative that the users (e.g., tourists in the case of AVANTI) can only access the underlying information system, rather than use the kiosk as a public Web access terminal.

From the preceding examples it becomes evident that the space of required interaction artifacts is directly related to the size and the diversity of the design space: The larger the number of end-user categories and the bigger the differences in their interaction needs, the more design alternatives will need to be developed.

The unified user interface design method requires that, in the process of identifying the instantiation styles (and style components) for different tasks, the design rationale behind alternative styles is captured, emphasizing the differences between instantiations. In the case of AVANTI, the rationale behind the selection/definition of each task, instantiation style, and style component, has been captured in respective documentation forms. These forms contain a general description of the (sub) task or style and its characteristics, certain design dimensions affecting dialogue design, the initiation parameters and the corresponding interaction objects, and the user and usage characteristics that have driven the definition of the style. In addition to capturing design rationale, these forms can be used in subsequent phases of the design process, in order to develop mock-ups and prototype versions of the defined tasks and styles, associating specific presentation attributes to each design alternative.

Tables 25.2 and 25.3 present the definition of two alternative instantiation styles for our exemplary open-location task. These are the direct-open-location and the indirect-open-location styles, respectively.

Polymorphic Task Decomposition.   The second stage in the unified user interface design process concerns the definition of a particular task hierarchy for each task. This includes a hierarchical decomposition of each task in subtasks and styles (dialogue patterns), as well as the definition of task operators (i.e., BEFORE, OR, XOR, *, and +; see also chap. 21, this volume) that enable the expression of dialogue control flow formulas for task accomplishment. The task hierarchy encapsulates the "nodes of polymorphism," that is, the nodes at which to proceed further down the task hierarchy one has to "select" an instantiation style. Polymorphism may concern alternative task subhierarchies, alternative abstract instantiations for a particular task, or alternative mappings of a task to physical interactive artifacts.

It should be noted at this point that, during the design of the AVANTI user interface, the definition/design of alternative interaction styles has been performed in parallel with the definition of the task hierarchy for each particular task. This entails the adoption of an iterative design approach in the polymorphic task decomposition phase, where the existence of alternative styles drives the decomposition process in subtask hierarchies, and the task decomposition itself imposes new requirements for the definition of alternative styles for each defined subtask. The latter is related to the requirement posed by the unified user interface design method, that "polymorphism" is introduced at those nodes in the task hierarchy that need to be differentiated due to user or usage characteristics that cannot be addressed through a single design artifact.

In practice, in the AVANTI user interface, we identified the specific nodes in the task hierarchies, where polymorphic decomposition was required (due to the

TABLE 25.2
Direct-Open-Location Style Description Form

Task Name: Open Location

Task Description

The user specifies a location (URL) for the system to access.

Style: *Direct Open Location (DOL)*

Dialogue Design Dimensions

Parallelism: modeless	Modalities
Component residency: fixed	**Task:** single
Confirmation: none	**Presentation:** single
Feedback type: implicit	**Lexical input:** single
Presentation design: main dialogue	

Task-Style Characteristics

Targets: speed, ease of use	Error-proneness
Duration: intermediate	**Semantic level:** normal
Difficulty: intermediate	**Syntactic level:** normal
Criticality: low	**Lexical level:** high

Task Initiation and Corresponding Interaction Objects

---

1. Selection of the corresponding interaction element	1. Text input object

Immediate Subtasks and Physical Instantiation of Initiating Interaction Objects

---

1. (Specify URL **BEFORE**	1. Dialogue object for text input
2. Indication of completion of input)	2. Standard method to indicate completion of input in the relevant interaction object
OR	3. Dialogue object indicating cancellation or initiation of any other task
3. Cancellation	

User Characteristics

Applicable to:	**Required Computer Experience:**
• fully able	frequent
• motor impaired	intermediate
	Required AVANTI Experience:
	frequent
	intermediate

Style Description

The user types in the URL of the target location. No feedback is provided, other than the one provided in the status field, and the successful or unsuccessful loading of the resource at the specified location. If the user initiates any other task, this task is being canceled.

TABLE 25.3

Indirect-Open-Location Style Description Form

Task Name: *Open Location*

Task Description

The user specifies a location (URL) for the system to access.

Style: *Indirect Open Location (IOL)*

Dialogue Design Dimensions

Parallelism: modal	Modalities
Component residency: on user request	**Task:** single
Confirmation: none	**Presentation:** single
Feedback type: implicit	**Lexical input:** single
Presentation design: input dialogue	

Task-Style Characteristics

Targets: ease of use, accuracy	Error-proneness
Duration: long	**Semantic level:** normal
Difficulty: intermediate	**Syntactic level:** normal
Criticality: low	**Lexical level:** high

Task Initiation and Corresponding Interaction Objects

---

1. Selection from command menu	1. Menu of choices
2. Input of acceleration key sequence	2. Keyboard (or replacement)

Immediate Subtasks and Physical Instantiation of Initiating Interaction Objects

1. (Specify URL	1. Dialogue object for text input
BEFORE	
2. Indication of completion of input)	2. Dialogue object indicating completion of input
OR	
3. Cancellation	3. Dialogue object indicating cancellation

User Characteristics

Applicable to:	**Required Computer Experience:**
• able	infrequent
• motor impaired	intermediate
	Required AVANTI Experience:
	infrequent
	intermediate

Style Description

The user invokes a subdialogue in which she or he types the URL of the target location. No feedback is provided, other than the successful or unsuccessful loading of the resource at the specified location.

user and/or usage parameters), and selected or defined alternative interaction styles were to accommodate these requirements. In the process of associating specific styles to each task/subtask, the rationale for making each decision (association) was depicted on the polymorphic task structure in the form of design parameters associated with specific "values," which were then used as "transition attributes," qualifying each branch leading away from a polymorphic task node. We see in the forthcoming section Adaptation Logic and Decision Making that these were then used to derive the run-time adaptation decision logic of the final interface. Figure 25.2 presents the result of the polymorphic task decomposition process for our exemplary open-location task.

As mentioned earlier, the open-location task allows the user to load a new document, by indicating its respective URL address. Two high-level styles have been defined for this task,[3] namely direct open location (DOL) and indirect open location (IOL). They are "compatible"; that is, they can be made concurrently available to the user, although, in practice, adaptation rules or user preferences may favor one over the other (e.g., arguably, the DOL style speeds up interaction, whereas IOL is better suited for novice users).

A "tip" is activated (adaptive prompting, see also the subsection Adaptations at the Syntactic Level of Interaction) if there is evidence that the user is unable to initiate the task (which would be indicated by the user-modeling component). For

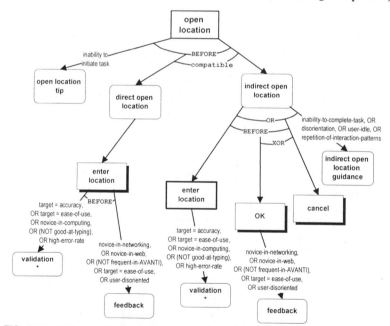

FIG. 25.2. Polymorphic task decomposition for the "open location" task.

[3]These styles are "high-level" ones, in the sense that they are further polymorphosed in the task hierarchy to cater, for example, to the users' physical abilities.

both styles, extensive feedback may be provided for the entire duration of the retrieval of the desired document. Implicit validation of the correctness of the URL may also be provided in both styles, for example, through the automatic completion of the URL address based on the "known" addresses (from bookmarked, or previously visited documents). Additionally, guidance may be provided for IOL concerning the sequencing of actions that need to be performed for the completion of the task. DOL is activated when the user selects the respective text input object, and ends on termination or cancellation. IOL initiates a modal dialogue, where the user enters the desired URL and then selects to load the document, or cancels the operation, by selecting the respective interaction objects.

Adaptation Rationale.   As already mentioned, the key factor that drives the polymorphic decomposition process is the user and usage-context characteristics that impose certain requirements in the definition/selection of alternative styles and style components, and provide information and constraints on the physical appearance and actual interactive behavior of the employed interaction objects.

An initial requirements analysis phase led to the definition of two separate sets of user-oriented characteristics, according to whether these characteristics were likely to permeate subsequent interaction sessions, or be confined within a single session. Both sets formed the basis of the decision parameters that drove the polymorphic decomposition process. The selection was made so as to ensure that adequate knowledge exists for the design of alternative styles for a wide range of user categories, taking into account not only possible disabilities, but also characteristics that differentiate individual users that may in general belong to the same broad category.

The first set was termed *static* user characteristics. The term static is used to denote that these characteristics are unlikely to change in the course of a single interaction session, and *not* that they never change. Moreover, these characteristics are assumed to be known prior to the initiation of interaction (retrieved from the user profile). This set comprises the following characteristics:

- *Physical abilities,* that is, whether the user is able-bodied, blind, or motor-impaired (in the last case the severity of the impairment is also taken into account).
- The *language* of the user (the system supports English, Italian, and Finnish).
- *Familiarity* of the user with computing, the Web in general, and the AVANTI system itself.
- The overall *interaction target:* speed, comprehension, accuracy, error tolerance.
- *User preferences* regarding specific aspects of the application and the interaction (e.g., whether the user prefers a specific style for a given task, or the preferred speech volume when links are read).

The second set of characteristics was termed *dynamic user states and interaction situations* to denote the fact that the evidence they hold is usually derived at run time, through interaction monitoring.[4] The following "states" and "situations" were included in this set:

- *User familiarity with specific tasks,* that is, evidence of the user's capability (or lack of it thereof) to successfully initiate and complete certain tasks.
- *Ability to navigate,* that is, ability to move within one document, or from one document to another in a consistent way.
- Error rate.
- *Disorientation,* that is, inability to cope with the current state of the system.
- *User idle time.*

As mentioned in the previous section, a set of user and usage-context characteristics, in the form of <decision parameters — value> pairs, is associated with each polymorphic alternative during the decomposition process, providing the mechanism for deciding on the need for, and assigning, different styles or style components. This set of <decision parameter - value> pairs constitute the adaptation rationale that is depicted explicitly on the hierarchical task structure (see Fig. 25.2).

Interaction Dialogue Design

In the previous sections we saw how the task decomposition phase led to the definition/selection of a number of styles and style components for each interaction task. In this section, we outline the next step of the design process, namely the design of the user interface interaction dialogue.

The AVANTI user interface was designed by a multidisciplinary team of experts, which included user interface designers, usability specialists, disability experts, and user interface software developers. The design of the alternative physical instantiations of styles and style components for each supported task followed the design rationale, as this was documented in the task decomposition phase. Figure 25.3 illustrates the design of the two styles that evolved from the decomposition of the exemplary open-location task (namely, the DOL—Fig. 25.3(a)—and the IOL—Fig. 25.3(b)—styles), as well as an additional style with encapsulated guidance—Fig. 25.3(c).

In addition to traditional dialogue design steps, the requirements for supporting accessibility and adaptation at the user interface level have led to a number of "nontraditional" design steps that are characteristic of the design of unified user interfaces. To start with, apart from the design of the alternative styles, it is also necessary to design adaptation "policies," or "strategies," for their activation and deactivation. In other words, part of the design must address the way in which different categories of adaptation can be introduced into the interface, as well as how

---

[4]Note that this is not always the case. For example, when long-term user modeling is employed, knowledge such as the user's familiarity with particular tasks is retained between sessions and is, thus, available to the system prior to the initiation of interaction.

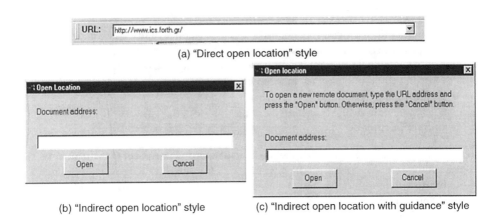

(a) "Direct open location" style

(b) "Indirect open location" style   (c) "Indirect open location with guidance" style

FIG. 25.3.   Alternative styles for the "open location" task.

the interface should choose between different policies and under what circumstances. An example of such a policy is discussed in the forthcoming section Instances of Adaptation. In the case of the AVANTI user interface, the design of adaptation "policies" was carried out through scenario-based techniques (Carroll, 1995) that "simulated" the behavior of the interface using different policies.

Another nontraditional design step concerned the selection/development of appropriate interaction techniques that would be used in association with the various "special" input and output devices supported by the browser. Input and output in AVANTI can involve any of the following devices/systems: keyboard (or any keyboard emulation device), mouse/trackball (or any mouse emulation device), touch screen, Braille display, touch tablet, binary switches, joystick, speech synthesis (output) and speech/command recognition (input), nonspeech audio output.

To facilitate the use of the special devices by disabled users, specific interaction techniques have been developed and subsequently used in the physical instantiation of each design style. Some examples of the developed special interaction techniques include:

- Switch-based interaction to be used by motor-impaired people is achieved through (automatic, or user-controlled) scanning and onscreen keyboards.
- Touch tablets can be used by blind users through demarcated areas (raised edges, Braille labels, etc.), each of which corresponds to specific functionality.
- Speech synthesis is used to present textual information to blind users and to signify attributes related to the possible hypermedia nature of the presented documents (e.g., links).

- Speech recognition can be used to allow blind users to issue vocal commands to the system, through a special set of control and navigation commands.
- Gesture recognition permits the use of a joystick by blind users, by coupling specific gestures to command sequences.
- Tactile presentation of hypertext in Braille is augmented with special symbolic annotations that facilitate the comprehension on the part of the user of the exact type of item being presented.

In addition to being accessible by a wide range of user categories, the user interface of the AVANTI Web browser implements features (interaction styles) new to Web/HTML-browsing applications that assist and enhance user interaction with the system. Some of these features have been used in hypermedia navigation systems and have proven to be of great assistance to users, whereas others have been developed specifically for the AVANTI system.[5] Such features include: enhanced history control for blind users, as well as linear and nonlinear (graph-based) history visualization for sighted users (Fig. 25.4); link review and selection acceleration facilities—Fig. 25.5(a); document review and navigation acceleration facilities—Fig. 25.5(b); and enhanced mechanisms for document annotation and classification (see, e.g., Fig. 25.6).

FROM UNIFIED DESIGN TO UNIFIED IMPLEMENTATION

In this section, we present the transition from the outcomes of the unified design of the AVANTI Web browser interface to their actual implementation. In particular, we focus on those parts of the implementation that are directly involved in the attainment of unification: (a) the coexistence of alternative dialogue patterns in a single interface, (b) the capability to monitor, and communicate information about, user interaction with the system, and (c) the capability to decide on and perform adaptations, based on statically, or dynamically derived information about the user and the context of use.

To this extent, we first briefly outline the overall architecture of the AVANTI Web browser, relating its modules to the functional components of the unified user interface architecture, described in chapter 20 of this volume. We then discuss the transformation of the adaptation logic (built into the polymorphic task hierarchy) into an adaptation decision-making module. Subsequently, we look at the different categories of adaptation that were defined and implemented in the browser's interface. Finally, we go through some exemplary instances of adaptation in the user interface, which illustrate some of the points made in the current and the previous sections.

---

[5]Many of the features have since been introduced in commercially available browsers. However, at the time of design and development, the AVANTI Web browser was the first to support them.

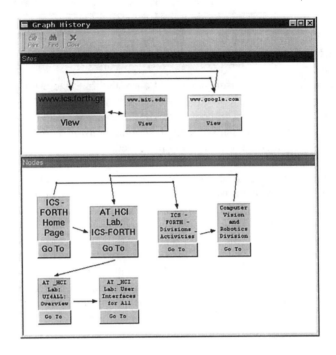

FIG. 25.4. Nonlinear representation of the document history.

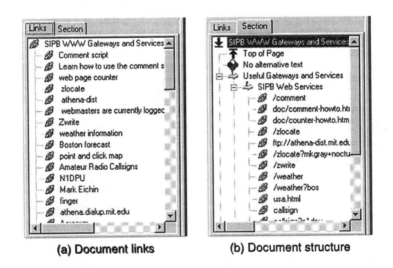

(a) Document links

(b) Document structure

FIG. 25.5. Link and document structure overview pane ("side-bar").

FIG. 25.6.   Unobtrusive visual feedback for bookmarked documents.

The AVANTI Web Browser Architecture

Figure 25.7 presents the overall architecture of the AVANTI Web browser, depicting the main software modules of the browser, the communication links with the other components of the AVANTI system, as well as the details of the unified user interface and the adaptation mechanism. The main modules and their role are described as follows (see also the overall AVANTI information system architecture discussed earlier in the section The AVANTI Information System):

- The *HTTP communication* module is used to communicate with the HSA and the MDI, to retrieve the information content. It can also be used to communicate with traditional HTTP servers, thus providing full, "standard" browser functionality.
- The *KQML communication* module enables the browser to communicate with the UMS, using the Knowledge Querying and Manipulation Language (Finin et al., 1993), in order to exchange interaction monitoring information, and inferences about user states and interaction situations respectively.
- The *monitoring* module, the role of which is to monitor user interaction and dispatch appropriate messages to the UMS. The information sent concerns both lexical and syntactic aspects of the interaction. The communication protocols between the UMS and the UI incorporate negotiation capabilities, so that, at any point in a session, the UMS is sent only information that is necessary for the inferences it attempts to make.
- The *adaptation mechanism* module is responsible for retaining and applying adaptation rules that concern syntactic and lexical adaptations at the level of the user interface, as well as for maintaining a knowledge space in which static user information and dynamically inferred (by the UMS) user states and interaction situations are stored.

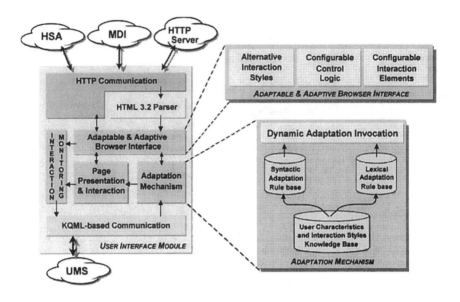

FIG. 25.7.   Architecture of the AVANTI Web browser.

- The *adaptable and adaptive browser interface* module is responsible for the presentation of the actual user interface. It instantiates the task decomposition and dialogue design by encapsulating the implementation of all the alternative tasks and styles. The different dialogue alternatives are selected for execution dynamically by consulting the adaptation mechanism and receiving appropriate decisions as a reply.
- The *page presentation and interaction* module is responsible for presenting the user with an HTML document and enabling interaction with the elements contained therein. Interaction modality, as well as other aspects of the presentation, is determined through user characteristics, with the assistance of the adaptation mechanism.
- The *HTML parser* module implements an HTML 3.2 parser, specifically developed to cater for the requirements of the AVANTI system. Special meta-tag syntax has been introduced, so that it is possible to affect the presentation of the user interface from within HTML documents [e.g., it is possible to enhance the command toolbar with new buttons and associated commands—see also Fig. 25.8(b)].

The development of the AVANTI browser's architecture was based on the unified user interface architectural framework (presented in chap. 20, this volume). This framework proposes a specific way of structuring the implementation of in-

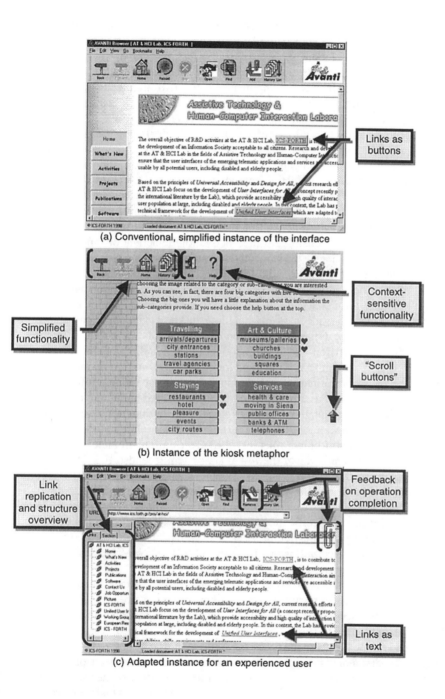

(a) Conventional, simplified instance of the interface

(b) Instance of the kiosk metaphor

(c) Adapted instance for an experienced user

FIG. 25.8. Adapting to the user and the context of use.

teractive applications by means of independent intercommunicating components, with well-defined roles and behavior. A unified user interface is comprised of (a) the dialogue patterns component, (b) the decision-making component, (c) the user information server, and (d) the context parameters server. The rest of this section presents the mapping between components in the two architectures (i.e., the AVANTI browser architecture and the unified user interface architecture) and outlines the adaptation mechanisms employed in the AVANTI browser.

First, the adaptable and adaptive browser interface (including the subcomponents: alternative interaction styles, configurable control logic, and configurable interaction elements), the interaction-monitoring component, and the page presentation and interaction component, in the AVANTI browser architecture, are the functional equivalent of the dialogue patterns component (DPC) in the unified user interface architecture. They encapsulate the implementation of the various alternative dialogue patterns (interaction styles) identified during the design process, and are responsible for their activation/de-activation, applying the adaptation decisions made by the respective module. Moreover, each style implementation has integrated functionality for monitoring user interaction and reporting the interaction sequence back to the user-modeling component of the AVANTI system (through a central—to the browser—communication module).

Second, the adaptation mechanism (including the knowledge space, the adaptation rule bases, and the dynamic adaptation invocation subcomponents) directly corresponds to the decision-making component (DMC) in the unified user interface architecture. It encompasses the logic for deciding on, and triggering adaptations, on the basis of information stored in its knowledge space (this information is, in turn, retrieved from user profiles and inferences based on interaction monitoring). Furthermore, adaptations can be triggered both during the initiation of interaction, or at run time.

The role of the user information server (UIS) in the unified user interface architecture is played by the UMS of the AVANTI system (see also Fig. 25.1). The UMS is implemented as a server, usually remotely located on the network, as it offers central, multiuser modeling functionality. The AVANTI implementation of the communication between the specific component and the browser/interface deviates from the protocol specification proposed in the unified user interface architecture, to cater to the particular needs of the project. Specifically, the communication is carried out using custom protocols encapsulated into KQML performatives (Finin et al., 1993).

The communication between the user interface and the UMS is bilateral: The interface sends messages regarding user actions at the physical and task levels (e.g., "the user pressed reload," or "the user successfully completed the loading of a new document," respectively). The UMS employs a set of stereotypes that store categorized knowledge about the users, their interactions, and environment, and a set of rules, programmed into it, to draw inferences on the current state of the user (making assumptions about the experience, abilities, and needs of the user), based on the monitoring information. It is important to note at this point that the communication between the two components can be tailored by either of them (as far as the type and detail of monitoring messages exchanged is concerned), through a separate negotiation protocol. Thus the UMS can, for example, "choose" to monitor only task-level events for the most part of the interaction, reverting to detailed, physical-action monitoring only when necessary, so as to avoid extraneous network traffic.

Finally, the context parameters server in the unified user interface architecture does not have an equivalent component in the AVANTI browser's architecture, as the needs of the user interface are very limited in that respect. In particular, the usage context characteristics that are taken into account in AVANTI concern only the interaction platform (i.e., if the browser is used on a personal computer, or on a public information kiosk) and the availability and characteristics of alternative input/output devices. This information is retrieved directly by the user interface from its device integration layer (see chap. 22, this volume, for a description of the role of the specific layer in a unified user interface).

Adaptation Logic and Decision Making

Of particular interest in the context of realizing the polymorphic task hierarchy into an adaptation-capable unified user interface is the transformation of the design parameters that guided the introduction of differentiated instantiation styles into a form appropriate for the construction of a decision-making component. As already discussed, this component is responsible for the application of the design knowledge in driving run-time adaptation.

The particular approach toward this transformation taken in the AVANTI Web browser was affected by two complementary aspects of the system: first, the adaptation logic itself, as this had been captured in the polymorphic task hierarchy, and, second, the capabilities of the UMS module. In particular, the UMS generates and communicates information in tuples of the form: <*user related hypothesis, probability that the hypothesis holds*>. This form is particularly suited to the development of "if ... then" rules, which combine the design-parameter-oriented rationale captured in the polymorphic task hierarchy, with the inference capabilities of the UMS (in the sense that each dynamically inferred parameter corresponds to a separate hypothesis).

Along these lines, the adaptation rationale was transformed into corresponding adaptation rules that are valuated at run time. The revaluation of any particular rule is triggered by the addition, or modification of inferred user- or context-related parameters in the interface's knowledge space (either from user profiles, or from the UMS).

Table 25.4 presents, in a simplified form, some actual rules used in the AVANTI browser where some of the following properties[6] are demonstrated:

---

[6]Note that these properties were mainly derived from representation requirements. Several alternative approaches could also prove appropriate for "expressing" adaptation logic.

TABLE 25.4
Simplified Adaptation Rules

Adaptation Rules	
if <*user novice in hypermedia*> then LinkType = Button	A
if <*user motor impaired*> and <*user is computer novice*> then ScanRate = Slow	B
if <*user is motor impaired*> and (<*user has high error rate*> or <*user unable to navigate*>) then ScanRate = Slow	C
if <*user is disoriented*> or <*user is idle*> then SpeechVolume = High	D

task "review bookmarks"   {
    if <*user novice is AVANTI novice*>
        then (Font = Large and ObjectSize = Large)
}
	E

task "review bookmarks"   {
    if <*user is unable to complete task*>
        then activate_style "review bookmarks with guidance"
        and SpeechSpeed = SpeechSpeed - SpeechSpeedUnit
}
	F

1.   Rules can trigger adaptations at the lexical level of interaction by directly modifying presentation or behavioral attributes of interaction objects; it is possible to apply modifications to all objects in the interface (e.g., Rules B, C, D, and E in Table 25.4), or only to objects belonging to a particular class (e.g., Rule A in Table 25.4).

2.   Rules can be "global"; that is, they can exist independently of particular user tasks (e.g., Rules A to D in Table 25.4). Rules in this category are usually addressing the lexical level of interaction and seek to apply adaptations that should affect all the objects in the interface (e.g., activating scanning-based interaction throughout the interface, if the user is severely motor impaired).

3.   Rules that are associated with specific tasks (by being encapsulated into a named "task" construct) can also trigger syntactic-level adaptations within that task by selectively activating and deactivating task instantiation styles (e.g., Rule F in Table 25.4).

4.   All rules can freely mix "static" and "dynamic" user characteristics and interaction states in their "if" clause (e.g., Rule C in Table 25.4).

5.   Rules that appear within a specific task context can also freely mix task-level and physical-level adaptations in their "then" clause (e.g., Rule F in Table 25.4).

To illustrate the relationship between the design logic in the polymorphic task hierarchy and the rules employed in the AVANTI system, we go back to our open-location exemplary task. Table 25.5 contains the actual rules derived from the polymorphic decomposition of the task presented in Fig. 25.2. Note that part of the design logic has been left out, as the respective design parameters were not included in the final user profile, or the dynamically inferred information (e.g., *<user is good at typing>*).

Categories of Adaptation

The design of alternative styles for the various tasks in the AVANTI browser user interface gave rise to a number of application-independent adaptation patterns, which are expected to be of general value in the design of adaptive user interfaces. This section discusses these patterns and examples of their application in the browser's interface. To facilitate the discussion, we follow the normative view of the user–computer interaction process, which identifies three levels at which interaction occurs (Hoppe, Tauber, & Ziegler, 1986):

- The *lexical* level of interaction (also referred to as *physical*), which concerns the structure, presentation attributes, and actual behavior of the input/output interaction elements that make up the user interface; it is at this level that interaction physically takes place.

TABLE 25.5
Adaptation Rules for the "Open Location" Task

```
task "open location" {
      if <user is unable to initiate task>
            then activate_style "adaptive prompting"

      if user_prefers_style "direct open location"
            then deactivate_style "indirect open location"
      else if user_prefers_style "indirect open location"
            then deactivate_style "direct open location"

      if (<interaction target> includes <accuracy> or
         <interaction target> includes <ease of use> or
         <user is computer novice> or <user has high error rate>)
            then activate_style "validation"

      if (<user is Web novice> or
         (not <user uses AVANTI frequently>) or
         <interaction target> includes <ease of use> or
         <user is disoriented>)
            then activate_style "extensive feedback"

      if (<user is unable to complete task> or
         <user is disoriented> or
         <user is idle>)
            then activate_style "guidance"
}
```

- The *syntactic* level of interaction, which concerns the structure and syntax of the dialogue between the user and the computer, through which the application semantics are made accessible to the user (e.g., specific interaction steps taken by the user, method of accomplishing tasks).
- The *semantic* level of interaction, which involves conveying the system functionality and domain-specific facilities to the end-user.

Adaptations at the Lexical Level of Interaction.   The lexical level of interaction is perhaps the most prominent level at which accessibility by persons with physical impairments can be treated. The adaptation patterns that have been employed to meet this goal in the AVANTI user interface, include:

- Incorporation of support for, and conditional activation/deactivation of, multiple interaction modalities, mostly based on the user profile. This implies that the interface is capable of integrating traditional, as well as special, input/output devices; supporting appropriate input/output interaction techniques for each of the supported devices; and selecting among the range of available devices and interaction techniques the appropriate ones, based on the user profile and the context of use. Exemplary adaptations at this level are: rendering of information content in speech, rather than on the VDU; activation of interface scanning to support switch-based interaction; activation of auditory cues to accompany, or replace visual feedback of initiated, ongoing, or completed system operations (e.g., loading of a new document); and so forth.
- Capability to automatically modify the presentation and conceived behavioral attributes of interactive elements. Presentation attributes that are subject to adaptation include size, color (combinations), font family and size, speech parameters (e.g., volume, gender, pitch, speed), spatial and temporal arrangement, grouping, and the like. Behavioral attributes of interaction objects (as conceived by the end-user) that are subject to adaptation include feedback offered when a user acts on them, representation of the state they are currently in, representation of contained components in the case of composite objects, and so on.

Adaptations at the Syntactic Level of Interaction.   The syntactic level of interaction has drawn considerable attention in user interface adaptation research in recent years. However, the corresponding literature does not yet provide enough empirical evidence to interrelate specific categories of syntactic-level adaptations with interaction situations in which they are mostly beneficial. This is due to the high dependency between syntactic adaptations and application context, as well as to the lack of appropriate methods for empirically comparing adaptable and adaptive system. The latter is true not only for comparing such systems between each other, or with their "static" counterparts, but also for comparing different adaptation policies within a single system. This section enumerates the syntactic-level adaptation patterns employed in the AVANTI user interface, cautioning, though, that the suitability of the techniques for specific application domains and interaction contexts needs to be assessed and thoroughly evaluated on a case-by-case basis:

- Support for *alternative task structures*. By possessing the capability to present the user with alternative task structures, the interface caters to situations where the user is incapable of dealing with the complexity of the interface, or the task domain (which is usually the case with people who have cognitive, or memory problems). On the other extreme, the particular adaptation pattern can be employed to provide access to advanced and rarely used system functionality of practical utility to only experienced users, or users with extensive domain expertise. Furthermore, the capability to incrementally introduce modifications in the task structure as needed allows the interface to retain high degrees of consistency and affordance within a single interaction session, but also across sessions. Some of the underpinning principles that support the requirement of presenting users with appropriately modified, personalized task structures have been well understood and articulated in relation to the concept of task orientation in the context of human–computer interaction (Ulich, Rauterberg, Moll, Greutmann, & Strohm, 1991), as well as explicitly related to the concept of interface adaptation (Paetau, 1994). A particular form of alternative task structures that has been introduced in the literature concerns the sequencing of subtasks (Brusilovsky, 1992).

- *Adaptable and adaptive help facilities*. Individual differences of users, as well as user expertise and domain knowledge, often necessitate the provision of help at different levels, and with varying degrees of detail (see, e.g., Smith & Tattersall, 1990). For example, the assistance required by a user who makes use of the application for the first time is entirely different from the assistance required by an experienced user seeking to make use of advanced system features. By employing adaptability and adaptivity techniques, the user interface attempts to provide the appropriate type and amount of information at all times, thus making sure that users are not intimidated, or put off by unnecessary detail, without, however, compromising the quality of provided assistance, or its appropriateness for the task at hand.

- *Task guidance (guided interaction)*. It is often the case that, when users are faced with new interactive tasks, they require assistance from the interface in the form of guidance. This assistance is usually explicitly requested by the user and provided through help facilities. However, it is possible for an adaptive system to detect cases where task performance is carried out in an erroneous, or inefficient way and automatically provide guidance to the user. Such a capability on the part of the interface becomes more important if the latter is also capable of task-structure adaptations; in this case, the user should be informed about the modifications, and guided through the modified, or newly available tasks. The notion of task guidance is similar in conception with that of intelligent tutoring systems (see, e.g., Benyon & Murray, 1993; Shute & Psotka, 1996), where users are guided through tasks; however, in this case, the aim is not to "teach" the users a particular concept, but to support users in accomplishing tasks. In the AVANTI user interface, such guidance is provided when the user is identified as unable to complete a particular task.

- *Awareness prompting.* The term awareness prompting is used to refer to adaptive prompting on the part of the system (Kühme, Malinowski, & Foley, 1993), intended to raise the users' awareness to features or functionalities of the interface, which they are not using, or are underutilizing. A nonadaptive form of such a facility has gained acceptance lately in the form of "tip-of-the-day" messages presented to users on application start-up. The goal is common in both cases: to stir the user's interest in portions of the interface that might be otherwise left unexplored (users tend to learn well and utilize only a small portion of the interface of large-scale applications and services, and rely on that to carry out common tasks). The main advantage to adaptively directing the user's attention to specific parts of the interface is that this only happens when a relevant need is detected. This increases the potential usefulness and the timeliness of providing such information, as well as the likelihood that newly acquired knowledge will be put into use soon after its provision, which would significantly enhance its memorability. In AVANTI, this adaptation pattern is activated when the user is identified as unable to initiate a particular task (see earlier subsections 'The AVANTI Web Browser Architecture' and 'Adaptation Logic and Decision Making' for discussions of how different user-related inferences are drawn in AVANTI).

- *Adaptive error prevention and correction.* The idea to use adaptation to handle users' errors in their interaction with computing systems is not a new one; in fact it has been argued that it constitutes one of the driving forces in developing adaptive systems (see, e.g., Browne, Norman, & Riches, 1990). User errors while interacting with an interface can be differentiated between the three levels of interaction: the lexical (e.g., user mistyping), the syntactic (e.g., user performed two interdependent actions in the reverse order than required), or the semantic level (e.g., user expects an action to have different results than it actually does). Whereas errors at the semantic level are hard to detect and compensate for, the same is not true for the first two levels. By monitoring users and learning the type of mistakes they usually make and the circumstances under which these mistakes are likely to be encountered, the interface can engage into a twofold process: first, attempt to prevent the user from erring by appropriately modifying the interaction syntax (e.g., by adaptively requesting explicit confirmation for an action); second, by "preprocessing" user actions before they are submitted to the interface dialogue management component and automatically substituting those that bear a high probability of being wrong, thus, in effect, correcting user errors after they have "happened." The AVANTI user interface employs the first of the two approaches, modifying aspects of the interaction syntax when the user is identified as exhibiting high error rate in a particular part of the dialogue.

Other syntactic-level adaptation patterns that were identified during the design process, but not implemented in the AVANTI user interface (due to the limited resources available in the context of the project) include:

- Support for *alternative syntactic paradigms*. By syntactic paradigms, we refer to collections of interaction steps, characterized by their well-established spatial and temporal arrangement and sequencing (e.g., "object-function" syntax vs. "function-object" syntax). Support for alternative syntactic paradigms would enable the interface to continue functioning appropriately under different contexts of use (e.g., different cultural environments may introduce different requirements in this respect), as well as meet specific user preferences, or requirements (potentially facilitating the exploitation of existing user experience in interacting with different interfaces).

- *Automatic replication of recurring interaction patterns.* Depending on the nature of the interface, it is possible that there will exist specific patterns that will recurrently appear in the user's interaction with the system (e.g., setting presentation preferences appropriately to view a specific type of document, bookmarking all open documents before exiting the browser). By detecting these patterns and possessing the capability to replicate them, the interface can relieve the user from potentially lengthy and tedious repetition of (possibly composite) tasks, not explicitly represented in the internal task structure of the interface. The net effect stemming from such system functionality would be highly personalized acceleration in the interaction of individual users with the system. The resulting acceleration would be highly beneficial in cases where interaction is decelerated by user difficulties/disabilities, as well as in cases where inexperience, tiredness, or disorientation might lead to errors in the performance of individual tasks.

Adaptations at the Semantic Level of Interaction.   Adaptations at the semantic level of interaction also suffer from the lack of strong empirical evidence to support their employment in given circumstances, and from the lack of an adequate number of experimental, or real-world systems that employ them.

It has been argued that interaction semantics relate to what the user perceives as the prevailing interactive embodiment of the computer (e.g., desktop, book, rooms), as well as the functionality made available through that embodiment

(Stephanidis & Akoumianakis, 1998). Adaptations at this level mainly concern the overall interactive metaphor(s) used to embody different functional properties of the system.[7] Recent advances in Web technologies (Dynamic HTML, Javascript, etc.) have made it possible for a metaphor to be either embedded in the interface of Web applications and services, or characterize the overall interactive environment. For instance, using a metaphor to develop a suitable visualization of a collection of related documents and the hyperlinks between them is an example of embedding metaphor in the user interface. In contrast, developing an interface that allows the user to exclusively interact with hyper-documents through book-rather than desktop-related concepts (such as table of contents, chapters, index, etc.) implies using a metaphor to characterize the overall interactive embodiment of the computer. Adaptation may be used not only to select and instantiate suitable interactive metaphors for different users and tasks, but also to individualize inter-action, by modifying only part of the metaphoric environment, in order to convey domain- or system-specific concepts in a more coherent, or easier to attain form.

In AVANTI, two moderately different metaphors are used to convey differences in the environment of use, namely desktop [see Fig. 25.8(a)], versus public information kiosk use [see Fig. 25.8(b)]. The two metaphors are differentiated both in terms of presentation, and in terms of the functionality they make available to the end-user. In particular, the desktop metaphor was presented as part of the overall windowing environment and offered the full range of functionalities to its users (subject, of course, to adaptations that originated from different parts of the system). In contrast, the kiosk metaphor presented itself as the unique embodiment of the system and removed part of the functionality that is available to the user (e.g., the user cannot explicitly indicate the address of a Web document to be viewed, and is thus unable to "exit" the information system).

Instances of Adaptation

To illustrate some of the categories of adaptation discussed in the previous section, as well as to demonstrate the practical application of the design issues discussed in the first part of the chapter, we now briefly review some instances of the AVANTI browser's user interface.

Figure 25.8 contains three instances of the interface that demonstrate adaptation based on the characteristics of the user and the usage context. Specifically, Fig. 25.8(a) presents a simplified instance intended for use by a user unfamiliar with Web browsing. Note the "minimalistic" user interface with which the user is presented, as well as the fact that links are presented as buttons, arguably increasing their affordance (at least in terms of functionality) for users familiar with windowing applications in general. The second instance, Fig. 25.8(b), presents the interface in "kiosk mode," that is, making use of the kiosk-based information system metaphor

---

[7]This use of the term *metaphor* should be distinguished from the case of embedding metaphoric properties into a user interface without modifying the overall interactive embodiment to be conveyed.

discussed earlier. Note that the typical windowing paradigm is abandoned and replaced by a simpler, content-oriented interface (even scrollbars, e.g., have been replaced by "scroll buttons," which perform the same function as scrollbars but have different presentation and behavioral attributes—e.g., they "hide" themselves if they are not selectable). Furthermore, note the presence of "content-sensitive" buttons in the interface's "toolbar." These are a result of the cooperation between the user interface and the HSA component of the AVANTI system architecture (see Fig. 25.1); the HSA places specific meta-tags into the HTML documents it sends to the browser, which are interpreted and presented as additional, dynamically introduced functionality into the user interface. Finally, in the third instance, Fig. 25.8(c), the interface has been adapted for an experienced user. Note the additional functionality that is available to the user (e.g., a pane where the user can access an overview of the document itself, or of the links contained therein; an edit field for entering the URLs of local or remote HTML documents).

Figure 25.9 contains some sample instances demonstrating disability-oriented adaptations in the browser's interface. The instance in Fig. 25.9(a) presents the interface with automatic scanning activated. Note the scanning highlighter over an image link in the HTML document and the additional toolbar that was automatically added in the user interface. The latter is a "window manipulation" toolbar, containing three sets of controls enabling the user to perform typical actions on the browser's window (e.g., resizing and moving). Figure 25.9(b) illustrates the three sets of controls in the toolbar, as well as the "rotation" sequence between the sets (the three sets occupy the same space on the toolbar, to better utilize screen real estate and to speed up interaction; the user can switch between them by selecting the first of the controls). Figure 25.9(c) presents an instance of the same interface with an onscreen, "virtual" keyboard activated for text input (interaction with the keyboard is also scanning based).

The single interface instance in Fig. 25.10 illustrates a case of adaptive prompting. Specifically, in this particular instance, the interface has received an inference from the UMS indicating that there exists high probability that the user is unable to initiate the open-location task (this would be the case if there was adequate evidence that the user is attempting to load an external document with unsupported means, e.g., using "drag and drop"). When this inference is added into the interface's knowledge space as a new piece of user-related information, the relevant rules are triggered. These trigger the activation of a "tip" dialogue, that is, a dialogue notifying the user about the existence of the open-location functionality and offering some preliminary indications of the steps involved in completing the task.

Figure 25.11 presents a more complex case of adaptation. In the first instance, Fig. 25.11(a), the user has opened the "history list" dialogue. The latter contains a list of previously visited documents, so that the user can choose one of them to retrieve and view. While the user is interacting with the dialogue, the UMS infers that the user is not capable of successfully completing the task (to arrive at this conclusion, the UMS might utilize monitoring knowledge collected during previous invocations of the dialogue). The inference is sent to the user interface and a

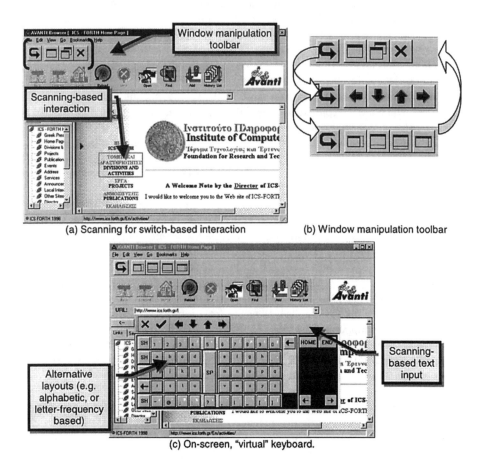

(a) Scanning for switch-based interaction   (b) Window manipulation toolbar

(c) On-screen, "virtual" keyboard.

FIG. 25.9.   Instances for motor-impaired users.

triggering process similar to that of the previous example takes place. At an abstract level, the adaptation triggered involves the provision of additional guidance
to the user for completing the task. However, the way in which it is applied in the
interface depends on the actual context of interaction. Figure 25.11(b) presents the
result of the adaptation if the user is still interacting with the "history list" dialogue; a separate dialogue is presented to the user, containing some brief guidance
on the task at hand. Figure 25.11(c) presents an alternative adaptation, that is, "embedding" the guidance in the dialogue, which can happen in the following two

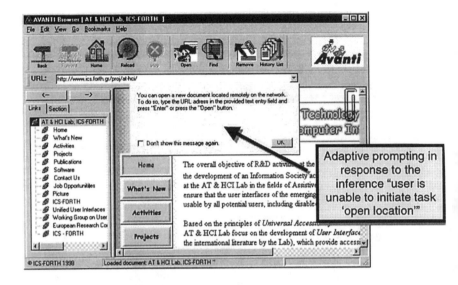

FIG. 25.10.   Awareness prompting.

cases: (a) if the user has already closed the "history list" dialogue, then the next invocation of the history list facility will bring up the new dialogue with embedded
guidance; and (b) the same thing will happen if the user disposes of both dialogues
in Fig. 25.11(b), but is still unable to complete the task successfully (as indicated
by the UMS).

EVALUATING THE UNIFIED USER INTERFACE

One of the key problems in the development of the AVANTI user interface has
been the inadequacy of available evaluation methods and techniques to be used for
the evaluation of unified user interfaces in particular, and adaptable and adaptive
interfaces in general (see also chap. 17, this volume). Specifically, the analysis of
existing evaluation methods revealed that these could not be used to assess the way
and extent to which the adaptation facilities of the interface affect interaction qualities, such as accessibility, usability, acceptability, and so on. Due to these shortcomings, the approach taken in the development of the AVANTI user interface has
been the introduction of a two-fold assessment process, which involved:

- Iterative, expert-based assessment cycles in the design of appropriate interaction styles, the definition of adaptation rules, and the development of
  the decision mechanism for materializing the required adaptable and
  adaptive behavior.

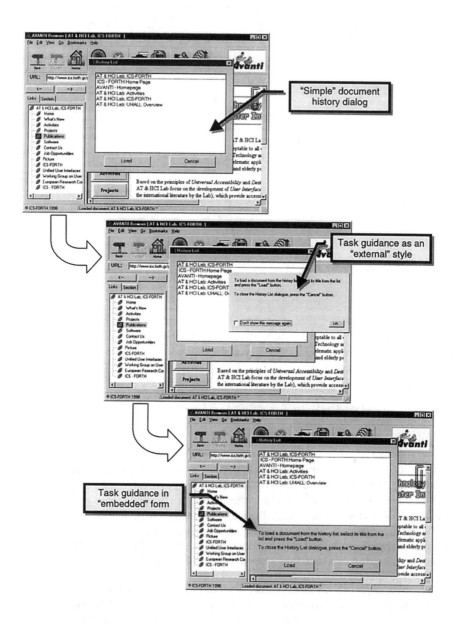

FIG. 25.11. Adaptive task guidance.

- End-user-based evaluation activities (using questionnaires, observations, and interviews), intended to assess the overall usability and accessibility of the user interface, independently of the latter's adaptation capabilities.

Before the evaluation activities are presented in more detail, it should be stressed that their scope and intention have been different. Expert-based assessment has been employed as a way of compensating for the lack of appropriate evaluation techniques for adaptation-capable user interfaces and the simultaneous lack of empirical evidence on which to base the design of adaptations. User-based evaluation has been employed as a means of assessing the system's acceptability, as a function of the user-perceived accessibility and usability of the system in general.

Expert Evaluation

Expert evaluation activities within the development of the user interface of the AVANTI Web browser aimed to employ accumulated knowledge and experience in the areas of user interface design, usability, and assistive technology, for (a) the design of alternative interaction styles that cater to the different user and usage-context requirements, as well as (b) the design of appropriate adaptation behavior to be built in the resulting interface.

Evaluating the Design of Interaction Styles.   The early evaluation activities that have been carried out as part of the design of the AVANTI user interface were intended to assess the appropriateness of the designed interaction styles for the specific interaction context and the particular user characteristics for which they were intended. Particular emphasis was placed in the evaluation of the accessibility features provided by the designed interaction patterns to the target disabled user categories (i.e., blind and motor-impaired). The nature of the user interface (i.e., ensuring accessibility to different categories of users in different contexts of use), as well as the complexity arising from its adaptation capabilities, necessitated the involvement of a multidisciplinary group of experts. The group of experts that have been involved in the early evaluation phases comprised user interface designers, disability experts, and usability specialists.

The evaluation process followed consisted of three phases, namely preparation, evaluation, and analysis. During the preparation phase, every interaction style was described in a uniform manner in documentation forms similar to those presented in Tables 25.2 and 25.3. These forms contained a general description of the interaction style and its characteristics, certain design dimensions affecting dialogue design, the initiation parameters and the corresponding interaction objects, and the user and usage characteristics that led to the definition of the style. Additional available material for each style (including, e.g., informal descriptions and paper-, or computer-based mock-ups) was collected and made available along with the respective style documentation form. Subsequently, the collected material was communicated to the participating experts.

During the evaluation phase, the experts reviewed each interaction style separately, based on established accessibility and usability guidelines and heuristics (e.g., Bergman & Johnson, 1999; IBM, 1999; ISO 9241-11, 1997; Microsoft, 1997; Story, 1998; Vanderheiden, 1994; Vanderheiden & Kaine-Krolak, 1995; World Wide Web Consortium, 1999). The experts were asked to identify potential accessibility, usability, or other problems of each interaction style (given the specified user characteristics and contexts of use), as well as to propose possible improvements in the design, based on their experience.

The outcome of these inspection activities was collected and analyzed in two steps. In the first step, the "agreement" of (i.e., lack of conflict between) the experts' recommendations was assessed. Based on that assessment, interaction styles were categorized into "conclusive" and "inconclusive" ones, according to whether there was any conflict in the identified problems and recommendations or not, respectively. The interactive styles identified as inconclusive became the subject of a group discussion involving all participating experts (second iteration of the evaluation phase). During the discussion, the rationale behind problems and recommendations was elaborated on and argued for, and the group reached decisions by consensus.

During the second step of the analysis, the collected material of the evaluation phase (including the outcome of the second iteration described previously) was further analyzed and categorized, eliminating duplicates and removing overlapped items. The final results were fed back into the design process and have led to three types of intervention to designed interaction styles: (a) redesign of styles, based on identified accessibility, usability, and other problems, or on contributed ideas toward the enhancement of the interaction styles, (b) elimination of redundant interaction styles (i.e., of styles that overlapped, due to the similarity in the end-user characteristics, or the characteristics of the usage context they were intended to cater to), (c) introduction of new interaction styles, to cover user characteristics and contexts of use that were not addressed adequately by existing styles.

In summary, the described process offered adequate evidence and support regarding the appropriateness of the different interaction styles to the user and usage-context characteristics they were intended for. Furthermore, it offered initial insight into the design of the related adaptation rules, the validation of which is discussed in the next section.

Validating the Adaptation Rules.   The development of adaptation rules (and the respective design activities in the construction of the polymorphic task hierarchy) has proven to be a very demanding task in the development of the AVANTI user interface, mainly due to the lack of adequate, empirically supported evidence in the literature, regarding the type of adaptations that need to be introduced as a result of changes in specific user characteristics and interaction parameters. To compensate for the lack of an adequate empirical foundation for the definition of the rules, two steps were taken. First, the rules were defined by a group of experts coming from the fields of user modeling, intelligent user interfaces, user interface design, disability, and usability. The definition of the rules involved several iterations, following each task-decomposition phase, as well as each stage in the definition/selection of alternative interaction styles. Second, a process was defined, intended to assess the design of adaptations by means of validating the resulting adaptation rules.

The validation of the adaptation rules has taken place in three consecutive phases, namely, evaluation of the rules by experts, verification of the adaptation mechanism on a per-rule basis, and verification of the adaptation mechanism across sets of rules. Each of these phases is briefly discussed next.

The evaluation of the rules involved a small group of human-factors and accessibility experts, who tested the adaptation rules in terms of potential inappropriate decisions, or introduction of usability problems. The process followed is similar to the one for the evaluation of individual styles, described in the previous section. The material provided to the experts in this case included: (a) the adaptation rules in a simplified form, similar to the one used in Tables 25.4 and 25.5, (b) indicative (fragments of) scenarios of use that might lead to the triggering of each rule, and (c) descriptions of the interface constituents that were affected by each rule (i.e., description of the affected presentation and behavioral attributes of interactive elements in the case of lexical-level adaptations, or description of the affected task instantiation styles in the case of syntactic-level adaptations).

As mentioned earlier, the experts were asked to assess the provided material across two dimensions: (a) appropriateness of individual adaptations triggered by each rule, that is, whether the rule introduced an adaptation appropriate to the newly identified characteristics of the user, or the context of use (e.g., when user confusion is identified, is it appropriate to provide guidance?), and (b) introduction of usability problems by (possibly conflicting) adaptations triggered by (sets of) rules (this question was approached through cross-situation analyses to ensure that there is no situation leading to another potentially problematic situation, e.g., disorientating users by trying to guide them through a task they might have found confusing).

The individual expert assessments were collected and analyzed in two steps. In the first step, rules were categorized as conclusive or inconclusive (see previous section for a description of the terms), based on the lack or presence of agreement, respectively, of the experts' recommendations. The rules identified as inconclusive were discussed by the experts collectively, and disagreements were resolved by consensus.

During the second step of the analysis, the collected material of the evaluation phase was further analyzed and categorized, and the results were fed back into the design process where they have led to four types of intervention to adaptation rules: (a) elimination of rules that were deemed inappropriate, or not sufficiently supported, (b) introduction of new rules (based on the recommendations of the experts), (c) modification of the rules' triggering conditions (e.g., adding, or removing a particular user characteristic from the description of the triggering situation), and (d) modification in the rules' decisions (e.g., addressing a particular situation only through guidance, instead of through guidance and extensive interim feedback). Furthermore, a side result of the previously described validation activities has been the introduction of modifications in the interaction styles, following respective observations of the experts. This type of input from the experts was voluntary and informal and should not be confused with the input derived from the dedicated interaction style evaluation activities described earlier. It is mentioned here because, in most cases, it was based on prototype instances of the interface and has, thus, affected significantly the final design of interaction styles.

The validation of adaptation rules was followed by the verification of the adaptation mechanism on a per-rule basis, and verification of the system's adaptation behavior across sets of rules. In particular, the mechanism used for valuating rules and carrying out the respective adaptation decisions was tested for consistent behavior, by: (a) testing that the triggering conditions for each individual rule (depending on the "static" user characteristics, the context of use, and the dynamically reported interaction situations) led to the desired (adaptation) behavior on the part of the user interface, and (b) testing sets of rules in combination, to assess the degree to which they affect each other from a functional, as well as from the user's, point of view.

The former procedure (i.e., testing rules individually) was performed by examining the defined rules one by one and verifying system behavior when the activation parameters were set, or changed. A "wizard of oz" technique was used to simulate the functionality of the user-modeling server.[8] The latter procedure (i.e., testing combinations of rules) was performed through the development of representative scenarios, where multiple activation parameters were set or changed simultaneously. The verification procedure resulted in the identification of conflicts in the activation of specific styles and inappropriate activation of certain rules in specific tasks. The main problem arose from the redundant activation of styles each time the dynamic situations reported by the user-modeling server were changed. The outcomes of the validation procedure initiated specific modifications in the predefined rules, as well as the adaptation mechanism itself.

User-Based Evaluation

Formal usability evaluation studies of the AVANTI Web browser with end-users (blind, motor-impaired, and able-bodied) have been carried out in the context of the experimental and field evaluations of the AVANTI system in the three user trial sites in Kuusamo, Siena, and Rome (Andreadis et al., 1998). It should be noted that the main goal of the usability studies carried out in the context of the AVANTI project has been to derive some initial results on the overall usability of the information systems. The evaluation activities that are relevant to our discussion focused on the provided support for accessibility, through the employment of adaptability and adaptivity, and the provision of alternative interaction techniques.

---

[8]For this purpose, a software module that simulates the functioning of a user model server has been developed and was used for "manually" generating the dynamic user situations.

The results of the evaluation were encouraging, both in terms of user acceptance of the characteristics of the interface, and in terms of the fulfillment of the initial goals that led to the employment of unification in the user interface. In particular, adaptability-addressing accessibility issues for the various end-user groups proved quite successful, as each user category conceived the interface as having been specifically developed to cater to their particular requirements. The results were similar for the non-disability-related categories in which users were classified (e.g., according to their computer expertise).

Adaptivity was assessed to a lesser degree than adaptability during the evaluation, due mainly to the following reasons: (a) adaptivity requires that interactive sessions are rather lengthy, so that adequate information about the user and the context of use is collected before any practically useful inferences can be made; and (b) existing user interface evaluation techniques do not offer themselves for the evaluation of dynamically changing, nondeterministic (from a user's perspective) systems. As far as the first of these issues is concerned, the typical duration of the interaction sessions performed during the experimental activities was not adequate for the extraction of dependable inferences on which dynamic adaptation could be based. As a result, users were aware of only a minimal set of adaptive features in the interface; however, their reaction to the features they did observe was positive.

As far as the availability of empirical methods and techniques for the evaluation of adaptivity is concerned, existing knowledge in the area of user interface evaluation is inadequate for the derivation of appropriate techniques and instruments to measure the effects of adaptive system behavior on interaction. Although there have been several attempts in the past to construct both objective and subjective expert- and user-based evaluation methods in this field (e.g., Grüniger & van Treeck, 1993; Höök, 1997; Totterdell & Boyle, 1990), the lack of understanding of the dynamic dimensions of adaptive user interfaces (as well as of the differences introduced by alternative approaches to achieving and "driving" adaptive behavior) compromises the applicability of solutions that have been suggested to date (for a more elaborate discussion, see chap. 17, this volume).

DISCUSSION AND CONCLUSIONS

This chapter has presented an overview of the employment of the unified user interface development approach in the construction of a Web browser that provides accessibility and high-quality interaction to a wide variety of user categories. The account of the different phases of development, or of the various issues involved in designing, implementing, and evaluating unified user interfaces has not been exhaustive. Rather, the chapter has focused on those aspects of the development process that are characteristics of the unification concept and of the implications this has in the development life cycle of interactive products. We now take a step back to make a few observations and point out some of the important lessons learned from the AVANTI case study.

First, looking at the design phase, one has to note that, due to the fact that there exist multiple target user categories and contexts of use (differentiated, potentially, across many dimensions), it is imperative to ensure a multidisciplinary approach to design. Specifically, it is necessary to involve end-users from each target category from the early phases of design, to ensure that design addresses their real needs and is differentiated adequately to meet their particular abilities, skills, requirements, and preferences. Furthermore, in the cases that the target population includes users with disabilities, it is important that experts in the respective disabilities are directly involved in the design process. Overall, it is argued that the design of unified user interfaces should make extensive use of user-centered techniques and practices (such as participatory design; Schuler & Namioka, 1993). Furthermore, the design process should be coordinated by participants who have a clear understanding of the desired nature and form of the outcomes, so that the individual design activities have a clear "goal" and deliverables, which are used as input in subsequent activities, or iterations.

Looking at the individual steps involved in the design of user interfaces, one can also observe that some of the traditional steps have an extended role in the case of unified user interfaces. Starting from requirements analysis, for example, one can see that in addition to identifying the characteristics of the users and of the general context of use that are relevant to the design of the user interface, an additional action is required. In particular, it is also necessary to identify those characteristics from the set discussed earlier that are candidates for differentiating the design between user categories, or contexts of use. This includes both "static" characteristics that can be determined prior to the commencement of interaction, and "dynamic" characteristics that can only be detected on the basis of interaction- (or overall interactive environment) monitoring information. Concerning the latter type of characteristics, it is also necessary to assess, at this point, a number of "practical" issues concerning their "detection." These include, but are not limited to, arriving at a precise description of the situation that each of the characteristics denotes, as well as the (approximate) interaction pattern (or, general "state" of interaction context) that would lead to an inference relating to the characteristic. This is necessary so that designers have a common and clear understanding of what the dynamic characteristics are, and decide on the feasibility of, and the resources that will be involved in, computationally deriving accurate "values" for the identified characteristics. For example, it may be deemed necessary to introduce adaptations in the interface if the user is "disoriented." That would entail the construction of an accurate definition of what "disorientation" denotes, in the context of the interface being designed, as well as an outline of how it would be detected (i.e., what interaction patterns would be considered as indicating its presence) at run time by the user-modeling component.

Moving on to the construction of the polymorphic task hierarchy, there are a number of issues that need to be carefully addressed, in order to arrive at a result that fully captures the design space and is also appropriate for implementation. The first point to note is that all polymorphic nodes in the hierarchy should represent real needs that have stemmed from differences in the user population, or in the context of use. In other words, polymorphic nodes should represent the real requirements for differentiation in the user interface and should be selected carefully, so that the hierarchy itself and the consequent implementation effort is not "bloated"; that, however, should not happen at the expense of compromising accessibility or interaction quality. Designers should only try to minimize the polymorphic nodes when that makes for better and more coherent design, rather than combine or remove polymorphic nodes so that implementation is facilitated.

A related issue in the construction of the task hierarchy is the differentiation between polymorphism at the abstract level of interaction, versus polymorphism that primarily concerns the physical instantiation of tasks. Care should be taken not to mix the two levels, or apply them in the wrong order. To better explain the implications of the previous statement, we consider the example of a task hierarchy that has to convey the existence of two alternative metaphors that will characterize the overall interactive environment.

First, assume two alternative versions of the desktop metaphor. These would probably have to be differentiated (polymorphosed) only at the physical level of interaction; the semantics and general task structure would probably be common in both cases. Now, assume two alternative metaphors, one based on the concept of the desktop, and one based on the concept of the newspaper. These would probably have to be differentiated at all three levels of interaction: semantic, as they would have to convey the system's functionality through significantly different real-world concepts; syntactic, as they might have to support tasks in a metaphor-consistent way; and physical, as they would employ different physical artifacts, with differing appearance and behavior, to compose the interface. In this case, polymorphism would have to be carefully planned, so as to avoid unnecessary replication of parts of the hierarchy. For example, if a particular task shared the same structure between the two metaphors, then it should not have two alternative abstract instances in the hierarchy, differentiated only by their physical attributes. The latter might happen if the arguably bad choice was made to polymorphose earlier in the hierarchy, on the basis of lexical-level differences.

By avoiding replication, the implementation of the unified interface is facilitated, while consistency is ensured in the design, without compromising coherence. Furthermore, following this approach design can be reused across user categories and contexts of use, where the possibility to employ similar, or related interaction dialogues, or techniques was not originally considered. This happened, for example, in the AVANTI user interface, where a particular enhancement was first introduced to facilitate interaction by people with light motor impairments, but it was later found to be of more general applicability. Specifically, interaction objects were enhanced so that they occupy an invisible ellipsoid area of screen real estate, in addition to the rectangular area they occupy by default (and within which they are "drawn"). User interaction outside the normal rectangular area of the object, but within the new invisible one, was associated to the object and treated as if it had actually happened "on" the object. This enhancement was introduced in order to facilitate users with light motor impairment in their upper limbs (e.g., coarse control of their movements, tremor), to interact with the system using pointing devices, lifting the requirement for high precision in pointing operations in the interface. However, it was later discovered that the technique was also useful for able-bodied individuals interacting with the interface through a touch screen (where achieving high precision in pointing operations is also a problem). Due to the fact that the particular technique was introduced into the polymorphic hierarchy (and consequently in the implementation) as a physical-level adaptation, it was possible (and in fact trivial) to reuse it across user categories and contexts of use. This would be unlikely if the technique had been introduced at a higher level in the hierarchy, before the distinction between the desktop and information kiosk metaphors of the AVANTI user interface were introduced.

In parallel to the construction of the polymorphic task hierarchy, the design team will normally have to consider the design of the dialogue between the user and the interface. It is important to point out again that, in addition to the conventional issues involved in dialogue design, two new items emerge from the fact that the interface is unified. First, unification is based on adaptation, making it necessary to design the adaptation "policies" to be used (i.e., the different ways in which adaptations are introduced in the interface, and the role of users in this process). There is yet no definitive empirical data concerning the appropriateness of different policies under different circumstances and for different categories of users. Therefore, the design team will need to employ mock-ups to identify the strong and weak points of different policies and assess their suitability in the context of the interface being designed. In general, these design efforts should not aim to arrive at a single policy that will be used throughout the interface, but, rather, to select "pools" of policies that are appropriate for given situations. Thus, consistency can be preserved, while there still exists adequate flexibility to cater to different requirements or preferences on the part of the user.

The second new aspect of dialogue design in unified user interface development is the design of dialogue techniques that are based on the employment of additional input/output devices. Such techniques need not be limited by the "known" ways of incorporating such devices in the interface. On the contrary, it is desirable that they are employed in novel ways that improve on the level of accessibility attainable, as well as offer the possibility of high-quality interaction to people with disabilities. This is a particularly important point: If the interaction techniques used are a mere one-to-one correspondence of standard direct manipulation techniques to special devices (e.g., using an integrated speech synthesizer for "screen reading"), the entire goal of unification is defeated and reduced to the coexistence of suboptimal accessibility solutions in the same interface.

The implementation of unified user interfaces also merits special attention in the context of the current discussion. Although certain portions of the implementation can be carried out using conventional tools (e.g., programming languages, interface builders), this requires careful planning and a lot of effort to go into the creation of software that satisfies the requirements introduced by unification. Consider, for example, meeting the implementation requirements put forward in chapter 22 of this volume, or creating mechanisms that will undertake the application of adaptations in the user interface, in a "policy-independent" way (i.e., without being bound to a particular adaptation policy). Such implementation issues are the very reason that led to the conception and development of specialized tools to assist and support the implementation of unified user interfaces (see also chaps. 23 and 24, this volume).

The unified user interface architecture, introduced in chapter 20, provides a roadmap for the development of the basic functional components that comprise a unified user interface. Of these components, we briefly focus on the decision-making component and the user information server (user-modeling component). These two modules undertake the transformation of design rationale into run-time logic that drives the adaptation behavior of the interface. In this chapter, we have presented a relatively simple but limited approach to decision making, based on simple rules directly derived from the polymorphic task hierarchy. Alternative, more powerful approaches can be employed (see, e.g., chap. 23, this volume, or Karagiannidis, Koumpis, & Stephanidis, 1997), although that might increase the complexity of the implementation, or introduce several levels of indirection between the actual design rationale as it appears in the task hierarchy and the actual adaptation logic, as this is applied in the final interface.

The evaluation of unified user interfaces is perhaps the most challenging of the development activities discussed so far. As already indicated in the section Evaluating the Unified User Interface (see also chap. 17, this volume), there do not exist today evaluation methods and techniques that adequately address the assessment of adaptive user interfaces. Because adaptation constitutes a major factor in achieving unification in the user interface, it is apparent that the evaluation of unified user interfaces must, therefore, be carefully planned and conducted on a case-by-case basis.

Specifically, evaluation of a unified user interface should aim to identify those aspects of the interface that have beneficial/detrimental effects on the accessibility and interaction quality offered by the interface for different categories of users and in different contexts of use. Two coarse evaluation dimensions can be derived from this goal. The first concerns the "appropriateness" of the different instantiation styles for the purpose they were developed. This entails the assessment of the styles themselves as individual interactive artifacts and as components of the overall interface, as well as the assessment of the design rationale/decision logic that activates (or deactivates) these styles, based on user and usage characteristics. The second dimension concerns the evaluation of the dynamic adaptation (adaptivity) in the interface. This is in fact the most difficult part to evaluate, as there are multiple "forces" that determine the various qualities of the interface. For example, an adaptation may be conceived as entirely dissatisfactory by the user if (a) the adaptation logic itself is flawed, (b) the "triggers" of the adaptation were wrongly inferred by the user-modeling component, (c) the adaptation was not "timely" (e.g., it came "too late" from the user's perspective), or (d) the adaptation policy is not satisfactory (e.g., because the user is not given enough control over it).

To counterbalance the inherent difficulties in evaluating dynamic adaptation in the interface, evaluators should plan the evaluation process carefully from the early design phases, and should actually base the evaluation plan on the unified design, where the different contributing forces discussed earlier are made explicit. Additionally, evaluation should not be restricted to summative activities; rather, it should proceed in parallel to the design of the user interface and should strive to identify deficiencies and possible problems as early as possible, informing and guiding the development process. The evaluation activities of the AVANTI user interface (described earlier) can be considered as the first preliminary steps toward generic methods and techniques for the evaluation of adaptation-capable user interfaces. However, a lot more research and practical experience are required in this direction, before we can derive valuable results that will be reusable across application domains, user categories, and contexts of use.

In summary, this chapter has presented the employment of unified user interface development in the construction of a large-scale, real-world user interface for a wide range of user categories interacting with the system in radically different situations. The practical applicability of the methodologies, techniques, and tools comprising the unified user interface development paradigm has been discussed, highlighting aspects of the development process that are expected to be of particular interest in the context of adopting the unified user interfaces approach toward achieving the goal of *user interfaces for all*.

REFERENCES

Andreadis, A., Giannetti, L., Marchigiani, E., Rizzo, A., Schiatti, E., Tiberio, M., Penttila, M., Perala, J., Leikas, J., Suihko, T., Emiliani, P.L., Bini, A., Nill, A., Sabbione, A., Sfyrakis, M., Stary, C., & Totter, A. (1998). *Global evaluation of the experiments* (ACTS AC042 AVANTI Project, Deliverable DE030). AVANTI Consortium.

Benyon, D., & Murray, D. (1993). Adaptive systems: From intelligent tutoring to autonomous agents. *Knowledge Based Systems, 6,* 197–219.

Bergman, E., & Johnson, E. (1999). *Designing for accessibility* [Online]. Available: http://www.sun.com/access/software.guides.html

Browne, D., Norman, M, & Riches, D. (1990): Why build adaptive systems? In D. Browne, P. Totterdell, & M. Norman (Eds.), *Adaptive user interfaces* (pp. 15–57). London: Academic Press.

Brusilovsky, P. (1992). A framework for intelligent knowledge sequencing and task sequencing. In C. Frasson, G. Gauthier, & G. I. McCalla (Eds.), *Intelligent tutoring systems. Proceedings of the Second International Conference* (pp. 499–506). Berlin: Springer-Verlag.

Byrne, M. D., John, B. E., Wehrle, N. S., & Crow, D. C. (1999). The tangled Web we wove: A taskonomy of WWW use. In M. G. Williams, M. W. Altom, K. Ehrlich, & W. Newman (Eds.), *Proceedings of the ACM Conference on Human Factors in Computing Systems: The CHI Is the Limit* (pp. 544–551). New York: ACM Press.

Carroll, J. M. (Ed.). (1995). *Scenario-based design: Envisioning work and technology in system development.* New York: Wiley.

Del Bianco, A. (1998, February). *Field trial network system in Italy* (ACTS-AVANTI AC042 Project Deliverable DE020).

Finin, T. W., Weber, J., Widerhold, G., Genesereth, M., Fritzson, R., McKay, D., McGuire, J., Pelavin, R., Shapiro, S., & Beck, S. (1993). *Specification of the KQML agent-communication language* [Online]. Available: http://www.cs.umbc.edu/kqml/ paers/ kqmlspec.ps

Fink, J. (1997). *Prototype content model* (ACTS-AVANTI AC042 Project Deliverable DE017).

Fink, J., Kobsa, A., & Nill, A. (1997). Adaptable and adaptive information access for all users, including the disabled and the elderly. In A. Jameson, C. Paris, & C. Tasso (Eds.), *Proceedings of the 6th International Conference on User Modeling* (pp. 171–173). New York: Springer-Verlag.

Fink, J., Kobsa, A., & Nill, A. (in press). Adaptable and adaptive information provision for all users, including disabled and elderly people. *New Review of Hypermedia and Multimedia.*

Fowler, C. J. H., Macaulay, L. A., & Siripoksup, S. (1987). An evaluation of the effectiveness of the adaptive interface module (AIM) in matching dialogues to users. In D. Diaper & R. Winder (Eds.), *Proceedings of the 3rd Conference of the British Computer Society Human Computer Interaction Specialist Group, "People and Computers III"* (pp. 345–359). Cambridge, England: Cambridge University Press.

Ghetti, C., & Bellini, A. (1998, March). *Rome system prototype* (ACTS-AVANTI AC042 Project Deliverable DE032).

Grüninger, C., & van Treeck, W. (1993). Contributions of a social science based evaluation for adaptive design projects. In M. Schneider-Hufschmidt, T. Kühme, & U. Malinowski (Eds.), *Adaptive user interfaces: Principles and practice* (pp. 319–330). Amsterdam: North-Holland, Elsevier Science.

Höök, K. (1997). Evaluating the usability of an adaptive hypermedia system. In *Proceedings of the ACM International Conference on Intelligent User Interfaces* (pp. 179–186). New York: ACM Press

Hoppe, H., Tauber, M., & Ziegler, J. (1986). *A survey of models and formal description methods in HCI with example applications* (ESPRIT Project 385 HUFIT, Report B.3.2.a). HUFIT Consortium.

IBM Corporation .(1999). IBM *accessibility guidelines* [Online]. Available: http://www.austin.ibm.com/ sns/guidelines.htm

ISO 9241-11. (1997). *Ergonomic requirements for office work with visual display terminals, Part 11: Guidance on usability* (Draft International Standard). Geneva, Switzerland: International Standards Organisation.

Karagiannidis, C., Koumpis, A., & Stephanidis, C. (1997). Modelling decisions in intelligent user interfaces. *International Journal of Intelligent Systems, 12*(10), 753–762.

Kobsa, A., & Pohl, W. (1995). The user modelling shell system BGP-MS. *User Modelling and User-Adapted Interaction, 4*(2), 59–106.

Kühme, T., Malinowski, U., & Foley, J. D. (1993). Adaptive prompting (Tech. Rep. No. GIT-GVU-93-05). Atlanta: Georgia Institute of Technology.

Microsoft Corporation. (1997). *The Microsoft Windows guidelines for accessible software design* [Online]. Available: http://www.microsoft.com/enable/dev/guidelines/software.htm

Nill, A. (1998, February). *Prototype of content adaptivity module* (ACTS AC042 AVANTI Project Deliverable DE024).

Paetau, M. (1994). Configurative technology: Adaptation to social systems dynamism. In R. Oppermann (Ed.), *Adaptive user support: Ergonomic design of manually and automatically adaptable software* (pp. 194–234). Hillsdale, NJ: Lawrence Erlbaum Associates.

Penttila, M., & Suihko, T. (1998, February). *Field trial network system in Finland* (ACTS-AVANTI AC042 Project Deliverable 021).

Perdue-Casali, S., & Chase, J. D. (1995). The effects of physical attributes of computer interface design on novice and experienced performance of users with physical disabilities. In G. Perlman, G. K. Green, & M. S. Wogalter (Eds.), *Human factors perspectives on human–computer interaction: Selections from proceedings of Human Factors and Ergonomics Society annual meetings, 1983–1994* (pp. 334–338). Santa Monica, CA: HFES.

Schreck, J., & Nill, A. (1998, February). *Prototype of user model server* (ACTS-AVANTI AC042 Project Deliverable 022).

Schuler, D., & Namioka, A. (Eds.). (1993). *Participatory design: Principles and practices.* Hillsdale, NJ: Lawrence Erlbaum Associates.

Shute, V. J., & Psotka, J. (1994). Intelligent tutoring systems: Past, present and future. In D. Jonassen (Ed.), *Handbook of research on educational communications and technology* (pp. 570–600). New York: Macmillan.

Smith, M., & Tattersall, C. (1990). Intelligent help: The results of the EUROHELP project. *ICL Systems Journal, 7*(2) 151–166.

Stephanidis, C., & Akoumianakis, D. (1998). Multiple metaphor environments: Issues for effective interaction design. In *8th ERCIM—DELOS Workshop on User Interfaces for Digital Libraries.* (5 pages), ERCIM, [On-line]. Available at http://www.ercim.org/publication/wsproceedings/DELOS8/index.html.

Story, M. F. (1998). Maximising usability: The principles of universal design. *Assistive Technology, 10*(1), 4–12.

Tauscher, L., & Greenberg, S. (1997). Revisitation patterns in World Wide Web navigation. In S. Pemberton (Ed.), *Proceedings of the ACM Conference on Human Factors in Computing Systems* (pp. 399–406). New York: ACM Press.

Totterdell, P., & Boyle, E. (1990). The evaluation of adaptive systems. In D. Browne, P. Totterdell, & M. Norman (Eds.), *Adaptive user interfaces* (pp. 161–194). London: Academic Press.

Ulich, E., Rauterberg, M., Moll, T., Greutmann, T., & Strohm, O. (1991). Task orientation and user-oriented dialog design. *International Journal of Human–Computer Interaction, 3*(2), 117–144.

Vanderheiden, G. C. (1994). *Application software design guidelines: Increasing the accessibility of application software to people with disabilities and older users* [Online]. Available: http://www.tracecenter.org/docs/software_guidelines/software.htm

Vanderheiden, G. C., & Kaine-Krolak, M. (1995). *Access to current and next-generation information systems by people with disabilities* [Online]. Available: http://www.tracecenter.org/docs/access_info_sys/full_doc.htm

World Wide Web Consortium. (1999). *User agent accessibility guidelines* (Working Draft) [Online]. Available: http://www.w3.org/TR/WAI-USERAGENT/

VII Support Measures

26 Making the Web Accessible

Daniel Dardailler
Judy Brewer
Ian Jacobs

The World Wide Web is fast becoming the de facto repository for on-line information. Yet, Web technology has inadvertently created barriers for people with disabilities. The World Wide Web Consortium co-ordinates the evolution of the Web core protocols (Hypertext Markup Language, Cascading Style Sheets, Extensible Markup Language, Hypertext Transfer Protocol, etc.) and has a mission to "lead the Web to its full potential". As of 1997, World Wide Web Consortium has taken on a new leadership role in removing these accessibility barriers and, to that effect, has launched the Web Accessibility Initiative. This chapter addresses the issue of accessibility to the Web by all potential users, and outlines current and on-going activities carried out by the World Wide Web Consortium - Web Accessibility Initiative in this direction.

The emergence of the World Wide Web has made it possible for individuals with appropriate computer and telecommunications equipment to interact as never before. The Web is the stepping stone, the infrastructure, that will pave the way for next-generation interfaces.

Before the advent of the graphical user interface (GUI), most human–computer interaction was text based; simple speech synthesizers, screen magnifiers, and other, well-known software, hardware, and firmware solutions enabled all but the most severely disabled to use computers. GUIs have introduced the graphical presentation of information, and have established direct manipulation as the primary method of interacting with computers, thus rendering most of the previous accessibility solutions obsolete and unusable. Along the same lines, "new" Web content freely combines graphics, audio, and video, severely limiting, in this way, efficient communication for a wide range of people with disabilities.

The current situation with Web accessibility is far from satisfactory and is abraded by the fact that many people "rush" into the Web, developing content without any awareness of the new limitations and frontiers they may create. No single disability population is unaffected. For example:

- People who are deaf cannot hear multimedia or audio events that do not contain captioning or audio descriptions.
- People who are blind struggle with the Web's inherent graphical interface, its graphic-based content, and any Web protocol or application that cannot be rendered or accessed easily using audio, Braille, large text, or synthetic voice.
- People with physical disabilities have difficulty using certain hardware devices or Web controls, including, for example, Web kiosks and WebTV™.[1]
- People with cognitive and visual impairments have difficulties interpreting most Web pages, because these pages have not been designed with this population in mind.

Worldwide, there are more than 750 million people with disabilities. A significant percentage of that population is affected by the emergence of the Web, directly or indirectly. For people without disabilities, the Web is a new technology that can help unify geographically dispersed groups. However, accessibility barriers put the Web in danger of disenfranchising people with disabilities in this emerging infrastructure, thus making the treatment of these barriers an absolute necessity.

Furthermore, even those without disabilities would benefit from many changes motivated by the needs of people with disabilities. When driving a car, for example, the driver may wish to browse the Web for information (movie schedules, etc.) using a voice-based interface similar to that used by the blind.

The position of the World Wide Web Consortium (W3C) is clear: Access to the World Wide Web could be significantly improved *for all* by changes to the Web's supporting protocols, applications and, most important, content. Therefore, all the protocols and languages issued as recommendations should meet or exceed established accessibility goals. In addition, the development of Web software and content that is accessible to most people with disabilities will be actively encouraged.[2]

THE W3C WEB ACCESSIBILITY INITIATIVE

Launching and Structure of the Web Accessibility Initiative

In order to fulfill its mission of leading the Web to its full potential, the W3C must promote a high degree of accessibility and usability for people with disabilities. To that effect, in coordination with other organizations, it launched the Web Accessibility Initiative (WAI) in 1997, to pursue accessibility of the Web through five primary areas of work:[3]

- *Technology reviews and development.* Centered on protocols and data formats, especially the Hypertext Markup Language (HTML), Cascading

---

[1]For more information, see: http://www.webtv.com

[2]For more information, see: http://www.w3.org

[3]For more information, see: http://www.w3.org/WAI/

Style Sheets (CSS), Extensible Markup Language (XML), Synchronized Multi-Media Interface Language (SMIL), and the Hypertext Transfer Protocol (HTTP).

- *Guidelines for use of the technology.* Guidelines targeted at content creators, browser vendors, and authoring tool vendors.
- *Education and outreach.* Raising the awareness of the content creation community to the needs of people with disabilities, as they relate to Web technologies.
- *Tools for evaluation and improvement of Web pages.*
- *Research and advanced development.* User interface design, novel devices, are all areas where additional work is required before standardization is appropriate.

Much like other W3C domains, the WAI coordinates several different activities through a series of mailing lists, Web areas, conference calls, and regular face-to-face meetings.[4]

Rationale and Mission of the WAI

In the area of software accessibility, education is one of the most important factors for success. One of the earliest roles of the W3C was educational; it has long acted as a repository of information about the World Wide Web for developers and users (especially technology specifications related to the Web).

To meet its accessibility goals, three additional roles are required with respect to accessibility:

1. *Act as a central point for setting accessibility goals for the Web.* This requires the W3C to coordinate with external organizations that represent people with disabilities to generate a widely accepted set of goals and guidelines, which take into account the needs of the user community, the details of technology, and engineering realities. The W3C already plays such a role in several other areas of technology, thus being the most appropriate actor to undertake the tasks of ensuring, supporting, and promoting Web accessibility.

2. *Act as an advocate for people with disabilities to the Web development community.* As the internationally acknowledged organization and leader for World Wide Web development, the W3C recognizes its responsibility for advocating Web accessibility for people with disabilities. As the Web user interface and infrastructure continues to evolve, the W3C will work to help its members become

---

[4]The Web site http://www.w3.org/WAI is continuously updated to refer to the latest and historical information about the project.

proactive in their efforts to design and develop the Web in a way that considers and addresses the special needs of this particular portion of the user population.

3. *Act as an advocate for people with disabilities to the Web content community*. The W3C already serves as a neutral party for distributing information about Web technology, and aims to extend this role to be proactive in explaining to content producers how to make optimal use of technology, in order to serve the needs of people with disabilities.

In order to adequately focus on education and outreach, as well as research and development, a WAI International Program Office (WAI-IPO) has been created that enables partnering and coordination among the many stakeholders in Web accessibility, that is, industry, disability organizations, government, and academic and research organizations.

One important role of the WAI-IPO is to sensitize content creators. We do not want to do just the technical part of the work and, in fact, we believe that without the content provider outreach aspect (through education and tools upgrade), the technical project alone is not worth running.

The WAI-IPO is separately funded from W3C member activities, through sponsorship of governments (U.S. National Science Foundation, U.S. Department of Education), European Commission support actions,[5] and corporations that have shown their leadership in Web accessibility and *universal design* (e.g., Microsoft, Lotus/IBM, NCR).

SCOPE OF WAI ACTIVITIES

This section briefly reviews the scope of the major areas of activity in WAI, involving (but not limited to): (a) the development of core Web technology, (b) the assembly and publication of guidelines, (c) the promotion and coordination of the development of tools for the evaluation and repair of Web sites (in terms of accessibility), and (d) activities related to education and awareness raising of Web content creators, technology developers, and users themselves.

Technology Development

This area is centered on Web protocols and data formats, especially HTML, CSS, SMIL,[6] RDF (Resource Description Framework), XML, and DOM (Document Object Model).[7] Because the WAI is intended to concentrate on Web (rather than

---

[5]The European Commission (Telematics Application Program, Disabled and Elderly Sector) funded the TAP DE 4105 project, "WAI—Web Accessibility Initiative," for a period of 18 months (January 1998–June 1999).

[6]For more information, see: http://www.w3.org/TR/REC-smil

[7]For more information, see: http://www.w3.org/DOM

general computer) accessibility, physical devices and related issues are not expected to be addressed by ongoing work. The work of the W3C, since its inception, has concentrated on precisely these technologies, and they can be expected to be at the core of W3C's focus and mission in the future. In this context, the following initial work items in the area have been addressed by WAI activities:

- HTML/CSS/SMIL: This is where most of the early technology development efforts in WAI have focused. The section Implementing Accessible Technologies later in this chapter is devoted to work carried out so far in this particular area.
- XML, DOM, HTTP, and others: As a follow-up to review work on the HTML and CSS specifications, several future activities are foreseen, mainly in reviewing different areas of Web technology, focusing on their incorporating sufficient support for accessibility.

Guidelines for Use of the Technology

There exist several guidelines for the use of HTML, aiming at the development of pages that are accessible by people with disabilities, but these are rapidly falling behind technology. Furthermore, there is confusion in the industry because the existing sets of guidelines are several, and, in some cases, incompatible. The industry needs a mechanism for (a) generating either a single set of guidelines, or, as a minimum, several compatible sets of guidelines, and, most important, (b) keeping the guidelines up to date as technology evolves. To attain both of these goals is far from trivial, as they heavily depend on a wide range of factors, including:

- Current Web technology.
- Desired use (e.g., conversion to Braille is different from speech output, which is, in turn, different from screen reading).
- Audience (different levels of accessibility have different connotations to a manager than to a developer).
- Disability being addressed.

The W3C does not, as a general rule, deal with producing style guidelines. Instead, it concentrates on "mechanism, not policy" and has allowed the market to shape the use of the technologies it defines. On the other hand, the W3C does attempt to provide mechanisms for this purpose (the "alt" attribute of the "IMG"—image—element is one example, allowing for descriptive text attached to an image) and does attempt to make it clear that this mechanism should be used to maximum advantage.

In addition to the popular markup guidelines, the WAI has also undertaken the development of two sets of guidelines for tool developers (concerning browser development and authoring tool development). These guidelines attempt to (a) identify accessibility features that need to be supported by user agents (e.g., browsers), through which users experience and interact with Web content, and (b) emphasize those features that need to be supported by development tools, so that content creators are facilitated and supported in the development of accessible content for people with disabilities.

The current state of the accessibility guidelines issued by WAI is briefly summarized in the final section of this chapter.[8]

Tools for Evaluation and Repair of Web Sites

Once the guidelines are stable, one needs to specify, coordinate, and foster the development of tools that will help users evaluate the accessibility of Web sites, as well as make any repairs necessary to improve their accessibility. Criteria for public and internal Web site accessibility must be developed, along with methods of applying these criteria via features that are either fully, or semiautomatically (in terms of human involvement) used to evaluate or repair Web sites.

Education (Sensitization) of Content Creators

As mentioned earlier, the primary issue is making sure that Web content is produced in a form accessible to people with disabilities. This process cannot be made completely automatic, even when given powerful authoring tools; it requires attention on the part of the designer to the needs of a community that is all too often ignored. The key to success is a combination of tools that make it easy to address accessibility requirements, and education that reinforces the importance of using the tools routinely and correctly.

To fulfill this mission, the WAI has created a special Working Group for Education and Outreach that will develop strategies and materials to increase awareness within the Web community of the need for, and the available solutions to achieve, Web accessibility.

Items included in the scope of work are:

- Planning and prioritizing education/outreach strategies and approaches.
- Identification and prioritization of audiences for dissemination/outreach/education.

---

[8]For additional and up-to-date versions of the guidelines, see: http://www.w3.org/WAI

- Compiling existing educational materials, and identifying event opportunities.
- Developing and refining outreach message and educational materials in a variety of formats.
- Promoting the implementation of accessibility improvements in Web technology.
- Coordinating with related educational activities.
- Coordinating the translation and localization of education and outreach materials.
- Assessing the impact of educational activities.

In order to reach our goal, we need to target several different audiences. The content providers, for instance, use, listen to, and are influenced by several other actors, including:

- *The authoring tools software vendors.* With increasing frequency, Web content is authored using specialized WYSIWYG tools and no longer textual editors "showing the tags" (i.e., editors that present their users with the actual markup used to structure, organize, and define the presentational attributes of content). By making sure that the providers of these tools take accessibility into account, the chances that the users of the tools will create accessible content are improved.
- *The Web site designers.* People "owning" the content are the content providers in the larger sense, but it often happens that those actually designing, or producing the content (i.e., implementing Web sites) are service companies, which can play a big role in advocating accessibility.
- *The Web-design educators.* When companies (usually big ones) set out to create a public or private (internal) Web space, they most of the time make use of educational services to train their employees on how to best take advantage of authoring tools. Organizations and companies providing these educational/formation services need to be made aware of the accessibility aspects of created content.
- *The press,* and in effect, the user base, can greatly influence content providers through their review of Web sites. It is important that accessibility becomes a regular criterion of choice in such reviews.

Of course, one other important actor is the W3C itself, due to its established role in the Web community; therefore, its undertaking of the type and range of activities described herein is of paramount importance and will hopefully act as a catalytic factor in the acceptance and employment of the advocated principles, methods, and techniques by the involved actors.

In order to reach all these communities, efforts need to be targeted along a series of events:

- Presentations/talks in major Web-related conferences.
- Organization of free seminars, either as part of conferences, or in isolation.
- Direct contact and awareness action with major Web site providers.
- Direct contact and lobbying with major authoring tool providers.
- Submission of papers in specialized and regular press.

Additionally, varied educational and outreach materials need to be created (in a variety of formats—Web-based, CD, hard copy, etc.), such as:

- Lists of frequently asked questions (FAQs) on accessibility improvements in W3C technical specifications.
- Code samples for accessible design.
- Sample accessible style sheets.
- Demonstrators of accessible and inaccessible design and innovations.
- Awareness and promotion materials.
- Policy references on accessibility.
- Business cases for accessibility and universal design *(design for all)*.
- Instructional modules for accessible design.
- Demonstration of Web sites on accessibility.
- Presentation and workshop packages.

Finally, an educational aspect that also needs to be explored is the education of the disability community itself, regarding the rights of people with disabilities with respect to accessing information like everybody else. This is particularly true and important in the Intranet context, where companies are already subject to existing legislation regarding access (see the U.S. Americans with Disabilities Ac, 1990; the U.K. Disability Discrimination Act, 1995; or the European Treaty of Amsterdam, 1997). Though we believe that educating all involved actors is perhaps the most important aspect of the work discussed so far, it is not something that falls easily within W3C's existing role. Clearly, part of this work should be included in the training programs that come with any Web-authoring tool. However, part of this work goes beyond individual tools, and touches on the traditional role of government: sensitizing the key players (content providers, in this case) to the needs of an important minority population with special needs.

This also explains why the education activity is the main externally funded activity within WAI, in contrast with usual W3C activities, which are usually exclusively funded by W3C core resources (i.e., membership fees). For the WAI and the IPO, given the extended nature of the work, we have sought and obtained external funding from the U.S. National Science Foundation, the U.S. Department of Education's National Institute on Disability and Rehabilitation Research, the European Commission's DG XIII, and industry.

IMPLEMENTING ACCESSIBLE TECHNOLOGIES

This section presents the features that have been introduced in the specification of HTML and CSS, in order to facilitate the development of Web content that is accessible by the widest possible population of users. It also discusses the positive impact these new accessibility features will have for the wider user community of the Web.

Built-In Accessibility Features in HTML 4.0

As part of its ongoing efforts to pursue and promote accessibility, the WAI joined forces with the W3C HTML Working Group in the design of HTML version 4.0,[9] which became a W3C Recommendation in December 1997. For this latest release of the World Wide Web's publishing language, the WAI group sought remedies for a number of authoring habits that cause problems for users of:

- *Screen readers.* Screen readers are software applications, which intercept output being sent to a monitor and direct it to speech synthesis devices, or refreshable Braille displays.
- *Audio browsers.* Audio browsers read and interpret HTML (and style sheets) and are capable of producing inflected speech output.
- *Text-only browsers.* Text browsers are used on some devices (including hand-held devices, with small character displays) that are only capable of displaying characters.

In particular, while reviewing the HTML 4.0 specification, the WAI group addressed: (a) unstructured pages, which disorient users and hinder navigation, (b) abuse of HTML structural elements for purposes of layout, or formatting, and (c) heavy reliance on graphical information (e.g., images, image maps, tables used for layout, frames, scripts, etc.) with no text alternatives.

---

[9]The HTML 4.0 recommendation is online available at: http://www.w3.org/TR/REC-html40

In the following sections, we look at how WAI contributions to HTML 4.0 (in conjunction with style sheets) allow authors to avoid accessibility pitfalls in their attempt to create more attractive, economical, and manageable Web pages.

Improved Structure.   Because highly structured documents are more accessible than those with little or no structure, HTML 4.0 has added a number of elements and attributes that enrich the authors' capabilities in designing and representing document structure.[10] The new constructs will also allow software tools (e.g., search robots, document transformation tools, etc.) to extract more information from these documents. The following structural elements are new in HTML 4.0:

- For structured text, the "ABBR" (abbreviation) and "ACRONYM" elements, in conjunction with style sheets and the "lang" (language) attribute, will provide assistance to speech synthesizers. The "Q" element identifies inline quotations, complementing the existing "BLOCKQUOTE" element.
- The "INS" and "DEL" elements identify new and deleted portions of a document, respectively (their existence is expected to offer great facilitation in document editors).
- To create structured forms, the "FIELDSET" and "LEGEND" elements organize form controls into semantically related groups. Respectively, the "OPTGROUP" element groups menu options into semantically related groups. Grouping menu options improves navigability and reduces the burden of browsing (and trying to remember) long lists of choices.
- Several new elements ("THEAD," "TBODY," "TFOOT," "COLGROUP," and "COL") have been added to group table rows and columns into meaningful sections. Several new attributes ("scope," "headers," and "axes") label table cells so that nonvisual browsers may render a table in a linear fashion, based on the semantically significant labels.

Style Sheets.   HTML was not designed with professional publishing in mind; its designers intended it to organize content, not present it. Consequently, many of the language's presentation elements and attributes do not always meet the needs of advanced page design. To overcome layout limitations, the W3C HTML Working Group decided not to add new presentation features to HTML 4.0, but instead to assign the task of presentation to style sheet languages, such as CSS.[11]

---

[10]The paper presentation of the new HTML 4.0 accessibility improvements in online available at: http://www.w3.org.WAI/References/html4-access

[11]Cascading style sheets (CSS), Level 2 is a W3C recommendation online available at: http://www.w3.org/TR/REC-CSS2

While style sheets are not part of the HTML proper, HTML 4.0 is the first version of the language to integrate them fully.

Why did the HTML Working Group adopt this strategy? For one thing, experience shows that distinguishing a document's structure and its presentation leads to more maintainable and reusable documents. Moreover, by extracting formating directives from HTML documents, authors may design documents for a variety of users and target media in mind, with minimal changes to their original HTML documents. The same HTML document, with different style sheets, may be tailored to color-blind users, those requiring large print, those with Braille readers, speech synthesizers, hand-held devices, text-only terminals, and so on. But style sheets have another significant impact on accessibility: They eliminate the need to rely on language "tricks" for achieving visual layout and formatting effects. In most cases, these tricks have the unfortunate side effect of making pages inaccessible.

For instance, HTML does not have an element, or attribute to indent a paragraph, so many authors have resorted to using the "BLOCKQUOTE" element to indent text (as many visual browsers indent the content of the element), even when there is no quotation involved. This is misleading to nonvisual users: When an audio browser encounters a "BLOCKQUOTE" element, it should be able to assume that the enclosed text is a quotation. More often than not, that assumption proves incorrect, because the element has been misused for a presentation effect.

The "BLOCKQUOTE" example demonstrates the misuse, for presentation purposes, of an element intended to provide logical information. Many similar traps can seduce HTML authors: They use tables and invisible images for layout; they use heading tags to change the font size of some text that is not a header; they use lists for alignment; they use the "EM" element to italicize text when, in fact, "EM" is meant to emphasize text (often presented with an italic font style by visual browsers, but rendered differently by a speech synthesizer), and so on.

Now, style sheets will give authors a richer palette for layout and formatting, eliminating at the same time the accessibility problems that arise from markup abuse. The section Built-in Accessibility Features in CSS2 later in the chapter discusses the accessibility improvements introduced in the cascading style sheets Level 2 (CSS2).

Alternate Content.   A picture may be worth a thousand words to some people, but others need at least a few words to get the picture. When developing content for the Web, authors should always complement nontextual content—images, video, audio, scripts, and applets—with alternate text content and textual descriptions. These are vital for visually impaired users, but also extremely useful to many others: those who browse with text-only tools, those who configure their browsers not to display images (e.g., because their modem is too slow, or they simply prefer nongraphical browsing), or those who are "temporarily disabled," such as commuters who want to browse the Web while driving to work.

In HTML 4.0, there are a host of new mechanisms for specifying alternate content and descriptions:

- As in previous versions of HTML, authors may use the "alt" attribute to specify alternate text. As of HTML 4.0, the specification of this attribute is mandatory for the "IMG" and "AREA" elements; it is optional for the "INPUT" and "APPLET" elements.
- The new "title" attribute gives a short description of an image and the like. This information is of vital importance to nonvisual browsers, but is equally useful for graphical browsers, which frequently pop up this information when a user pauses over an element with the mouse (tool tips).
- The new "longdesc" attribute designates an external document that gives a long description of an image and so on.
- The new "CAPTION" element and "summary" attribute (of the "TABLE" element) describe a table's purpose.
- In addition to the new frame elements, the "NOFRAMES" element specifies content to be rendered when a browser cannot render a document with frames.
- In addition to the "SCRIPT" element, the "NOSCRIPT" element specifies content to be rendered when a browser cannot render script content.

The "title" attribute, in particular, has many accessibility-related applications. For instance, with the new "ABBR" and "ACRONYM" elements, it may indicate the expanded text of an abbreviation, or provide a short description of an included sound clip, or provide information about why a horizontal rule (signified with the "HR" element) has been used to convey a structural division (although authors should use structural markup as well, such as the "DIV"—block-level division element—or "SPAN"—inline text container element—elements).

But of all the new elements, the "OBJECT" element (used to include images, applets, or any type of object in a document) is the most important for specifying alternate content. With it, authors may specify rich alternate content (containing markup, impossible with attribute values), at the same location where they specify the object to be included. When a browser cannot render the image, applet, and so forth, included by an "OBJECT" element, it renders the (marked-up) content instead.

One important application of this feature involves client-side image maps. In HTML 4.0, the content model of the "MAP" element has been expanded to allow marked-up anchor ("A") elements, that give the geometry of the map's active regions. When placed inside of an "OBJECT" element, the textual version of the image map will be rendered only if the graphical version cannot be. Thus, authors may create graphical and nongraphical image maps at the same location in their documents.

Easier Navigation and Orientation.   Visually impaired users have tremendous difficulties browsing pages in which navigation options rely largely on graphical cues. For instance, image maps with no textual alternatives are next to impossible to navigate. Similarly, link text that offers no context (e.g., a link that simply reads "click here") is as frustrating as a road sign that reads only "Exit"—exit to where? On a different level, adjacent links not separated by nonlink characters confuse screen readers, which generally interpret them as a single link.

HTML 4.0 includes several features to facilitate navigation:

- For client-side image maps, the "MAP" element may contain anchor elements ("A") that simultaneously specify the active regions of the image map and provide detailed textual explanations of links. As mentioned earlier, these links may be marked up with rich alternate descriptions of the image map.
- The "title" attribute, with the "A" element, can describe the nature of a link, so that users may decide whether to follow it.
- The "accesskey" attribute allows users to activate links or form controls from the keyboard.
- The "tabindex" attribute allows users to use the keyboard to navigate the links or form controls on a page in a logical sequence.
- The "LINK" element (specified in the header of a document) together with the "media" attribute allow user agents (e.g., browsers) to load appropriate pages for a specific target medium automatically, making user navigation to those pages unnecessary.

Impact of Accessibility Features Beyond the Disabled Community.
Investing in physical-world accessibility modifications (wheelchair ramps, curb cuts, etc.) has benefited a much larger community than those with disabilities: How often have parents with baby carriages or cyclists appreciated these same improvements? The benefits from accessibility innovations can similarly be generalized to situations that involve nondisabled population. The following are considered indicative examples:

- Highly structured documents convey more information (e.g., for use by search engines).
- Using style sheets with HTML instead of images for aesthetically pleasing text and fancy layout can reduce the combined size of Web documents and, once widespread, is likely to improve overall Internet performance because of faster download times.

- Smoother browsing can be achieved in situations where the user's eyes or hands are occupied, or in the case of display units with limited capabilities (e.g., browsing the Web in your car, on hand-held devices, etc.).

Built-In Accessibility Features in CSS2

In the area of CSS2,[12] WAI's efforts have shown again that what benefits the disabled community benefits the entire community. As they are supported by more browsers and authoring tools, style sheets will make life much easier for Web users with disabilities, but also for:

- Authors, as style sheets are simple and powerful design tools.
- Site managers, because pages may be shared, reused, and cached.
- The entire Internet community, because style sheets, in tandem with other technologies, have been shown to speed up downloads and reduce Internet traffic.

CSS may be used with HTML 4.0 as well as applications of XML 1.0.[13] In fact, XML authors will have increased needs for style sheets to describe presentation, because XML applications have no generally accepted rendering.

No More Cumbersome Tricks.   Style sheets eliminate the need for many of the cumbersome font and layout "tricks" that page designers have devised to overcome HTML's limitations. However clever these tricks may be, they often have the unfortunate side effect of making pages inaccessible. Prior to style sheets, authors wishing to display text with a fancy font had to capture the effect in an image. Images not only require more bits than text, but they are also not searchable and cannot be rendered by speech synthesizers. CSS2's powerful capabilities for font manipulation and presentation (collectively termed WebFonts) let browsers download or even synthesize fonts, so that authors can retain text in textual form and still achieve the desired effect.

Resourceful authors have also relied on images to lay out their pages. Scattering invisible images around a document makes it very difficult for nonvisual browsers to identify actual content reliably. Similar obstacles to nonvisual browsers include tables used for alignment, and "PRE" (preformatted) or "BLOCKQUOTE" elements for indenting and positioning. CSS2 offers several positioning and alignment mechanisms so that content is not littered with spaces or images that challenge blind users:

---

[12]The paper presentation of the new CSS Level 2 accessibility improvements is online available at: http://www.w3.org/WAI/References/css-2-access

[13]The XML 1.0 recommendation is online available at: http://www.w3.org/TR/REC-XML

- Spaces between letters confuse blind users. Authors can use the "word-space" and "font-stretch" properties to achieve spacing and stretching effects on "normal" text.
- Paragraphs can be indented with the "text-indent" property and centered and justified with the "text-align" property.
- The "position" and "float" properties allow authors to place content anywhere on the screen or page, whereas the text still appears in the file in a logical order for reading. In audio presentation environments, it is important for the browser to receive the text in logical reading order; it must be easily "linearized" because it will be heard as a single, sequential stream. The same properties may be used to create margin notes (which may be automatically numbered), side bars, framelike effects, simple headers and footers, and more.
- The "empty cells" property means that authors can get better looking tables without filling them with blank spaces or " " characters (which force the presence of whitespace characters in the rendered document).

All of these mechanisms emphasize the separation of a document's content from its presentation. When authors design with this principle in mind, everyone benefits because documents become smaller (no redundant presentation tags, fewer images), easier to maintain (external style sheets may be shared, altered without changing the main document, and cached), searchable, faster to download, and usable by speech synthesizers, Braille readers, text-only display terminals, cell phones, and so on.

User Control.   CSS2 surpasses CSS Level 1 (CSS1) in user control over styles, audio rendering of documents, positioning, downloadable fonts, table formatting, numbering, and generated text, as a result of contribution from, and coordination with WAI.

The term *cascading* is used to denote the fact that style sheets from a variety of sources—the browser, the author, and the user—are interwoven and sorted through a mechanism called "the cascade." In CSS1, the cascade mechanism gave authors final control over the rendering of an element. In CSS2, this has been reversed, specifically so that users with accessibility needs can be sure that their requirements will be met. Any rule in a user style sheet indicated as "!important" overrides any competing author or user agent rule, so that the user always retains final control over the presentation of any element. Thus, for example, people requiring large print, or certain backgrounds or color combinations, will be able to control the appearance of the document when browsing. The new "inherit" value that every CSS2 property may take, when combined with "!important," gives users even tighter control of appearance, should it be required. The cascade allows users, but not authors, to override such rules for special cases if necessary.

Orientation and Navigation.   Users who are blind, or engaged in eyes-busy activities (such as driving), or using a device with limited or no display screen, can find it difficult to stay oriented in the midst of unstructured textual information, or to navigate links and forms on a page. One way to facilitate orientation is to number lists, paragraphs, sections, and so on. CSS2's lists, markers, generated text, and automatic numbering mechanisms allow numbers to be generated for almost any sequence of elements in a document, in a variety of international numbering styles, and with flexible style control.

Additionally, CSS2's aural properties provide information to nonsighted users, much in the same way fonts provide visual information. The following example shows how various CSS2 aural properties (including "voice-family," which is conceptually similar to an "audio font") can let a user know that the spoken content is a header:

```
H1 {

    voice-family: paul;

    stress: 20;

    richness: 90;

    cue-before: url("ping.au")

}
```

The "cue-before" property is just one of CSS2's "generated content" properties. In the preceding example, it is used to signify the fact that an important header ("H1" denotes top-level headings in a document) is about to be spoken.

When an author employs header elements correctly—using them to introduce sections, and not to change font size—the headers can provide significant information about a document's structure. CSS2's powerful font mechanism will let authors use header elements as intended, allowing browsers to exploit this structural information to generate navigation tools automatically, such as tables of contents and outline views of a document.

Access to Alternative Representations of Content

Though CSS2 eliminates the need for some bit-mapped images, it is not meant to eliminate graphical or pictorial representation of content. Such content, however, also needs to be represented textually (through attributes). CSS2 gives users direct access to alternate text and other attribute values thanks to attribute selectors, the "attr()" function, and the "content" property. Attribute selectors match elements that have a certain attribute (such as "name") or a specific value for that attribute (such as "name = submit"). Thus, it is possible to make rules applicable, not globally, but only in selected problematic cases. Attribute values may also be inserted into a document with the "content" property, making it possible, for example, to retrieve the value of the "alt" attribute for an "IMG" element, and render it immediately following the image.

CONCLUSIONS

At the beginning of 1999, the technical accessibility work carried out so far (review and improvement to the HTML/CSS/SMIL specifications) has already given very good results. Participation in the WAI technical working group has been very rewarding and we have received very strong support by, and heavy participation from the experts in the HTML and CSS groups.

We are looking forward to more achievements in the area of XML, DOM, and presentation style specification languages (e.g., embedding direct support for presentation in Braille).

The guidelines working groups are moving at fast paces. There currently exist public draft versions of the following sets of guidelines:

1. The Page Author Guidelines,[14] intended to be followed by content creators in order to make their pages more accessible for people with disabilities, as well as more useful to other users, new page-viewing technologies (mobile and voice), and electronic agents (such as indexing robots).

2. The User Agent Guidelines,[15] intended for user agent developers for making their products more accessible to people with disabilities and for increasing usability for all users. In the context of these guidelines, the term *user agent* is used to denote browsers (graphic, text, voice, etc.), multimedia players, and assistive technology products (such as screen readers, screen magnifiers, and voice input software).

3. The Authoring Tools Guidelines,[16] intended to assist developers in designing authoring tools that generate accessible Web content, as well as in creating an accessible user interface to the authoring tool itself.

The education program has started, and we think it is going to be one of the most exciting tasks in this project. Our first experiences and contacts are showing that most content providers are very willing and ready to make their Web space accessible; they just need to be told how. An example of the activities of the education program are the "QuickTips Card,"[17] which is a business card containing con-

---

[14]The latest version of the Page Author Guidelines can be obtained from: http://www.w3.org/WAI/GL/WD-WAI-PAGEAUTH

[15]The latest version of the User Agent Guidelines can be obtained from: http://www.w3.org/TR/WD-WAI-USERAGENT

[16]The latest version of the Authoring Tools Guidelines can be obtained from: http://www.w3.org/WAI/AU/WD-WAI-AUTOOLS

[17]More information about the QuickTips card are online available from: http://www.w3.org/WAI/References/QuickTips

densed accessibility guidelines for quick reference; guidelines cover accessibility issues in: images and animations, page organization, image maps, hypertext links, graphs and charts, audio and video, scripts, applets and Web browser plug-ins, frames, tables, and validation/verification of the overall accessibility of Web sites with the use of automated tools.

Above all, we would like to stress the following two points: (a) new information technologies have the potential to eliminate isolation effects of the past, through appropriate design and planning, taking thoroughly into consideration the needs of the full spectrum of their potential users, and, in particular, of people with various forms and severity levels of disabilities; and (b) improving access to the Web by people with special needs, has the potential to also improve comprehension of, and interaction with, Web content for all users, irrespective of disability. The well-known "curb-cut effect" in the field of physical space accessibility is a good example of what can be achieved in the latter case.

REFERENCES

Americans with Disabilities Act. (1990). U.S. Department of Justice [Online]. Available: http://www.usdoj.gov/crt/ada/adahom1.htm

Disability Discrimination Act. (1995) (Chapter 50). U.K. Acts of Parliament [Online]. Available: http://www.hmso.gov.uk/acts/acts1995/1995050.htm

Treaty of Amsterdam Amending the Treaty on European Union, the Treaties Establishing the European Communities and Certain Related Acts. (1997). [Online]. Available: http://ue.eu.int/Amsterdam/en/amsteroc/en/treaty/treaty.htm

27 Industrial Policy Issues

**Constantine Stephanidis,
Demosthenes Akoumianakis,
Nikolaos Vernardakis, Pier Luigi Emiliani,
Gregg Vanderheiden, Jan Ekberg,
Juergen Ziegler, Klaus-Peter Faehnrich,
Anthony Galetsas, Seppo Haataja,
Ilias Iakovidis, Erkki Kemppainen,
Phill Jenkins, Peter Korn, Mark Maybury,
Harry J. Murphy, and Hirotada Ueda**

This chapter explores some industrial policy issues related to User Interfaces for All. *In particular, it addresses two key questions, namely: (a) who are likely to be the "willing recipients" of a contemplated effort to transfer* User Interfaces for All *technologies towards the mainstream software industry; and (b) what type of policy interventions would be useful in this direction. With regards to the former issue, we examine the conditions for success across different sectors of the industry to determine the characteristics of potential recipients. With regards to the latter issue, we describe two policy variables, namely legislation and standardization, which are known to influence the generation and diffusion of social innovation. The main conclusion of the chapter is that despite recent progress, there is still much to be done in order to establish an industrial environment favorable to* User Interfaces for All.

Following a description of the rationale, technologies, and challenges involved in achieving the goal of *user interfaces for all,* this chapter focuses on some policy issues, in an attempt to shed light into the issue of transferring technical know-how and know-why to industry, so as to promote wider adoption and internalization of the respective principles and practices. Specifically, the objective is to briefly review the possible pathways through which the benefits of user interfaces for all can be made visible to, and appropriated by, the mainstream software industry into socially desirable and market-wise successful innovations. In this endeavor, we are equally interested in the sectors of the industry involved, their composition, and on the conditions for successful uptake and diffusion.

The chapter is structured as follows. The following section focuses on two themes. The first is the likelihood that different sectors will adopt, and make use of, advanced technologies for user interfaces for all. The second is an analysis of suitable technology transfer mechanisms, given some distinctive characteristics of technologies, products and services complying with the requirements of user interfaces for all. Subsequently, we review alternative strategies for introducing user interfaces for all into prevailing computing paradigms. In particular, three alternatives are explored: (a) the market for user interface development tools characterizing the interactive applications running on the vast majority of graphical desktops in the current office environments, (b) distributed and component-based technologies, and (c) the Internet and the World Wide Web (WWW). The analysis is followed by a discussion on the relative merits of each strategy. The chapter ends with a summary and conclusions.

TECHNOLOGY TRANSFER

The themes and objectives of user interfaces for all have influenced both the research community (e.g., precompetitive collaborative research and development [R&D] efforts), as well as the industry (e.g., see chaps. 16, 17, this volume). Though the respective technologies have improved and matured substantially since their conception, there are still issues pertaining to their adoption and use by the wider software community. A recent study on the diffusion of Java,[1] for example, emphasizes this point by indicating that, at present, Java is still building momentum.

A relevant question, therefore, is "who" can adopt, internalize, and further develop technologies for user interfaces for all, as well as "how" this can be achieved.

A recent study on the diffusion of Java, for example, emphasizes this point by indicating that, at present, Java is still building momentum. This is evident from the fact that, at the time the quantitative measures were taken, the quantity of Java resources were continuing to increase in the official directory for Java, Gamelan. This implies that as more people learn the Java programming language, the individual categories of resources containing useful development tools are also growing (Regan, 1998). The turning point for the wider adoption of Java, however, is likely to be the size, range and scope of powerful applications built with Java, so as for the technology to continue its upward climb.

Relevant questions, therefore, are "who" can adopt, internalize, and further develop technologies for user interfaces for all, and "how" such a result can be achieved. In other words, we are not only interested in establishing potential sources and recipients of innovation, but also in the mechanisms which need to be in place, in order to initiate, sustain and facilitate successful technology transfers.

---

[1] It should be noted that reference to Java here implies reference to the concepts that underpin the language (e.g., platform independence) and constitute an advance toward user interfaces for all.

The rest of this section seeks to answer these two questions. Specifically, in the first part of the section we consider "who" may be the sources and recipients of transfers of technologies for user interfaces for all, while in the second part we examine critical mechanisms to facilitate such transfers and their relative advantages and disadvantages.

Sources and Recipients

With regard to the first question, there are three main alternatives for the wider adoption of user interfaces for all, beyond the technological feasibility stage. The first option is that industries servicing consumer markets, which are currently deprived from uninhibited access to the information-intensive environment (i.e., assistive technology industry), decide to lead the way toward either the provision of tools for user interfaces for all, or the integration of user interfaces for all principles into available user interface software and technology.

A second alternative is that the mainstream information technology (IT) industry, and in particular the sector concerned with the development of user interface software tools, progressively adopts the principles of user interfaces for all and delivers them as commercial components—either low-level components (e.g., libraries of interaction elements, toolkits) or higher level ones (e.g., tools for interface development, reusable repositories).

Finally, a third plausible option would be that the concept of user interfaces for all is adopted and further pursued in research organizations, or research consortia, and delivered to the industry as a de facto standard, or encapsulated in public domain software. This has been a very successful strategy characterizing the diffusion of several innovations in the software industry (cf. the Mosaic browser). In what follows, we consider each option in turn.

The Assistive Technology Industry.   The assistive technology industry is an engineering sector concerned with the design and manufacturing of products intended for use by disabled and elderly people. This industry can be characterized as being dominated by small- and medium-size enterprises, addressing local, regional, sometimes national, and occasionally international markets. Despite their small size, the vast majority of these firms are multiproduct firms and usually do not limit themselves to assistive technology goods, but produce them along with conventional ones.[2] Thus, a degree of complementarity between the assistive technology goods produced and conventional/traditional product lines may be expected, although this need not necessarily be so. From this point of view, specialization in production is not very widespread.

The companies tend to specialize in particular sectors of the assistive technology market, and usually address rather narrow market segments. As a result, the

---

[2]This does not mean that there are no companies that are exclusively focused on producing assistive technology products.

assistive technology market is very fragmented, which is a suboptimal situation for all parties concerned. Competition among producers is limited, and tends to concern the product differentiation, rather than the price. The oligopolistic structure of the market seems to predominate the operation and behavior of the relatively small number of companies that are active in the field. Besides the small size of the market, a possible reason why it remains oligopolistic may be attributed to the interventions of third-party payers (e.g., insurance organizations), or other intermediary organizations. These are individuals or organizations that have been given by law the task to assess disabled or elderly people and prescribe solutions. Such organizations prefer to deal with a few of the larger companies rather than with many small ones; hence, the oligopoly.

We consider it highly unlikely that this industry will engage in the effort required to develop and launch products and services based on the premises of user interfaces for all. This is due to several structural characteristics prevailing in this industry, which have been identified and elaborated elsewhere (Vernardakis, Stephanidis, & Akoumianakis, 1995, 1997). We provide only a brief account of the more important of these factors here.

One critical determinant of the capability of a sector to adopt and internalize the principles of user interfaces for all is associated with the propensity of the actors involved to innovate. The previous studies, among others, have concluded that not only is this condition not met in the vast majority of the assistive technology actors, but, more important, the assistive technology industry is generally characterized by an unfavorable environment for industrial innovation. Traditionally, it has occupied a position of a technology recipient rather than a technology-producing sector, and has been found to have both a low innovation-generation rate and a low innovation-adoption rate.

The industry's capacity to identify innovation opportunity, and subsequently turn it into new products and services, is constrained by several structural characteristics. These include the small size, underfinancing conditions, and multiproduct nature of firms, the oligopolistic structure of the industry, the prevalent competitive strategies, which foster a posteriori adaptations rather than product differentiation, and third-party interventions. The latter turns out to be a distinctive characteristic of the assistive technology industry and a serious impediment to innovation, as it results in a revealed (or filtered) demand, with severe implications on the behavior of firms. For a more detailed account of these factors, the reader is referred to Vernardakis et al. (1995).

From the preceding analysis, it can be concluded that the likelihood of assistive technology industries providing tools for and supporting user interfaces for all is substantially hindered, given the current composition and structure of this sector.

The Mainstream Industry.   An alternative would be that mainstream industry either delivers *design for all* tools (in the form of libraries, toolkits, or other type of product) or integrates design for all principles into emerging technologies (i.e., either middle-ware components, or high-level tools). Indeed, this would be a plausible option, as the mainstream IT industry is known to be highly innovative and technically capable to advance the current generation of user interface software and technology. However, the likelihood of this option materializing is again low, unless vendors obtain access to relevant research results and commit to embedding the principles of user interfaces for all into subsequent releases of existing product lines (e.g., operating systems, tools, programming environments).

Currently, this is evidenced by the presence of major software vendors such as Microsoft, IBM, and Sun in the field of assistive technology. These companies, by being actively involved in the *universal design* movement, seek to provide their new technologies with reliable build-in support for *universal access*.

One way toward this state of affairs is by upgrading a de facto industry standard through collaborations across vendors. This option is currently being materialized in the context of various accessibility initiatives, such as the Active Accessibility and Java Accessibility initiatives, launched by major software vendors (Microsoft and Sun, respectively). It is important to mention that recent studies (e.g., Regan, 1998) reveal that such an option is not only a valid approach, but also meets all the conditions for successful innovations. For example, the study by Regan, which investigates the diffusion of Java 1 year after its introduction, provides both qualitative and quantitative evidence as to why Java is favorably received, as well as why it its deemed to be a successful innovation.

Research Institutes or Consortia.   The third alternative is the undertaking of the responsibility for the wider adoption of user interfaces for all by research institutes, or research consortia of partners involving academic/research organizations and industry. An example of this type of exploitation in the past has been the X Consortium and the development of the X Windows System, which introduced a de facto standard in the mainstream IT. A similar approach for user interfaces for all would require that the technical features of the resulting products and accompanying services would be of the level and quality needed to facilitate a new de facto industry standard.

This is in fact a realistic alternative that could lead toward the uptake of user interfaces for all, if supported appropriately. To this end, the role of nonmarket institutions, in the broader sense of the term, including government and transnational decision-making bodies (e.g., European Union, National Science Foundation) is crucial. Their role is primarily in funding developments, promoting awareness, as well as establishing directives, legislative acts, and standards for ensuring the diffusion of technology and its adoption by a wide proportion of the industry. It follows, therefore, that nonmarket institutions can contribute consciously and explicitly toward the creation of an environment favorable for *designing for all,* by acting as a catalyst for overcoming the technological threshold factors associated with the availability of appropriate infrastructure (i.e., networks, terminals, etc.) to communities of end-users.

Mechanisms for Technology Transfer

It is evident from the previous discussion that, if the concept of user interfaces for all is to materialize and serve the design for all principle in human–computer interaction (HCI), this is likely to involve a purposeful plan of action characterized by interaction- and collaboration-intensive technological development, and bidirectional transfers among capable and willing parties. In this context, the objective should be in the direction of transfer of know-how and know-why, as opposed to mere embodied formations of technology. In general, such technology transfers may be facilitated through various mechanisms, some of which are relatively simple (e.g., licensing, technical advice, technical support, contract of R&D), whereas others are more advanced and effort demanding (e.g., cooperative R&D, joint venture R&D agreements, joint ventures aiming at keeping partners informed, large–small firm agreements).

Thus, it is worth considering which mechanisms from the preceding range stand a good chance of delivering the intended solutions (e.g., tools for user interfaces for all). We propose that this is done, first by characterizing the type of products foreseen under user interfaces for all (by assessing them against a small set of technological criteria) and, subsequently, by considering the suitability of each mechanism against the requirements posed by user interfaces for all. The criteria to be considered are listed in Table 27.1. A summarizing account of the user interfaces for all case for each of these criteria is depicted in Table 27.2.

An important conclusion from Table 27.2 is that any contemplated technology transfer effort toward user interfaces for all should involve recipients that are both willing and committed to adopt, internalize and deploy the technology. It follows that the transfers required to facilitate user interfaces for all require intensive interaction and collaboration among the relevant parties, toward learning by doing, transfer of know-how and know-why. This, by itself, renders relatively simple forms of technology transfer unsuitable for user interfaces for all, because they do not contribute toward the aforementioned requirements (Vernardakis et al., 1997).

Instead, continuous and persistent reliance on such mechanisms may result in permanent technological disadvantage. Therefore, the plausible range of mechanisms, in the present context, includes only those that are classified under the category of advanced technology transfer mechanisms (cooperative R&D, joint

TABLE 27.1
Basic Analytic Criteria

C1:	Level of technology involved
C2:	Type of technological innovation involved
C3:	Size and technological capabilities of firms to be involved
C4:	Nature and degree of technical character of products to be developed
C5:	Role of nonmarket institutions

TABLE 27.2

Analysis of User Interfaces for All Based on the Selected Set of Criteria

Criterion	The case of User Interfaces for All for Each of the Selected Criteria
C1	In the case of user interfaces for all, the technology involved is primarily high and radically changing. To develop tools for user interfaces for all, it is required that a basic research phase has been executed, demonstrating clearly the technical feasibility of the results to be delivered.
C2	The innovation concerned is mainly in software tools. Although small- and medium-size enterprises (SMEs) have been found to be capable of pursuing innovation opportunities, when the latter are in software, the structure and composition of the assistive technology industry renders the size of firms involved a prohibitive factor, necessitating that large firms become engaged in the collaboration effort.
C3	Both the source and the recipient should be above a certain technological threshold and competent in the technology under consideration. The recipient should be a large firm, active in the mainstream user interface tool development industry, and capable of investing the resources required to integrate technology into mainstream software components. Although we do not foresee any reason restricting the range of organizations that may act as a source, it is likely that R&D institutions are the best candidates for the reasons discussed in the previous sections. Notwithstanding this, there could be a possibility for a high-technology SME to act as the source (large–small firm agreements).
C4	With regard to the type of products to be produced through the collaboration, they are typically strongly standardized. It is also the case that there should exist a strong degree of complementarity with existing product lines.
C5	Extremely important and necessary.

venture R&D agreements, joint ventures aiming at keeping partners informed, and large–small firm agreements). All of these perform well against the identified requirements of user interfaces for all. The selection of the most suitable option is rather an attribute of the individual case.

Strategies for Integrating Principles of User Interfaces for All Into Mainstream User Interface Architectures

The success of user interfaces for all and the development of spin-off technologies will critically depend on the strategy chosen for integrating accessibility functionality with the target environment or application. The scope of potential approaches ranges from a full integration of accessibility functions into the application, to add-on tools, or to the distribution of responsibilities among applications, interaction platforms, and intermediaries. In what follows, we present three alternative exploitation strategies that cover the integration of the principles of user interfaces for all with (a) prevalent user interface development techniques, (b) componentware technologies, and (c) Internet technologies. Subsequently, we discuss relative advantages and disadvantages.

Integration With Prevalent User Interface Development Techniques.
The first exploitation strategy targets the user interface tools industry, and aims to consider the likelihood of integrating user interfaces for all principles in prevalent mainstream user interface development techniques. Some of the techniques considered have already been supported in commercially available user interface tools, others are embodied in tools that are currently available as public domain software, whereas yet another cluster is still in prototype versions. In considering these techniques, our objective is to assess their suitability for designing for all in HCI. This means that we do not seek to compare them against any particular implementation approach underpinning user interfaces for all (see chaps. 13 and 16, this volume; see also Part VI "Unified User Interfaces"). Instead, we wish to consider whether or not, and to what extent, the specific techniques considered make provisions for the fundamental requirements of *encapsulation* of alternative interactive behaviors and *abstraction*, which are necessary to facilitate designing for all in HCI.

Currently, there are various interface development techniques and tools (Myers, 1995). For the purposes of our assessment, these are classified under six distinctive categories, namely presentation-based, physical-task-based, demonstration-based, model-based, abstract objects and components, and declarative fourth-generation languages (4GLs). Presentation-based techniques include graphical construction tools, such as VisualBasic™ and TAE plus™ (TAE Plus, 1998).[3] Physical-task-based techniques lead to a specific sequence of user actions and include TAG (Reisner, 1981) and UAN (Hartson, Siochi, & Hix, 1990). Demonstration-based techniques are similar to graphical construction methods with the exception that they allow the interactive definition of a physical interface instance through an example, or a demonstration. Demonstration-based techniques have been embedded in systems such as Peridot (Myers, 1988), Pavlov (Wolber, 1996), and DemoII (Fischer, Busse, & Wolber, 1992). Declarative 4GL methods are typically found in some user interface management systems (UIMSs), such as SERPENT (Bass, Hardy, Little, & Seacord, 1990) and HOMER (Savidis & Stephanidis, 1998). Abstract objects/components are techniques that support alternative physical realizations through either object abstractions, such as metawidgets (Blattner, Glinert, Jorge, & Ormsby, 1992) or componentware technologies, such as Active X™ by Microsoft and JavaBeans™ by SunSoft. Finally, model-based techniques represent a more recent effort based on the notion of generating interactive behaviors from suitable models about users, tasks, platforms, environment, and so on. A comprehensive retrospective account of model-based technology can be found in Szekely (1996). Examples of model-based tools include HUMANOID (Szekely, Luo, & Neches, 1992) and MASTERMIND (Szekely, Sukaviriya, Castells, Muthukumarasamy, & Salcher, 1995).

---

[3]For more information, see: http://www.cen.com/tae/

Because the key ingredients for designing for all in HCI are encapsulation of alternative interactive behaviors and abstraction (see chap. 19, this volume), it follows that development methods closer to the physical level of interaction (e.g., presentation based, physical task based, and demonstration based) are less effective. On the other hand, techniques that focus on higher level dialogue properties and offer mechanisms for articulating alternative interactive components stand a better chance and could be considered as candidates for integrating principles of user interfaces for all (see Fig. 27.1).

Integration in Emerging Distributed and Component-Based Technologies.
A second market strategy could be seen in the integration of the principles of user interfaces for all in emerging distributed and component-based technologies (see abstract objects/components option in Fig. 27.1). At present, there are several competing technologies for distributed and component-based computing including JavaBeans™ (by SunSoft), DCOM™/Active X™ (by Microsoft), and CORBA (by OMG).

It is important to mention that these technologies, in their current version, hardly interoperate and most of them are still subject to further development. Moreover, some of them are explicitly targeted to supporting distributed computing (e.g., CORBA), whereas others are more suited to component-based development (e.g., JavaBeans™).

The main difference between distributed and component-based technologies is that, whereas the former offer a model for breaking down software into concrete components, which may offer remotely accessible services, the latter

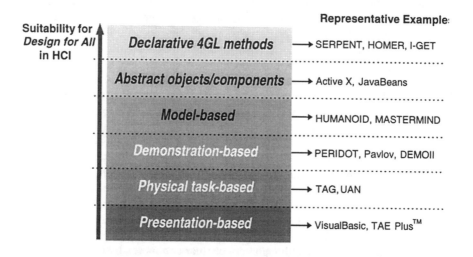

FIG. 27.1.   Integration of alternative access methods in the UIMS model.

focus on breaking down the interactive aspects of applications into meaningful pieces of binary code, which may be combined into a single software package. In the recent past, there has been only one research effort in the area of user interfaces, namely Fresco (X Consortium, 1994), which combined distributed computing and componentware to provide support for distributed component-based interfaces.

Integration With Internet Technologies.   A third and equally interesting exploitation strategy is to deploy user interfaces for all in such a way so as to facilitate Internet accessibility. As the WWW is progressively becoming a pervasive communication infrastructure, both within and across enterprises, the provision of user interfaces for all is of highest relevance, in order to facilitate access to the large information spaces distributed across the network. In the paradigm of user interfaces for all, the WWW represents yet another platform to be integrated. One way for improving the accessibility of WWW-based interactive applications is described in chapter 25 of this volume. In addition to considering accessibility in the context of specific Internet applications (such as the WWW), it is equally important to deploy accessibility features in emerging software technologies likely to have a global impact. One such technology is Java, which is currently being extended to facilitate a range of accessibility services. Java Accessibility refers to a collection of tools that are provided as part of the Java Development Kit (JDK) to facilitate the construction of accessible Java applications. Java Accessibility resembles many of the properties of user interfaces for all: for example, Microsoft's Active X and the pluggable look-and-feel architecture of Java, built on concepts similar to those that have given rise to the *dual user interface* concept in the HOMER UIMS (Savidis & Stephanidis, 1995) and the transformable user interface architecture of the unified user interface development paradigm (see chap. 20, this volume). All of them foster the principle that accessibility should no longer be considered as a mere translation of available visual manifestations, but as a requirement pending a generic solution.

Assessment.   The three integration approaches just outlined can be evaluated with respect to a small set of exploitation-oriented criteria, depicted in Table 27.3.

From Table 27.3, it follows that user interfaces for all constitutes a realistic approach to facilitating design for all in HCI, as it can be effectively deployed to all three sectors examined. Any effort to address the mainstream user interface tool development sector should aim to ensure compliance with existing de facto standards. Moreover, as already pointed out, such an effort should be targeted toward techniques that exhibit abstraction and encapsulation capabilities, as these are closer to the user interfaces for all development paradigm. By implication, this recommendation precludes alternatives, such as presentation-based and demonstration techniques, which although very popular commercially (e.g., Visual Basic, TAE Plus), are less suitable for designing for all in HCI.

TABLE 27.3
Assessment of the Alternative Exploitation Strategies

Exploitation Criterion/Strategy	User Interface Tools Industry	Distributed/ Componentware	Internet Accessibility
Market size addressed	Large	Medium/Increasing	Rapidly increasing
Development effort	High	Medium/High	Low/Medium
Tool/Application variety to be served	Low	Few	Low
Dependency on technology provider	High	Low/Medium	Low

The case of radically changing and emerging technologies, such as, for example, distributed computing and componentware, offer a better opportunity for user interfaces for all. This is largely due to the fact that mature solutions have yet to be consolidated, as the case may be in the tools development sector. As a result, there is a window of opportunity in the distributed and componentware market, provided that user interfaces for all principles exercise the required influence on subsequent releases of evolving and emerging de facto standards.

Finally, the third exploitation strategy relates more to emerging Internet technologies that are likely to deliver a new interaction paradigm for a wider range of functionality, targeted at communication and cooperation in the context of the emerging information society. As in the case of distributed and componentware computing, Internet technologies (e.g., HTML, scripting languages) offer very promising opportunities for appropriating the benefits of user interfaces for all. This is not only due to the fact that the driving technologies are still emerging, and thus can be influenced, but also due to the target consumer base, which extends far beyond the traditional "business" users, as well as the anticipated level of residential demand for products and services.

CREATING A FAVORABLE ENVIRONMENT

Thus far in this chapter, we have provided an account of "who" can adopt the principles of user interfaces for all and "how." We have discussed the opportunities offered by alternative industry sectors, the mechanisms needed to facilitate technology transfer, as well as possible exploitation strategies. A catalyst, however, for the success of the options elaborated so far is the degree to which the overall industrial environment is favorable to a shift toward user interfaces for all. We propose to address this issue by briefly examining two policy variables, namely legislation and standardization, that are known to be extremely important in establishing favorable conditions for the diffusion of social innovations. In the context of our treatment, such policy variables are also important for another reason, which is the fact that, under certain conditions, they can provide prime means for induced innovation (Vernardakis et al., 1995).

Legislation

Legislation can be used either to induce innovation, or to ensure sustained diffusion of an innovation. An example of legal intervention to induce innovation is the case of the Americans with Disabilities Act in the United States, which has influenced major software vendors and communications providers to adopt strategies toward increased accessibility of core technologies and services. On the other hand, an example of legislation being used to facilitate diffusion of an innovation is the case of alarm telephones service diffusion in Finland (see also chap. 28, this volume).

In the case of user interfaces for all, as well as the more general case of design for all, legislation seems to be relevant insofar as it can initiate and sustain developments toward technological improvements (e.g., induced innovation) that facilitate these objectives. As pointed out, this has already been the case in the United States with the introduction of the Americans with Disabilities Act in 1993 and the Telecommunications Act of 1996. However, elsewhere in the world similar progress is not observed. Before elaborating the international scene, it is perhaps appropriate to briefly review what type of legislation is in question and what the end results that can be expected are.

Europe. In Europe there are very few clauses that are in effect and explicitly aim to advance the perspective of design for all in the context of information and telecommunications technologies (ITT). As an example of such legislation, we may refer a new open network provision (ONP)/voice telephony directive, where it is stated that "Member States shall, where appropriate, take specific measures to ensure equal access to and affordability of fixed public telephone services, including directory services, for disabled users and users with special social needs." (Article 8, Directive 98/10/EC of the European Parliament and of the Council on the Application of Open Network Provision to Voice Telephony and on Universal Service for Telecommunications in a Competitive Environment).

In addition, there is an important "soft law" example. The Commission has issued a communication entitled "Equality of Opportunity for People with Disabilities—A New European Community Disability Strategy." It is related to the Resolution of the Council and the Representatives of the Governments of the Member States meeting within the Council of December 20, 1996 on equality of opportunity for people with disabilities. According to the communication, the Commission is actively interested in exploring the possibilities for harnessing all aspects of the information society in the achievement of equal opportunities for people with disabilities and in improving their living and working conditions. It should also be noted that these questions are generally discussed in the Commission's Green Paper on Living and Working in the Information Society: "People First."

Another example of useful, though limited legislation at the European level is the Amsterdam Treaty, according to which the Council may take appropriate action to combat discrimination based on gender, racial or ethnic origin, religion or belief, disability, age, or sexual orientation. Although the meaning and interpretation of the new rule is open, it expresses what a European social model should be in this respect.

In addition to the aforementioned attempts at introducing legislation in Europe, we should also point out the efforts and ongoing work of COST 219,[4] which has tried to promote the principle of accessibility so that it is accounted for in relevant European directives. COST 219 has also put forward and continues to support the recommendation for including the disabled and elderly people as target user groups in universal service obligations in telecommunications.

Asia.   In Asia the situation is similar to Europe. Though there are various examples of legislation, very few of them address specific requirements in the context of the emerging information society. The law of the People's Republic of China on the Protection of Disabled Persons gives an overview of how disability is understood in the region. The law stipulates both general rights and specific assistance. According to Article 3, disabled persons shall enjoy, on an equal basis with other citizens, rights in the political, economic, cultural, and social fields, in family life and other aspects. The citizenship rights and personal dignity of disabled persons shall be protected by the law. Discrimination against and insult of disabled persons and commission of harmful acts against disabled persons shall be prohibited. According to Article 4, the state shall provide disabled persons with special assistance by adopting supplementary and supportive methods, with a view to alleviating or eliminating the effects of their disabilities, as well as removing social and built environment barriers and ensuring the fulfillment of their rights.

In Japan, a comprehensive legislation includes the Law for Employment Promotion for people with disabilities. The legislation defines specific measures for its implementation. There is also an employment quota rate for the physically disabled, calculated in a complex way. The law defines payment of an adjustment allowance for employing physically disabled people. Moreover, policy measures seeking to provide livable environments for all have been declared by the Ministry of Construction along with a new Accessible and Usable Buildings Act (1994). Although it is not yet obligatory to abide by the Act (because it does not replace the existing Building Standard Law), local governments are asking building owners to observe the concepts, both for new construction and remodeling.

Another crucial legislation has been initiated by the Japanese government. During the past 2 years, the Japanese Ministry of International Trade and Industry (MITI) has been working to set up guidelines for the universal design of products. They need further development to be workable, but MITI recognizes their potential as a basis for standardization at international level, and has proposed to the

---

[4]For more information, see: http://www.stakes.fi/COST219/

Committee on Consumer Policy (COPOLCO) of the International Standards Organisation (ISO) that the concept of guidelines should be established as an international standard. The proposal was accepted and discussion has already begun.

The United States. A landmark of antidiscrimination legislation is the Americans with Disabilities Act (ADA). It was signed into law by President Bush on July 26, 1990. ADA should be seen in the context of other legislation. In particular, Section 504 of the Rehabilitation Act, enacted in 1973, prohibits programs supported by the federal government from discrimination against persons with disabilities. Other important legislation includes the Hearing Aid Compatibility Act, which requires telephones to allow inductive coupling to hearing aids, the Telecommunications Accessibility Enhancements Act, which requires the Federal Telecommunications System to be accessible to hearing-impaired, speech-impaired, and deaf persons, and provides for relay service and text-to-text communication, and Section 508 of the Rehabilitation Act, which requires governmental computer systems to be accessible to disabled users.

On February 8, 1996, the President of the United States signed the Telecommunications Act of 1996. Section 255 provides that a manufacturer of telecommunications equipment or customer premises equipment shall ensure that the equipment is designed, developed, and fabricated to be accessible to and usable by individuals with disabilities, if readily achievable. Whenever either of these is not readily achievable, a manufacturer or provider shall ensure that the equipment or service is compatible with existing peripheral devices or specialized customer premises equipment commonly used by individuals with disabilities to achieve access, if readily achievable.

The Architectural and Transportation Barriers Compliance Board (Access Board) is responsible for developing accessibility guidelines in conjunction with the Federal Communications Commission (FCC) under Section 255(e) of the Act for telecommunications equipment and customer premises equipment. The Access Board has issued final guidelines for accessibility, usability, and compatibility of telecommunications equipment and customer premises equipment (published in the *Federal Register,* February 3, 1998).

Remarks and Reflections. The preceding brief review is indicative of the kind of legislation that is currently in effect around the world. One prominent remark with regard to the clauses relevant to ITT is that they do not always seek to attain the same objective. Thus, generalizing across regions, we can classify existing legislation as aiming to improve accessibility through: (a) the promotion of design for all, which requires that accessibility is taken into consideration in the initial product design, or (b) providing adaptations or alternatives as standard options, available on request from a normal vendor, or (c) attaching into an existing product special assistive devices from a third-party vendor, or (d) customizing the product (Thorén, 1993).

In the future, the objective should be to establish legal interventions that are more explicitly targeted to the first option, which seeks to promote design for all.

However, it is important to note that the existence of legislation is not sufficient. It must be implemented and made to "live." There are different traditions and models as to how this can be done. One approach is to establish general principles, rules designating goals, or framework-type legislation. The shortcoming of such an approach is that the result may be of limited use because it can be so abstract that it is not binding or meaningful in individual cases. On the other hand, it can be used to raise awareness, establish a basis for a more detailed regulation, or provide a guide for interpretation in different cases. An alternative option is to strive for detailed and binding clauses that require limited (if any) interpretation. Detailed legislation can be effective, but it needs sufficient enforcement, monitoring, and supervision to become an agent of change. Such conditions may not be met in sectors of the industry characterized by radical technological change and innovation.

Standardization

The setting of standards is not associated in the literature with the encouragement of innovation, quite the opposite in fact (Vernardakis et al., 1995). In general, standards tend to reduce the range of technological options explored at a given moment. They may be instrumental in the establishment of situations of "lock in" where future development options are preempted by the dominance of a single trajectory (OECD, 1990). However, it may be argued that in contrast with the general case, under certain conditions and specific situations, not only do the negative impacts of standards on innovation disappear, but also innovation is actually supported and induced. The basic condition is the definition of functional standards. Such standards ensure the requirement for technological variety and thus safeguard all promising technological development. However, functional requirements have to be defined in ever greater detail and are thus bound to run against the wishes of manufacturers who would want standards as narrow as possible so that they are very close to match the specifications of their own products.

In the area of ITT, standards are usually developed either through processes of consensus building, or through industrial market forces. They seek to create market opportunities through harmonization and interoperability, providing the basis for achieving critical masses required for a broad market penetration. Standards can improve the acceptance of products in the market, because buyers can better predict product characteristics or quality features, or can expect that products with those features will be available for some foreseeable period of time. This predictability is an important prerequisite for protecting customers' investments in equipment, but also for the investments related to introducing the technology in organization and for building up the human resources and qualifications required to operate it. This is particularly true for ITTs due to their high level of complexity and their pervasive use in practically all fields of work and private life.

In the area of user interfaces and HCI, there are mainly two kinds of standards, namely those developed by industry (i.e., de facto standards) and those developed by official standardization organizations. Whereas industry user interface standards (in the form of user interface style guides) are often related to concrete software components or programming interfaces, providing concrete rules for the application of those components, official standards in this field mainly provide principles and guidelines that are largely technology-independent (cf. Dzida, 1997).

There have been several ongoing efforts aiming to establish national or international standards. Examples of international standards include the ISO FDIS 9241 (1998) (Ergonomic Requirements for Office Work With Visual Display Terminals), ISO CD 13407 (1996) (Human-Centered Design), ISO CD 14915 (1998) (Multimedia User Interface Design), and so on. A review of the relevant published material indicates that none of these standards makes any explicit provisions for accessibility. Only recently has a new work item proposal been accepted within ISO TC159 SC4 committee (ergonomics of human–system interaction), which will produce a technical report as a basis for a future standard on accessibility. Additionally, a special chapter on accessibility has been defined within the HFES/ANSI (1997) usability standard, which is currently being developed.

Outside the ISO, there are a number of standards and guidelines, usually at national levels, that address accessibility. The Nordic Guidelines for Computer Accessibility (Thorén, 1993) are a well-known Scandinavian standard that addresses both hardware and software aspects. In terms of application design, it focuses on facilitating the use of assistive technologies, such as screen readers, as well as the adaptability and individualization of the user interface. Further accessibility-oriented activities can be found in Bergman (1997).

Particularly in the United States, various sets of accessibility guidelines have been developed especially by large IT manufacturers, partially due to the legal requirements that have been established in the United States over the past years (e.g., ADA). IT manufacturers have also developed accessibility-enabled technologies (e.g., accessibility application programming interfaces [APIs]) that allow third-party developers to construct assistive tools that make use of the APIs provided by the operating system, or a programming language. Examples include Microsoft's Active Accessibility technology and Sun's Java Accessibility.

With the evolution of the Web, accessibility guidelines (see also chap. 26, this volume), mainly for writing accessible Web pages, have been developed by universities, user organizations, and individuals to fulfill the requirement of the community of users with disabilities. The World Wide Web Consortium (W3C)[5] is an organization with numerous members from industry and academia, which, among other things, develops and promotes standards related to Web technologies. The W3C launched the Web Accessibility Initiative (WAI)[6] in April 1997, with the aim to collect, consolidate, and further develop guidelines for Web accessibility. Within this initiative, individual working groups have developed three sets of accessibility guidelines for Web-related issues, namely *Page Authoring*, *User Agents*, and *Authoring Tools Guidelines* (see World Wide Web Consortium, 1998).

---

[5]For more information, see: http://www.w3.org

[6]For more information, see: http://www.w3.org/WAL

Having briefly reviewed the state of the art in standardization and accessibility, it can be concluded that, although some important initiatives have emerged, they are still at an infant stage. Thus, the overall environment for accessible interactive software, though improving, is far from being considered as favorable. The challenges are many and remain complex. A compelling requirement is to broaden the scope of accessibility and exercise influence not only on user interface standardization committees, but also on other bodies.

Table 27.4 summarizes the results of a preliminary analysis of the potentially relevant standardization bodies and committees that could be encouraged to address accessibility as a main theme. Introducing accessibility-related aspects into these standards may be achieved either through refinement of existing drafts, or by initiating new work items, such as that recently introduced in the context of ISO 9241/TC 159/SC 4.

In any case, it is important to note that future efforts focusing on accessibility and standardization should progressively adopt the view that accessibility of interactive computer-based systems should be considered as a nonfunctional quality attribute, just as portability, reusability, scalability, modifiability, and so forth. Such a view is useful because it provides a global perspective on accessibility, thus realizing the whole range of implications of accessibility on the process of designing, the techniques, and the architectural underpinnings of the resulting interactive system.

From the preceding discussion, it follows that, despite the potentially useful role of standardization in creating an industrial environment favorable to user interfaces for all, this is not currently the case, due to the lack of eminent principles from major standards.

SUMMARY AND CONCLUSIONS

This chapter has investigated some policy issues that shed light into the question of "who" is likely, capable, and willing to adopt, internalize, and further advance

TABLE 27.4
Standardization Committees Where New Items on Accessibility Could Be Proposed

- ISO TC159 Ergonomics
- ISO/IEC JTC 1 Information Technology/SC7 Software Engineering
- IEEE P2001 Web Page Engineering for Intranet/Extranet Environments (well-engineered Web page guidelines)
- Special Joint Working Group IEC/TC3-ISO/TC10 (SJWG/13) Future standardization needs in the field of documentation
- ISO TC46 Information and Documentation
- IEEE P1063 Software User Documentation
- ISO/IEC JTC 1 Information Technology/SC 34 Document Description and Processing Languages
- ETSI STC HF 2 Human Factors for People With Special Needs

technologies in the direction of user interfaces for all, and "how" this is likely to be achieved. Such a discussion could not be informed by empirical or statistical evidence, as we are still very early in the life cycle of such technologies. Instead, our intention was to build on existing experiences and examine some of the conditions that are needed to push these technologies out of the research laboratories and into the development processes of the wider software industry.

From our analysis, several conclusions may be drawn. First of all, technologies supporting the aims and objectives of user interfaces for all in HCI are progressively being adopted by an increasing number of the mainstream industry. Initial studies on the diffusion of these technologies indicate that the principles being advanced by user interfaces for all (e.g., portability, platform integration, specification-based user interface development techniques, emphasis on designing rather than programming interactions, etc.) are being received very favorably by the respective social systems.[7] Such evidence, supported by the fact that, in the case of software technologies, the social system welcomes innovation, leads to the conclusion that, in the near future, and as the technologies mature, more users (of the technology) will progressively become accustomed to, and familiar with, their application.

This can be facilitated through suitable mechanisms for technology transfer. It is important to mention that such transfers should aim toward transfer of both codified and tacit knowledge. Codified knowledge, embodied into suitable tools, will enable users to easily grasp and become productive with a particular technology. Tacit knowledge will contribute to developing a design- as opposed to a programming-oriented focus, which is needed in order to deliver products and services of high user-perceived quality.

Equally important as the technological threshold is the capability of the technology vendors to influence wider communities, such as, for example, international standardization organizations. This has been the case with unified user interface development and both ISO and W3C-WAI. Sun is also pursuing a commitment to standards, which will eventually open up Java development to all interested parties, and ensure that tools using this technology will maintain adherence to Sun's stand on platform independence, security, and robustness. In the context of user interfaces for all, these efforts, which are still ongoing, raise an additional requirement; this is the compelling need for international cooperation to ensure knowledge sharing and exchange of experience and practice. The role of nonmarket institutions in this direction is critical.

Finally, and in order to facilitate the aforementioned targets, it is important that the available wisdom on user interfaces for all in HCI is consolidated in a manner that is accessible, sharable, and useful to the industry when it comes to internalizing and applying the respective technologies. Although this is clearly pointed out in the research agendas presented in Part VIII of this volume, it is also mentioned here as it constitutes a prerequisite for any success stories.

---

[7]For a definition of the term see Rogers (1982), where it is defined as "the boundaries within which an innovation diffused and which is composed of individuals, groups and/or organisations"

REFERENCES

Bass, L., Hardy, E., Little, R., & Seacord, R. (1990). Incremental development of user interfaces. In G. Cockton (Ed.), *Engineering for human–computer interaction* (pp. 155–173). Amsterdam: North-Holland, Elsevier Science.

Bergman, E. (1997). The role of accessibility in HCI standards. In G. Salvendy, M. J. Smith & R. J. Koubek (Eds.), *Proceedings of HCI International '97* (pp. 441–444). Amsterdam: North-Holland, Elsevier Science.

Blattner, M., Glinert, E., Jorge, J., & Ormsby, G. (1992). Metawidgets: Towards a theory of multimodal interface design. In *Proceedings of COMPSAC '92 Conference* (pp. 115–120). New York: IEEE Computer Society Press.

Dzida, W. (1997). International user-interface standardization. In A. Tucker (Ed.), *The computer science and engineering handbook* (pp. 1474–1493). Boca Raton, FL: CRC Press.

Fischer, G., Busse, D., & Wolber, D. (1992). Adding rule-based reasoning to a demonstrational interface builder. In *Proceedings of ACM Symposium on User Interface Software and Technology* (pp. 89–97). New York: ACM Press.

Hartson, H. R., Siochi, A., & Hix, D. (1990). The UAN: A user-oriented representation for direct manipulation interface design. *ACM Transactions on Information Systems, 8*(3), 181–203.

HFES/ANSI. (1997). *Draft HFES ANSI 200 Standard. Section 5: Accessibility.* Santa Monica, CA: Human Factors and Ergonomics Society.

IEC PT 61997. (1998). *New project: Guidelines for the user interface in multimedia equipment for general purpose use.* Geneva, Switzerland: International Standards Organisation.

ISO/IEC 9126. (1991). *Software product evaluation: Quality characteristics and guidelines for their use.* Geneva, Switzerland: International Standards Organisation.

ISO 9241-10. (1996). *Ergonomic requirements for office work with visual display terminals: Dialogue principles.* Geneva, Switzerland: International Standards Organisation.

ISO 9241-10 (1997). *Ergonomic requirements for office work with visual display terminals: Dialogue principles.* Geneva, Switzerland: International Standards Organisation.

ISO 9241-14. (1997). *Ergonomic requirements for office work with visual display terminals: Menu dialogues.* Geneva, Switzerland: International Standards Organisation.

ISO CD 13407. (1996). *User centred design process for interactive systems.* Geneva, Switzerland: International Standards Organisation.

ISO CD 14915. (1998). *Multimedia user interface design.* Geneva, Switzerland: International Standards Organisation.

ISO DIS 9241-13. (1997). *Ergonomic requirements for office work with visual display terminals: User guidance.* Geneva, Switzerland: International Standards Organisation.

ISO FDIS 9241-11. (1998). *Ergonomic requirements for office work with visual display terminals: Guidance on usability.* Geneva, Switzerland: International Standards Organisation.

ISO FDIS 9241-12. (1998). *Ergonomic requirements for office work with visual display terminals: Presentation of information.* Geneva, Switzerland: International Standards Organisation.

ISO FDIS 9241-15. (1997). *Ergonomic requirements for office work with visual display terminals: Command dialogues.* Geneva, Switzerland: International Standards Organisation.

ISO FDIS 9241-16. (1997). *Ergonomic requirements for office work with visual display terminals: Direct manipulation dialogues.* Geneva, Switzerland: International Standards Organisation.

ISO FDIS 9241. (1998). *Ergonomic requirements for office work with visual display terminals*. Geneva, Switzerland: International Standards Organisation.

ISO/IEC 14598, Parts 1–6. (1997). *Information technology—Evaluation of software products*. Geneva, Switzerland: International Standards Organisation.

Myers, B. (1988). *Creating user interfaces by demonstration*. Boston: Academic Press.

Myers, B. (1995). User interface software tools. *ACM Transactions on Computer–Human Interaction, 2*(1), 64–103.

OECD. (1990, October). *Technology and the process of internationalisation—Globalisation* (Technology/Economy Programme, Chapter 9, SG/TEP/1).

Regan, B. (1998). *An innovation's diffusion: The Java programming language one year after release* [Online]. Available: http://www.webreference.com/content/java/index.html

Reisner, P. (1981). Formal grammar and human factors design of an interactive graphics system. *IEEE Transactions on Software Engineering, 7*(2), 229–240.

Rogers, E. M. (1982). *Diffusion of innovations*. New York: The Free Press.

Savidis, A., & Stephanidis, C. (1995). Developing dual user interfaces for integrating blind and sighted users: The HOMER UIMS. In *Proceedings of ACM Conference on Human Factors in Computing Systems* (pp. 106–113). New York: ACM Press.

Savidis, A., & Stephanidis, C. (1998). The HOMER UIMS for dual user interface development: Fusing visual and non-visual interactions. *Interacting With Computers, 11*(2), 173–209.

Szekely, P. (1996). Retrospective and challenges for the model-based interface development. In F. Bodard & J. Vanderdonckt (Eds.), *Proceedings of 3rd Eurographics Workshop on the Design, Specification and Validation of Interactive Systems* (pp. 1–27). Berlin: Springer-Verlag.

Szekely, P., Luo, P., & Neches, R. (1992). Facilitating the exploration of interface design alternatives: The humanoid model of interface design. In *Proceedings of the ACM Conference on Human Factors in Computing Systems* (pp. 152–166). New York: ACM Press.

Szekely, P., Sukaviriya, P., Castells, P., Muthukumarasamy, J., & Salcher, E. (1995). Declarative interface models for user interface construction tools: The Mastermind approach. In L. Bass & C. Unger (Eds.), *Proceedings of Engineering for Human–Computer Interaction* (pp. 120–150). London: Chapman & Hall.

Thorén, C. (Ed.). (1993). *Nordic guidelines for computer accessibility* (Nordiska Nämnden för Handikappfrågor, NNH 4/93). Vällingby, Sweden: Nordic Committee on Disability.

Vernardakis, N., Stephanidis, C., & Akoumianakis, D. (1995). On the impediments to innovation in the European assistive technology industry. *International Journal of Rehabilitation Research, 18*(3), 225–243.

Vernardakis, N., Stephanidis, C., & Akoumianakis, D. (1997). Transferring technology toward the European assistive technology industry: Mechanisms and implications. *Assistive Technology, 9*(1), 34–36.

Wolber, D. (1996). Pavlov: Programming by stimulus-response demonstration. In *Proceedings of ACM Conference on Human Factors in Computing Systems* (pp. 252–259). New York: ACM Press.

World Wide Web Consortium. (1998). *WAI accessibility guidelines: User agent* [Online]. Available: http://www.w3.org/TR/WD-WAI-USERAGENT

X Consortium. (1994). *FRESCO™ Sample implementation reference manual* (X Working Group Draft, Version 0.7). Mountain View, CA: Silicon Graphics & Japan: Fujitsu.

28 Economics and Management of Innovation

Nikolaos Vernardakis
Demosthenes Akoumianakis
Constantine Stephanidis

This chapter aims to address some of the issues pertaining to the economics and management of technologies underpinning the development of User Interfaces for All. The focus is on characterizing the innovation involved and identifying the parameters that are likely to determine subsequent take-up and diffusion. In discussing the diffusion of User Interfaces for All, we will be guided by innovation diffusion theory, essentially Mansfield's models, as related to software innovations. Reference to comparable studies will offer an enlightening view on what has been empirically derived, while pointing out the aspects needing further investigation. The chapter concludes with an account of the accompanying measures needed to facilitate a shift towards adopting socially desirable innovations, such as User Interfaces for All.

Previous parts of this book have addressed a broad range of issues with immediate or future impact on the design and implementation of *user interfaces for all*. These contributions have followed different pathways, offering alternative perspectives toward user interfaces for all. Previous chapters offered either enlightening views on critical dimensions in the study of user interfaces for all (see Part II "Dimensions"), or perspectives on underpinning theoretical science base (see Part III "Design"), or technological insights into the current state of the art and/or emerging technological paradigms (see Part IV "Software Technologies and Architectural Models" and Part V "Evaluation"). In this chapter, we are concerned with some issues pertaining to the economics and management of *universal design* in human–computer interaction (HCI). To facilitate this, we first try to characterize the type of innovation constituted by universal design in HCI and summarize some of the benefits for the intended user communities. This helps us identify not only the characteristics pertaining to such innovations, but also some of the conditions that are critical for their uptake and diffusion.

Characterizing the Innovation

In addressing universal design in HCI, our focus is on innovations in user interface software technologies that advance our current conceptions of HCI toward universal design. Innovation in this area accounts for a broad range of challenges, including, but not limited to:

- HCI design (e.g., new methods and techniques for inquiring, unfolding, understanding, capturing, representing, and articulating design constraints) aiming to improve the process, the means for, and deliverables of design activities.
- Development tools for building integrated interactive environments (e.g., new capabilities to integrate, extend, augment, and abstract from platform-specific interactive behaviors) accessible by anyone, anytime, anywhere.
- The design, conduct, and feedback of evaluation so as to provide timely input into design.

Inevitably, such an account extends our treatment of innovation beyond technological aspects that can be articulated or embodied into software tools and environments, and claims that the shift toward universal design in HCI is critically dependent on a broad range of issues, including: (a) codifiable elements of technology and "soft science" in the form of tools, principles, and methodologies, and (b) tacit knowledge, ranging from awareness of the need for universal design, to commitment to support it, and other uncodifiable components, that cannot be immediately integrated into user interface software technologies. Such knowledge is transmitted through alternative channels (i.e., publications, tutorials, manuals, examples of best practice, interpersonal communication), but most important, through learning by doing.

The primary social group of such innovations is the Internet community, which includes application developers (including those developing digital content), tool developers, and the general user interface programming community. Moreover, the demand for such innovations, which is currently not well articulated, may be expressed through additional social groups, such as the end-users and/or nonmarket institutions (e.g., governments). It follows that, in this early phase, universal design in HCI seeks to deliver both process and product innovations, which can be subsequently improved by change agents to meet specific industry and end-user requirements.

In the past, there have been several studies aiming to characterize innovations in software. Some of these studies have investigated determinants of innovative activity in the software sector, such as the size of firms, whereas others have inquired into what the respective communities perceive as being innovative, and under what conditions they decide to adopt a particular innovation. For example, since the early period of graphical user interfaces (GUIs), commercially successful tools for user interface development were those that exhibited strong compatibility and compliance with de facto standards, such as:

- The C++ programming language in the case of popular user interface development toolkits.
- The Basic programming language in the case of Visual Basic.
- The Pascal programming language in the case of Delphi.

Such path dependencies are also evident in more recent developments, such as Sun's Java programming environment, which follows very tightly the C++ programming conventions. This leads to the observation that even though the software design and programming communities have been receptive to changes, they have developed certain standards in the form of general conceptions that determine the adoption of a particular innovation. Later on in this chapter, we try to unfold some of them in order to develop an understanding of the conditions for successful innovations in this industry.

Focus of the Chapter

Previous chapters in this book, as well as other supporting studies (e.g., Müller, Wharton, McIver, & Kaux, 1997; National Research Council, 1997; Story, 1998), provide evidence to conclude that designing interactions for the broadest possible end-user population, including people with special needs, is not only desirable, but indeed possible. Consequently, what is of interest to the present study is to unfold the factors and conditions for its adoption by, and diffusion toward, the academic and industrial communities.

To this end, the chapter seeks to build on the existing evidence and provide an alternative but complementary line of argumentation in favor of user interfaces for all and *design for all* in HCI. This is attained by discussing a range of issues, which, though nontechnological in nature, come to bear substantially on the adoption and diffusion of socially desirable software innovation, such as user interfaces for all. In doing so, we are not concerned with a particular product (i.e., Java vs. Unified User Interface development tools) or type of innovation (i.e., whether in hard/soft science, design methodology, or technology). After all, competing alternatives examined in different chapters of this book share many common characteristics. Instead, we aim to address how the accumulated technological background can facilitate the diffusion and adoption of the principles, technologies, and practices of user interfaces for all, and facilitate a "culture" for designing for the broadest possible end-user population in the context of HCI. To this end, whenever possible and appropriate, we refer to specific evidence and cases.

THEORETICAL LINKS

Economics and management of new technologies provide a disciplined approach and a consolidated body of knowledge for studying a broad range of issues pertaining to innovation, adoption, and diffusion of new technologies. In this chapter, such a perspective serves the purpose to extend the domain of inquiry into user interfaces for all beyond technological feasibility—which has been the focus of attention in previous chapters of this book—to account for additional aspects that are likely to determine both the "race" between alternative/competing technological perspectives toward user interfaces for all, and the adoption of the principles of designing for all in HCI by academic and industrial communities.

Diffusion theory is a research approach that measures how an innovation is adopted among a population (Rogers, 1982). After an innovation is introduced, it is adopted by a somewhat eccentric and/or entrepreneurial group called the innovators. This group, being slightly outside of the norm, does not possess the weight necessary to drive adoption. Change agents or opinion leaders among the social system will step in next, thereby legitimizing the innovation and opening the potential for adoption to all members of the system. The next stage in an innovation's adoption is characterized by widespread adoption until such point that the innovation has saturated the social system and growth tapers off. This process is plotted by diffusion researchers as an S-shaped growth curve (see Fig. 28.1).

In discussing the diffusion of user interfaces for all, we are guided by innovation diffusion theory, essentially Mansfield's (1968) models, as outlined in more recent publications such as the book *Diffusion of Innovations* by Rogers (1982). For a more involved account, the reader should see Paul Stoneman's (1983) book or the collection edited by Dosi et al. (1988). An innovation is defined as an idea, practice, or object that is perceived as new by an individual or other unit of adop-

The Innovation Adoption Curve

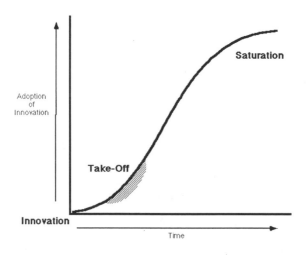

FIG. 28.1.   The S-shaped innovation diffusion curve.

tion. The model of diffusion contains five characteristics that help project an innovation's rate of adoption. Those innovations possessing most of the identified characteristics have a greater chance of acceptance by the social system. The five characteristics are summarized in Table 28.1.

Innovations diffuse through *communication channels.* A communication channel is the means by which messages get from one individual to another. The effectiveness of a communication channel decides the fate of an innovation's adoption. Mass media channels figure prominently in many diffusion processes because of their ability to carry information to a wide audience in a short amount of time. Another effective communication channel in the diffusion process is interpersonal communication. Information and advice from peers often carries more weight in an adoption decision than technical specifications or product documentation.

Equally important to diffusion is the *social system* of an innovation. The social system constitutes the boundaries within which an innovation is diffused, and is composed of individuals, groups, and/or organizations. All members of a social system (innovators, change agents, opinion leaders, early to late adopters) are joined in the common objective of seeking and spreading information about the innovation. Varying relationships within the social system occur among its members that can affect the diffusion process. Change agents and opinion leaders directly influence the course of the diffusion.

The final element in the model is *time.* Time is crucial to an innovation diffused within a social system, because the rapid communication rate speeds up the diffusion process and shrinks the amount of time necessary for sharing knowledge and for making a decision about the innovation. The innovation, then, has a small time frame within which to establish itself.

TABLE 28.1

The Five Characteristics That Determine the Diffusion of Innovations

Characteristic	Definition
Relative advantage	"The degree to which an innovation is perceived as being better than the idea it supersedes"
Compatibility	"The degree to which an innovation is perceived as being consistent with existing values, past experiences, and needs of potential adopters"
Complexity	"The degree to which an innovation is perceived as difficult to understand and use"
Trialability	"The degree to which an innovation may be experimented with on a limited basis"
Observability	"The degree to which the results of an innovation are visible to others"

DIFFUSION THEORY AND USER INTERFACES FOR ALL

The choice of the theoretical frame of reference just discussed is justified by the type of analysis allowed by the early stage of universal design in HCI. In particular, with reference to the S-curve of Fig. 28.1, which is generally accepted to depict the diffusion of innovations, user interfaces for all technologies have just started to make a presence with an expectation for growth to reach maturity. Consequently, one can hardly expect reliable quantitative measures in advance (Twiss, 1992) to determine the exact shape and type of the diffusion curve. Due to these circumstances, comprehensive studies on the diffusion of relevant technologies are hardly available. Consequently, in the present chapter, Rogers' (1983) account of innovation diffusion is used as an analytical and qualitative aid for the treatment of the diffusion process and the numerous elements involved.

Another point to mention is that traditional innovation diffusion models, including Rogers' (1983), make certain assumptions on what determines diffusion, which has been criticized in the recent literature. The first is that adopters (i.e., the social system in Rogers' terminology) constitute a homogeneous social group with similar objectives, aims, and values. Though we recognize that this may not be a general case for all type of innovations, we accept this assumption, given the characteristics of the potential adopters discussed in the introduction.

A second point regarding Rogers' (1983) model is that it makes no reference to cost as a determinant of an innovation's diffusion. Nevertheless, for the purposes of our study and given the patterns that have been followed with previous innovations in software (e.g., free distribution of the Mosaic browser, free distribution of JDK by Sun, etc.), the zero cost difference assumption (Katz & Shapiro, 1994) can be maintained. This means that we assume a zero cost difference with regard to the adoption of user interfaces for all versus competing technologies.

The final remark on Rogers' (1983) model is that the five characteristics of Table 28.1 are all technological and supply-side characteristics. Demand is loosely addressed by the notion of the social system. In this chapter, we extend this view to address some of the demand-side conditions that are believed to determine the adoption of user interfaces for all versus competing technologies. This is done by pointing out some of the technological performance thresholds that user interfaces for all should meet in order to be preferred by adopters in the long run.

The Five Characteristics and User Interfaces for All

Having made the preceding clarifications, we now examine how Rogers' (1983) model applies in the case of user interfaces for all. To this effect, we draw parallels with the study by Regan (1998), which builds on innovation diffusion theory to measure how Java stacks up against the five key characteristics of successful innovations. Such an account provides a comparative measure that reveals what can be done to facilitate the diffusion of innovations in user interface software technologies aiming to provide support for design for all in HCI. The comparison has face validity, as innovations in the direction of user interfaces for all possess similar characteristics to those of Java. In particular, they both constitute innovations in software, thus they can be compared against the five characteristics of successful innovations of Table 28.1.

Moreover, they both are transmitted through similar (if not identical) communication channels and have similar (if not identical) social systems. Consequently, the qualitative measures used in Regan's (1998) study and the main results provide a useful yardstick for assessing critical elements of innovations in user interfaces for all software technologies. The important features that should characterize user interface software technologies supporting user interfaces for all are as shown in Table 28.2.

In particular, with regard to relative advantages of user interfaces for all technologies, these can be found in the areas of improved interactivity (e.g., in Web browsing), advanced technology features (e.g., extended APIs, simplicity of programming principles and style, object orientation, portability, architecture neutrality, security, platform integration), and emphasis on distributed and networked computing. As discussed in various chapters in this volume, all major initiatives satisfy the majority of these features, thus rendering user interfaces for all technologies quite strong in terms of the relative advantage characteristic.

TABLE 28.2

Desirable Features in Technologies for User Interfaces for All Versus Rogers'
Characteristics of Successful Innovations

Characteristic	Feature
Relative advantage	a) Improved interactivity in Web browsing b) Advanced technology features (e.g., simple programming language, object-orientation, portability, robustness, architecture neutrality, platform integration, security) c) Distributed/network computing
Compatibility	d) Compliance with existing values in the object-oriented environment e) Similarity to prevailing and widely used programming languages (e.g., C++) f) Compatibility with the open standards of the Internet community g) Compatibility with the expectation for free distribution of knowledge and tools through the Internet (cf. Mosaic, JDK)
Complexity	h) Compliance with prevailing knowledge and practices of intended social system i) Availability of tools (e.g., visual development, debugging) j) Resource availability (e.g., documentation, examples, applications, use experiences)
Trialability	k) Tools to promote wide experimentation l) Reference examples m) Availability of source and binary code
Observability	n) Visibility to members of the social system o) Compliance with other de facto standards (e.g., Netscape, Internet Explorer) p) Availability of archives

The compatibility characteristic stresses the necessity for user interfaces for all technologies to comply with existing values and standards in the software-engineering and user interface tool development sectors. In particular, compliance with prevailing object-orientedness of popular programming languages (e.g., C++) and tools (e.g., UIMSs) is a crucial element, and something that is consistently supported by the two approaches described in this book, namely Sun's Java Accessibility and unified user interface development. In addition, the expectation for free distribution of knowledge and tools through the Internet turns out to be a critical success factor, as shown by recent innovations. For example, in the recent past, the X-Consortium has distributed freely the Mosaic browser, as well as its source code. Similarly, today, both Sun and Microsoft distribute key technologies, such as the Java Development Kit (JDK) and the Active Accessibility Software Development Kit (SDK), respectively, free of charge.

Complexity is the third technological characteristic of successful innovations in Rogers' (1983) theory of diffusion. It generally refers to compliance with prevailing knowledge and practices of intended users, and the availability of tools and resources (e.g., documentation, examples, reference material). Among the three, the availability of tools currently seems to stand out as the main obstacle. In the past, the availability of tools has had a major impact on the diffusion of software technologies. For example, the emergence of visual programming environments changed radically the capacity, productivity, and practices of programmers (see chap. 9, this volume). Similarly, it was not until tools became available that GUIs really took off to become popular. In the case of user interfaces for all, the situation is as follows. With the exception of unified user interface development, which realized the need for tools and emphasized tool development, neither Active Accessibility nor Java Accessibility are currently being supported by powerful and well-behaved tools. This is in line with the more general observation of Regan (1998), which indicated the lack of Java tools as the primary impediment in the diffusion of the programming language in the early period.

Trialability of software technologies refers to wide experimentation and availability of reference examples, source, and binary code. None of the technologies reviewed in this book can be said to have passed the trialability test, especially when it comes to testing properties that are critical to user interfaces for all. For example, platform integration is an important feature that needs to be supported, but very few (if any) developments have been devoted to testing this property with the technologies available. One explanation is that the technologies are still maturing to reach stable versions and, therefore, cannot be subject to an exhaustive testing of their underlying technical properties. Another explanation is that the need for testing some of these properties (such as platform integration), is not yet well articulated due to the limited attention by mainstream developers.

Finally, observability refers to making the technology and the respective advantages visible to the members of the social system. This can be achieved by emphasizing adherence and compliance to de facto standards (e.g. Netscape, Internet Explorer) or by making available source code, design guidelines, example templates, and so on. Both these techniques have been used to make Active Accessibility SDK and the Sun JDK observable by the respective communities of practice.

Communication Channels

For a successful innovation, information must be easily available to all potential adopters. In the case of Java, these communication channels are staggering in their diversity and in their ability to instantaneously spread information to millions of potential adopters. Traditionally, information on an innovation spreads through interpersonal channels, and only reaches the mass media once a sizable number of adopters warrant the media's attention. The communication channels involved in this diffusion process contain three different mass media channels (mainstream, trade, and Web sites) and at least five interpersonal channels (newsgroups, list servers, personal Web sites, user groups, and traditional face-to-face communication). In the years to come, all of these channels should be exploited to raise awareness regarding the merits and outcomes of user interfaces for all.

To this effect, one can be informed from Sun's communication strategy about Java as presented in Regan's (1998) study. In the year since Java's release, 50 newspaper articles reporting specifically on the programming language have appeared in mainstream newspapers, including *The Wall Street Journal, The New York Times, USA Today,* the *Los Angeles Times,* the *Boston Globe,* and the *Chicago Tribune.* Additionally, industry trade publications and other media offering more in-depth insight into the innovation, as well as future projections, have been widely used by Sun. In particular, from May 1995 to March 1996, 218 articles with Java as the main topic appeared in computer periodicals. An additional 970 articles made mention of, or discussed Java within the scope of other related information.

In addition to the aforementioned, Web sites can constitute mass media channels in their ability to convey information to a wide audience in a short amount of time. The Javasoft site, for example, documents its innovation, offers online tutorials, publicizes the company, and announces new developments. Established Web sites (such as Yahoo[1] or Gamelan[2]) serve a similar purpose. Users know that these sites contain desired information and seek them out on a regular basis.

Finally, information about the innovation is also transmitted through interpersonal channels in the form of newsgroups, list servers, individual Web sites, and user groups. This type of communication is in electronic form, resulting in a great expansion of the traditional face-to-face interpersonal communication channels. Interpersonal channels also include informal face-to-face communications among professional and student colleagues. Such a wide information network increases the amount of information about the innovation, exploration into the innovation, and the amount of time necessary for the dispersal of that information. These communication channels are a major factor in the Java hype that has guaranteed momentum.

---

[1] http://www.yahoo.com

[2] http://www.gamelan.com

The Social System and Technology Performance

Users.   The user interfaces for all social system is made up of all members of the Internet community and in particular, Web designers, user interface programmers and end-users, regardless of their level of experience or technical knowledge. This user base constitutes the potential adopters, ultimately accepting or rejecting user interfaces for all as an innovation.

Within this system, varying relationships and statuses exist. The majority of the user base contains average Web users receiving information about the innovation coincidentally, through exploration of the Web. These users, although potential adopters, are more casual bystanders of the user interfaces for all innovation. Because of the coincidence of their knowledge, this user base is less likely to influence ultimate adoption, though important in an environment where hype equals momentum. Communication maintains a collective consciousness about the innovation within the social system. This group will most likely follow the opinion leaders' decisions to adopt, or reject.

Another set of average users actively seeks out information through Web sites, media channels, and interpersonal communication. This subset of potential adopters has not yet formed a decision to either adopt or reject the innovation. Some members of the social system seek out information and try to learn the technology or use already developed applications and prototypes. This set of potential adopters has made an initial decision to adopt on a limited basis.

Other members of the social system are experienced developers who have received enough information to discuss the various reasons for, or against, the adoption of the innovation. These members actively engage in trading viewpoints throughout the diffusion process, and are the most likely to develop products using the innovation. The decisions made by this set of potential adopters are critical to the success of the diffusion process. However, at the present time, there are many issues still pending, thus rendering immature any tentative conclusion. In particular, a critical turning point in the adoption of user interfaces for all by this social group is the realization of the compelling need for "anyone, anytime, anywhere" systems. Criteria such as portability, toolkit interoperability and nomadic access are likely to determine their choice. Consequently, the user interfaces for all community should actively engage in the provision of reference examples that clearly demonstrate the relative advantages along these dimensions. This brings about the issue of technological performance threshold as a demand-related determinant of diffusion. In other words, it is argued that user interfaces for all will not be adopted unless the technology exceeds certain thresholds.

Technological Performance Thresholds.   Technology performance focuses on what enterprises actually do in technological terms during the various stages of production (e.g., the knowledge, skills, equipment, and materials employed). When studying technology performance, it is important to consider the long-term target that needs to be sustained, rather than performance at a given point in time (Daly, 1998). Thus, people, equipment, facilities, and processes in the enterprise must embody sufficient technological knowledge and flexibility to wisely select among alternative techniques available for each step in the process, to adapt technology to meet local circumstances and changes in them, and ideally to develop new technology as and when appropriate and necessary.

Though technological performance in enterprises comprises diverse elements, what is relevant to the present study are the criteria or thresholds that characterize the capability of an enterprise to adopt, internalize and progressively improve the technology. Figure 28.2 depicts technological performance thresholds (from lower to higher) and associates them to technologies for which they are adequate.

As far as user interfaces for all are concerned, it is argued that higher order criteria than technological feasibility, such as technical reliability, economic efficiency, and efficacy, are likely to be some of the critical success factors that, in turn, will determine both adoption and diffusion in the long run. Though technological feasibility is a necessary condition toward user interfaces for all, it is not sufficient, introducing a compelling need for support measures intended to create an overall environment that is favorable to designing for the broadest possible end-user population.

Technological feasibility refers to establishing initiatives aimed at demonstrating technical possibility and viability, by offering proof-of-concept prototypes

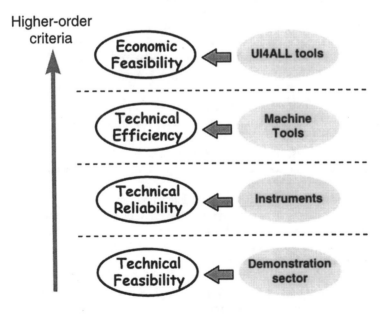

FIG. 28.2.   Technological performance thresholds for user interfaces for all in the IT&T sector.

and implementations. Thus, for example, technological feasibility is the typical outcome of precompetitive collaborative research and development projects in which emerging technologies are involved. The TIDE-ACCESS and the ACTS-AVANTI projects (see Part VI "Unified User Interfaces" of this volume and the Acknowledgments section) and their respective results are examples of early attempts toward demonstrating the technical feasibility of (some of) the principles of user interfaces for all. In a similar manner, FRIEND21 (chap. 13, this volume), though oriented explicitly toward basic research, undertook development efforts to demonstrate the principles underpinning its methodological grounds.

Whether or not technical feasibility is sufficient depends on several industry-specific parameters. For example, though it may be an attribute sufficient to satisfy the needs of what DeBresson (1991) called the "demonstration" sector (meaning state research laboratories, military research, or the medical sector), it is far from being sufficient to satisfy the requirements of the information technology and telecommunications (IT&T) sector, in terms of its capacity to provide user interfaces for all.

Technical reliability is an attribute of higher order than technical feasibility for it presupposes the latter and adds to it. Nevertheless, although it may be a sufficient criterion for a sector such as "instruments," it falls short of the expectations arising by the intention to facilitate computer-based products and services accessible and highly usable by the broader possible target end-user population across different contexts of use. Strictly speaking, even technical efficiency, which presupposes both technical feasibility and technical reliability, and would be appropriate for a sector such as "machine tools," would still be an insufficient attribute. What is really needed is economic feasibility in the long run (see Fig. 28.2), leading to versatility and economic efficiency (see chaps. 29 and 30, this volume).

Besides the technologies involved, the nature of the products and services to be offered is also important. For instance, within the broad category of consumer goods, different attributes can be used to qualify products within subcategories. Thus, in luxury goods, new utility is addressed; in mass luxury goods, user friendliness is expected; in durables, standardization is offered, whereas in subsistence goods, cost advantage is a feature.

Though econometric studies on the determination of demand behavior for IT&T products and services complying to the principles and requirements of user interfaces for all are still to be performed, one would hardly err by saying that accompanying measures should address the issues of new or enhanced utility, conviviality (user friendliness), and standardization.

In conclusion, what is important to note is that reaching higher technological performance thresholds is a long and challenging task. Studies in several countries, such as Brazil and some Asian countries, have monitored enterprise-learning processes that have taken some 20 years. Mastering a technology may require even more time. Additionally, the development of the technological capacities of entire industries is likely to be even more demanding.

Time

Time is important for the diffusion of an innovation for several reasons. First of all, it may be used as a yardstick to measure the adoption of an innovation. This was the case in Regan's (1998) study, where the time factor was used to measure the current rate of adoption as a reference point in Java's diffusion process. Additionally, time can improve the potential of an innovation, either by attracting new innovators (e.g., all those who develop professional applications that further enhance or add to the innovation) or by attracting the attention of early adopters (e.g., all those who experiment with the innovation or commit to it one way or another). Successful innovations usually exhibit both these characteristics. For example, in the case of Java, early adopters had been major actors such as Netscape Communications Corporation, Symantec, Oracle, and Borland.

What is important to mention is that in a fast-paced environment, relative newness does not last very long, thus giving early adoption a small window of opportunity. Technologies supporting the development of user interfaces for all must be able to capitalize on such a window so that they become the innovation used as the underlying architecture for interactive products and services, rather than the innovation that merely shifted the industry's focus.

IN SEARCH OF APPROPRIATE INTERVENTIONS

Based on the previous discussion, what would be appropriate intervention for the IT&T sectors to attain the required technological performance so as to make user interfaces for all a reality? This process is likely to require purposeful action, commitment, and time. The development of appropriate social and economic policies, and necessary social and economic institutions, to permit enterprises to undertake and benefit from improving technological performance appear vital to the success and benefit of any long-term efforts at technological improvement. One theory of "induced technological innovation" suggests that, if the institutions are built and adequate policies are adopted, the improvement of technological performance will surely follow. In the best case, improvement in technological performance will be the result of correct predictions about what technological change will be appropriate at each point in the process of sustainable development, and of hard work aiming to achieve those changes. It has alternatively been suggested that the opportunities provided by technological progress may in turn stimulate the development of appropriate sustainable development policies and institutions. It seems most likely that progress in technology, institutions, and policies are all inextricably interlinked, and that an explicit and energetic effort to improve technological performance will be a valuable and even a critical component of a successful development policy.

Given the paramount importance of policy setting, in the rest of this chapter we are concerned with the type, range, and scope of support measures required for IT&T to progressively attain economic efficiency in the area of HCI design for the broadest possible end-user population and, in particular, user interfaces for all.

ACCOMPANYING MEASURES

Given the preceding insights, the diffusion of user interfaces for all raises a compelling need for a targeted set of support measures intended to develop an industrial environment favorable to social innovation toward user interfaces for all. The type, scale, and scope of the support measures may vary depending on firm- and industry-specific factors, to include best practice experiments, incentives for industry to adopt and apply proven technology, as well as policy instrumentation. All of them, however, relate to the issue of technology transfer and how tacit knowledge in the form of know-how and know-why, and subsequently tools for building interactions, can progressively become embedded into novel products and services.

Types of Accompanying Measures

In a recent study (Vernardakis, Stephanidis, & Akoumianakis, 1997a, 1997b), the likelihood of design for all was assessed against different technology transfer mechanisms. The main outcome from that investigation was that design for all is likely to proliferate when advanced mechanisms are involved. The types of such mechanisms that were considered appropriate include cooperative research and development (R&D) (M_1), joint venture R&D agreements (M_2), joint ventures that aim at keeping partners informed (M_3), cross-licensing—referring to separate product markets (M_4), and large–small firm agreements (M_5).

Table 28.3 projects each of these alternative mechanisms against a small set of criteria. The criteria include the level of technology involved (C_1), the size and technological capabilities of the firms involved (C_2), and the nature and degree of technical character of products to be developed (C_3). Table 28.3 summarizes the assessment and identifies the type of products and services that could be expected under different regimes toward design for all.

In recent years, the field of HCI has experienced attempts toward design for all building directly on some of the aforementioned mechanisms or calling on such mechanisms to meet pressures introduced by legislation or other policy initiatives. Though these alternative pathways toward effective support measures may share common objectives, they differ in a number of dimensions. Some of these dimensions are depicted as criteria in Table 28.4, which compares cooperative R&D against policy-oriented interventions, such as legislation and standardization, toward user interfaces for all. As shown, each alternative may potentially turn out to be ineffective, as there are strong conditions placed on them. Though a detailed review of each of those is beyond the scope of this chapter, it is perhaps appropriate to give some examples.

Collaborative R&D.   Collaborative R&D aiming to advance technologies for user interfaces for all has been a popular mechanism for technological advancement. In particular, projects such as ACCESS and AVANTI are examples of collaborative R&D funded by the European Union, whereas FRIEND21 and MERCATOR

TABLE 28.3

Assessment of Technology Transfer Mechanisms

	C_1	C_2	C_3
M_1	High	Both source and recipient are very large	Strongly standardized; existence of significant technological complementarities
M_2	High	Large firm, international oligopolies	Not very standardized, with significant technological complementarities
M_3	Same as above	Same as above	Same as above
M_4	High	Oligopoly	Technological complementarities and/or special sectors in submarkets
M_5	High and rapidly changing	The small firm is knowledge intensive	Same as above

(Mynatt & Weber, 1994) constitute similar cases funded by Japan's Ministry of International Trade and Industry (MITI) and the National Science Foundation (NSF) in the United States, respectively. It is important to mention that cooperative R&D has its limits and does not always guarantee success (Pavitt, 1998). Though it may facilitate the consolidation of a sound R&D base, initiate proactive technological intervention, raise awareness, and establish a common ground, it seldom exceeds the technological feasibility stage. In the past, cooperative R&D projects funded by the European Union have demonstrated their capability to be at the technological frontiers by being multidisciplinary, collaborative, user-involved, transnational, and precompetitive R&D initiatives, where technological feasibility is demonstrated in selected application domains, or pilot case studies.

Though such attributes constitute necessary ingredients for socially desirable innovation, in the sense that without them the generation of such innovation may be hard to attain, they are clearly not sufficient to assure diffusion. This is due to several reasons, the most important being the fact that technological feasibility, though very important, does not necessarily imply economic efficiency and efficacy, which is what is really needed in the long run. A second aspect is that adoption and diffusion of socially desirable innovation depends on additional parameters, such as the degree to which the environment is favorable to such innovation, the commitment of firms to appropriate the resulting benefits, sectional characteristics, such as industry composition, competitive strategies, the nature of the sector (e.g., producer vs. recipient of technology), as well as prevalent practices in regulation and in particular standardization (e.g., de facto industry standards) and legislation.

Consequently, as shown in Table 28.4, the conditions considered as prerequisites for success in cooperative R&D are quite substantial. Partner selection, recip-

TABLE 28.4

Comparative Assessment

Criteria	Policy Instruments		
	Legislation	Standardization	Cooperative R&D
Target	√ Reinforcement	√ Consolidation of knowledge √ Guidance	√ Establish common R&D agenda √ Provide a solid basis for R&D √ Promote cohesion
Prerequisites	√ Demand already articulated √ Commitment	√ Solid R&D base √Timely intervention	√ Cross-industry focus √ Reciprocal investments √ Willingness and commitment √ Favorable conditions for transfers
Potential shortcomings	√ Difficult to guarantee compliance, due to • industry opposition • tendency to bypass • lack of user demand • lack of awareness	√Lock-on effect √ Appropriate recommendations √ User involvement √ Industrial participation √ Not possible in highly competitive industries	√ Exploitation capability √ Technology must be emerging √ Conditions for sources and recipients √ Conditions on mechanism used
Role of nonmarket institutions	√ Initiate and sustain √ Monitoring application	√ Funding √ Disseminate knowledge	√ Funding R&D work √ Facilitating collaboration √ Offering guidance √ Undertaking technological forecasting √ Provision of incentives √ Establishing favorable conditions

rocal investments, cross-industry focus, consortium composition and responsibilities, as well as management practices are critical determinants of a project's technical outcomes and the corresponding exploitation.

Standardization. Standardization is an alternative policy-oriented mechanism that may be used to promote use of technologies for user interfaces for all. In recent years, standardization bodies have embarked on an effort to provide guidelines and ergonomic recommendations for designing user interfaces for different user groups. Examples include the work carried out in the context of HFES/ANSI,[3] W3C, and W3C-WAI,[4] as well as international standardization bodies (e.g., the new work item on accessibility by ISO 9241/TC 159/SC 4/WG 5). All these efforts follow conventional standardization processes whereby a consolidated body of research is translated into ergonomic standards. Though final results are still pending, it is important to underline some of the effects of the sequential process that is currently in place in the vast majority of standards bodies. Typically, this results in extensive time horizons before a standard can be issued (cf. time taken for ISO 9241, 1998, and ISO 13407, 1997, to reach their final state). In certain industries, this is an acceptable life cycle. In others, such as the IT&T industry, which is characterized by rapid technological change, it may result in lengthy revisions of the original drafts, amendments, and extensions, in order to avoid rendering the standard obsolete by the time it is finalized.

Legislation. Legislation is yet another policy-oriented alternative. In Europe, there have been very few and rather naive attempts at introducing legislation as an instrument to facilitate user interfaces for all. On the other hand, in the United States, there has been notable progress by means of the Americans with Disabilities Act in 1992 and the 1996 Telecommunications Act. The Americans with Disabilities Act (ADA), in particular, introduced explicit clauses to the effect of meeting or accounting for the requirements of people with disabilities. The initial formulation was subsequently revised and refined in the Telecommunications Act of 1996. The implications of such initiatives on national industries, though not radical, have been very beneficial and stimulating. For example, both Microsoft's Active Accessibility efforts and Sun's Java Accessibility Initiative have been, at least partially,[5] influenced by the effect of ADA.

Despite the aforementioned, it is important to point out that the case of legislation offers no trouble-free alternative. In cases where there has been a solid R&D basis and a well-articulated demand for products and services, legislation may provide an effective policy instrument. An example of this was reported in Vernardakis, Stephanidis, and Akoumianakis (1995a, 1995b), which describe an econometric study on the diffusion of the alarm telephone service in municipalities in Finland (see Fig 28.3 for the S-shaped diffusion curve). The introduction of legislation in 1987 was found to have a positive accelerating effect on service diffusion.

It can, therefore, be concluded that policy initiatives by themselves cannot always guarantee economic efficiency and acceptability in the long run. In industry sectors where this has been the case, there was usually a solid R&D basis (see, e.g.,

---

[3]See also HFES/ANSI (1997).

[4]World Wide Web Consortium—Web Accessibility Initiative (1997). For more information, please refer to: http://www.w3.org/WAI/

[5]It should be noted that both these companies and others have been conscious of the accessibility issue for a number of years now.

the case of alarm telephones service diffusion in Finland, where state interventions led to wide technological adoption for the benefit of disabled and elderly people). In contrast, where a solid R&D basis is lacking (or being formed), as in the case of designing for all in HCI, policy initiatives are unlikely to bring about the expected results by themselves. In such cases, it is likely that efforts to introduce legislation will be confronted by strong industrial opposition, resulting in postponement or cancellation. Even when legislation is brought into effect, it may be by-passed or ignored, especially if demand for the type of products and services foreseen has not been articulated (Kodama, 1992; see also chap. 30, this volume). A possible explanation is the fact that the industry is consolidated and, as a consequence, policy instruments may be hard to formulate, agree on, and introduce. Under such conditions, policy instruments are likely to be broad, frequently vague, and open to interpretation, thus enabling unsatisfied industries to find means to either oppose or by-pass the legislation, or, in the best of cases, meet the absolute minimum requirements.

Conditions for Success

From the discussion thus far it follows that, though each alternative has its pros and cons, neither seems to be by itself capable of facilitating the intended objective of

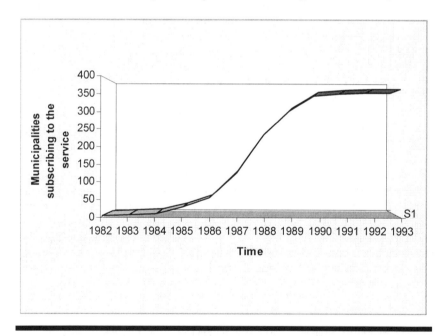

FIG. 28.3.   S-shaped curve of the diffusion of alarm telephones in Finland—effects resulting from the introduction of legislation in 1987. From Vernardakis, Stephanidis, and Akoumianakis (1995b). Copyright 1997 by RESNA. Reprinted by permission.

user interfaces for all in the IT&T industry. It can therefore be concluded that only a sound R&D basis complemented with the required policy initiatives may bring about the desired outcomes. These should aim at introducing both suitable standards, where missing, and legislation. With regard to standardization, it is important to point out the compelling requirement for speeding up current standardization processes, as well as for international coordination of standards in the long term.

To this end, actions are needed to facilitate coordination across efforts initiated in the context of research consortia (e.g., the W3C-WAI project and the ERCIM Working Group on "User Interfaces for All"[6]), as well as in national (e.g., HFES/ANSI) and international standardization bodies (e.g., the new work item on accessibility by ISO 9241/TC 159/SC 4/WG 5). Coordination in this context should also involve the exchange of input so as to avoid incompatible standards. In this context, the requirements of mainstream industries need to be carefully studied so as not to impede adoption of design for all principles and recommendations. Another important issue in this line of action is the establishment of suitable assessment and certification measures for accessibility and usability of new computer-based interactive products and services.

Legislation is also needed to provide the framework of operation and the required incentives for both the consumer base and the industry. To this effect, recent experience in the United States with the ADA of 1992 and the Telecommunications Act of 1996 should be assessed, and similar actions should be encouraged internationally. Such efforts could also draw upon general rules and recommendations compiled by industrial consortia (e.g., The Telecommunications Policy Roundtable in the United States), technical committees (e.g., the USACM-Public Policy of ACM), and international organizations (e.g., the United Nations General Assembly Standard Rules of 1995). The type of actions that are expected in this area are accompanying/support measures (or horizontal actions).

IMPLEMENTING THE SUPPORT MEASURES

The Need for International Collaboration

There are many ways in which the support measures just discussed could be introduced to effect. One possible option in the direction of attaining the aforementioned objectives would be for national governments and funding bodies to develop the required research and technological development (RTD) policy mechanisms, in order to facilitate a favorable environment for social innovation and user interfaces for all. However, many of the items and recommendations that are needed to this effect either have an explicit international dimension, or can be more effectively addressed at an international level. As a result, what is needed is international collaboration and cooperation, which, in any case, would provide the necessary input to national and

---

[6]For more information, please refer to: http://www.ics.forth.gr/proj/athci/UI4ALL/index.html

transnational RTD policy fora, as well as establish liaisons with the relevant special interest groups, consortia, and scientific committees.

To this end, there is a requirement for a continuous engagement in ongoing international collaboration programs, tasked with the goal to undertake all necessary actions to promote visionary goals in the direction of user interfaces for all and the more general case of an information society for *all,* across academia, industry, and policy levels. At the core of these efforts is the undertaking of work toward the implementation of the R&D agendas that have been formulated in Stephanidis et al. (1998) and Stephanidis et al. (1999), included in this volume as chapters 29 and 30 respectively. Moreover, it is necessary to initiate and sustain cooperation between international initiatives (e.g., international standardization bodies), in order to ensure that medium- and long-term targets are accomplished.

Types and Role of Nonmarket Institution

To attain the objectives outlined previously, nonmarket institutions can be the catalyst for success. In particular, their role is deemed crucial in funding developments, promoting awareness, establishing directives, legislative acts, and standards, assuring the diffusion of technology and its adoption by a wide proportion of the industry, and, finally, contributing consciously and explicitly toward the creation of an environment favorable for designing for all in HCI, by acting as a catalyst for overcoming the technological threshold factors associated with the availability of infrastructure (i.e., networks, terminals, etc.) to communities of end-users.

In recent years, a number of studies have converged to the preceding conclusion, pointing out that nonmarket institutions can play a catalytic role toward universal design, especially in sectors of the industry that are known to be recipients, as opposed to natural generators of innovation (chap. 29, this volume; Vernardakis et al., 1997b). In general, there is a broad range of actions that such organizations could undertake. Some of them have been identified in Stephanidis et al. (1998) and are reproduced in Table 28.5.

In the recent past, there have been several modalities through which interventions by nonmarket institutions have come into effect. A traditional type of nonmarket institution is that of governmental bodies involved in legislation, policy setting, and so on. More recent forms of nonmarket institutions have provided the means for undertaking research toward demonstrating technological feasibility, as well as for ensuring economic efficiency and efficacy in the long run.

This has been achieved by introducing initiatives and funding schemes that, in addition to targeted RTD for the purpose of technological feasibility, include support measures for technology transfer, exploratory awards, best practice and experience, process improvement experiments, and policy interventions. Moreover, nonmarket institutions may also be formed with an explicit transnational focus. For example, the European Commission and the U.S. NSF constitute a particular type of transnational organization facilitating technology development and transfer. Usually, they employ various mechanisms to attain the set objectives. Table

28.6 provides examples of such mechanisms implemented in various RTD programs of the European Commission.

Another type of nonmarket institution can be formed by industrial consortia (e.g., X Consortium), academia-industry consortia (e.g., W3C), as well as standards organizations. As already shown, the type of interventions facilitated by such nonmarket institutions may range from public domain technologies (e.g., the Mosaic browser) to policy instrumentation in the form of legislation and/or standardization.

SUMMARY AND CONCLUSIONS

The main argument in this chapter is that technological feasibility, though necessary, is not a sufficient condition to facilitate the diffusion and adoption of user in-

TABLE 28.5

Potential Role of Nonmarket Institutions

- Establish effective strategies and legal frameworks
- Act as catalyst in bringing about a shift in perspectives
- Provide funding for R&D initiatives
- Provide incentives to both large industries and SMEs
- Regulatory role aiming to support accessibility in scientific, technical, and social terms
- Identify technological potential
- Promote coherence in the market
- Support actions for standardization and legislation
- Adopt policy measures ensuring that industries design for all
- Guide industry toward design for all
- Promote the generation of the required knowledge and facilitate access to research results
- Encourage strategic alliances
- Act as technology transfer intermediary
- Develop standards and guidelines
- Establish an international usability assurance scheme to facilitate consumer confidence in product usability

TABLE 28.6

Example Mechanisms Used by Transnational Nonmarket Institutions

Target	Mechanism
Technology development	Collaborative RTD
Awareness raising	Horizontal actions
Technology transfer	Horizontal actions
	Best practice experiments

terfaces for all. In particular, what is additionally needed is a conscious effort to create an environment favorable to social innovation, by advancing the available results beyond technological feasibility. To this effect, the complementary role of cooperative R&D and policy initiatives, such as legislation and standardization, was discussed in the light of recent experiences from the United States and Europe. The main outcome from this review is that neither is sufficient by itself to guarantee the designated objectives.

The main conclusions from the present chapter are twofold. First of all, it becomes evident that user interfaces for all, though a tangible goal, poses several challenges for the IT&T sector of the industry. In other words, the currently available know-how has reached a level of maturity that provides evidence of technological feasibility in the area of computer-based products and services. Nevertheless, technological feasibility is not enough. What is really needed is economic efficiency and efficacy in the longer term. To this end, the role of nonmarket institutions such as the European Commission in Europe, MITI in Japan, and the NSF in the United States can play a critical role in initiating and establishing suitable measures to promote awareness, facilitate adoption and diffusion, and, in general, create an industrial environment favorable to social innovation.

The second conclusion is the compelling need for international collaboration in a wide area of technological and nontechnological fields. Such collaborations should aim to establish a common ground, and a holistic approach covering scientific, technological, and policy issues, for the advancement of user interfaces for all. Moreover, efforts in this direction need to be carried out in a coordinated and timely fashion, so as to provide input to current and future initiatives across the world, such as the Fifth Framework Program of the European Commission and the National Information Infrastructure in the United States.

Finally, it is important to point out that, despite the recent success stories in the practice and application of universal design in the area of HCI, there is much more R&D work needed to cover the remaining challenges (see chap. 29, this volume, for an R&D agenda). Such efforts are needed not only to provide a deeper understanding of human behavior in computer-mediated human activities, but also to enable the development of tools that will ease designers in constructing the new virtual spaces to characterize an information society for all, including people with special needs.

REFERENCES

Daly, A. J. (1998). *Improving technology performance in small and medium enterprises* [Online]. Available: http://www.worldbank.org/html/fpd/technet/sme-daly.htm

DeBresson, C. (1991). Technological innovation and long wave theory: Two pieces of the puzzle. *Journal of Evolutionary Economics, 1,* 241–272.

Dosi, G., Freeman, C., Nelson, R. R., & Soete, L. (Eds.). (1988). *Technical change and economic theory.* London: Pinter Publishers.

HFES/ANSI (1997). *Draft HFES/ANSI 200 standard, section 5: Accessibility.* Santa Monica, CA: Human Factors and Ergonomics Society.

ISO 9241. (1998). *Ergonomic requirements for office work with visual display terminals (VDTs).* Geneva, Switzerland: International Standards Organisation.

ISO 13407. (1997). *Human-centered design processes for interactive systems.* Geneva, Switzerland: International Standards Organisation.

Katz, M., & Shapiro, S. (1994). Systems competition and network effects. *Journal of Economic Perspectives, 8*(2), 93–115.

Kodama, F. (1992). Technological fusion and the new R&D. *IEEE Engineering Management Review, 20*(2), 6–11.

Mansfield, E. (1968). *Industrial research and technological innovation.* New York: Norton.

Müller, M. J., Wharton, C., McIver, W. J., & Kaux, L. (1997). Towards an HCI research and practice agenda based on human needs and social responsibility. In *Proceedings of the ACM Conference on Human Factors in Computing Systems* (pp. 155–161). New York: ACM Press.

Mynatt, E., & Weber, G. (1994). Non-visual presentation of graphical user interfaces: Contrasting two approaches. In *Proceedings of the ACM Conference on Human Factors in Computing Systems* (pp. 166–172). New York: ACM Press.

National Research Council. (1997). *More than screen deep: Toward every-citizen interfaces to the nation's information infrastructure* [Online]. Washington, DC: National Academy Press. Available: http://www.nap.edu/readingroom/books/screen/index.html

Pavitt, K. (1998). The inevitable limits of EU R&D funding. *Research Policy, 26*(6), 559–568.

Regan, B. (1998). *Java: One year out* [Online]. Available: http://www.webreference.com/content/java/

Rogers, E. M. (1982). *Diffusion of innovations.* New York: The Free Press.

Stephanidis, C. (Ed.), Salvendy, G., Akoumianakis, D., Arnold, A., Bevan, N., Dardailler, D., Emiliani, P-L., Iakovidis, I., Jenkins, P., Karshmer, A., Korn, P., Marcus, A., Murphy, H., Oppermann, C., Stary, C., Tamura, H., Tscheligi, M., Ueda, H., Weber, G., & Ziegler, J. (1999*).* Toward an information society for all: HCI challenges and R&D recommendations. *International Journal of Human–Computer Interaction, 11*(1), 1–28.

Stephanidis, C. (Ed.), Salvendy, G., Akoumianakis, D., Bevan, N., Brewer, J., Emiliani, P-L., Galetsas, A., Haataja, S., Iakovidis, I., Jacko, J., Jenkins, P., Karshmer, A., Korn, P., Marcus, A., Murphy, H., Stary, C., Vanderheiden, G., Weber, G., & Ziegler, J. (1998). Toward an information society for all: An international R&D agenda. *International Journal of Human–Computer Interaction, 10*(2), 107–134.

Stoneman, P. (1983). *The economic analysis of technological change.* Oxford, England: Oxford University Press.

Story, M. F. (1998). Maximising usability: The principles of universal design. *Assistive Technology, 10*(1), 4–12.

Twiss, B. C. (1992). *Managing technological innovation* (4th ed.). London: Pitman Publishing.

Vernardakis, N., Stephanidis, C., & Akoumianakis, D. (1995a). On the impediments to innovation in the European assistive technology industry. *International Journal of Rehabilitation Research, 18*(3), 225–243.

Vernardakis, N., Stephanidis, C., & Akoumianakis, D. (1995b). The use of analytical tools in analysing the demand for assistive technology products: The case of alarm telephones in Finland. In *Proceedings of the 2nd TIDE Congress* (pp. 211–214). Amsterdam: IOS Press.

Vernardakis, N., Stephanidis, C., & Akoumianakis, D. (1997a). The transfer of technology towards the European assistive technology industry: Current impediments and future opportunities. *International Journal of Rehabilitation Research, 20*(2), 189–192.

Vernardakis, N., Stephanidis, C., & Akoumianakis, D. (1997b). Transferring technology toward the European assistive technology industry: Mechanisms and implications. *Assistive Technology, 9*(1), 34–46.

VIII Looking to the Future

29 Toward an Information Society for All: An International Research and Development Agenda

Constantine Stephanidis, Gavriel Salvendy,
Demosthenes Akoumianakis, Nigel Bevan,
Judy Brewer, Pier Luigi Emiliani,
Anthony Galetsas, Seppo Haataja,
Ilias Iakovidis, Julie A. Jacko, Phil Jenkins,
Arthur I. Karshmer, Peter Korn,
Aaron Marcus, Harry J. Murphy,
Christian Stary, Gregg Vanderheiden,
Gerhard Weber, and Juergen Ziegler

This article introduces the visionary goal of an Information Society for all, *in which the principles of universal access and* quality in use *prevail and characterize computer-mediated human activities. The paper is based on the outcome of the first meeting of the International Scientific Forum "Towards an Information Society for All", which took place during the Seventh International Conference on Human Computer Interaction (HCI International '97). The objective of this meeting was to define a short-, medium-, and long-term international R&D agenda in the context of the emerging Information Society, based on the principle of designing for all users. The proposed agenda addresses technological and user-oriented issues, application domains and support measures, which are necessary for the establishment of a favorable environment for the creation of an Information Society acceptable to all citizens.*

The radical technological changes in the information technology and telecommunications (IT&T) sectors of the industry have contributed toward a more information- and interaction-intensive paradigm for computer-mediated human activities. This has been a direct derivative of continuous and evolutionary changes that rapidly transform society, from one based on the production of physical goods, to one where the main emphasis is on the production and exchange of information. Such a trend, which is expected to continue, raises a whole new range of human, social, economic, and technological considerations regarding the structure and content of societal activities at the turn of the 21st century. In this context, the requirements for universal access and quality in use for the broadest possible user population progressively emerge into first-order objectives for an information society for *all* citizens.

The term *information society,* although attributed with different meanings and connotations, is frequently used to refer to the new socioeconomic and technological paradigm likely to occur as a result of an all-embracing process of change that is currently taking place. This process is expected not only to alter human interaction with information, but also to affect individual behavior and collective consciousness (Danger et al., 1996). The information society is neither the mere effect of radical technological progress brought about by research and technological development (RTD) work, nor the result of incremental demand-driven innovation in a particular sector of the industry. Instead, it is a product of a *technology fusion* (Kodama, 1992) of IT&T.[1]

This far-reaching effect of combining incremental technical improvements from several previously "separate" fields of technology brings about radically new opportunities and market windows. As Kodama (1991) reported, marrying optics and electronics technologies produced optoelectronics, which gave birth to fiber optics communications systems; fusing mechanical and electronics technologies produced the mechactronics revolution, which has transformed the machine tool industry. In a similar fashion, the fusion of IT&T technologies is expected to introduce radical changes in the society, as well as far-reaching organizational and institutional changes in all aspects of human activity (e.g., workplace, leisure, shopping, commerce, education).

Such a progressive transformation introduces new challenges and requirements (European Commission, 1994; National Research Council, 1997) regarding the content of information being communicated, the range of information services available, the telecommunication services for access to information, the media used for the presentation of information, and so on. These, in turn, have direct implications on the structure and organization of human activities at the turn of the 21st century (European Commission, 1997), and necessitate new means for computer-mediated human activities.

One important issue, in this context, is that the goods likely to proliferate in the emerging information age (i.e., information-based commodities) should be made available to anyone, anywhere, and at anytime (Stephanidis, 1995b). This challenge not only applies to computers and their interfaces, but also to information itself, and how it is created, collected, represented, stored, transferred from one place to another, and used. The information society has the potential to improve the quality of life of citizens, and increase the efficiency of our social and economic organization. At the same time, it may lead to a "two-tier" society of "haves" and "have-nots," in which only part of the population has access to the new technology, or is comfortable using it, and can thus fully enjoy the benefits (European Commission, 1994). It is in this context, that the principle of *design for all* becomes an important vehicle toward ensuring social acceptability of the emerging information society.

In many ways, the term design for all (or universal design; the terms are used interchangeably) is not entirely new. It is well known in several engineering disci-

---

[1]Some also include the media sector (see Spectrum Strategy Consultants, 1997).

plines, such as, for example, civil engineering and architecture, with many applications in interior design, building and road construction, and so on. This is not to say that the built environment we all live in has been designed for *all*, but merely points to the fact that universal design is not specific to information society technologies. However, although the existing knowledge may be considered sufficient to address the accessibility of physical spaces, this is not the case with information society technologies, where universal design is still posing a major challenge.

In the context of this chapter, design for all in the information society is the conscious and systematic effort to proactively apply principles, methods, and tools in order to develop IT&T products and services that are accessible and usable by *all* citizens, thus avoiding the need for a posteriori adaptations, or specialized design.

This chapter reports the process used to collect and consolidate experts' opinion on the issue of universal access in the emerging information society. The result of this process is an international research and development (R&D) agenda toward the development of an information society accessible and socially acceptable to *all* citizens. The current effort, which aims to provide a first milestone in this direction, was undertaken by a group of experts sharing the concerns regarding the issues of accessibility and social acceptability of the emerging information society.

The chapter is structured as follows. The next section reviews and describes some of the prominent characteristics of the information society, the new research challenges already in place, and the requirements that need to be addressed by future work. Then, design for all in the context of the emerging information society is introduced and discussed, in terms of the underlying rationale and of different views regarding its feasibility and cost justification. The subsequent section describes the process and instruments used to develop an international R&D agenda to facilitate the creation of an information society *for all,* and discusses the agenda in detail. The chapter concludes with a summary of results and plans for future activities.

MAIN CHALLENGES AND REQUIREMENTS
IN THE INFORMATION SOCIETY

The Shift of the Techno-Economic Paradigm

The emergence of the information society is associated with radical changes in both the demand and the supply of new products and services, resulting from the fusion of IT&T. The changing pattern in demand is due to a number of characteristics of the customer base, including (a) increasing number of computer users characterized by diverse abilities, requirements, and preferences, (b) product specialization to cope with the increasingly knowledge-based nature of tasks, resulting from the radical changes in both the nature of work and the content of tasks, and (c) increasingly diverse contexts of use.

On the other hand, one can clearly identify several trends in the supply of new products and services, aiming to develop the required know-how and know-why to meet the changing pattern in demand. These can be briefly summarized as follows:

(a) increased scope of information content and supporting services, (b) emergence of novel interaction paradigms (e.g., virtual and augmented realities, ubiquitous computing), and (c) shift toward group-centered, communication-, collaboration-, and cooperation-intensive computing.

Figure 29.1 summarizes this shift in computing paradigms and illustrates the new role of computers in the emerging information society. In this context, one of the problems that need to be addressed is the *design* (as opposed to programming) of information artifacts that inherently support the expanded range of computer-mediated human activities (Kapor, 1991; Winograd, 1995).

The Varieties of Context

The notion of *context* has traditionally been a crucial design factor for any type of computational system (Norman & Draper, 1986). Despite this, existing practice indicates that, in the majority of cases, the study of context is ignored, or bound to what is normally encountered by "typical" users (Nardi, 1996). This narrow and implicit view of context no longer suffices, given the broad range of computer-mediated human activities in the emerging information society. Instead, what is needed is a new ground for the study of context in the information age, based on a multidisciplinary protocol of exchange between research and technological development, as well as practice and experience.

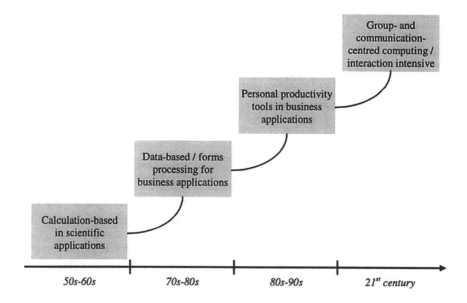

FIG 29.1.   Shift in computer paradigms and forecast of trends for the 21st century.

Variety in the Target User Population and the Nature of Work. The radical innovation in the IT&T sectors and the increasing number of emerging application domains has broadened the range and type of the "target user population." As suggested by Figure 29.1, from the early calculation-intensive nature of work that was prevalent in the early 1960s, computer-based systems are progressively becoming a tool for communication, collaboration, and social interaction. From a specialist's device, the computer is being transformed into an information appliance for the citizen in the information society. It follows, therefore, that designers will increasingly have to provide information artifacts to be used by diverse user groups, including people with different cultural, educational, training, and employment background, novice and experienced computer users, the very young and the elderly, and people with different types of disabilities.

Variety in the Context of Use. Another dimension of variation is the *context of use*. In particular, the "traditional" use of computers (i.e., *scientific* use by the specialist, *business* use for productivity enhancement) is increasingly being complemented by *residential* and *nomadic* use, thus penetrating a wider range of human activities, in a broader variety of environments, such as the school, the home, the marketplace, and other civil and social contexts. As a result, information artifacts should embody the capability to interact with the user in all those contexts, and independently of location, machine, or run-time environment. Usability in such "nontraditional" usage contexts is likely to prove a harder target to meet than in the case of the workplace (Stephanidis & Akoumianakis, 1996).

Variety in the User Access Medium. In addition to the preceding, there will be also variation in the systems or devices used to facilitate access to the community-wide pool of information resources. These devices include computers, standard telephones, cellular telephones with built-in displays, television sets, information kiosks, special information appliances, and various other "network-attachable" devices. Depending on the context of use, users may employ any of these to review or browse, manipulate and configure information artifacts, at any time.

Universal Access and Quality in Use

Given the techno-economic paradigm shift and the variety in the context of computer-mediated human activities, it is important that *universal access* and *quality in use* are considered as prerequisites for an information society where *all* citizens have equal opportunities for interpersonal communication, education, vocational training, employment, and so on.

In the context of this chapter, universal access in the information society signifies the right of *all* citizens to obtain equitable access to, and maintain effective interaction with, a community-wide pool of information resources and artifacts.

Universal access implies more than *direct access* or access through *add-on (assistive) technologies* (Vanderheiden, 1990), because it emphasizes the principle that accessibility should be a design concern, as opposed to an afterthought. In other words, it is claimed that universal access entails the development of systems that can be used effectively, efficiently, and enjoyably by *all* users. To this end, it is important that the needs of the broadest possible end-user population are taken into account in the early design phases of new products and services.

The notion of quality, on the other hand, has various meanings and connotations (Bevan, 1995; Garvin, 1984), and there are also different approaches to achieving product quality as part of the production process (e.g., International Standards Organisation [ISO]: ISO 9001, 1987; ISO 8402, 1994). In particular, quality in use is the high-level design objective for a system to meet the real-world needs of its intended users (Bevan & Azuma, 1997; ISO/IEC 14598-1, 1998) and entails the consideration of a broad range of functional and nonfunctional attributes, that characterize the use of information artifacts by humans in their various problem-solving, information-seeking, and communication-intensive computer-mediated activities.

This notion of quality goes beyond the "traditional" concept of usability (i.e., ease of use and learnability, etc.), to include aspects (such as usefulness, suitability for the task, tailorability, etc.) that are not easily measurable by current approaches based on performance criteria, such as effectiveness, efficiency, satisfaction, and the like.

DESIGNING FOR ALL IN THE INFORMATION SOCIETY

Introducing Design for All

As already pointed out, design for all in the information society is the conscious and systematic effort to proactively apply principles, methods, and tools in order to develop IT&T products and services that are accessible and usable by *all* citizens, thus avoiding the need for a posteriori adaptations or specialized design. The rationale behind universal design is grounded on the claim that designing for the "typical" or "average" user, as the case has been with "conventional" design of IT&T products, leads to products that do not cater to the needs of the broadest possible population, thus excluding categories of users (Bergman & Johnson, 1995). Contrasting this view, the normative perspective of universal design is that there is no "average" user and, consequently, design should be targeted toward *all* potential users.

The vision of universal design has underpinned recent work, predominantly in the fields of human–computer interaction (HCI) and assistive technology. The main results today vary in context, scope and applicability across application domains. Nevertheless, they constitute a useful repository of experience and best practice that can influence future developments.

In particular, recent advances toward universal design in HCI have provided a design wisdom in the form of universal design principles and guidelines (TRACE R&D Center, 1997); platform-specific accessibility guidelines, for example, for graphical user interfaces (GUIs) or the World Wide Web (WWW) (Gunderson,

1996); or domain-specific guidelines, for example, for text editing or graphic manipulation (Human Factors and Ergonomics Society, 1997). The systematic collection, consolidation, and interpretation of these guidelines is currently pursued in the context of international collaborative initiatives, such as the Web Accessibility Initiative (WAI) of the World Wide Web Consortium (W3C; 1997) and the ISO TC159/SC4/WG5 (Stephanidis, Akoumianakis, Ziegler, & Faehnrich, 1997).

In addition to the aforementioned, in recent years, several technical research and development projects have provided insights toward new user interface development frameworks and architectures that account (explicitly or implicitly) for several issues related to accessibility and interaction quality. Examples include the European Commission–funded projects TIDE-ACCESS TP1001 (Stephanidis, Savidis, & Akoumianakis, 1997) and ACTS-AVANTI AC042 (Stephanidis, Paramythis, Karagiannidis, & Savidis, 1997), as well as the Japanese FRIEND21 initiative (FRIEND21, 1995).

Moreover, efforts toward universal design in the fields of IT&T have met with wide appreciation by an increasing proportion of the research community, including (see also Müller, Wharton, McIver, & Kaux, 1997): (a) research consortia in the context of various Programmes of the European Commission, such as the ERCIM[2] Working Group on "User Interfaces for All" (Stephanidis, 1995a); (b) industry, such as the U.S. Telecommunications Policy Roundtable, Microsoft Active Accessibility, and Java Accessibility; (c) scientific and technical committees, such as Association for Computing Machinery (ACM's) USACM public policy committee; (d) legislative acts, such as the Americans with Disabilities Act, 1993, and the U.S. Telecommunications Act, 1996—Sec.255; and (e) the United Nations General Assembly Standard Rules (United Nations, 1995).

Universal Design Deliberations

In contrast to the supporting initiatives and efforts just discussed, there is also skepticism concerning the practicality and cost justification of universal design. In particular, there is a line of argumentation raising the concern that "many ideas that are supposed to be good for everybody aren't good for anybody" (Lewis & Rieman, 1994). However, universal design should not be conceived as an effort to advance a single solution for everybody, but as a user-centered approach to providing environments designed in such a way that they cater to the broadest possible range of human needs, requirements, and preferences.

Another common argument is that universal design is too costly (in the short-term) for the benefits it offers. Though the field lacks substantial data and comparative assessments as to the costs of designing for the broadest possible population, it has been argued that (in the medium to long term) the cost of inaccessible systems is comparatively much higher, and is likely to increase even more, given the current statistics classifying the demand for accessible products (Bergman & Johnson, 1995; National Council on Disability, 1996; Vanderheiden, 1990).

---

[2]ERCIM: The European Research Consortium for Informatics and Mathematics (www.ercim.org).

It is important, however, to underline that any particular technology (in the broad sense of the term) aiming toward universal design should satisfy much more than mere demonstration of technical feasibility, in order to be acceptable. Strictly speaking, even technical efficiency, which presupposes both technical feasibility and technical reliability, may still not be sufficient. What is really needed is economic feasibility in the long run, leading to versatility and economic efficiency (Vernardakis, Stephanidis, & Akoumianakis, 1997). In this context, considering universal design practices within a user-centered process of system development (Bevan & Azuma, 1997) is likely to provide a successful business case for universal design, and a framework for realizing its promises in an effective and cost-efficient manner.

The Need for a New R&D Agenda

It follows then that there is a compelling requirement for greater awareness of the premises and challenges of universal design; in order to advance universal design toward new frontiers and promote its practice, a closer international collaboration is necessary between the fields that drive the development of information society technologies. Figure 29.2 provides an indicative illustration of the levels of concern relevant to universal design, emphasizing HCI, the telecommunications infrastructure, as well as the content level.

To this effect, there is a compelling need to critically review recent accomplishments in the relevant disciplines, with a renewed focus on the issues that prevail as a result of the new requirements. It is important that the existing inventory of wisdom is assessed against new criteria that characterize computer-mediated human activities in the context of the information society. The normative perspective of this effort lies in the notion that any advances toward an information society for *all* citizens are likely to be made as a result of a purposeful, multidisciplinary approach to the appropriation of reciprocal investments in cross-sector RTD work, leading to fusion-type innovation.

In this context, an international R&D agenda is required for an information society acceptable to *all* citizens; such an agenda would facilitate the fusion of theoretical perspectives into a common ground, so as to inform and improve design practice, and promote a deeper understanding of how humans interact and communicate with other humans and with information artifacts. The multidisciplinary focus necessitates not only cross-sector commitment, but also a conscious effort to ensure a broad scope of the envisioned R&D activities empowered with new concepts, tools, and techniques from different (and up to now rather dispersed) scientific disciplines, technological strands, and socioeconomic and policy perspectives.

Bridging the Gap Across Relevant Scientific Disciplines. At the scientific level, the intention is to establish a cross-discipline agenda for collaborative research, fostering potential synergies among relevant disciplines and the

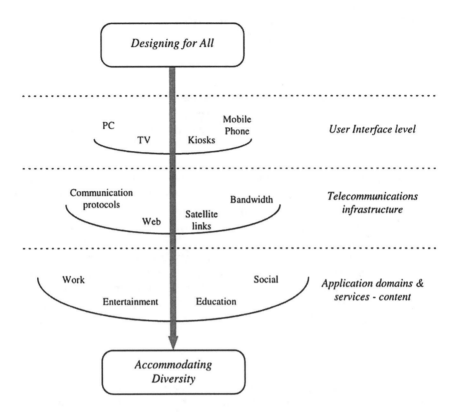

FIG. 29.2.   Levels of concerns and implications of design for all.

appropriation of the resulting benefits. Such a requirement is now widely recognized and, in some cases, it is also actively pursued.

For example, in the field of HCI there have been efforts in the direction of revisiting existing approaches to design (in view of the new requirements), such as the human factors evaluation paradigm (O'Brien & Charlton, 1996) and cognitive science (Norman, 1986). The intention has been to (a) specialize the HCI science base, (b) facilitate better utilization of information-processing psychology, (c) extend the scope of information-processing psychology, or (d) to broaden the range of psychology being applied (Carroll, 1991). Advocates of these proposals claim that enriching the available HCI theories, design methods, and tools with suitable concepts from the social sciences (e.g., developmental approaches to psychology, anthropology, sociology) can provide the means to overcome well-known problems associated with more traditional approaches, and facilitate more powerful frameworks for HCI design. Examples of recent efforts to this end include the development of an *action-science* for HCI, advocated by Carroll and colleagues (Carroll & Rosson, 1992), as well as new frameworks for HCI based on *activity theory* (Bødker, 1989, 1991), *situated action models* (Suchman, 1987), and *distributed cognition* (Norman, 1993).

Synergies, such as those driving recent developments in the field of HCI, can be promoted across other scientific disciplines, such as information theory, behavioral and organization science, and so on, as well as the natural sciences, in order to bring about a new conceptualization of computer-mediated human activities within the information society.

Bridging the Gap Across Relevant Technological Fields.   The role of technology is equally critical, as it is the sole provider of the required tools through which humans will interact with information artifacts. Though much can be achieved with the current pool of technological wisdom, it is expected that the convergence postulated in the previous section, and a more effective linkage with design theory and practice, will help technology provide solutions that meet pragmatic requirements in various contexts of use. To this effect, it is imperative that contributing technologies cover the whole range of information management activities, including information generation and extraction, storage and tracing, retrieval and presentation, communication, as well as the broad range of issues pertaining to human interaction with information artifacts encountered in various contexts, collaborative structures, and virtual spaces.

Bridging the Gap Across Social, Economic, and Industrial Policy.   Socioeconomic and policy issues are relevant to the extent to which they cover RTD planning, industrial policy and innovation, assessment of the products and services envisioned in the information age, cost factors, diffusion and adoption patterns, standards, legislation, technology transfer, and so forth. An important requirement to be observed is that of cross-industry collaborative research and development, which, in turn, involves reciprocal investment decisions and joint R&D efforts toward fusion-type innovations (Kodama, 1991). Policymakers should take appropriate measures to promote and facilitate RTD work, grounded on promising technology synergies and identified and selected on the basis of suitable criteria and assessment protocols. To achieve this, however, requires deep and detailed knowledge of possible synergies, how they are to be appropriated, the nature of the collaboration needed, and the type of collaborators who should be involved. In this context, "traditional" economic theory no longer suffices (Vernardakis et al., 1997). Instead, a broader view of economic, social, and technological factors and their interrelationships is required, in order to enable the specification of the type and nature of the envisaged products and services, the requirements that they should satisfy, as well as the way in which industry should respond to the changing patterns of demand. These issues reflect on a complex of research challenges, related to the economics of innovation, technology management, and industrial policy.

THE INTERNATIONAL SCIENTIFIC FORUM

The call for closer collaboration across disciplines and a holistic approach covering scientific, technological, and policy issues advocated in the previous section, would facilitate a common research and practice agenda for the advancement of an information society that is acceptable to *all* citizens. Such an effort needs to be carried out in a coordinated and timely fashion, so as to provide input to current and future initiatives across the world (e.g., the Fifth Framework Program of the European Commission and the National Information Infrastructure in the United States).

The International Scientific Forum "Towards an Information Society for *All*" held its first meeting with the aim of addressing these challenges, as well as exploring and exploiting the new opportunities in this direction. The following sections outline the preparatory work leading to the meeting, and present the consolidated results of this meeting.

Objectives and Themes of the Meeting

The overall objective of the meeting was to elaborate an international R&D agenda that would contribute toward the establishment of a favorable environment for the creation of an information society accessible and acceptable to *all* citizens. Such an agenda should address short-, medium-, and long-term issues, based on the consolidation of recent progress in the identified scientific fields, and should specify the support measures necessary to stimulate future R&D activities at the international level. In addition to this aim, during the meeting, participants unanimously agreed on the potential of the International Scientific Forum to continue as a *network* for collaboration and exchange of experience between researchers, industrialists, and policymakers, aiming to advance the existing wisdom and stimulate new developments toward universal access and social acceptability of the emerging information society.

The driving themes of the first meeting of the Forum centered around three major topics: *technological and user-oriented challenges, critical application domains and services,* and *support measures.* Participants contributed toward the identification of the relevant issues in each of the themes, the assessment of the available scientific and technological knowledge providing the foundations for further work, and the definition of an R&D agenda covering specific areas of critical importance for the short, medium, and long term.

The technological challenges included for discussion in the agenda of the meeting were, among others, assessment of current design approaches and practice, advanced interaction technologies, user interface software, architectural models for interactive systems, supporting tools, and novel interactive environments. Some of the user-oriented challenges included for discussion were user participation in the design process, frameworks of evaluation and assessment of end-user opinion, user-modeling approaches, experimentation, reports on practice and experience, and frameworks for studying computer-mediated human activities.

The theme of application domains and services covered critical areas to be addressed, such as health care, education, social interaction, banking, electronic commerce, public-access information kiosks, as well as application domains and services built on top of the WWW.

Finally, the theme of support measures identified issues related to standardization, legislation, and policy that are necessary to establish and sustain a favorable environment for industrial innovation, technology transfer, awareness raising, consensus, networking, diffusion, and so on, toward an information society for *all*.

Preparatory Activities

A number of preparatory activities were undertaken leading up to the meeting. These facilitated more focused and targeted discussions during the meeting, as well as the formulation of a common ground for discussion. In particular, a short questionnaire was used, together with accompanying background material, to facilitate targeted response and the collection of preliminary data on a number of issues pertaining to the agenda of the meeting. The questionnaire and the background material were circulated to participants prior to the meeting. The background document included the meeting's objectives, rationale, structure, agenda, as well as specific guidelines.

The questionnaire was primarily intended to elicit "high-level" responses from the experts. It intentionally did not include technical questions, as the meeting was not intended to provide a discussion forum advocating specific technical solutions, but instead, to provide a medium for knowledge exchange, experience sharing, identification of challenges, and formulation of targets. The questionnaire collected expert opinions on the present state of the art, current impediments, and industrial challenges, aiming to identify and prioritize alternatives, and to develop an action plan in the form of an R&D agenda for the short, medium, and long term. The questionnaire and the accompanying documentation were subject to a small trial run with three experts, whose comments and remarks were incorporated in a final version. This final version was subsequently circulated to all participants electronically, who were given 2 weeks to study the documents, respond to the questionnaire, and provide position statements related to the themes of the meeting.

The participants' expertise was varied and included HCI, assistive technology, user interface tool development, human factors, usability evaluation, standardization, legislation, and technology transfer. The affiliated organizations of the participants comprised three large industrial organizations (IBM Corporation, NOKIA Mobile Phones, and SUN Microsystems); one high-technology small-medium enterprise (SME) (Aaron Marcus and Associates), one international consortium (W3C), representatives of the European Commission (the Information Technology ESPRIT Program—DG III, and TELEMATICS APPLICATION Program— DG XIII), five universities, and five research institutes.

Roundtable Discussion and Consolidation

The preliminary results of the analysis of the questionnaires were the basis for a roundtable discussion at the beginning of the meeting (see Tables 29.1–29.4). Several of the issues appearing in these tables were also considered as short-, medium-, and long-term targets. The remainder of this section outlines the responses and data gathered for each question.

Sufficiency of Existing Know-How. The responses gathered on this question are summarized in Table 29.1. An important conclusion regarding the sufficiency of available know-how was that, in certain cases, the existing design wisdom is sufficient to facilitate a number of developments with clear benefits for the population at large. For example, several guidelines for universal design are available today, which, if followed, will bring about improvements in the accessibility and usability of popular information environments, such as the WWW. However, it was also pointed out that further work is needed toward experimentation as well as technological developments. As an example of the former, one can point to the need for better design foundations and the much needed experimental evidence of what constitutes "good enough" design. On the other hand, further technological developments are needed to provide suitable tools for different human needs.

Some of the participants raised the concern that, in some cases, it is difficult to assess what is already available, because there is a lack of coordination of international efforts. This is further complicated by the inadequacy of traditional channels of dissemination of research results, practice, and experience, to facilitate knowledge sharing and exchange in a timely fashion. As a result, in many cases, it is hard to know what has been accomplished, so as to avoid duplication of efforts.

Impediments in Developing an Information Society for All Citizens. In this question, participants provided a broad view as to the possible obstacles to achieving an information society for *all* citizens (see Table 29.2). Some of the impediments identified pertain to the current state of the mainstream industry and the assistive technology field; others relate to consumers and their willingness and capacity to actively participate and articulate demand for accessible and usable products. In spite of these widely accepted shortcomings, it was also acknowledged that, as a result of recent initiatives, a number of these targets have begun to gain attention.

TABLE 29.1

Responses Related to Sufficiency of Existing Know-How

In some cases, the existing know-how is sufficient; in other cases further work is needed.

Lack of coordination of international efforts impedes a detailed understanding of what is available.

Packaging and broader dissemination of available information is necessary to improve awareness of the existing know-how.

TABLE 29.2

Impediments in Developing an Information Society for *All* Citizens

Current Status of the Mainstream Industry	Assistive Technology Prevalent Practice	Attitude of Consumers
Large installed base of inaccessible building blocks	Lack of well-designed and user-friendly systems	Lack of user participation
Lack of appropriate user interface software architectures	"Looking at the tree and not the forest"	Need for usability as product differentiation determinant, or as purchase determinant
Lack of cost-effective technology to deliver products that comply to design for all principles	Small and fragmented markets	Lack of appropriate education mechanisms for consumers to that they can participate in advanced technology discussion
Limited view of the user; traditional focus on average user	Lack of understanding of user interaction with new technologies	Lack of appropriate usability and user-centered design techniques to elicit effectively consumer opinion
Prevailing competitive strategies	High cost of accessibility in currently prevailing practices	
Limited awareness and appreciation of diversity	Reactive nature of the assistive technology field	
Lack of knowledge about specific disabilities and people with multiple handicaps	Lack of coordination of efforts/legal/regulatory frameworks	
Currently prevailing view that existing technology, if used carefully, can provide effective solutions to all problems	Lack of reference materials, documentation, and research data	
Lack of usability training in computer science courses	Lack of internationally coordinated, comprehensive, cross-disability research programs	
Insufficiency of currently available design methods and tools		

Role of Nonmarket Institutions.   In recent years, a number of studies have reached the conclusion that nonmarket institutions can play a catalytic role toward universal design, especially in sectors of the industry that are known to be recipients, as opposed to natural generators of technology (Vernardakis et al., 1997). This point was even more emphasized in the replies of the respondents. To this effect, the possible actions that such organizations could undertake are summarized in Table 29.3.

Industrial Impediments.   Table 29.4 summarizes the responses of the participants regarding specific impediments that restrict the capacity and capability of the industry to develop products and services that comply to the principles and requirements of design for all. From these responses, a point that was raised earlier on

TABLE 29.3
Potential Role of Nonmarket Institutions

Establish effective strategies and legal frameworks.

Act as catalyst in bringing about a shift in perspectives.

Provide funding for R&D initiatives.

Provide incentives to both large industries and SMEs.

Regulatory role aiming to support accessibility in scientific, technical, and social terms.

Identify technological potential.

Promote coherence in the market.

Support actions for standardization and legislation.

Adopt policy measures ensuring that industries design for all.

Guide industry toward design for all.

Promote the generation of the required knowledge and facilitate access to research results.

Encourage strategic alliances.

Act as technology transfer intermediary.

Develop standards and guidelines.

Establish an international usability assurance scheme to facilitate consumer confidence in product usability.

TABLE 29.4
Industrial Impediments

Lack of internal knowledge as to how to design for all.

Lack of legal framework to ensure/reinforce design for all.

Lack of awareness of potential longer term benefits.

Lack of effective/efficient technology transfer mechanisms.

Cost–benefit justification/lack of perceived market demand.

Fierce competition and pace of development.

Limitations of prevailing systems development practice.

Attitude that universal design is cost prohibitive.

Lack of standards.

Lack of consensus by disability groups (e.g., as to what is "good enough")

Lack of skills and methods.

Lack of qualified people.

Reluctance to change existing practice.

Lack of corporate commitment.

becomes even more evident, namely that any technological push toward universal design should satisfy the criterion of economic efficiency, rather than mere technical feasibility and reliability, otherwise it will not be acceptable.

THE PROPOSED INTERNATIONAL R&D AGENDA

Following an analysis of the results of the questionnaires, participants were subsequently involved in consolidating the available material for each of the three themes of the meeting. More specifically, participants were requested to contribute with research targets, a brief explanation of content, scope, and purpose, as well as examples for reference.

Technological and User-Oriented Issues

The specific recommendations that were raised for this theme are summarized in Table 29.5 and are discussed next.

Design Processes, Methods, and Tools for Computer-Mediated Human Activities. It was recommended that further research is needed to provide enriched frameworks, theories, and methodologies to facilitate the design of systems that exhibit a range of qualities that render them accessible and usable by the broadest possible end-user population (e.g., user-centered goals to design). Indicative quality attributes include: *software ergonomic criteria*, such as ease of learning and understanding, ease of use, adaptability, error tolerance, suitability to the task, and so on (ISO 9241, 1995); *performance criteria,* such as effectiveness, efficiency, satisfaction, and the like (Bevan & Macleod, 1994); and *nonfunctional quality attributes,* such as accessibility, scalability, reliability, reusability, portability, and so forth.

An immediate action should therefore be targeted towards the definition of criteria for design processes that take full account of accessibility and the desired quality in use attributes. The availability of such criteria would guide future developments and inform and improve practice. In addition, as another short-term target, it is recommended that universal design advocates gather data to confirm, consolidate, and expose the current design principles and guidelines known to be valid, so as to facilitate greater awareness and wider adoption by designers. As a result, it is expected that industry will become more sensitive to delivering products and services for diverse contexts of use, and will eventually adopt the required science base to facilitate it.

In the short to medium term, emphasis should be on disciplines that focus on people and communication. More specifically, prescriptive frameworks are needed for deriving technical features for the study of activity orientation, collaboration, social awareness, and social immersion, as well as the diversity of context resulting from cultural, physical, and cognitive user characteristics, or technological proliferation. Equally important is that validated design criteria for user interfaces supporting interaction in the emerging information society are derived from such empirically confounded frameworks. To this effect, end-user participation is

TABLE 29.5

Critical Areas of Research and Development Related to Technology
and User-Oriented Issues

	Time Horizon		
Issues	Short Term	Medium Term	Long Term
Design process, methods, and tools	Consolidate and disseminate available wisdom on universal design Establish criteria that take account of accessibility and usability	Use methods, tools, and results from disciplines focusing on people and communication Promote use of methods within a user-centered design process	Advance new prescriptive instruments to facilitate design for scalability and modality independence
User-oriented challenges	Study variety of user contexts and address requirements of tail population Build upon and expand existing techniques from participatory design and ethnography to promote user involvement	Develop standardized methods for eliciting end-user requirements	Develop an empirical science base for the study of users
Input/Output Technology	Speech input Natural-language input and processing Multimodal language input Flexible, portable, high-resolution, compact displays Haptic devices Voice and synthesized sound Multimodal output generation Computer-supported cooperative work (CSCW) Hypermedia	Computer vision Gesture processing High-resolution, full-page tactile displays Advanced/alternative interaction platforms Embedded intelligence and agents	Tools for abstraction, integration, and expansion of interaction platforms Building new forms of cooperative structures and virtual spaces Integrated agency models
User interface architectures	Enhancements to build new interaction facilities such as scanning into existing platforms (e.g., WINDOWS95) Distributed object environments Transportable software	Requirements engineering methods and tools that incorporate accessibility issues System development methodologies that take full account of accessibility and usability	Tools for effort and cost-effective integration of access into products Appropriate user interface architectures Architectures for collaboration Toolkit interoperability Architectures for collaboration

651

crucial and should be strongly encouraged by the methods and tools to be employed, or developed.

In the medium term, the existing inventory of methods, techniques, and tools for user-centered design (ISO 13407, 1997) should be suitably applied and enhanced as needed, by drawing on accumulated knowledge and results in social sciences to promote and facilitate the use of more developmental approaches to the study of computer-mediated human activities. In all cases, the tight evaluation loop advocated by user-centered design should provide the primary channel for timely feedback into the design process, so as to assure that design deficiencies are corrected at an early stage, while updates are less costly to make (ISO/IEC 14598, 1996).

In the long term, efforts should be mainly targeted toward gaining an improved understanding of basic principles to support upward scalability to new technologies and modality and medium independence. Scalability, in this context, refers to the requirement that interactive devices should be extensible, so as to be capable of incorporating the next generation of technology. On the other hand, modality and medium independence imply that design artifacts are represented and stored in a form that is not bound to the specifics of any particular technology platform. Such representations can subsequently provide alternative instantiations to different modalities through different media.

User-Oriented Issues.   In order to improve current design practice and guide the evolution toward new interaction and communication paradigms, it is necessary, in the short term, to employ available methodologies from the fields of sociology, psychology, anthropology, physiology, and other contributing disciplines. These would provide a more in-depth understanding of human interaction and communication with other humans and information, in the context of the emerging information age. It is argued that the user's experience with new information society technologies is more than likely to be different from that associated with the currently prevailing desktop embodiment of the computer. As a result, interaction with such technologies is bound to follow different metaphors and involve multiple human sensory channels. To this end, participation of diverse user groups is critical and should be encouraged, in order to provide a direct account of the range and implications of the combination of user abilities, constraints, skills, requirements, preferences, and so on.

In the medium to long term, the emphasis should be on the development of a rigorous experimental science base pertaining to how diverse users interact with information artifacts. Such an endeavor should necessarily take account of users with disabilities and elderly people, as they frequently constitute the "cutting edge" in the assessment of the usability of a particular technology. Moreover, part of what is currently missing from this science base is an empirical characterization of how the physiology of specific user impairments dictates the interaction strategies that can be employed by users with disabilities. There is also insufficient data concerning how (combinations of) impairments affect preferences for, and performance with, various types of technologies; understanding such relationships will allow a more systematic approach to matching individual end-user abilities, skills, requirements, and preferences with features of interactive systems.

Input/Output Technologies.   Investment is also needed in further R&D in emerging technologies, which will carry the power of computing to people and environments not currently serviced. A broad range of technological developments is recommended in the short, medium, and long term. These include technologies for alternative input and output, expanding the current horizon of computer-mediated human activities beyond the conventional visual embodiment of the desktop metaphor, as well as technologies that improve the process and means by which individuals and groups of people communicate, cooperate, and collaborate to accomplish common objectives.

Critical areas of technologies to facilitate alternative input include speech input, natural-language input and processing, computer vision, gesture recognition, and their combination with "traditional" technologies to facilitate multimodal input. On the other hand, technologies facilitating a wider range of output modalities include: flexible, portable, high-resolution and compact displays; haptic devices; high-resolution full-page tactile displays; voice and synthesized sound and multimodal output generation; and so on. A critical issue in undertaking such technological developments will be the integration of the resulting hardware and software components into mainstream operating systems, using suitable and reliable facilities. This would ensure the wide utilization and interoperability of such technologies.

In addition to the aforementioned, technologies contributing to the creation of new information spaces for experience sharing, exchange of ideas, and social communication should also be supported. Such technologies need to be advanced toward pragmatic needs and requirements of individual members in groups, thus appropriating the benefits of social engagement and collective decision making. Additionally, their combination with other emerging technologies such as hypermedia, interactive multimedia, and alternative interaction environments (e.g., virtual/augmented environments) should provide a new ground for supporting a wide range of computer-mediated human activities.

User Interface Architectures.   Research on user interface architectures will be a crucial contributing factor toward a new generation of user interface software and technology for the broadest possible population (Müller et al., 1997; Stephanidis & Akoumianakis, 1997). The critical issue in the short to medium term is to provide the components that will allow the development of integrated environments, which, among other things, are adaptable, personalized, cooperative, easy to learn, error tolerant, and responsive to a changing environment and context of use. Such architectures should extend the current conception of the visual desktop metaphor, or the metaphors proliferating in Web-based applications. To this end, they should progressively enlarge the scope of computer-mediated human activities to facilitate the attainment of communication-oriented goals, in addition to "traditional" task-oriented activities.

An important issue in this context is to enable human interaction with other humans and information artifacts through alternative access media, such as the telephone or the television, which are becoming information appliances. Therefore, it is likely that neither the conventional desktop, nor the prevalent Web metaphors will suffice as a single solution to designing for the broadest possible end-user population (Winograd, 1997). Having said this, it is important to clarify that new interface architectures should not aim to replace visual user interfaces, but instead, to empower them with new capabilities, through the technological development efforts identified previously.

In the medium and long term, research on advanced user interface architectures should also be complemented by efforts to advance the tools currently available for requirements engineering, designing, implementing, and evaluating interactive systems in a cost-effective and efficient manner. As an example, recent trends indicate that, in order to advance the current generation of cross-platform environments, research is needed to facilitate multiple toolkit platforms, supporting multiple/alternative interactive embodiments of the functions in a source application domain (e.g., the office, the marketplace).

Application Domains and Services

The theme on application domains and services identified, and provided an account of, critical areas to be addressed. The significance of the application domains not only reflects their role in establishing a coherent and socially acceptable information society, but also the diverse range of human activities likely to be penetrated, as a result of the fusion in the IT&T sectors of the industry. Table 29.6 summarizes the consolidated results, which are briefly discussed next.

Life-long learning is a critical area where emphasis should be placed in the "knowledge" society of the future. It entails a continuous engagement in the acquisition of knowledge and skills to facilitate and sustain equitable participation in the information society (European Commission, 1997). New technologies may play a

TABLE 29.6
Critical Application Domains and Services

Life-long learning.

Public information systems, terminals, and information appliances (e.g., kiosks, smart home environments).

Transaction services (e.g., banking advertisement).

Electronic commerce applications and services.

Social services for the citizens (e.g., administration, elderly, transport, health care, awareness).

Tools to allow for added-value information services (e.g., creation storage, retrieval, and exchange of user experiences, traces, and views).

Security.

catalytic role in providing new educational mechanisms and structures, thus allowing learning to become an inseparable part of life-long human activities in the context of knowledge-intensive *learning communities,* and social interaction among groups of people.

Another important application area and a critical short-term target is the development of general-purpose *public information systems, terminals,* and *information appliances* (e.g., information kiosks for access to community-wide information services). These are expected to be used in increasingly different contexts, including public places, homes, classrooms, and the like, and provide the means for ubiquitous and nomadic access. *Environmental control* will also become increasingly important. *Smart environments* will progressively penetrate a wide range of human activities in hospitals, hotels, public administration buildings, and so on. Tele-operation of such environments will also gain increasing attention to facilitate responsiveness to unforeseen events, enhanced mobility, and security.

Finally, a broad range of *transaction services* (e.g., banking, advertising, entertainment), *social services for the citizens* (e.g., administration, health care, education, transport), and *electronic commerce* applications will become increasingly important in reshaping business and residential human activities. These should also be addressed within a short- to medium-term time horizon.

Independently of any particular application domain or service, there are certain quality attributes and added-value functionality that should be accommodated into future services. For instance, *security, privacy,* and *control* are central themes in the evolution of a socially acceptable information society and should receive immediate attention. At the same time, they will increasingly constitute more complex targets to accomplish, as they span across different levels of the telecommunications infrastructure, from network services to application services (such as business transactions and entertainment), terminals and information appliances.

Additionally, important quality attributes to be observed include accessibility, intuitive operation, and ease of use, as well as functionality to allow users to create, store, and tailor available data into added-value information, and leave traces of their experience. Such quality attributes, and the extent to which they are satisfied, are likely to be important determinants of diffusion and adoption of emerging information services by *all* citizens. These factors are also likely to provide the primary explanatory variables of early versus late adoption rate by different user groups.

Support Measures

The theme on support measures identified a broad range of issues related to policy that are required to provide a favorable environment for industrial innovation, technology transfer, awareness raising, consensus, networking, and diffusion, for the evolution of an information society accessible and acceptable to *all* citizens. The results consolidated at the roundtable discussion are summarized in Table 29.7 and are elaborated next.

TABLE 29.7

Support Measures

Issues	Time Horizon		
	Short Term	Medium Term	Long Term
Articulating demand for universal design	Awareness raising	New product concepts	Developments and delivery of new products and services
Supporting the industry	Provide incentives for universal design International standards coordination Legislation	Provide incentives for universal design Internationnal standards coordination Legislation	Provide incentives for universal design International standards coordination Legislation
Awareness and knowledge dissemination	Encourage qualified universal design practice Consolidate and disseminate knowledge known to be true Devise suitable international dissemination strategy	Encourage qualified universal design practice Consolidate and disseminate knowledge known to be true Devise suitable international dissemination strategy	Encourage qualified universal design practice
Technology transfer	Promote advanced mechanisms for technology transfer Establish a coordinating body to guide, facilitate transfers Consider policy variables required for networking	Consider policy variables required for networking	Consider policy variables required for networking

Articulating Demand for Universal Design.   In the short term, support efforts should be devoted to the articulation of a demand for design for all. In the recent literature (see Kodama, 1991), articulating demand has been defined as a two-step process: First, translate market data into a product concept; and second, decompose the concept into a set of development projects. It should be mentioned that such a two-step process is sufficient only in cases where a market already exists and has reached a level of maturity whereby it can react and respond to the needs of its customer base. However, in the case of the emerging information society it is argued that there is an additional short-term need for creating awareness as to the new challenges likely to emerge, as well as the new opportunities offered.

Consequently, support measures are needed in the direction of raising consumer awareness, education, and training. These should cover a broad range of the population, including "tail" populations, such as people with disabilities, and should be aimed toward building a *public expectation* for usable products and services, and intolerance of inaccessible forms of technology. Additionally, until a certain maturity level is reached and more effective end-user input into the design process can be attained, procurement guidance is necessary with regard to accessible and usable technology.

Regarding subsequent steps required for demand articulation, namely the translation of market data to product concepts and the decomposition of product concepts to development projects, it is important that a range of questions are addressed. These include what product types are needed in the market, how they could be produced, through which technologies, and what other characteristics should the envisaged technologies exhibit.

Supporting the Industry.   Another line of action should be targeted toward the creation of an environment favorable to industrial innovation. At the core of such activities should be the provision of incentives toward design for all. Industrial incentives need not necessarily be of a financial type, though this would be critical for SMEs. Incentives should also include access to research results that would be difficult to obtain otherwise, provision of a suitable infrastructure, collaborative R&D activities for technology transfer (see also later discussion), as well as a favorable legal framework.

Additionally, participants recognized the compelling requirement for speeding up current standardization processes, as well as the need for internationally coordinated standards in the long term. To this end, it is recommended that work already initiated in the context of research consortia (e.g., the W3C WAI Project, http://www.w3c.org/WAI/, and the ERCIM Working Group on "User Interfaces for All," http://www.ics.forth.gr/at-hci/UI4ALL/) as well as in national (e.g., HFES/ANSI) and international standardization bodies (e.g., the new work item on accessibility by ISO 9241—http://scitsc.wlv.ac.uk/~c9584315/iso9241.html; ISO/TC 159/SC 4/WG 5) should be coordinated and should exchange input so as to avoid incompatible standards. Another recommendation along the same lines is to establish suitable assessment and certification measures for accessibility and usability of new products and services.

Legislation is also needed to provide the framework of operation and the required incentives for both the consumer base and the industry. To this effect, recent experience in the United States with the Americans with Disabilities Act of 1993 and the Telecommunications Act of 1996 should be assessed, and similar actions should be encouraged internationally. Such efforts could also draw on general rules and recommendations asserted by industrial consortia (e.g., The Telecommunications Policy Roundtable in the United States; ACM, 1994), technical committees (e.g., the USACM-Public Policy Committee of the ACM; http://jafar.ncsa.uiuc.edu/usacm) and international organizations (e.g., the United Nations General Assembly Standard Rules of 1995; United Nations, 1995).

Awareness and Knowledge Dissemination. One of the critical impediments to the adoption of universal design practice is the lack of qualified practitioners who understand what is required to achieve universal access and quality in use. To overcome this, it is recommended that, in the short term, usability is introduced as a mandatory component of university education; additionally, existing knowledge about the benefits of, and how to achieve accessibility and usability should be systematically disseminated. This will progressively encourage the adoption of universal design principles by professional bodies and individuals in the medium to long term. To this effect, an international dissemination strategy to raise awareness and provide access to information and training would speed up the transfer of knowledge into design practice.

Technology Transfer. Effective and efficient technology transfer is another critical target, requiring a range of support measures to be effected. Technology, in this context, includes both embodied and disembodied forms. Embodied technology is evident in new products and services, machines, tools, and research equipment. Disembodied technology appears as learning-by-doing, documentation, know-how, and know-why. To facilitate successful transfers of technology, suitable mechanisms are needed in the short term, to the effect of targeted and purposeful exchange of knowledge, know-how, and know-why. It is recommended that, from the broad range of technology transfer mechanisms that can be considered, emphasis is on advanced measures (such as cooperative R&D, joint venture R&D agreements, joint ventures aimed at keeping partners informed, large–small firm agreements), rather than simpler ones (such as licensing, technical advice, technical support, contract of R&D). This is because the former cluster is better suited for the type of knowledge transfer that is required (Vernardakis et al., 1997). In this context, it is important to mention that collaborative, interdisciplinary, multinational, multicultural, and cross-industry R&D activities, involving industry and institutions, are of primary importance.

Moreover, given the broad range of technologies that will drive the emergence of the information society, it is recommended that any contemplated technology transfer effort should closely and carefully consider potential sources and recipients, and assess alternative technological performance thresholds. In principle, it is recommended that design for all should be the aim of all emerging technologies. The newer a technology, the greater are the chances of being influenced in the direction facilitating design for all. However, for such a condition to materialize, a "monitoring system" for critical emerging technologies should be established; this could be part of an extensive collaborative network, and should aim to identify potential synergies and possibilities for international collaborative R&D efforts.

SUMMARY AND CONCLUSIONS

The emergence of the information society creates new opportunities and challenges for *all* citizens. The progressive shift from physical goods to information-based products and services is likely to introduce new patterns for demand and supply of such products and services. One important issue in this transition is the extent to which the emerging information society advances in a manner that ensures nondiscrimination and social and economic inclusion of the broadest possible end-user population. This question may be rephrased to explicitly point out the requirement for a new society for *all* citizens. This chapter has outlined the rationale for the argument that universal access and quality in use should be integral components of future developments in this direction, and that design for all provides a means to this end.

To this effect, the chapter discusses new and forthcoming requirements for such a society, and proposes an R&D agenda based on the consolidated opinion of experts within the International Scientific Forum "Towards an Information Society for *All*." The agenda was advanced in the course of a 1-day roundtable discussion, which was preceded by several preparatory activities. The objective of the roundtable was to identify short-, medium-, and long-term activities that would contribute toward the visionary goal of an information society acceptable to *all* citizens. Additionally, the roundtable addressed critical issues and necessary steps to facilitate the implementation of the R&D agenda at an international level.

The agenda points out a broad range of required actions relevant to three main themes, namely technology and user-oriented issues, critical application domains and services, and support measures. Each recommended action was discussed in terms of target objective, tentative scope, and likely time horizon. Under the theme of technology and user-oriented issues, the agenda highlights the need for additional work covering the development of critical technologies, the advancement of suitable design frameworks, and the evolution of powerful user interface architectures. Critical application domains and services include life-long learning, public information systems, terminals and information appliances, transaction services, social services and electronic commerce, as well as global issues such as security, reliability, and the like. Support measures that would facilitate a favorable environment toward an information society for *all* should cover the articulation of demand for universal design, support to industry, awareness raising, and knowledge dissemination, and technology transfer.

In addition to these short-, medium-, and long-term targets, the roundtable discussed alternatives in the direction of establishing a favorable environment for supporting and implementing the proposed agenda. One possible option in this direction would be for national governments and funding bodies to develop the required RTD policy mechanisms, in order to integrate and support future research along the lines suggested in this chapter. However, as already pointed out in the previous sections, many of the items and recommendations of the R&D agenda have either an explicit international dimension, or can be more effectively addressed at an international level. As a result, what is needed is international collaboration and cooperation that, in any case, would provide the necessary input to national and transnational RTD policy forums, as well as establish liaisons with the relevant special interest groups, fora, consortia, and scientific committees.

To facilitate this effect, the roundtable discussions concluded that a possible path for realizing the proposed R&D agenda and supporting its implementation could be through the continuous engagement in ongoing international collaboration efforts, tasked with the goal to undertake all necessary actions to promote the visionary goal of an information society for *all,* across academia, industry, and policy levels. Such a strategy is currently being further investigated, and is expected to attract wider attention. At the core of these efforts should be the undertaking to implement several parts of the proposed agenda and to initiate and sustain cooperation with other international efforts (e.g., standardization bodies), to ensure that medium- and long-term targets are accomplished. In the short term, a number of the recommendations identified earlier can provide the ground for such international collaboration.

For instance, the International Scientific Forum has initiated, and is planning to define activities aiming to consolidate current practice and experience in the area of universal design, and make it widely available as reference material, or provide contributions to ongoing national and international standardization activities. Similarly, in the short to medium term, work can be directed toward the identification of key accessibility criteria or requirements to be met by products and services. These efforts can be consolidated into an appropriate form (e.g., accreditation scheme) that would guide subsequent efforts by both industry and academia toward information products and services accessible and usable by the broadest possible end-user population. Furthermore, such activities can help industry to gain a renewed focus on the issue of universal design, and facilitate justification for the costs and benefits of alternative technologies. Additionally, it can stimulate new developments, and establish the ground whereby universal design informs and improves practice.

The preceding are only some of the goals of international collaboration aiming to promote an information society for *all* citizens. They demonstrate the benefits resulting from an international, multidisciplinary effort to advance the concepts and principles of universal design in the context of information society technologies, as well as the added value of undertaking this effort through a network of partners cooperating and collaborating toward a shared vision. Finally, it is also important that, for such a committee to be established and efficiently operated, support is provided by individuals, organizations, nonmarket institutions, national governments, and international bodies.

In summary, this chapter has combined the opinions of experts regarding the research requirements that should be addressed to facilitate the emergence of an information society for *all* citizens. The proposed R&D agenda is considered an important step toward this visionary goal, and it is presented in the hope that significant coordinated efforts will ensue. It is anticipated that such efforts will promote awareness, stimulate interest, and initiate international cooperative activities, and make inroads toward establishing a favorable environment for the creation of an *information society accessible, usable, and acceptable to all citizens.*

ACKNOWLEDGMENTS

This chapter is a reprint from Stephanidis et al. (1998). *International Journal of Human–Computer Interaction, 10*(2), 107–134. Copyright 1998 by Lawrence Erlbaum Associates and IJHCI. It is based on the roundtable discussion of the first meeting of the International Scientific Forum "Towards an Information Society for All" that took place during the Seventh International Conference on Human Computer Interaction (HCI International '97). The meeting was organized by Constantine Stephanidis of FORTH-ICS, Greece, and Gavriel Salvendy of Purdue University, United States. The contributors are listed in alphabetical order.

We acknowledge IBM Corporation for sponsoring the 1-day roundtable meeting. The support of the European Commission is also acknowledged for partially funding initial work leading to the adoption, realization, and promotion of the *design for all* concept.

REFERENCES

ACM. (1994). Renewing the commitment to a public interest telecommunications policy. *Communications of the ACM, 37*(1), 106–108.

Bergman, E., & Johnson, E. (1995). Towards accessible human–computer interaction. In J. Nielsen (Ed.), *Advances in human–computer interaction* (Vol. 5, pp. 87–113). Norwood, NJ: Ablex.

Bevan, N. (1995). Measuring usability as quality of use. *Software Quality Journal, 4,* 115–130.

Bevan, N., & Azuma, M. (1997). Quality in use: Incorporating human factors into software engineering lifecycle. In C. Walnut (Ed.), *Proceedings of the 3rd IEEE International Software Engineering Standards Symposium and Forum* (pp. 169–179). Los Alamitos, CA: IEEE Computer Society.

Bevan, N., & Macleod, M. (1994). Usability measurement in context. *Behavior and Information Technology, 13*(1 & 2), 132–145.

Bødker, S. (1989). A human-activity approach to user interfaces. *Human–Computer Interaction, 4*(3), 151–196.

Bødker, S. (1991). *Through the interface: A human activity approach to user interface design.* Hillsdale, NJ, Lawrence Erlbaum Associates.

Carroll, J. (1991). Introduction: The Kittle house manifesto. In J. Carroll, (Ed.), *Designing interaction: Psychology at the human–computer interface* (pp. 1–16). New York: Cambridge University Press.

Carroll, J., & Rosson, M. B. (1992). Getting around the task-artefact framework: How to make claims and design by scenario. *ACM Transactions on Information Systems, 10*(2), 181–212.

Danger, S., Huizing, N., Walker, A., Rowland, A., Anderson, R., & Sciaccaluga, R. (1996). *EU information society guide.* Brussels: The EU Committee on the American Chamber of Commerce in Belgium.

European Commission. (1994). *Recommendations to the European Council: Europe and the global information society—The Bangemann report* [Online]. Available: http://www.ispo.cec.be/infosoc/backg/bangeman.html

European Commission. (1997). *Building the European Information Society for us all: Final policy report of the high level expert group, April 1997* [Online]. Available: http://www.ispo.cec.be/hleg/Building.html

FRIEND21. (1995). *Human interface architecture guidelines.* Tokyo: Institute for Personalised Information Environment.

Garvin, D. A. (1984, Fall). What does "product quality" really mean? *Sloan Management Review, 26*(1), 25–48.

Gunderson, J. (1996). *World Wide Web browser access recommendations.* Urbana: University of Illinois Press.

Human Factors and Ergonomics Society. (1997). *Draft HFES/ANSI 200 Standard, Section 5: Accessibility.* Santa Monica, CA: Author.

ISO 8402. (1994). *Quality vocabulary.* Geneva, Switzerland: International Standards Organisation.

ISO 9001. (1987). *Quality systems—Model for quality assurance in design, development, production, installation and servicing.* Geneva, Switzerland: International Standards Organisation.

ISO 9241. (1995). *Ergonomic requirements for office work with visual display terminals.* Geneva, Switzerland: International Standards Organisation.

ISO 13407. (1997). *Human-centred design processes for interactive systems.* Geneva, Switzerland: International Standards Organisation.

ISO/IEC 14598. (1996). *Information technology—Evaluation of software products, Part 1: General overview.* Geneva, Switzerland: International Standards Organisation.

ISO/IEC 14598-1. (1998). *Information technology—Evaluation of software products, Part 1: General guide.* Geneva, Switzerland: International Standards Organisation.

Kapor, M. (1991). A software design manifesto. In T. Winograd, Stanford University, & Interval Research Corporation (Eds.), *Bringing design to software* (pp. 1–9). New York: ACM Press.

Kodama, F. (1991). *Analysing Japanese high technologies: The techno-paradigm shift.* New York: Pinter.

Kodama, F. (1992). Technological fusion and the new R&D. *IEEE Engineering Management Review, 20*(2), 6–11.

Lewis, C., & Rieman, J. (1994). *Task-centred user interface design: A practical introduction* [Online]. Available: http://www.syd.dit.csiro.au/hci/clewis/contents.html

Müller, M. J., Wharton, C., McIver, W. J., & Kaux, L. (1997). Towards an HCI research and practice agenda based on human needs and social responsibility. In S. Pemberton (Ed.), *Proceedings of the ACM Conference on Human Factors in Computing Systems* (pp. 155–161). New York: ACM Press.

Nardi, B. (1996). *Context and consciousness: Activity theory and human computer interaction.* Cambridge, MA: MIT Press.

National Council on Disability. (1996). *Achieving independence: The challenge of the 21st century.* Washington, DC: Author.

National Research Council. (1997). *More than screen deep: Toward every-citizen interfaces to the nation's information infrastructure.* Washington, DC: National Academy Press.

Norman, D. (1986). Cognitive engineering. In D. Norman & W. Draper (Eds.), *User-centered design: New perspectives on human–computer interaction* (pp. 31–61). Hillsdale, NJ: Lawrence Erlbaum Associates.

Norman, D. (1993). *Things that make us smart.* Reading, MA: Addison-Wesley.

Norman, D., & Draper, W. (Eds.). (1986). *User-centered design: New perspectives on human–computer interaction*. Hillsdale, NJ: Lawrence Erlbaum Associates.

O'Brien, T. G., & Charlton, S. G. (Eds.). (1996). *Handbook of human factors testing and evaluation*. Mahwah, NJ: Lawrence Erlbaum Associates.

Spectrum Strategy Consultants. (1997). *Development of the information society—An international analysis* [Online]. Available: http://www.isi.gov.uk/isi/dotis/index.html

Stephanidis, C. (1995a). Editorial. In *Proceedings of the 1st ERCIM Workshop on "User Interfaces for All"* [Online]. Available: http://www.ics.forth.gr/proj/at-hci/UI4ALL/

Stephanidis, C. (1995b). Towards user interfaces for all. In *Proceedings of the 6th International Conference on Human Computer Interaction* (pp. 137–143). Amsterdam: Elsevier, Elsevier Science.

Stephanidis, C., & Akoumianakis, D. (1996). Usability requirements for advanced IT products. In A. Mital, H. Krueger, M. Menozzi, & J. E Fernandez (Eds.), *Advances in occupational ergonomics and safety I* (pp. 145–149). Amsterdam: IOS Press.

Stephanidis, C., & Akoumianakis, D. (1997). *Designing for all in the emerging information society: A utopia or a challenge?* Unpublished manuscript.

Stephanidis, C., Akoumianakis, D., Ziegler, J., & Faehnrich, K.-P. (1997). User interface accessibility: A retrospective of current standardisation efforts. In *Proceedings of the HCI International '97 Conference* (pp. 469–472). Amsterdam: Elsevier, Elsevier Science.

Stephanidis, C., Paramythis, A., Karagiannidis, C., & Savidis, A. (1997). Supporting interface adaptation: The AVANTI Web browser. In *Proceedings of the 3rd ERCIM Workshop on "User Interfaces for All"* [Online]. Available: http://www.ics.forth.gr/proj/at-hci/UI4ALL/

Stephanidis, C., Savidis, A., & Akoumianakis, D. (1997). Unified interface development: Tools for constructing accessible and usable user interfaces: Tutorial No. 13. in the *7th International Conference on Human–Computer Interaction* [Online]. Available: http://www.ics.forth.gr/proj/at-hci/html/tutorials.html

Suchman, L. A. (1987). *Plans and situated actions: The problem of human machine communication*. Cambridge, MA: Cambridge University Press.

TRACE R&D Center. (1997). *Designing an accessible world* [Online]. Available: http://www.trace.wisc.edu/world/world.html.

United Nations. (1995). *General Assembly standard rules on universal accessibility*. New York: Author.

Vanderheiden, G. C. (1990). Thirty-something million: Should they be exceptions? *Human Factors, 32*, 383–396. [On-line]. Available: http://www.trace.wisc.edu/text/univdesn/30_some/30_some.html

Vernardakis, N., Stephanidis, C., & Akoumianakis, D. (1997). Transferring technology toward the European assistive technology industry: Mechanisms and implications. *Assistive Technology, 9*, 34–46.

Winograd, T. (1995). From programming environments to environments for designing. *Communications of the ACM, 38*(6), 65–74.

Winograd, T. (1997). Interspace and an every-citizen interface to the national information infrastructure. In *More than screen deep: Toward every-citizen interfaces to the nation's information infrastructure*. Washington, DC: National Academy Press. [Online]. Available: http://www.nap.edu/readingroom/books/screen

World Wide Web Consortium. (1997). *Web accessibility initiative* [Online]. Available: http://www.w3.org/WAI/

30

Toward an Information Society for All: HCI Challenges and R&D Recommendations

Constantine Stephanidis, Gavriel Salvendy, Demosthenes Akoumianakis, Albert Arnold, Nigel Bevan, Daniel Dardailler, Pier Luigi Emiliani, Ilias Iakovidis, Phill Jenkins, Arthur I. Karshmer, Peter Korn, Aaron Marcus, Harry J. Murphy, Charles Oppermann, Christian Stary, Hiroshi Tamura, Manfred Tscheligi, Hirotada Ueda, Gerhard Weber, and Juergen Ziegler

This article reports on the results of the second meeting and workshop of the International Scientific Forum "Towards an Information Society for All", that took place in Crete, Greece, June 15-16, 1998. In particular, it elaborates on the international Research and Development (R&D) agenda (Stephanidis et al., 1998), which resulted from the first meeting and workshop of the Forum in San Francisco, California, USA, August 29, 1997, in the context of the HCI International '97 Conference. The present document elaborates on the proposed R&D agenda by identifying Human Computer Interaction (HCI) challenges and clusters of concrete recommendations for international collaborative Research and Technological Development (RTD) activities. Four clusters of recommendations are proposed, of which the first three are intended to facilitate reaching technological targets, while the fourth comprises accompanying measures. The three technological clusters are related to the corresponding transitions from: (a) productivity tools to environments of use; (b) individual users to communities of users; and (c) computer-assisted business tasks to computer-mediated human activities. The fourth cluster covers support (horizontal) actions needed to establish a favorable environment for the creation of an Information Society acceptable to all citizens. Each cluster is elaborated by means of specific recommendations, plausible RTD objectives and likely or expected outcomes.

The International Scientific Forum (ISF) "Towards an Information Society for All" is a network for collaboration, discussion, and exchange of experience and practice on the broad range of issues related to the accessibility, usability, and ultimately, the acceptability of the emerging information society. The objective of the ISF is to promote the establishment of a favorable environment for the creation of an information society acceptable to *all* citizens.

Information society refers to the new status quo and the new socioeconomic and technological paradigm likely to occur because of the current all-embracing process of change. It is expected to affect the interaction in computer-mediated human activities, individual human behavior, the collective consciousness, and the economic and social environment (Stephanidis et al., 1998). The emergence of the information society signifies the transition toward a new form of society based on the production and exchange of information (see Figure 30.1), as opposed to physical goods. Its evolution is likely to introduce new virtual spaces (Winograd, 1997) and a whole range of computer-mediated human activities (Nardi, 1996).

In this context, the ISF aims to promote *universal design* in information society technologies, emphasizing *accessibility* and *high quality of interaction* by the broadest possible end-user population, including people with special needs. *Information society technologies* refers to innovative technologies that drive the emergence of the information society, as a result of either incremental demands on behalf of the customer base, technological breakthroughs, or fusion-type innovation. The primary industry sectors that generate and push the development of these technologies are the information technology, telecommunications, and media sectors.

Universal design or *design for all* (here used interchangeably) has different connotations. For some individuals, it is considered as a new politically correct term, referring to efforts intended to introduce "special features" for "special us-

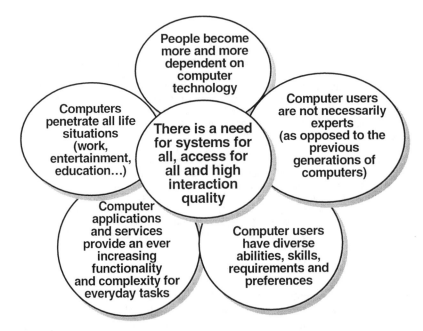

FIG. 30.1.   Critical trends toward the emergence of an information society.

ers" during the design of a product. To others, universal design is a deeply meaningful and rich topic that elevates what designers like to call "good user-based design" to a more encompassing concept of addressing the needs of all potential users. In this article, the phrase is used to reflect a new concept, or philosophy for design that recognizes, respects, values, and attempts to accommodate the broadest possible range of human abilities, skills, requirements, and preferences in the design of all computer-based products and environments. Thus, it promotes a design perspective that eliminates the need for "special features" and fosters individualization and end-user acceptability (Stephanidis et al., 1998; Story, 1998). As already pointed out, universal design is used interchangeably with design for all users. This does not imply a single design solution suitable for all users. Instead, this interchangeability should be interpreted as an effort to design products and services in such a way that they suit the broadest possible end-user population. In doing so, different solutions for different (categories of) users and different contexts of use are more likely to emerge.

Accessibility, on the other hand, traditionally has been associated with disabled and elderly people (Bergman & E. Johnson, 1995; Müller, Wharton, McIver, & Kaux, 1997) and reflects the efforts devoted to the task of meeting prescribed code requirements for use by people with special needs (Story, 1998). However, due to recent technological developments (e.g., proliferation of interaction platforms, such as wireless computing, wearable equipment, and user terminals), the range of different categories of users that may gradually be confronted with accessibility problems extends beyond the population of disabled and elderly users (National Research Council, 1997). In this chapter, *accessibility* denotes the global requirement for access to information by individuals with different abilities, skills, requirements, and preferences, in a variety of contexts of use (International Standards Organisation [ISO], ISO 13407, 1997). Its meaning is intentionally broad so it encompasses accessibility challenges that *diversity* poses in (a) the target user population (including people with special needs) and the individual and cultural differences, (b) the scope and nature of tasks (especially as related to the shift from business tasks to communication- and collaboration-intensive computer-mediated human activities), and (c) the technological platforms and associated devices through which information is accessed.

Finally, the notion of quality in use is typically associated with various meanings and connotations (Bevan, 1995; Bevan & Azuma, 1997; Garvin, 1984), and the ways it can be achieved as part of the production process also vary (e.g., ISO 9001, 1987; ISO 8402, 1994). In this article, quality entails the consideration of a broad range of functional (e.g., domain-specific qualities, such as interoperability and search efficiency) and nonfunctional attributes (e.g., portability, scalability, modifiability) that affect the use of information artifacts by humans in their various problem-solving, information-seeking, and communication-intensive computer-mediated activities. This notion of quality goes beyond the traditional concept of usability (i.e., measurable attributes based on performance criteria such as effectiveness, efficiency, satisfaction, etc.), to include aspects (such as usefulness, suitability for task, tailorability, etc.) that may not be measurable with currently available means.

The aims of the first ISF meeting and workshop in San Francisco (Stephanidis et al., 1998) were to (a) assess the state of the art in the area of accessibility and universal design, (b) stimulate new developments, and (c) advance the existing wisdom toward universal access and high-quality interaction in the emerging information society. The specific objectives were:

- To identify, consolidate, and characterize progress in the relevant scientific fields and develop a short-, medium-, and long-term research and practice agenda toward an information society accessible and acceptable by *all* citizens.
- To stimulate support measures required at national and international levels in order to raise awareness, achieve consensus, promote work in the relevant fields, and establish a favorable environment for the creation of an information society for *all* citizens.

Following this meeting, the ISF produced a White Paper entitled "Toward an Information Society for All: An International R&D Agenda" (Stephanidis et al., 1998), which introduced the goal of an information society for *all* citizens, and proposed a short-, medium-, and long-term international research and development (R&D) agenda, based on the principle of designing for *all* users. The proposed agenda addresses technological and user-oriented issues, application domains, and support measures that are deemed as necessary for the establishment of a favorable environment for the creation of an information society acceptable to *all* citizens.

The focus of the second ISF meeting[1] in Crete was more narrow in scope, aiming to elaborate, review, and consolidate the outcome of the first ISF meeting, as reported in Stephanidis et al. (1998). This was approached in the context of four themes related to *HCI design, usability, user interface software technologies,* and *standardization,* as well as in connection to their relevance with respect to accessibility and universal design. For each of these topics, the goal of the workshop was to develop concrete recommendations that would result in influential interventions, anticipated to expand the current scope of R&D activities and support mea-

---

[1] The meeting was organized as a roundtable discussion coordinated by the chairman. Secretarial support was also available. A draft agenda was provided in advance a part of a background document. Minutes of the meeting were kept by the secretary and were distributed to all participants shortly after the meeting. Each participant actively contributed to the selected thematic topics of the meeting. Each thematic topic was explored in a sequence of four phases: (a) review, (b) short presentations (position statements), (c) user issues, and (d) discussion and consolidation. Participants came from Europe (12), the United States (7), and Japan (2). The research community was represented by academics (9) and researchers (7). Industrial participation included four (4) large information technology actors and one (1) small-size enterprise. All participants are senior professionals within their organization and some of them participate in international standards committees and are heavily involved in transnational collaborative activities.

sures in the direction of *universal access* and *high quality of interaction* for *all* citizens in the information age. The deliberate intention to address these HCI-related topics is based on the critical role of this technology in shaping the type, range, and scope of computer-mediated human activities in the emerging information society. It is anticipated that in future, follow-up meetings, the ISF will progressively address additional relevant topics, as emerging technologies mature and become embedded into new products and services.

We wish to acknowledge the European Commission and FORTH-ICS for sponsoring the second meeting of the International Scientific Forum (ISF) "Towards an Information Society for All" in Crete, Greece, June 15–16, 1998.

R&D ROADMAP—AN OVERVIEW

Consolidation

As already pointed out, the second ISF meeting focused on accessibility as related to four primary themes, namely HCI design, usability, user interface software technologies, and standardization. In this context, accessibility was defined as the global requirement for access to information, communication, and social interaction by individuals with different abilities, requirements, and preferences, in a variety of contexts of use. Thus, the meaning of the term is intentionally broad to encompass accessibility challenges as posed by diversity in:

- The target user population (including people with special needs) and the individual and cultural differences.
- The scope and nature of tasks (especially as related to the shift from computer-assisted business tasks to residential and social computer-mediated human activities).
- The technological platforms and associated devices likely to penetrate the emerging broad range of computer-mediated human activities.

Following reviews by participants and discussions pertaining to each one of these topics, several recommendations were compiled regarding future R&D activities. In total, over 50 recommended research items were identified in the context of the four themes of the workshop. These items interrelate and give rise to certain clusters that highlight the group's visionary targets to be met en route to an information society. Four main clusters were identified and include recommendations regarding *software technologies, users,* the *scope of usage,* and *accompanying* or *support measures.*

Figure 30.2 highlights some of the targets that R&D has facilitated until now, as well as the new potential targets that should be facilitated in the future. The clusters reflect the progressive shift from an industrial society toward an information society, as well as the expected/desirable transition in terms of the objectives for each of the three dimensions. In addition to the three clusters, and the corresponding targets depicted in the diagram of Fig. 30.2, the ISF stressed the importance of accompanying measures (or supporting actions) as a fourth dimension relevant to the evolution of an information society for *all* citizens. This fourth dimension is orthogonal to any particular sector, or application domain, and reflects global requirements for the establishment of a favorable environment for the envisioned information society.

High-Level (Clusters of) Recommendations

In the light of the schematic of Fig 30.2, the group arrived at the following high-level recommendations:[2] (a) promote the development of environments of use, (b) support communities of users, (c) extend user-centered design to support new virtualities (and novel usage contexts), and (d) establish suitable accompanying measures.

The four topics are interrelated. Thus, recommendations under one topic link with recommendations under a different topic. The type of actions envisaged, with the exception of the accompanying measures cluster, cover all phases of technological development, ranging from feasibility studies, to basic and applied research, and demonstration. Following are brief descriptions of the meaning and rationale for the

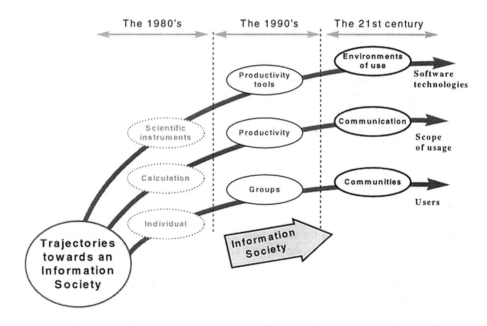

FIG. 30.2.   Trajectories towards an information society.

---

[2]The first three are proposed as RTD topics, whereas the fourth is proposed as a support measure, or "horizontal" activity.

four high-level recommendations, and the next section elaborates on specific recommended research and technological development (RTD) activities.

Promote the Development of Environments of Use. Environments of use imply integrated systems shareable by communities of users. They should, in contrast to the traditional notion of computers as productivity tools, allow for richer communications, and signify the progressive integration of the computing environment with the physical environment (see Fig. 30.3). Moreover, in contrast to tools, which enhance the productivity of individuals, environments of use would promote the concept of loveable systems suitable for a broad range of communication and collaboration intensive activities among groups of people. Such environments should be characterized by sympathy and care for users and nonusers[3] and should be accessible by anyone, anytime, anywhere. Finally, they should provide unobtrusive means for supporting social activities. As depicted in Fig. 30.3, environments of use are likely to become integral components of daily activities among communities of users and facilitate the establishment of new forms of social endeavors. Consequently, they should be conceived and designed as community centered, shareable, expandable, cooperative, collaborative, and responsive media, catering through user and environment monitoring to a broad range of human needs for both users and nonusers. Additionally, they should offer voluntary and context-specific user support, and facilitate error-tolerant behavior and preventive actions against unforeseen circumstances and/or misuse.

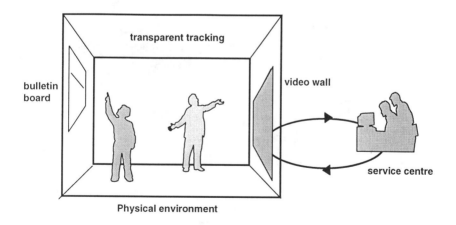

FIG. 30.3.   From productivity tools to environments of use.

[3]*Nonuser* refers to a member of a community who, although not interacting with the environment him or herself at a particular point in time, is affected by this environment, or its use by other active users.

Support Communities of Users.   Another critical trajectory en route to an information society is the one that progressively shifts the focus of attention from individual users to communities of users. The important element in this trajectory is the emphasis on social interaction in virtual spaces. To design interactions in such virtual worlds, it is pertinent to enhance the currently prevailing interaction paradigms (e.g., graphical user interfaces [GUIs] and the World Wide Web [WWW]) to support the broad range of group-centric and communication-intensive computer-mediated human activities (see Fig. 30.4). Such a community-wide design perspective requires that activities among members of communities of users become the primary unit of analysis, as opposed to an individual's keystrokes or performance measures. Moreover, the design focus should be on the cumulative experiences of the communities of users with the shared resources, as well as on the way in which communities move from early formation to maturity. To this end, there is a compelling need to study and understand how such communities (e.g., the virtual city) are formed, evolve, grow, and intra-/interoperate in order to synthesize methods that facilitate the design of suitable virtualities and computer-mediated activities for all potential community users.

Extend User-Centered Design to Support New Virtualities.   To facilitate the design of *new virtualities* likely to be encountered in the information age, the existing inventory of methods, techniques, and tools for user-centered design

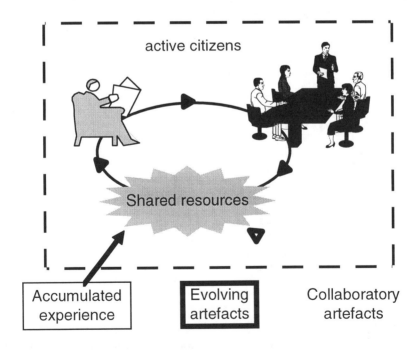

FIG. 30.4. Supporting community-wide experiences.

should be suitably applied and enhanced (see Fig. 30.5). To this end, attention should be drawn to the accumulated knowledge and results in the social sciences (e.g., human communication theories, language theories, action theories, etc.), to promote and facilitate the use of more *developmental* approaches[4] to the study of computer-mediated human activities. In all cases, the tight *evaluation-feedback* loop advocated by user-centered design should provide the primary channel for timely input into design processes, so as to ensure that deficiencies are corrected at an early stage, while updates are less costly to make.

Establish Suitable Accompanying Measures.   Support measures cover a whole range of multidisciplinary and cross-sector actions that are needed to facilitate the development of an industrial environment favorable to an information society for the broadest possible end-user population. Actions are needed to promote and facilitate the adoption and diffusion of good practice in the areas of accessibility and usability, so as to ensure quality in the use of products and services. To this

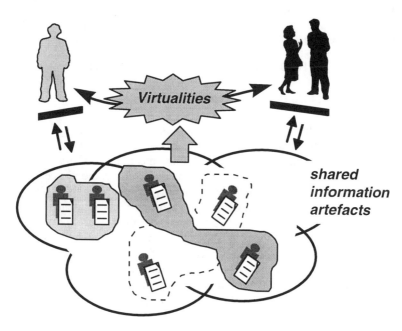

FIG. 30.5.   New virtual spaces for individual and collective experience.

[4]The term *developmental approach* to studying computer-mediated human activities is used to refer to various established theoretical strands within the social sciences and/or psychology that take explicit account of and model development in human behavior and capability. Such developmental approaches have recently started to progressively find their way into HCI. Examples include *activity theory* (Bødker, 1989, 1991; Nardi, 1996), *situated action plans* (Suchman, 1987), *distributed cognition* (Hutchins, 1995), *language/action theory* (Winograd, 1987), and so on.

end, it is important that accompanying measures are initiated to articulate demand (Kodama, 1992) for universal design, support the industry in adopting novel methods and practices, raise awareness, promote knowledge dissemination, and transfer technology in the form of know-how and know-why.

Specific Recommendations

The high-level recommendations discussed earlier are derived from a collection of specific proposals for future R&D actions that could be implemented in the context of an international agenda for collaborative RTD work. A summarizing account of the recommendations is depicted in Table 30.1. In what follows, each of the proposed items is elaborated through a brief description of the item, identification of key objectives, identification of required types of actions (e.g., basic research, applied research, technological development and demonstration, input to standardization activity), examples of plausible RTD activities, and a summary of some of the expected outcomes. It should be noted that this structure is not followed for the recommended research actions falling under "accompanying/support measures," as this thematic topic is a "horizontal" action, rather than an RTD cluster.

R&D ROADMAP—SPECIFIC ACTIONS

Environments of Use

Properties of Environments of Use.   *Environments of use* constitute integrated systems shareable by communities of users (e.g., students, teachers, researchers) and facilitate a broad range of communication, collaboration, and social activities. Such environments (e.g., the virtual university, the virtual theater, the virtual marketplace) are likely to be substantially different in their architectural underpinnings from conventional interactive software. To facilitate the construction of such environments, studies are needed to identify their respective properties and characteristics, as well as the prevailing norms that characterize their operation.

Actions in this area should aim to (a) identify requirements of different user groups (including people with special needs) and nonusers, (b) determine quality attributes such as loveability, usability, accessibility, unobtrusiveness, and so on, that characterize environments of use, and (c) map user requirements to technical characteristics.

The type of actions envisaged include basic research, applied research, and demonstration. Typical RTD activities may include, but are not limited to, the following:

- Undertaking of experimental studies to determine requirements of, and suitable interaction mechanisms among, members of online communities.
- Identification and assessment of desirable properties of environments of use (e.g., accessibility, usability, cooperativity, adaptability, individual-

Table 30.1

Proposed R&D Roadmap

Promote the Development of Environments of Use	Support Communities of Users	Extend User-Centered Design to Support New Virtualities	Establish Suitable Accompanying Measures
Determine desirable properties of environments of use (e.g., augmented capabilities on user's demand, multimodality, cooperativity, intelligence, adaptation, etc.)	Individual/collective intelligence and community knowledge management	Develop suitable foundations for design, by applying, integrating, and extending existing user-centered design methods to facilitate the design of new virtual spaces	Articulating demand for design for all
Novel architectures for interactive systems for managing collective experiences of users and nonusers	Methodologies for collecting/analyzing requirements and understanding virtual communities	Develop metrics for important quality attributes (e.g., usability, accessibility, adaptation, intelligence, etc.)	Supporting the industry
Architectures for multiple-metaphor environments	Provide means to access community-wide information resources	Provide computation support for usability engineering (e.g., computer-supported usability platforms)	Awareness and knowledge dissemination
Multiagent systems and components to support cooperation and collaboration	Develop models to support social interaction among members of online communities	Extend existing requirements engineering methods to facilitate the elicitation of requirements in novel contexts of use and different user groups	Technology transfer
Support individualization and user interface adaptation (e.g., adaptability and adaptivity) of environments of use		Promote user involvement and develop protocols for effective user participation in design activities	
		Investigate and provide design recommendations for alternative interaction modalities and their combinations	

ization, capability to provide augmented services on user's demand, intelligence).

- Development of demonstrators of innovative, integrated, shareable, and interoperable systems for online communities of users facilitating social interaction and exchanges among members.

Expected outcomes may include empirical evidence, requirements, guidelines, and demonstrator prototypes.

Novel Architectures for Interactive Systems for Managing Collective Experiences of Users and Nonusers.

Traditional interactive software architectures do not account for several of the desirable or envisioned properties of environments of use, such as interoperability, adaptation, cooperation, intelligence, and so forth (Stephanidis et al., 1998). Actions are needed to refine/extend/revise existing, as well as introduce new architectural models that can address the new requirements of such types of systems and facilitate the broad range of computer-mediated human activities that will emerge.

Actions in this area should strive to introduce and validate new architectural models for interactive software, define desirable architectural properties (e.g., adaptation, cooperation, collaboration, portability, interoperability, scaleability, modifiability), and produce guidelines on how they can be met (e.g., for adaptation, these would include the issues of platform abstraction, integration, augmentation, etc.). These actions should span basic and applied research, technological feasibility studies, as well as technology demonstration. Examples of RTD activities include, but are not limited to:

- Exploration of novel architectural models for interactive software (covering both interactive and noninteractive elements).
- Illustration of technical feasibility of new architectural models.
- Development of systems complying to the proposed architectural models.
- Evaluation and usability assessment of demonstrators.
- Provision of guidelines for integrating the proposed architectural models into new systems and services.

The expected outcomes may include new architectures, feasibility studies, prototypical demonstrators, guidelines, and tools.

Architectures for Multiple-Metaphor Environments.

The notion of a multiple-metaphor environment implies a particular computer-based embodiment of an integrated system, capable of performing context-sensitive mapping between concepts from a source domain to symbols in a (target) presentation domain (or metaphor), and vice-versa. Alternatively, it may be conceived as an integrated platform of multiple and concurrently available interaction toolkits, capable of context-sensitive mapping. The fusion of multiple metaphors into an integrated system will allow such systems to adapt to different user requirements and contexts of use. Actions are needed to determine how multiple-metaphor environments can be constructed and how they can provide the technology for building systems exhibiting desired usage properties.

Actions in this area should aim to (a) explore alternative metaphors for interaction, (b) develop interactive computer embodiments of metaphors, (c) determine properties of multiple-metaphor environments (e.g., adaptive change, user awareness, context-sensitive processing, intelligence), (d) provide experimental evidence and illustrate face validity of proposals, (e) develop demonstrators of multiple-metaphor environments, and (f) compile recommendations and guidelines for building and supporting multiple-metaphor environments. These actions could take the form of basic and applied research and should accommodate a strong element of technological demonstration. Examples of RTD actions may include, but are not limited to:

- Investigation of alternative theories of metaphor[5] and analysis of how they could influence the construction of environments of use.
- Exploration of novel metaphors suitable for new virtual spaces.
- Development of systems that exhibit characteristic properties of multiple-metaphor environments and tools for developing multiple-metaphor environments.
- Undertaking of evaluation and usability assessment of demonstrators.
- Provision of guidelines for developing new metaphors and building multiple-metaphor environments.
- Establishment of quality criteria for multiple-metaphor environments.

Expected outcomes include, but are not limited to novel interaction metaphors, methodologies for developing interaction metaphors, architectural abstractions for multiple-metaphor environments, demonstrators, experimental evidence, design and development guidelines, and tools.

Multiagent Systems and Components to Support Cooperation and Collaboration.

Two of the important dimensions of environments of use are expected to be (a) a further shift in the computing paradigm, departing from the desktop embodiment of the computer to distributed "intelligence" in the living environment, which, in turn, will necessitate a departure from the traditional direct manipulation approach in using a computer environment, to more delegation-oriented activities; and (b) cooperation and collaboration, which will need to be further facilitated and actively supported, so that humans will be able to seamlessly perform joint activities, independent from geographic location, specific characteristics of the software and hardware used, differences in language and culture, and so on. In this context, and in order to address the two dimensions together, it is necessary to cater for environments

---

[5]For a review of metaphor theories, see Lakoff and M. Johnson (1980).

that will be jointly inhabited by human and software agents, and that will interact and cooperate with each other toward common goals and under a multitude of social and other circumstances.

Actions in this area should strive to support and facilitate the cooperation and collaboration between humans in the new environments of use. Of particular interest is the construction of systems and components that will facilitate the coexistence of humans and software entities, in diverse contexts of use, allowing them to communicate, to effortlessly share knowledge and activities, and to evolve. The type of actions envisaged include basic research, applied research, and demonstration efforts. Examples of illustrative RTD activities include, but are not limited to:

- Investigation of suitable agent architectures and frameworks for constructing computational environments that exhibit voluntary domain-specific support and guidance, as well as interoperability with other tools and systems.
- Development of agent-based communication languages with embedded constructs facilitating social interaction (e.g., multiple ontology references, ability to employ different "social" contracts, communication across heterogeneous media/platforms).
- Experimentation with cooperative intelligent agents, personal assistants, and service integration agents.
- Development of communities of agents (e.g., formation of virtual communities of agents based on thematic interests).
- Construction of multiagent environments (e.g., support for different architectures and different languages).

The expected results include, but should not be limited to, architectural models, languages for agent-based communications, technology demonstrators, and guidelines.

Individualization and User Interface Adaptation.   A critical property of environments of use will be their capability for adaptation and individualization. This is necessary to ensure accessibility by *all* users to community-wide information and communication resources, as well as to satisfy experiences in the use of systems that carry out a broad range of social activities. In this context, adaptation refers both to the system's capability to tailor aspects of its interactive behavior prior to an interactive session in anticipation of a user's requirements (adaptability), as well as to run-time dialogue enhancements on the basis of dynamically acquired and maintained knowledge regarding the user. Individualization is a term that is typically associated with adaptation (adaptivity). In the present context, it implies additional capabilities on the part of the system—which need not be covered by adaptation—such as transparency and modifiability of a system's knowledge about the user, as well as the broad range of issues pertaining to ethics, privacy, and security.

Actions in this area should aim to (a) provide methods to facilitate the design of adaptations, (b) construct tools to support adaptable and adaptive behavior, (c) evaluate and assess adaptation strategies, (d) develop instruments for evaluating adaptable and adaptive behavior, (e) explore alternative architectural models for user interface adaptations, and (f) assess user's opinion toward individualized interactions. The type of actions envisaged include basic research, applied research, development, and demonstration efforts. Examples of RTD activities include, but are not limited to:

- Enrichment of the currently prevalent concept of adaptation to accommodate changes in the overall community-based environment (e.g., introducing new members and new services, revised social norms).
- Determination of suitable strategies to facilitate novel adaptation objectives, such as customization through the combination of interactive pluggable modules, interpreting the intentions of user operations, and so on.
- Investigation of the acceptability of adaptations and the social implications for individuals and communities of users.
- Demonstration of economic, social, and personal value of adaptable and adaptive systems.
- Development of prototypical demonstrators of adaptable and adaptive systems.
- Assessment and exploration of alternative technological options toward improved adaptable and adaptive behavior.

Some of the expected outcomes include, but are not limited to, demonstrators of adaptable and adaptive environments of use, enriched theoretical models of adaptation rooted in social sciences and substantiated into suitable interactive environments, and guidelines for developing adaptable and adaptive systems.

Supporting Communities of Users

Individual/Collective Intelligence and Community Knowledge Management.   Recent advances in telecommunications and, in particular, networking have broadened the scope and type of computer-mediated human activities. Increasingly, users find themselves associated with various virtual/online communities to attain professional and social goals. Knowledge, therefore, no longer constitutes an individual's asset, but a community-wide resource that can be shared and articulated by members of that community. Actions are needed to support the life cycle of communities and community-based knowledge management.

Actions in this area should strive to: facilitate capturing community-generated wisdom and collective experiences; support social awareness, collaboration, knowledge sharing and persistence, and the exchange of experiences. The type of actions envisaged include basic research, applied research, and technology demonstration. Examples of RTD activities include, but are not limited to:

- Development of tools for managing large information spaces (e.g., knowledge ontologies, evolutionary knowledge repositories, recommender systems).

- Development of interaction mechanisms to cope with new virtual spaces (e.g., collaborative filtering, virtual and augmented realities, etc.).
- Development of agent-based communication languages for knowledge sharing.

Expected outcomes may include demonstrator prototypes, tools for building collaborative systems, interoperable environments, or any other component that fits the preceding.

Methodologies for Collecting/Analyzing Requirements and Understanding Virtual Communities.
Traditional models and tools of information-processing psychology, focusing on individual users, need to be enhanced to provide a broader view of HCI, accounting for small groups and communities of users. Novel methodologies making use of analytical and developmental approaches to human communication need to be developed to provide prescriptive frameworks for the study of communities of users and to support the broad range of computer- mediated human activities. Suitable models should facilitate effective protocols for collecting/analyzing requirements and understanding online virtual communities.

Actions in this area should aim to (a) develop an understanding of virtual community life cycles and (b) investigate how online communities are formed, operate, and grow. The type of actions envisaged include basic research, applied research, and technological development and demonstration. Examples of RTD activities include, but are not limited to:

- Exploration of alternative models from the social sciences for the study of virtual communities.
- Development of novel methods for capturing requirements of virtual communities.
- Development of novel design methods and tools for mapping requirements to technical specifications.
- Experimentation with virtual communities.

The expected outcomes include, but should not be limited to methodologies for eliciting/analyzing requirements and studying virtual communities of users, guidelines for setting up virtual communities, as well as experimental demonstrators.

Accessing Community-Wide Information Resources.
Information generated and captured by virtual communities of users should be stored and accessed in a manner that is effective, efficient, and satisfactory for the individual members of the community. As this information is likely to expand rapidly and have a long life cycle (extending beyond a particular generation of users), it needs to be accessible through different media and from a variety of devices and locations. Actions are needed to facilitate human interactions with large information spaces (and concurrently with other humans) and to provide technological solutions that will make

community-wide information resources accessible, scaleable to new generations of technology, persistent, and secure. For example, the WWW provides enormous potential to enrich people's lives in many spheres, but understanding how to exploit the potential of the Web, requires, at present, considerable skills and experience. Current WWW tools and access methods are general purpose, like an "operating system" interface. New methods and techniques are required to support a much wider range of users and to facilitate access to resources related to specific tasks in areas including education, training, leisure, contacting people with common interests, supporting local communities, and so on. This needs to be integrated with existing consumer devices, such as cable TV and telephone, without necessarily demanding the skills needed to operate a personal computer.

Actions in this area should aim to (a) ensure accessibility and usability of community-based information resources by *all* potential users, (b) develop suitable interaction techniques that meet the requirements of individual members of communities of users, and (c) establish demonstrators of good practice. The type of actions envisaged include basic research, applied research, and technological development and demonstration. Examples of RTD activities include, but are not limited to:

- Investigation of and studies on user requirements for novel interaction technologies.
- Development of advanced three-dimensional domain-specific visualizations.
- Analysis and experimentation with new metaphors for interaction in social settings.
- Development of multimodal interaction mechanisms (e.g., gestures, natural-language understanding, tactile, and their combination).
- Exploration of alternative designs for nonvisual modalities to facilitate interaction in radically different contexts of use.
- Development of novel interaction techniques based on emerging technologies (e.g., wearable computing, virtual/augmented realities).
- Development of advanced content-based retrieval engines.
- Development of technologies for managing large information spaces (e.g., digital libraries).

The expected outcomes include, but should not be limited to, prototypical demonstrators, empirical studies, new implemented interactive metaphor environments, tools for building multimodal, or alternative-modality interactions, guidelines for human interfaces, interoperable interactive components, and tools to present and support navigation and searching in large information spaces.

Social Interaction Among Members of Online Communities.   A primary characteristic of the emerging broad range of computer-mediated human activities in the information age is their inherently group-centric and social nature. In order to facilitate the development of "sociable" interactive environments, it is important to enrich the currently prevailing practice with concepts that have a social focus. New models are needed to facilitate social awareness, social immersion, and social navigation in large virtual spaces. Moreover, such models should be validated in real contexts of use to determine their suitability as prescriptive instruments for new design foundations.

Actions in this area should aim to explore novel concepts for embodying information spaces (e.g., the virtual theater, the virtual city, the virtual marketplace), assess their relevance to the design of virtual communities, and provide experimental evidence to support novel design concepts for community-based activities (e.g., collaborative concept creation, community-based learning). The type of actions envisaged include basic research, applied research, and technological development and demonstration. Examples of RTD activities include, but are not limited to:

- Construction and validation of conceptual/prescriptive frameworks for envisioning virtual communities.
- Development and validation of novel community-based navigation concepts.
- Investigation of methods for evolutionary and persistent knowledge management in virtual communities.
- Development and validation of guidelines for constructing virtual communities.
- Development of demonstrators of "sociable" interactive environments for virtual communities.
- Development of tools for interpersonal communication between humans within and across the same virtual communities.

The expected outcomes include, but should not be limited to, theoretical and conceptual models, experimental foundations, guidelines, and technical and prototypical demonstrators.

Extend User-Centered Design to Support the Construction of New Virtualities

Foundations for Designing Computer-Mediated Human Activities in the Information Age. User-centered design (Norman & Draper, 1986) has surfaced as the primary design approach to facilitate usable interactive systems. It offers a broad collection of tools and methods for planning, iterative development, and evaluation, while it fosters a tight evaluation feedback loop to assure that deficiencies are identified and corrected at an early stage of the development life cycle, when the cost of refinement is not prohibitive. In view of the trends in technology, it becomes evident that in order to provide the required support for the design of the broad range of computer-mediated human activities in the emerging virtual spaces, user-centered design, as a philosophy, should be extended to provide a more prescriptive design framework. To this end, actions are needed to apply, refine, and extend existing techniques and tools of user-centered design with concepts from the social sciences, so as to provide a broader foundation for HCI design.

Actions in this area should strive to assess potential HCI design contributions rooted in disciplines that focus on human communication in social contexts (e.g., developmental psychology, the social sciences, the humanities, etc.) and extend existing analytical design approaches (e.g., design space analysis techniques) with social constructs to provide new methods for studying virtual spaces. The type of actions envisaged include basic research, applied research, and technology demonstration. Examples of RTD activities include, but are not limited to:

- Exploration and assessment of emerging, alternative approaches to HCI design based on activity theory (Bødker, 1989, 1991), language/action theories (Winograd, 1987), situated action models (Suchman, 1987), distributed cognition (Hutchins, 1995), and other frameworks from the social sciences (e.g., structure-functionalistic theory).
- Development of new tools for studying social, group, and individual behaviour and informing design practice (e.g., taxonomies of requirements, capabilities, and design alternatives).
- Compilation and validation of measurable yardsticks (e.g., metrics) for evaluating new virtualities and assessing social impact.
- Development of prototypical implementations of concepts (e.g., the virtual city) using novel design techniques and approaches and compilation of guidelines and examples of good practice.

The expected results include, but should not be limited to novel design methods and tools, guidelines, experimental and empirical evidence, as well as technology demonstrators.

Metrics for Important Interaction Quality Attributes.   Metrics provide a powerful instrument for measuring different aspects of an interactive system. In the past, the field of HCI has attempted to provide metric-based techniques in the form of usability scales for measuring qualities of interactive systems. Examples include usability scales based on performance criteria (e.g., effectiveness, efficiency, and satisfaction), cognitive workload, and so on. These techniques usually take the form of a questionnaire (e.g., for measuring user satisfaction and cognitive workload) or user tests (e.g., in the case of measuring effectiveness and efficiency). Despite the fact that such techniques have been around for several years, their adoption has been rather slow. Consequently, actions are needed to (a) extend the available range of metrics to cover additional quality attributes such as accessibility, adaptation, intelligence, and so forth, likely to determine the outcome of computer-mediated human activities in the emerging information age, (b) embed such metrics into tools for automatic evaluation and measurement, and (c) establish (technology-independent) protocols for measuring quality attributes of systems, taking account of the various contexts of use and the new virtualities that such systems are intended to support. The type of actions envisaged include basic research, applied research, and technological development and demonstration. Examples of RTD activities include, but are not limited to:

- Development of metric-based instruments for quality attributes, or design targets, including accessibility, cooperativity, social awareness, social immersion, intelligence, adaptability, adaptivity, and so on.
- Development of life-cycle metrics that allow usability, accessibility, and quality in use to be included in the requirements specification, and to be monitored during early development activities, in order to provide confidence that final validation of the designated quality attributes of the finished system will meet the specified requirements.
- Studies intended to correlate product attributes for usability and accessibility with quality in use to address a range of questions such as: Which usability and accessibility features need to be designed into a product so that it provides quality in use for specific user groups and tasks? How can these be specified and evaluated without the expense of laboratory usability testing? What is the correlation between the inclusion of specific features and the user-perceived quality of the final system for particular user groups?
- Demonstration of the validity of the aforementioned techniques by applying them across design cases, contexts of use, and application domains.
- Development of tools for automatic evaluation and measurement of software quality based on metrics.
- Provision of guidance as to how different quality attributes may be attained.
- Formulation and validation of guidelines on the basis of experimental evidence.

The expected results include, but should not be limited to new instruments, tools for community-centered design, examples of good practice, and experimental evidence.

Computational Tools for Usability Engineering.   Usability engineering has been traditionally conducted by experts without the assistance of computational environments or tools. This bears on cost factors and rates of adoption of specific techniques. Computational environments to support usability engineering have the potential to lead to both cost justification and improved usability practices, as they may automate certain tasks, guide designers toward usability targets, or provide extensible environments for capturing, consolidating, and reusing previous experience. Actions in this area should try to provide interoperable components covering the broad range of usability engineering tasks within a user-centered design protocol. The intention could range from attempts to fully automate specific and well-defined stages, to augmenting the capabilities of human designers or usability practitioners to undertake and carry out effectively a collection of tasks.

Actions in this area should aim to investigate characteristic properties of, and to provide for computer-supported usability engineering platforms comprising interoperable software components. The type of actions envisaged include basic research, applied research, and technological development and demonstration. Examples of RTD activities include, but are not limited to:

- Development of tools for a range of design and evaluation tasks; for example, working with guidelines, facilitating inspections, critiquing tentative designs, capturing/reusing past experience, evaluating designs, capturing design rationale, embedding rationale into designs, providing computational support for metric-based techniques, generating specifications that meet predetermined usability targets, cost–benefit analysis, and so forth.
- Development of usability support environments to integrate existing usability/accessibility tools and new tools for design for all; such platforms should allow for the accumulation of results from usability and accessibility assessment activities, impact analysis, and cost–benefit analysis; moreover, they should offer support for the resultant modification-implementation decisions.
- Development of planning tools to assist with selecting the best methods and techniques for user-centered design, usability, and accessibility, and integrating them into existing development activities; the tools should include information on advantages, disadvantages, and cost benefits of each instrument or method, and the skills required.
- Development of interoperable architectures linking usability platforms to design environments and user interface development systems.
- Development of requirements elicitation tools on widely available platforms (e.g., WWW).
- Development of responsive prototyping media tools addressing alternative and combination of modalities.

Expected outcomes include, but should not be limited to, computer-based usability platforms and improved methodologies for user-centered design and usability engineering.

Requirements Engineering Methods to Facilitate the Elicitation of Requirements in Novel Contexts of Use and Different User Groups.   The study of requirements in the design of computer-based interactive systems has always been a challenge for system designers and developers. With the advent of user-centered design, several tools have emerged to facilitate requirements elicitation, capture, and/or specification. However, existing techniques (e.g., brainstorming, scenarios, prototyping) have been used only in traditional contexts of use to elicit requirements of average/typical users. It is, therefore, important that existing techniques are refined and extended to facilitate requirements engineering in novel contexts of use and for user groups with radically different requirements.

Actions in this area should strive to provide improved means for eliciting, capturing, and consolidating requirements for a broad range of computer-mediated human activities in the information age, including the development of tools to facilitate the mapping of requirements to design concepts. The type of actions envisaged include basic research, applied research, and technological development and demonstration. Examples of RTD activities include, but are not limited to:

- Development of software environments for requirements engineering.
- Development of tools for collecting/documenting requirements in novel contexts of use.
- Development of tools for collecting/documenting requirements of different user groups, including people with special needs.
- Development of software tools for integrating requirements engineering into iterative prototyping environments.
- Development of taxonomies of human abilities versus interaction requirements.

The expected outcomes include, but should not be limited to, improved requirements for engineering instruments (e.g., questionnaires, protocols) and tools to facilitate the transition from requirements specification to iterative prototyping.

Protocols for Effective User Participation in Design Activities.   User involvement in the design of computer-based interactive systems has long been a challenging issue. Despite its potential value, it needs to be carefully planned and assessed in different phases of a product's life cycle. Participatory design has provided useful insights into how user involvement may be managed in practice and offers several tools and guiding principles. However, the existing wisdom offers very little in the direction of involving different user groups with diverse abilities, skills, requirements, and preferences. Therefore, actions should be undertaken to refine and extend the available instruments in such a way that they can effectively guide the design of new computer-mediated human activities.

Actions in this area should aim to (a) establish new methods and tools for managing user participation in design projects that are intended to be accessible to the broadest possible end-user population, including people with special needs, and (b) promote practice and experience of participatory design and develop suitable models. The type of actions envisaged include basic research, applied research, and technological development and demonstration. Examples of RTD activities include, but are not limited to:

- Development of conceptual models and guidance for participatory design in suitable selected application domains and design cases (e.g., accessibility).
- Assessment of the cost and benefits of participatory design in real-world case studies.
- Provision of guidance in the use of different techniques for participatory design.
- Establishment of links between participatory design and user-centered design activities.

The expected outcomes include, but should not be limited to, novel methods for participatory design, cost–benefit studies, practice and experience, and guidelines.

Design Recommendations for Suitable/Plausible Interaction Modalities and Combinations. In the recent history of HCI, the visual modality has been predominant in the systems and tools that have been developed for humans to work with.[6] However, with the advent of multimedia and the new capabilities that are being offered, it becomes pertinent to enrich, rather than replace, the visual modality to facilitate "broad-band" interactions between humans and artifacts. Such media-rich environments are increasingly needed due to the variety in the context of use that may render certain presentations inappropriate. Thus, it is important to investigate how to design for alternative modalities and how to combine modalities into integrated environments. This would not only facilitate more effective computer-mediated communication, but it would also substantially reduce the problems faced by users with special needs. This line aims to promote R&D activities that would facilitate the creation of a design corpus for constructing purely multimedia interactions, as well as the development of the required tools that would ease the task of constructing and building such interactions.

Actions in this area should aim to establish a basis for designing for alternate interaction modalities and combinations of modalities, as well as to demonstrate the benefits of developing multimodal and multimedia systems for communities, groups, and individual users. The type of actions envisaged include basic research, applied research, and technological development and demonstration. Examples of RTD activities include, but are not limited to:

- Experimental studies on design for alternative modalities.
- Development and validation of a taxonomy of modality combinations in computer-based interactive software (detailing issues, such as modality compatibility, modality expressiveness, redundancy) to guide toward effective multimodal interaction design.
- Assessment of the usefulness of multimedia in specific application domains and contexts of use.
- Experimentation with alternative interaction design (e.g., nonvisual).
- Demonstrators of advanced multimodal interaction techniques (e.g., gestures, speech recognition, tactile interaction).

Expected outcomes include, but should not be limited to, experimental evidence, guidelines for designing for alternative modalities and modality combinations, novel interaction techniques, and technology demonstrators.

Establish Suitable Accompanying Measures

Articulating Demand for Design for All. In the short term, support efforts should be devoted to the articulation of demand for design for all. Articulating de-

---

[6]Among the very few exceptions is interpersonal communication, which makes use of the auditory modality.

mand has been defined as a two-step process: First, translate market data into a product concept; and second, decompose the concept into a set of development projects. It should be mentioned that such a (two-step) process is sufficient only in cases where a market already exists, and has reached a level of maturity whereby it can react and respond to the needs of its customer base. However, in the case of the emerging information society, it is argued that there is an additional short-term need for creating awareness as to the new opportunities offered, as well as to the new challenges likely to emerge.

Consequently, support measures are needed in the direction of raising consumer awareness on the value of accessibility and usability, educating consumers and producers in the need to include requirements for usability and accessibility in product specifications, as well as helping them evaluate usability and accessibility when making design, or purchase, decisions. Such actions would help toward building a public expectation for accessible and usable products and services, and intolerance of inaccessible forms of technology. Additionally, until a certain maturity level is reached and more effective end-user input into the design process can be attained, procurement guidance is necessary with regard to accessible and usable technology.

Regarding subsequent steps required for demand articulation, namely the translation of market data to product concepts and the decomposition of product concepts to development projects, it is important that a range of questions are addressed. These include: (a) what product types are needed in the market, (b) how they could be produced, (c) through which technologies, and (d) what other characteristics should the envisaged technologies exhibit. The type of actions that area expected in this area are accompanying/support measures (or horizontal actions).

Supporting the Industry.   This line of action should be targeted toward the creation of an environment favorable to industrial innovation. At the core of such activities should be the provision of incentives toward design for all. Industrial incentives need not necessarily be of a financial type, though this would be critical for small- and medium-size enterprises (SMEs). They should also include access to research results that would be difficult to obtain otherwise, provision of a suitable infrastructure, collaborative R&D activities for technology transfer (see also later section, Technology Transfer), as well as other policy initiatives, such as the establishment of an accessibility/usability certificate.

There is also a compelling requirement for speeding up current standardization processes, as well as for more intensive international coordination of standards in the long term. To this end, actions are needed to facilitate coordination across efforts initiated in the context of research consortia (e.g., the W3C-WAI[7] project and the ERCIM Working Group on "User Interfaces for All"[8]), as well as in national (e.g.,

---

[7]World Wide Web Consortium—Web Accessibility Initiative (1997). For more information, please refer to http://www.w3.org/WAI/

[8]For more information please refer to http://www.ics.forth.gr/proj/at-hci/UI4ALL/index.html

HFES/ANSI[9]) and international standardization bodies (e.g., the new work item on accessibility by ISO 9241/TC 159/SC 4/WG 5). Coordination in this context should also involve exchange of input so as to avoid incompatible standards. To this effect, the requirements of mainstream industries need to be carefully studied so as not to impede adoption of design for all principles and recommendations. Another important issue in this line of action is the establishment of suitable assessment and certification measures for accessibility and usability of new products and services.

Legislation is also needed to provide the framework of operation, and the required incentives for both the consumer base and the industry. To this effect, recent experience in the United States with the Americans with Disabilities Act of 1993 and the Telecommunications Act of 1996 should be assessed, and similar actions should be encouraged internationally. Such efforts could also draw upon general rules and recommendations compiled by industrial consortia (e.g., The Telecommunications Policy Roundtable in the United States), technical committees (e.g., the Association of Computing Machinery [ACM] Public Policy Committee), and international organizations (e.g., the United Nations General Assembly Standard Rules of 1995). The type of actions that are expected in this area are accompanying/support measures (or horizontal actions).

Awareness and Knowledge Dissemination.   One of the critical impediments to the adoption of universal design practice is the lack of qualified practitioners who understand what the requirements for universal access and quality in use are. To overcome this, it is recommended that, in the short term, accessibility, usability, and quality in use are introduced as mandatory components of university education.

Additionally, efforts should be devoted to the collation and dissemination of comprehensive information on the practical resources available for user-centered design, usability, and accessibility. This would include information on the available methods, techniques, and tools for user-centered design, usability, and accessibility, the skills required to adopt, internalize, and appropriate the benefits of the methods, as well as their socioeconomic benefits and costs. Such efforts would necessarily build upon the accumulated wisdom[10] collected through past and ongoing collaborative project work in the context of transnational projects.

---

[9]See also Human Factors and Ergonomics Society (1997).

[10]The European Commission has funded several projects aiming to collect, consolidate, and disseminate available knowledge and experience. Examples include: INUSE ("European Usability Support Centres," Telematics Applications Programme, Telematics Engineering Sector, 1996–1998); RESPECT ("Requirements Engineering and Specification in Telematics," Telematics Applications Program, Telematics Engineering Sector, 1996–1998); MEGATAQ ("Methods and Guidelines for the Assessment of Telematics Application Quality," Telematics Applications Program, Telematics Engineering Sector, 1996–1998); BASELINE ("Data for User Validation in Information Engineering," Telematics Applications Program, Telematics Engineering Sector, 1996–1998) INCLUDE ("INCLUsion of Disabled and Elderly People in Telematics," Telematics Applications Programme, Telematics Engineering Sector, 1996–1998); ACCESS ("Development Platform for Unified Access to Enabling Environments," Technology for the Disabled and Elderly [TIDE] Program—Bridge Phase, 1994–1996); USER ("User Requirements Elaboration in Rehabilitation and Assistive Technology," Technology for the Disabled and Elderly [TIDE] Program—Bridge Phase, 1994–1996).

Dissemination methods could include guidance and reference documents, lists of resources, and provision of tutorials and workshops for potential consortium partners. Ideally, such activities could be supported through the sponsorship of a network of excellence that would adopt the aforementioned targets as part of its global function. Such a network could employ specific means to desired ends. For instance, the dissemination targets could be promoted by organizing and running on predetermined time intervals a multidisciplinary conference with the intention to bring together the previously disparate communities, and to help them develop a shared understanding of the common problems and goals. This will progressively encourage the adoption of universal design principles by professional bodies and individuals in the medium to long term. The type of actions that are expected in this area are accompanying/support measures (or horizontal actions).

Technology Transfer.   Effective and efficient technology transfer is another critical target, requiring a range of support measures to be effected. Technology, in this context, includes both embodied and disembodied forms (Vernardakis, Stephanidis, & Akoumianakis, 1997a). Embodied technology is evident in new products and services, machines, tools, and research equipment. Disembodied technology appears as learning-by-doing, documentation, know-how, and know-why. To facilitate successful transfers of technology, suitable mechanisms are needed in the short term, to the effect of targeted and purposeful exchange of knowledge, know-how, and know-why. It is recommended that, from the broad range of technology transfer mechanisms that can be considered, emphasis is on advanced measures (such as cooperative R&D, joint venture R&D agreements, joint ventures aimed at keeping partners informed, large–small firm agreements), rather than simpler ones (such as licensing, technical advice, technical support, contract of R&D). This is because the latter cluster is better suited for the type of transfers (e.g., know-how and know-why) that is required. In this context, it is important to mention that collaborative, interdisciplinary, multinational, multicultural, and cross-industry R&D activities, involving industry and research institutions, are of primary importance.

Moreover, given the broad range of technologies that will drive the emergence of the information society, it is recommended that any contemplated technology transfer effort should closely and carefully consider potential sources and recipients and assess alternative technological performance thresholds (Vernardakis, Stephanidis, & Akoumianakis, 1997b). In principle, it is recommended that design for all should be the aim of all emerging technologies. The newer a technology, the greater the chances are of being influenced in the direction of facilitating design for all. However, for such a condition to materialize, a "monitoring system" for critical emerging technologies should be established; this could be part of an extensive collaborative network, and should aim to identify potential synergies and possibilities for international collaborative R&D efforts. The type of actions that are expected in this area are accompanying/support measures (or horizontal actions).

SUMMARY AND CONCLUSIONS

The emergence of the information society creates new opportunities and challenges for *all* citizens. The progressive shift from physical goods to information-based products and services is likely to introduce new patterns for demand and supply of such products and services. One important issue in this transition is the extent to which the emerging information society advances in a manner that ensures nondiscrimination and social and economic inclusion of the broadest possible end-user population, thus posing the requirement for a society caring for *all* citizens.

This chapter presents an R&D roadmap for activities that could be undertaken in the context of international collaboration in R&D to contribute toward the advancement of an information society accessible to the broadest possible end-user population.

To this effect, this document discusses research items under four main thematic topics, namely, *promote the development of environments of use, support communities of users, extend user-centered design to support new virtualities,* and *establish suitable accompanying measures and supporting actions.* Though all four of them have a cross-sector nature, the first three are RTD clusters, whereas the fourth is intended as a (horizontal) support action. Each thematic topic has been elaborated in terms of a contextual description and a cluster of specific recommendations for collaborative R&D activities. The description of the topic provides an insight as to how this topic emerged and the rationale for including it as a main research target. The specific recommendations that relate to each topic exemplify the context and scope of the proposed R&D activities.

The four topics are interrelated. Thus, recommendations under one topic link with recommendations under a different topic. For example, the recommendation for suitable methods and tools to support the design of new virtualities relates to the recommendation for novel architectures for adaptable, adaptive, multimodal, and cooperative interactive systems. It is important to note that such interrelations do not imply any particular preference, or priority. Instead, they indicate possible pathways for the diffusion of innovative action across, or within sectorial programs of national, European, and transnational nonmarket institutions, such as, for example, the European Commission of the European Union and the National Science Foundation in the United States. This means that some of the recommendations may be considered in the context of specific key sectorial actions, or as components of a cross-sector theme that spans across key actions and domains.

Finally, it should be mentioned that the International Scientific Forum "Towards an Information Society for All" is in the process of becoming an international association aiming to promote the evolving objective of an information society accessible and acceptable to *all* citizens, through a series of activities. Some of these activities will aim to strengthen transnational cooperation at various levels, as well as to establish an international conference and an archival scientific journal. The next ISF meeting[11] is scheduled to take place in Munich, Germany, August 22–23, 1999, in the context of the 8th International Conference on Human–Computer Interaction (HCI International '99).

ACKNOWLEDGMENTS

This chapter is a reprint from Stephanidis et al. (1999). *International Journal of Human–Computer Interaction, 11*(1), 1–28. Copyright 1999 by Lawrence Erlbaum Associates and IJHCI.

REFERENCES

Bergman, E., & Johnson, E. (1995). Towards accessible human–computer interaction. In J. Nielsen (Ed.), *Advances in human–computer interaction* (Vol. 5, pp. 87–113). Norwood, NJ: Ablex.

Bevan, N. (1995). Measuring usability as quality of use. *Software Quality Journal, 4,* 115–130.

Bevan, N., & Azuma, M. (1997). Quality in use: Incorporating human factors into the software engineering life-cycle. In C. Walnut (Ed.), *Proceedings of the 3rd IEEE International Software Engineering Standards Symposium and Forum* (pp. 169–179). Los Alamos, CA: IEEE Computer Society.

Bødker, S. (1989). A human-activity approach to user interfaces. *Human–Computer Interaction, 4*(3), 151–196.

Bødker, S. (1991). *Through the interface: A human activity approach to user interface design.* Hillsdale, NJ: Lawrence Erlbaum Associates.

Garvin, D. A. (1984). What does "product quality" really mean? *Sloan Management Review, 26*(1), 25–48.

Human Factors and Ergonomics Society. (1997). *Draft HFES/ANSI 200 standard, Section 5: Accessibility.* Santa Monica, CA: Author.

Hutchins, E. (1995). *Cognition in the wild.* Cambridge, MA: MIT Press.

ISO 8402. (1994). *Quality vocabulary.* Geneva, Switzerland: International Standards Organisation.

ISO 9001. (1987). *Quality systems—Model for quality assurance in design, development, production, installation and servicing.* Geneva, Switzerland: International Standards Organisation.

ISO 13407. (1997). *Human-centred design processes for interactive systems.* Geneva, Switzerland: International Standards Organisation.

---

[11]The specific objectives of the third ISF meeting are three-fold. First, the workshop will aim to review recent advances in HCI and other related fields and assess their impact in the context of selected emerging information society technologies (e.g., healthcare, disabled and elderly, transport, digital libraries) and novel application domains (e.g., electronic commerce). Second, the workshop will aim to consolidate the group's exchanges into meaningful recommendations for future collaborative activities in the involved fields (e.g., guidelines for new HCI paradigms in the context of the emerging information society). Finally, the workshop will devise a suitable dissemination mechanism and plan for reaching a wider range of target audience including industry, nonmarket institutions, research and academic community, user organizations, standardization organizations, and so on. These objectives are expected to advance the notion of user acceptability in information society technologies beyond the traditional fields of inquiry (e.g., HCI, assistive technologies, housing, consumer electronics) and into the core of emerging information society technologies and application domains.

Kodama, F. (1992). Technological fusion and the new R&D. *IEEE Engineering Management Review, 20*(2), 6–11.

Lakoff, G., & Johnson, M. (1980). *Metaphors we live by.* Chicago: University of Chicago Press.

Müller, M. J., Wharton, C., McIver, W. J., & Kaux, L. (1997). Towards an HCI research and practice agenda based on human needs and social responsibility. In S. Pemberton (Ed.), *Proceedings of the ACM Conference on Human Factors in Computing Systems* (pp. 155–161). New York: ACM Press.

Nardi, B. (1996). *Context and consciousness: Activity theory and human computer interaction.* Cambridge, MA: MIT Press.

National Research Council. (1997). *More than screen deep: Toward every-citizen interfaces to the nation's information infrastructure.* Washington, DC: National Academy Press.

Norman, D. A., & Draper, W. S. (Eds.). (1986). *User-centered system design: New perspectives in human–computer interaction.* Hillsdale, NJ: Lawrence Erlbaum Associates.

Stephanidis, C., (Ed.), Salvendy, G., Akoumianakis, D., Bevan, N., Brewer, J., Emiliani, P. L., Galetsas, A., Haataja, S., Iakovidis, I., Jacko, J., Jenkins, P., Karshmer, A., Korn, P., Marcus, A., Murphy, H., Stary, C., Vanderheiden, G., Weber, G., & Ziegler, J. (1998). Toward an information society for all: An international R&D agenda. *International Journal of Human–Computer Interaction, 10*(2), 107–134.

Story, M. F. (1998). Maximising usability: The principles of universal design. *Assistive Technology, 10*(1), 4–12.

Suchman, L. A. (1987). *Plans and situated actions: The problem of human machine communication.* Cambridge, England: Cambridge University Press.

Vernardakis, N., Stephanidis, C., & Akoumianakis, D. (1997a). The transfer of technology towards the European assistive technology industry: Current impediments and future opportunities. *International Journal of Rehabilitation Research, 20*(2), 189–192.

Vernardakis, N., Stephanidis, C., & Akoumianakis, D. (1997b). Transferring technology toward the European assistive technology industry: Mechanisms and implications. *Assistive Technology, 9*(1), 34–46.

Winograd, T. (1988). A language/action perspective on the design of co-operative work. *Human–Computer Interaction, 3*(1), 3–30.

Winograd, T. (1997). Interspace and an every-citizen interface to the national information infrastructure. In *More than screen deep: Toward every-citizen interfaces to the nation's information infrastructure.* Washington, DC: National Academy Press. [Online]. Available: http://www.nap.edu/readingroom/books/screen/index.html

Author Index

J

Jacko, J., 109, 223, 363, 628, 665, 666, 667, 668, 676
Jacob, R., 492, 495
Jacobson, I., 390, 394, 397
Jameson, A., 271
Jenkins, P., 109, 223, 359, 363, 364, 628, 665, 666, 667, 668, 676
John, B.E., 154, 529
John, O.P., 27
Johnson, E., 9, 558, 640, 641, 667
Johnson, H., 417, 421, 428
Johnson, J., 174
Johnson, M., 175, 677
Johnson, P., 390, 394, 397, 417, 421, 428
Johnson, R., 408
Johnson-Laird, P.N., 160
Jones, S., 178
Jorge, J., 13, 461, 493, 596
Joy, D., 67, 77
Just, M.A., 147, 148

K

Kaasinen, E., 105
Kahler, S., 301
Kahn, R., 210
Kaine-Krolak, M., 558
Kalyuga, S., 31
Kamada, T., 492
Kaplan, C., 344
Kapor, M., 166, 638
Kaptelinin, V., 185, 190, 192, 193, 194, 195
Karagiannidis, C., 105, 240, 444, 565, 641
Karat, C.M., 355
Karshmer, A., 109, 223, 359, 363, 364, 628, 665, 666, 667, 668, 676
Karunanithi, N., 290
Kasabach, C., 92
Kasday, L.R., 51
Katz, M., 614
Kautz, H., 203
Kaux, L., 109, 611, 641, 653, 667
Kay, J., 278, 279, 469
Keil, M., 355
Kellogg, W.A., 192, 255, 442
Kensing, F., 6, 349
Kernal, H., 304

Kieras, D.E., 154, 340, 341
Kiesler, S., 28, 160, 298, 312
Kimball, R., 442
King, W.J., 298, 312
Kirakowski, J., 349, 359
Kline, R.L., 23
Kobsa, A., 105, 273, 279, 282, 283, 287, 394, 398, 469, 526, 527
Kochanek, D., 104, 108
Koda, T., 298, 304, 312
Kodama, F., 626, 636, 644, 656, 674
Kolojejchick, J., 70
Kono, Y., 282
Konstan, J.A., 215, 290
Koons, D.B., 69
Korn, P., 109, 223, 359, 363, 364, 628, 665, 666, 667, 668, 676
Koskinen, M., 105
Kosslyn, M.S., 66
Koumpis, A., 240, 565
Kouroupetroglou, G., 105, 379, 408, 444, 482
Kozulin, A., 187
Kühme, T., 12, 38, 271, 393, 394, 408, 550
Kuhn, S., 171, 177
Kuutti, K., 185, 187, 190, 191, 193, 195
Kyng, M., 178, 192, 284

L

Laaksolahti, J., 302
Lai, K., 469
Laird, J.E., 145, 147
Lakoff, G., 175, 677
Lamp, M., 491
Landauer, T.K., 142, 143, 147
Lanier, J., 312
Laroche, C., 32
Latour, J.C., 32
Laurel, B., 171
Lave, J., 186
Lawrence, D., 154
Lazar, R., 145
Lee, E., 190
Legge, G.E., 23
Leheureux, J.M., 473
Leikas, J., 560
Leont'ev, A.N., 186, 187, 189
Lester, J., 301
Leventis, A., 105, 444
Levinson, D.J., 210

Subject Index